Capillary Electrophoresis of Nucleic Acids
Volume I

METHODS IN MOLECULAR BIOLOGY™

John M. Walker, Series Editor

177. **Two-Hybrid Systems:** *Methods and Protocols,* edited by *Paul N. MacDonald, 2001*
176. **Steroid Receptor Methods:** *Protocols and Assays,* edited by *Benjamin A. Lieberman, 2001*
175. **Genomics Protocols**, edited by *Michael P. Starkey and Ramnath Elaswarapu, 2001*
174. **Epstein-Barr Virus Protocols,** edited by *Joanna B. Wilson and Gerhard H. W. May, 2001*
173. **Calcium-Binding Protein Protocols, Volume 2:** *Methods and Techniques,* edited by *Hans J. Vogel, 2001*
172. **Calcium-Binding Protein Protocols, Volume 1:** *Reviews and Case Histories,* edited by *Hans J. Vogel, 2001*
171. **Proteoglycan Protocols,** edited by *Renato V. Iozzo, 2001*
170. **DNA Arrays:** *Methods and Protocols,* edited by *Jang B. Rampal, 2001*
169. **Neurotrophin Protocols,** edited by *Robert A. Rush, 2001*
168. **Protein Structure, Stability, and Folding,** edited by *Kenneth P. Murphy, 2001*
167. **DNA Sequencing Protocols,** *Second Edition,* edited by *Colin A. Graham and Alison J. M. Hill, 2001*
166. **Immunotoxin Methods and Protocols**, edited by *Walter A. Hall, 2001*
165. **SV40 Protocols**, edited by *Leda Raptis, 2001*
164. **Kinesin Protocols,** edited by *Isabelle Vernos, 2001*
163. **Capillary Electrophoresis of Nucleic Acids, Volume 2:** *Practical Applications of Capillary Electrophoresis,* edited by *Keith R. Mitchelson and Jing Cheng, 2001*
162. **Capillary Electrophoresis of Nucleic Acids, Volume 1:** *Introduction to the Capillary Electrophoresis of Nucleic Acids,* edited by *Keith R. Mitchelson and Jing Cheng, 2001*
161. **Cytoskeleton Methods and Protocols**, edited by *Ray H. Gavin, 2001*
160. **Nuclease Methods and Protocols,** edited by *Catherine H. Schein, 2001*
159. **Amino Acid Analysis Protocols,** edited by *Catherine Cooper, Nicole Packer, and Keith Williams, 2001*
158. **Gene Knockoout Protocols,** edited by *Martin J. Tymms and Ismail Kola, 2001*
157. **Mycotoxin Protocols,** edited by *Mary W. Trucksess and Albert E. Pohland, 2001*
156. **Antigen Processing and Presentation Protocols,** edited by *Joyce C. Solheim, 2001*
155. **Adipose Tissue Protocols,** edited by *Gérard Ailhaud, 2000*
154. **Connexin Methods and Protocols,** edited by *Roberto Bruzzone and Christian Giaume, 2001*
153. **Neuropeptide Y Protocols**, edited by *Ambikaipakan Balasubramaniam, 2000*
152. **DNA Repair Protocols:** *Prokaryotic Systems*, edited by *Patrick Vaughan, 2000*
151. **Matrix Metalloproteinase Protocols,** edited by *Ian M. Clark, 2001*
150. **Complement Methods and Protocols,** edited by *B. Paul Morgan, 2000*
149. **The ELISA Guidebook**, edited by *John R. Crowther, 2000*
148. **DNA–Protein Interactions:** *Principles and Protocols* **(2nd ed.)**, edited by *Tom Moss, 2001*
147. **Affinity Chromatography:** *Methods and Protocols*, edited by *Pascal Bailon, George K. Ehrlich, Wen-Jian Fung, and Wolfgang Berthold, 2000*
146. **Mass Spectrometry of Proteins and Peptides,** edited by *John R. Chapman, 2000*
145. **Bacterial Toxins:** ***Methods and Protocols***, edited by *Otto Holst, 2000*
144. **Calpain Methods and Protocols**, edited by *John S. Elce, 2000*
143. **Protein Structure Prediction:** *Methods and Protocols*, edited by *David Webster, 2000*
142. **Transforming Growth Factor-Beta Protocols**, edited by *Philip H. Howe, 2000*
141. **Plant Hormone Protocols**, edited by *Gregory A. Tucker and Jeremy A. Roberts, 2000*
140. **Chaperonin Protocols**, edited by *Christine Schneider, 2000*
139. **Extracellular Matrix Protocols**, edited by *Charles Streuli and Michael Grant, 2000*
138. **Chemokine Protocols**, edited by *Amanda E. I. Proudfoot, Timothy N. C. Wells, and Christine Power, 2000*
137. **Developmental Biology Protocols, Volume III**, edited by *Rocky S. Tuan and Cecilia W. Lo, 2000*
136. **Developmental Biology Protocols, Volume II,** edited by *Rocky S. Tuan and Cecilia W. Lo, 2000*
135. **Developmental Biology Protocols, Volume I,** edited by *Rocky S. Tuan and Cecilia W. Lo, 2000*
134. **T Cell Protocols:** *Development and Activation*, edited by *Kelly P. Kearse, 2000*
133. **Gene Targeting Protocols,** edited by *Eric B. Kmiec, 2000*
132. **Bioinformatics Methods and Protocols,** edited by *Stephen Misener and Stephen A. Krawetz, 2000*
131. **Flavoprotein Protocols,** edited by *S. K. Chapman and G. A. Reid, 1999*
130. **Transcription Factor Protocols,** edited by *Martin J. Tymms, 2000*
129. **Integrin Protocols,** edited by *Anthony Howlett, 1999*
128. **NMDA Protocols,** edited by *Min Li, 1999*
127. **Molecular Methods in Developmental Biology:** Xenopus *and Zebrafish*, edited by *Matthew Guille, 1999*
126. **Adrenergic Receptor Protocols,** edited by *Curtis A. Machida, 2000*
125. **Glycoprotein Methods and Protocols:** *The Mucins*, edited by *Anthony P. Corfield, 2000*
124. **Protein Kinase Protocols,** edited by *Alastair D. Reith, 2001*
123. ***In Situ* Hybridization Protocols (2nd ed.),** edited by *Ian A. Darby, 2000*
122. **Confocal Microscopy Methods and Protocols,** edited by *Stephen W. Paddock, 1999*
121. **Natural Killer Cell Protocols:** *Cellular and Molecular Methods,* edited by *Kerry S. Campbell and Marco Colonna, 2000*
120. **Eicosanoid Protocols,** edited by *Elias A. Lianos, 1999*
119. **Chromatin Protocols,** edited by *Peter B. Becker, 1999*
118. **RNA–Protein Interaction Protocols,** edited by *Susan R. Haynes, 1999*
117. **Electron Microscopy Methods and Protocols,** edited by *M. A. Nasser Hajibagheri, 1999*
116. **Protein Lipidation Protocols,** edited by *Michael H. Gelb, 1999*

METHODS IN MOLECULAR BIOLOGY™

Capillary Electrophoresis of Nucleic Acids

Volume I: Introduction to the Capillary Electrophoresis of Nucleic Acids

Edited by

Keith R. Mitchelson

Australian Genome Research Facility, University of Queensland, Brisbane; and Walter & Eliza Hall Institute, Parkville, Australia

and

Jing Cheng

Biochip Research and Development Center, State Key Laboratory for Biomembrane and Membrane Biotechnology, Tsinghua University, Beijing, China; and Aviva Biosciences Corporation, San Diego, CA

Humana Press **Totowa, New Jersey**

Dedication

To Nanne and Moon
Each is unique.

999 Riverview Drive, Suite 208
Totowa, New Jersey 07512

This publication is printed on acid-free paper. ∞
ANSI Z39.48-1984 (American Standards Institute) Permanence of Paper for Printed Library Materials.

Cover design by Patricia F. Cleary.

For additional copies, pricing for bulk purchases, and/or information about other Humana titles, contact Humana at the above address or at any of the following numbers: Tel.: 973-256-1699; Fax: 973-256-8341; E-mail: humana@humanapr.com; Website: http://humanapress.com

Printed in the United States of America. 10 9 8 7 6 5 4 3 2 1

Library of Congress Cataloging in Publication Data

Main entry under title:

Methods in molecular biology™.

Capillary electrophoresis of nucleic acids/edited by Keith R. Mitchelson and Jing Cheng.
p. cm. -- (Methods in molecular biology ; 162-163)
Includes bibliographical references and index.
Contents: v. 1. Introduction to the capillary electrophoresis of nucleic acids -- v. 2. Practical applications of capillary electrophoresis.
ISBN 0-89603-779-7 (v. 1 : alk. paper) -- ISBN 0-89603-765-7 (v, 2 : alk. paper)
1. Nucleic acids--Separation. 2. Capillary electrophoresis. I. Mitchelson, Keith R. II. Cheng, Jing. III. Series.
QP620.c36 2000
572.8--dc21 00-038911
For Library of Congress CIP

Preface

The development of PCR, which enables extremely small amounts of DNA to be amplified, led to the rapid development of a multiplicity of analytical procedures that permit use of this new resource for the analysis of genetic variation and for the detection of disease-causing mutations. The advent of capillary electrophoresis (CE), with its power to separate and analyze very small amounts of DNA, has also stimulated researchers to develop analytical procedures for the CE format. The advantages of CE in terms of speed and reproducibility of analyses are manifold. Furthermore, the high sensitivity of detection, and the ability to increase sample throughput with parallel analysis, has led to the creation of a full range of analysis of DNA molecules, from modified DNA adducts and single-strand oligonucleotides through PCR-amplified DNA fragments and whole chromosomes. *Capillary Electrophoresis of Nucleic Acids* focuses on analytical protocols that can be used for detection and analysis of mutations and modification, from precise DNA loci through entire genomes of organisms. Important practical considerations for CE, such as the choice of separation media, electrophoresis conditions, and the influence of buffer additives and dyes on DNA mobility, are discussed in several key chapters and within particular applications. The use of CE for the analysis of drug–DNA interactions and for examination of the metabolism of therapeutic oligonucleotides and modified nucleosides illustrate the burgeoning applications of CE to the analysis of a wide range of medical and molecular diagnostic problems. The study of interactions between nucleic acids and diverse ligands by CE is also a new and rapidly developing area. Microanalytical devices, which include capillary electrophoretic separations, have resulted from the fusion of semiconductor devices and microdiagnostic analysis. Several seminal papers on the application of microchip-based capillary electrophoresis for genotyping and DNA sequence determination signal the revolution that is occurring in both capillary electrophoresis and laboratory analysis.

Capillary Electrophoresis of Nucleic Acids comprises two volumes divided into twelve parts, each part containing chapters that address particular general goals or experimental approaches. Broadly, Volume I addresses instrumentation, signal detection, and the capillary environment, as well as the integration of mass spectrometry and CE for the analysis of small oligonucleotides and modified nucleotides, whereas Volume II addresses techniques for high-throughput analysis of DNA fragments of less than 1 kb, employing SNP detection, mutation detection, DNA sequencing methods, and DNA-ligand interactions.

Volume I

Volume I, Part I presents basic CE instrumentation and the theoretical background to the separation of DNA by CE. The choice of different modes of CE and detection systems, the characteristics of sieving matrices, sample preparation, and quantitation of measurement are emphasized. DNA sequencing by CE is discussed. Methods for the manufacture of microchip devices are illustrated.

Volume I, Part II presents important factors, such as choice of capillary coatings and separation media, which offer new possibilities for separations. The theoretical background for the selection and composition of sieving polymers for optimal separation is illustrated.

Volume I, Part III presents choices of separation media and the CE environment for size-based (fragment length) separations of DNAs. Emphasis is given to new copolymer matrix materials and liquid agaroses.

Volume I, Part IV discusses the fast separations of large DNA molecules and chromosomes using pulsed-field capillary electrophoresis formats. Methods of collection of separated CE fractions for the analysis of supercoiled DNAs are also presented.

Volume I, Part V presents practical choices for the analysis of small therapeutic oligonucleotides, nucleosides, ribonucleotides, and DNA metabolism products by CE. Practical applications for the quality control of nucleotides and oligonucleotides are illustrated.

Volume I, Part VI presents choices for the sensitive CE analysis of nucleotides and DNA metabolism products arising from environmental and cellular damage through disease. Practical applications of mass spectrometry and capillary zone electrophoresis for the detection of oncogenic change are also presented.

Volume II

Volume II, Part I presents practical considerations necessary for very rapid and accurate separations of linear DNAs by CE using short capillaries.

Volume II, Part II presents details of the high sensitivity detection of single nucleotide polymorphism in DNA fragments by CE. Particular emphasis is given to the use of denaturing gradient environments. Techniques include SNuPE, heteroduplex analysis, ARMS analysis of DNA, and SSCP analysis.

Volume II, Part III details various modes of genotyping by CE. Particular emphasis is given to the use of multiplex separations. The techniques include parallel analysis of multiple simple repeat loci in human and CE analysis of gene markers following degenerate oligonucleotide amplification from single cells. Methods for the application of chemical cleavage of mismatched DNA for mutation detection are illustrated.

Volume II, Part IV presents techniques for the quantitative estimation of gene expression using CE. Both quantitative RT-PCR and differential display analyses are highlighted.

Volume II, Part V presents practical choices of separation media and the capillary electrophoresis environment for DNA sequencing by CE. Rapid sequencing at elevated temperature and sequencing on micro- and array-CE devices are discussed. Methods for sequencing with selective primers are also illustrated.

Volume II, Part VI presents the analysis of DNA protein and DNA ligand interactions using CE. Techniques discussed include mobility shift assays and capillar affinity-gel electrophoresis.

Keith Mitchelson
Jing Cheng

Contents

Contents of the Companion Volume

Capillary Electrophoresis of Nucleic Acids

Volume II: Practical Applications of Capillary Electrophoresis

Contributors

DONALD H. ATHA • *Biotechnology Division, National Institute of Standards and Technology, Gaithersburg, MD*
IRIS BARMÉ • *Drug Metabolism and Pharmacokinetics, Novartis Pharma AG, Basel, Switzerland*
JOHN BASHKIN • *Fremont, CA*
PETR BOČEK • *Institute of Analytical Chemistry of the Czech Academy of Sciences, Brno, Czech Republic*
EWAN BOOTH • *Toxicology Department, Shell Research and Technology Center, Shell International Chemicals BV, Amsterdam, Netherlands*
GERARD J. M. BRUIN • *Drug Metabolism and Pharmacokinetics, Novartis Pharma AG, Basel, Switzerland*
TREVOR CARNELLEY • *Department of Public Health Sciences, Faculty of Medicine and Dentistry, University of Alberta, Alberta, Canada*
HONG-YUAN CHEN • *Department of Chemistry, Nanjing University, Nanjing, China*
JING CHENG • *Biochip Research and Development Center, State Key Laboratory for Biomembrane and Membrane Biotechnology, Tsinghua University, Beijing, China; and Aviva Biosciences Corporation, San Diego, CA*
MARCELLA CHIARI • *Institute of Biocatalysis and Molecular Recognition, CNR, Milano, Italy*
BENJAMIN CHU • *Department of Chemistry and Department of Materials Sciences and Engineering, State University of New York at Stony Brook, Stony Brook, NY*
MARINA CRETICH • *Institute of Biocatalysis and Molecular Recognition CNR, Milano, Italy*
OLIVIA DE CARMEJANE • *Department of Chemistry, University of Michigan, Ann Arbor, MI.*
LAWRENCE A. DEDIONISIO • *Lynx Therapeutics, Hayward, CA*
DIETER L. DEFORCE • *Laboratory for Pharmaceutical Biotechnology, University of Ghent, Ghent, Belgium*

CLAUDE DESRUISSEAUX • *Department of Physics, University of Ottawa, Ottawa, Canada*
JOSEPH M. DEVANEY • *Transgenomic, Gaithersburg, MD*
NORMAN J. DOVICHI • *Department of Chemistry, University of Washington, Seattle, WA*
CHRISTOPH HELLER • *Max-Planck-Institute for Molecular Genetics, Berlin, Germany*
JOS HOOGMARTENS • *Laboratory for Pharmaceutical Chemistry and Drug Analysis, Katholieke Universiteit Leuven, Leuven, Belgium*
SYLVAIN J. HUBERT • *Department of Physics, University of Ottawa, Ottawa, Canada*
HALEEM J. ISSAQ • *Analytical Chemistry Laboratory, National Cancer Institute at Frederick, Frederick, MD*
YONGSEONG KIM • *Division of Chemistry and Chemical Engineering, Kyungnam University, Kyungnam, South Korea*
KAREL KLEPÁRNÍK • *Institute of Analytical Chemistry of the Czech Academy of Sciences, Brno, Czech Republic*
CHERYL KOERTJE • *Department of Chemistry, Northeastern University, Boston, MA*
ROY M. A. LAUTAMO • *Agilent Technologies/J & W Scientific, Folsom, CA*
X. CHRIS LE • *Department of Public Health Sciences, Faculty of Medicine and Dentistry, University of Alberta, Alberta, Canada*
JANE LEE • *Experimental Oncology, Cross Cancer Institute, Alberta, Canada*
RAINER LEHMANN • *Abteilung Innere Medizin IV, Endokrinologie, Stoffwechselerkrankungen und Pathobiochemie, Medizinische Klinik und Poliklinik der Universität Tübingen, Tübingen, Germany*
DEHAI LIANG • *Department of Chemistry, State University of New York at Stony Brook, Stony Brook, NY*
HARTMUT M. LIEBICH • *Abteilung Innere Medizin IV, Endokrinologie, Stoffwechselerkrankungen und Pathobiochemie, Medizinische Klinik und Poliklinik der Universität Tübingen, Tübingen, Germany (Tel. 0049-7071 2982715 Fax:0049-7071-293188 Email:Hartmut.Liebich@med.uni-tuebingen.de)*
TIANBO LIU • *Department of Chemistry, State University of New York at Stony Brook, Stony Brook, NY*
YINGJIE LIU • *Biochip Systems, Motorola, Tempe, AZ*
SOFFIA MAGNÚSDÓTTIR • *Facolta di scienze, Istituto Policattedra, Universita degli studi di Verona, Strada le Grazie, Verona, Italy*

ZDEŇKA MALÁ • *Institute of Analytical Chemistry of the Czech Academy of Sciences, Brno, Czech Republic*
DAVID T. MAO • *Agilent Technologies/J & W Scientific, Folsom, CA*
MICHAEL A. MARINO • *Transgenomic, Gaithersburg, MD*
LISA A. MARZILLI • *Department of Pharmacology, Johns Hopkins University Medical School, Baltimore, MD*
BRUCE MCCORD • *Department of Chemistry, Clippinger Laboratories, Ohio University, Athens, OH*
STEVE MENCHEN • *Applied Biosystems Corporation, Foster City, CA.*
KEITH R. MITCHELSON • *Australian Genome Research Facility, University of Queensland, Brisbane; and Walter & Eliza Hall Institute, Parkville, Australia*
MICHAEL D. MORRIS • *Department of Chemistry, University of Michigan, Ann Arbor, MI*
SYLVIA MÜLLER-HAGEDORN • *Abteilung Innere Medizin IV, Endokrinologie, Stoffwechselerkrankungen und Pathobiochemie, Medizinische Klinik und Poliklinik der Universität Tübingen, Tübingen, Germany*
ZEENA E. NACKERDIEN • *Rockefeller University, New York, NY*
STACEY A. NEVINS • *Department of Chemistry, University of Michigan, Ann Arbor, MI*
ANDERS K. PALM • *Molecular Sciences, AstraZeneca R&D Lund, Lund, Sweden*
MARTIN J. PEARCE • *Detection, Defence Evaluation and Research Agency, CBD Porton Down, Salisbury, Wiltshire, UK*
ROYA POURMAND • *Analytical Chemistry and Quality Control, Development Chemistry & Pharmaceutics, Isis Pharmaceuticals, Carlsbad, CA*
MARK A. QUESADA • *Science & Technology, Corning Incorporated, Corning, NY*
RANDOLPH L. RILL • *Department of Chemistry and Institute of Molecular Biophysics, Florida State University, Tallahassee, FL*
JAN SAEVELS • *Algemene Pharmaceutische Bond, Brussel, Belgium*
JEFFREY J. SCHWINEFUS • *Department of Biochemistry, University of Minnesota, St. Paul, MN*
PHILLIPE SERGOT • *Laboratoire de Physico-Chimie Structurale et Macromoléculaire, ESPCI, Paris, France*
BARBARA A. SILES • *University of Denver Research Institute, Denver, CO*
GARY W. SLATER • *Department of Physics, University of Ottawa, Ottawa, Canada*
G. SUSAN SRIVATSA • *ElixinPharma, Encinitas, CA*
ELFRIDE G. VAN DEN EECKHOUT • *Laboratory for Pharmaceutical Biotechnology, University of Ghent, Ghent, Belgium*

ANN VAN SCHEPDAEL • *Laboratory for Pharmaceutical Chemistry and Drug Analysis, Katholieke Universiteit Leuven, Leuven, Belgium*

JEAN-LOUIS VIOVY • *Laboratoire de Physico-Chimie, Section de Recherche, Insitut Curie, Paris, France*

PAUL VOUROS • *Department of Chemistry and the Barnett Institute, Northeastern University, Boston, MA*

NIGEL D. WATSON • *Forensic Science Unit, Department of Pure and Applied Chemistry, University of Strathclyde, Glasgow, UK*

WILLIAM P. WATSON • *Toxicology Department, Shell International Chemicals BV, Shell Research and Technology Center, Amsterdam, Netherlands*

TIM WEHR • *LC Resources, Walnut Creek, CA*

MICHAEL WEINFELD • *Experimental Oncology, Cross Cancer Institute, Alberta, Canada*

STEVEN WINTERS • *Analytical Chemistry and Quality Control, Development Chemistry & Pharmaceutics, Isis Pharmaceuticals, Carlbad, CA*

CHUNHUNG WU • *Chemistry Department, Tamkang University, Taipei Hsien, Taiwan*

WENZHANG XIE • *Biochip Research and Development Center, School of Life Sciences and Engineering, Tsinghua University, Beijing, China*

JAMES Z. XING • *Department of Public Health Sciences, Faculty of Medicine and Dentistry, University of Alberta, Alberta, Canada*

WANLI XING • *Biochip Research and Development Center, School of Life Sciences and Engineering, Tsinghua University, Beijing, China*

DAN-KE XU • *Beijing Institute of Radiation Medicine, Chinese National Biomedical Analysis Center, Beijing, China*

GUOWANG XU • *National Chromatographic Research and Analysis Centre, Dalian Institute of Chemical Physics, Chinese Academy of Sciences, Dalian, China*

JUNQUAN XU • *Biochip Research and Development Center, School of Life Sciences and Engineering, Tsinghua University, Beijing, China*

RONG YANG • *Biochip Research and Development Center, School of Life Sciences and Engineering, Tsinghua University, Beijing, China*

EDWARD S. YEUNG • *Ames Laboratory-USDOE and Chemistry Department, Iowa State University, Ames, IA*

JIANZHONG ZHANG • *Molecular Dynamics, Sunnyvale, CA*

LIANG ZHANG • *Biochip Research and Development Center, School of Life Sciences and Engineering, Tsinghua University, Beijing, China*

MINGDE ZHU • *Life Sciences Group, Bio-Rad Laboratories, Hercules, CA*

I

Analytical Considerations for Capillary Electrophoresis

1

Overview

The Application of Capillary Electrophoresis for DNA Polymorphism Analysis

Keith R. Mitchelson

1. Introduction

The development of capillary electrophoresis (CE) technology has been rapid over the past three years for application to the analytical separation in a variety of biopolymers such as proteins, polysaccharides, and DNA *(1–3)*. CE offers high throughput and high resolution, automatic operation and on-line detection with automatic data acquisition, and this has stimulated its particular application to the analysis of DNA mutations for genetic analysis, and medical diagnosis. These advantages have also provided the impetus to the recent miniaturization of CE equipment to silicon-chip based devices *(3–12)*, which provide all of the above facilities, as well as a significant improvement in the speed and degree of automation of analysis. Significant developments of other miniaturized electro-separation devices including molecular electrophoresis sieves and dielectric trapping using microelectrodes *(13,14)* have been described, which may be able to be integrated with CE to create micro-analytical or preparative devices. This chapter reviews the development of mutation-detection assays for use with CE.

CE instrumentation (*see* **Fig. 1**) consists of two electrolyte chambers linked by a thin capillary, typically of 50–100 µm id. The temperature gradients and distortions that can affect the resolution of bands in conventional gels are virtually absent in CE as the microcapillary facilitates rapid heat dissipation, despite the application of large electric fields. Data on fractionated molecules is acquired automatically by an on-line detector positioned close to the outlet of the capillary. DNA may be detected using the natural UV absorption, although its low sensitivity may limit the detection of samples at low DNA concentrations or low-abundance molecules. Laser-induced fluorescence (LIF) *(15)* provides extremely high sensitivity (approx 100 times UV absorption) through induced detection of additives attached to the DNA, and greatly extends the lower limit of

From: *Methods in Molecular Biology, Vol. 162:*
Capillary Electrophoresis of Nucleic Acids, Vol. 1: Introduction to the Capillary Electrophoresis of Nucleic Acids
Edited by: K. R. Mitchelson and J. Cheng © Humana Press Inc., Totowa, NJ

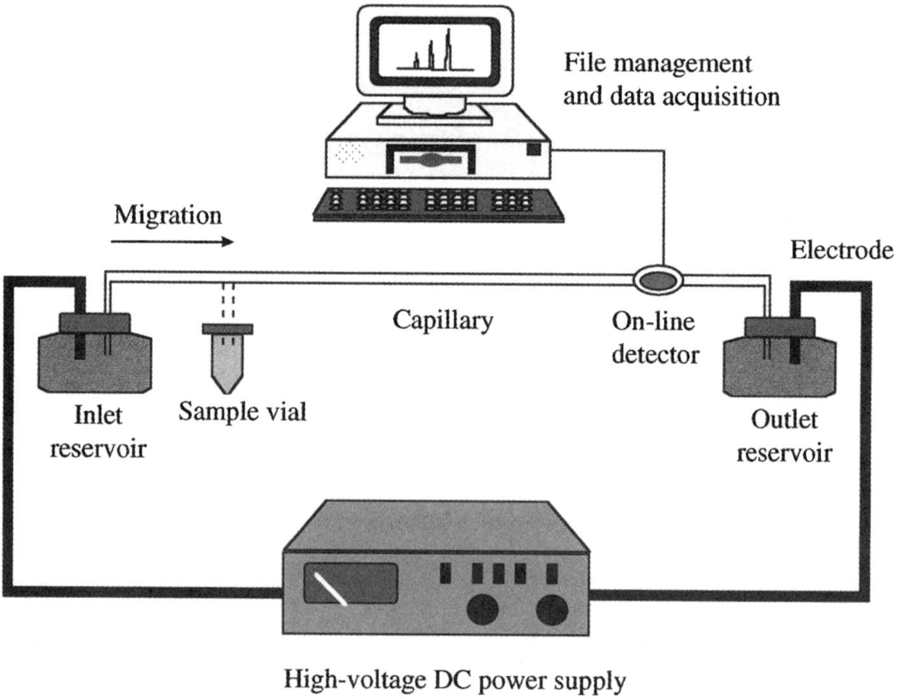

Fig. 1. Diagrammatic representation of a capillary electrophoresis apparatus. The analyte is passaged under an electric field through sieving matrix held within a fine capillary column. The passage of the analyte past a window is detected using a photometric device. The high surface to volume ratio of the capillary aids dissipation of Joule heat. Reprinted from Mitchelson, K. R., Cheng, J. and Kricka, L. J. (1997) Use of capillary electrophoresis for point mutation screening. *Trends in BioTechnology* **15**, 448–458. Copyright (1997), with the permission of Elsevier Science.

concentration at which DNA may be detected. New instrument designs for more efficient laser excitation and signal detection have been described by Yeung ***(16)***, in which the laser light propagates through an array of immersed square capillaries, without undergoing a serious reduction in power. The excitation scheme can be potentially scaled up to hundreds of capillaries to achieve high speed and extremely high throughput.

2. Sieving Media, Electrolytes

As important as the development of new protocols and applications, new electrophoretic sieving-media show potential for high resolution of DNA molecules based on shape ***(17)*** and length ***(18–20)***. Novel sieving media, with copolymers between acrylamide and β-D-glucopyranoside and glucose producing a medium with high-resolving capacity and low viscosity ***(18)*** improve media exchange and handling. Similarly, media comprising block copolymers ***(21–23)*** allow high-resolution sieving matrices to be formed in capillary from low viscosity precursors. New electrophoretic procedures have also been developed that allow better separation of short DNA molecules (<150 bp) by the use of a histidine buffer *(24)*, although fresh buffer must be used to maintain high resolution and DNA-histidine complexes may form under some ionic conditions *(25)*. The application of stepwise increments in the electric field, which improves the theoretical plate number and decreases the run times *(20)*, can result in the improved resolution of longer dsDNA and ssDNA molecules (>500 bp).

An electrophoresis protocol such a "temperature programmed electrophoresis" *(26,27)*, which produces temperature microenvironments in the capillary, allows for efficient separation of heteroduplex DNA molecules based on DNA conformation. Another procedure, "variable-field electrophoresis" *(20)* in which both electric field (and temperature) may be modulated during a run provides for improved separation of single-strand molecules during high-resolution DNA sequencing. Stepwise electric field gradients are useful for both sizing experiments *(17)* and for DNA sequencing of longer fragment *(20)*. Combinations of novel media, buffers, and electrophoresis procedures will continue to provide new paradigms for resolution of DNA and oligonucleotides. An example is the new type of grafted copolymer medium, poly(*N*-isopropylacrylamide)-g-poly(ethylene oxide) (PNI-PAM-g-PEO) solution, which self-coats the capillary tubing *(28)*. A ϕX174/*Hae*III digest could be separated within 24 s using an 8% w/v PNIPAM-g-PEO solution in a 1.5-cm long column with a field voltage of 2400 V.

3. DNA Mutation Detection

The detection of DNA mutations and natural variation has become central to the characterisation and diagnosis of human genetic diseases and is a core to many aspects of molecular biology and medicine/genetics. Several recent reviews provide detailed descriptions of CE applications developed for the detection of point mutations *(3,4,26,29,30)*. Most methods of mutation detection (*see* **Table 1**) can be classified into two general categories: (1) methods to detect known mutations and (2) methods to detect unknown mutations.

Known mutations in target loci are detected by employing various techniques including DNA sequencing *(15,20)*, DNA mini-sequencing and allele-specific amplification (ASA) *(31–33)*, selective primer sequencing *(34–36)*, amplification-refractory-mutation assay (ARMS) *(37)*, and the ligase chain reaction (LCR) *(38,39)*. Other methods that examine changes in defined DNA regions such as polymerase chain reaction-restriction fragment length polymorphism (PCR-RFLP) *(40,41)* and short tandem repeat (STR) length polymorphism *(42–47)* are also used to identify mutations.

The methods for detecting unknown mutations in DNA fragments include DNA sequencing, single-strand-conformation polymorphism (SSCP) *(48–51)*, heteroduplex-polymorphism assay (HPA) *(52,53)*, constant denaturant *(54,55)* and denaturing gradient gel electrophoresis (DGGE) *(29,56)* and chemical or enzymatic cleavage of mismatches (CMC or EMC) *(57–64)*. Frequently, combinations of several complementary techniques are employed to characterize an unknown mutation.

3.1. Polymorphism Detection by DNA Sequencing, Sizing, and Quantification

CE can size-fractionate DNA fragments up to several kilobases in less than 20 min. It has been successfully adapted to standard analytical techniques, in which multiple

Table 1
Abbreviations and Mutation Detection Methods

Summary guide to abbreviations	Reference to conventional mutation-detection methods	Reference to analysis using capillary electrophoresis
ASA, allele-specific amplification	*31–33*	
AFLP, amplified fragment length polymorphism	*34,36*	
ARMS, amplification refractory mutation system	*4*	*37*
ACE, array capillary electrophoresis		*8,16,76,106*
Block co-polymer sieving media		*21,22,28*
CAGE, capillary affinity gel-electrophoresis	*97*	*97–100,105*
CAE, capillary array electrophoresis		*43–46,76,80,89*
Capillary coating materials		*19*
CDGE, constant denaturant capillary gel electrophoresis		*54,55*
CGE, non-denaturing capillary gel electrophoresis		*24–27,29,37*
CEMSA, capillary electrophoresis mobility shift assay		*101–104*
CE/MS, capillary electrophoresis/ mass spectrography		*110–119*
Chip capillary electrophoresis		*3–13,105–108*
CMC, chemical mismatch cleavage	*63,64*	*62*
Collection of capillary electrophoresis sample fractions		*125–127*
Co-polymer sieving media		*18,23,81,85*
CZE, capillary zone electrophoresis		*24–27,29,37,71*
DGGE, denaturing gradient gel electrophoresis	*29*	*29*
DD RT-PCR, differential display reverse-transcriptase-polymerase chain reaction		*67–69,74*
DNA sequencing by capillary electrophoresis	*1,2*	*15,20,76–82*
DOP-PCR, Degenerate oligonucleotide primed-polymerase chain reaction		*10*
Electric field strength	*1,2,29*	*1,2,20,26*
EMC, enzymatic mismatch cleavage	*57–59*	*58,60,61*
ESCE, entangled-solution capillary electrophoresis		*21–24,39,46,53,85*
HPA, heteroduplex DNA polymorphism assay		*52,53*
Integrated micro-analytical device		*3–6,105,107,108*
Isothermal DNA amplification	*90–92*	*93*
LCR, ligase chain reaction		*38,39*
LIF, laser-induced fluorescence		*15,16,38*
LIFP, laser-induced fluorescence polarization		*100*
Microfabrication inside the capillary		*128*
Minisequencing/ primer extension	*31*	*32,33*
MIDAS, mismatch cleavage DNA analysis system		*60,61*
PCR-CE, automated polymerase chain reaction and capillary electrophoresis assay		*6,10,96*
PF-CE, pulsed-field capillary electrophoresis	*1,2*	*1,2,81,83–87*
Pyrosequencing	*124*	
Q RT-PCR, quantitative reverse-transcriptase-polymerase chain reaction	*67*	*66–73*

Table 1 ***(continued)***

Summary guide to abbreviations	Reference to conventional mutation-detection methods	Reference to analysis using capillary electrophoresis
Quantification of genomic alleles		*37,71*
Radial capillary array electrophoresis microplate		*127*
Selective primer amplification analysis	*34,35*	*34–36*
Sieving media		*17–23,83–85*
SSCP, single-stranded DNA conformation polymorphism	*30*	*48–51*
STR, short tandem (microsatellite) repeat	*4*	*12,43–46*
TPCE, temperature-programmed capillary electrophoresis		*26,27,29,48,49*
Ultra-fast capillary electrophoresis		*88,89*

DNA species are size fractionated and parallel analyses are compared for differences in fragment profiles. Such parallel analyses include PCR-RFLP of gene loci ***(40,41)***, for characteristic repeated DNA length polymorphism's such as the bacterial terminal RFLP (T-RFLP) of ribosomal genes ***(47)***, RAPD polymorphisms ***(65)***, amplified fragment length-polymorphism (AFLP) genetic markers ***(34,36)*** and for the analysis of simple tandem repeats (STR) ***(43–46)***. With pressurized or careful electrokinetic loading of samples, CE can be used for quantification of relative amounts of an individual DNA species within a mixture of DNAs ***(37,66–71)***. The direct estimation of the concentration of analytes during CE can be used for the quantification of PCR fragments and applied to the estimation of allele frequency in genome analysis ***(37,71)***.

3.1.1. DNA Sequencing by CE

DNA sequencing by CE is increasingly reported ***(1,15,20,77–80)*** to offer high reproducibility and greatly increased speed compared to planar gels, with elimination of problems associated with electrophoretic distortion and lane tracking ***(75,76)***. Array capillary sequencing allows for simple handling of multiple sample changeovers and very high throughput with sequence reads of more than 1000 bases within 80 min using ACE ***(76,77,80)***. In addition, several technical advances such as, thermal ramping programs ***(20)***, pulsed-field electrophoretic separation ***(81)*** have resulted in improved base-calling and higher resolutions particularly for long DNA fragments, resulting in cost saving through longer sequence reads ***(80)***. An on-column method of sample concentration for capillary-based DNA sequencing was achieved simply by electrokinetic injection of hydroxide ions ***(82)***. Field focusing occurs upon neutralization of the cationic Tris buffer, resulting in a zone of lower conductivity. Even unpurified products of dye-primer sequencing reactions are concentrated at the front of this low-conductivity zone allowing sample injection times as long as 360 s at 50 V/cm. Both resolution and signal strength are excellent relative to highly purified samples and a resolution of at least 0.5 can be generated for fragments up to 650 nt long.

3.1.2. Pulsed-Field CE

Pulsed-field gel electrophoresis formats have found wide application for the improved separation of large (20–100 kb) and very large (several Mb) DNA molecules, however separations are slow because of the low field strength and low mobility of large molecules in solid gels. In a CE format, both entangled solution sieving media ***(83–85)*** and field inversion capillary electrophoresis (FICE) ***(86)*** have been applied for separation of both large and very large DNA molecules with an improvement in speed by 1–2 orders of magnitude ***(1,2,87)***. FICE has also been found to improve the resolution of ssDNA fragments for DNA sequencing, particularly improving resolution of longer fragments and increasing the length of sequence read ***(20,81)***. Pulsed-field CE methods would be suited to a microdevice format where very substantial gains in speed of separation would also be realized.

3.1.3. Mini-Sequencing and Single-Nucleotide Primer Extension

Mini-sequencing is an assay in which a probe is extended by single labeled dideoxy terminator nucleotide if the correct allele is available as template, and incorporation of specific labeled nucleotides can simultaneously identify several SNP alleles at a locus ***(31)***. The application of capillary electrophoresis-laser-induced fluorescence (CE-LIF) for detecting single-nucleotide primer extension (SNuPE) products detected three different point mutations in human mitochondrial DNA ***(32)***. SNuPE analysis using CE-LIF provides high speed and has the potential for multiplexing with the provision of differentially labeled primers.

3.1.4. Selective Primer Amplification and Sequencing

Kambara and colleagues ***(34–36)*** have developed a selective polymerase chain reaction (PCR) using two-base anchored primers to improve the amplification specificity and eliminate base-mispair amplification. This selectivity has been applied to the improvement of genetic marker technology, specifically to the AFLP assay with high fidelity, which when coupled with CE analysis, provides for rapid genotyping and for identification of linked gene markers. This selective amplification primer approach can also be used for amplifying one fragment from a DNA fragment mixture ***(34,35)*** which may then be classified by CE analysis according to its terminal-base sequences and its length. Fragments produce characteristic electropherograms, which may be used to select PCR reaction primers for any fragment in a digestion mixture. Comparison of the electropherograms of two different DNA strands allows selective amplification and specific sequencing of several kilobases of DNA without subcloning, which dramatically simplifies DNA fragment analysis.

3.2. Refractory Amplification Systems

3.2.1. Amplification Refractory Mutation System

ARMS is a PCR-based assay for mutations at known loci in which PCR primers fully complementary to particular alleles amplify a defined product, whereas other alleles are refractory to amplification. Rapid diagnosis of the classic form of human 21-hydroxylase deficiency is achieved by the simultaneous detection of common point mutations in the *P450c21 B* gene by nested PCR-ARMS in conjunction with capillary zone electrophoresis (CZE) in sieving liquid polymers *(37)*. The common mutations in the *CFTR* gene are detected using ARMS in conjunction with entangled solution capillary electrophoresis (ESCE). In the first PCR, genes are selectively amplified, then in the nested reaction ARMS-detected wild-type and mutated alleles are separately pooled and resolved by CZE and detected by the fluorescent dye SYBR Green I using LIF detection. The PCR reaction products could be separated without desalting of samples using the CZE and detected with LIF, without sample preconcentration *(37)*.

3.2.2. Ligase Chain Reaction

Ligase chain reaction (LCR) is a thermocycler-based assay for known mutations in which oligonucleotide primers fully complementary at loci to particular alleles amplify a defined ligation product, whereas on other alleles, primers do not align fully and are thus refractory to cyclic-ligation. Since the dsDNA ligation products are short (typically ~ 50–100 bp) and can be rapidly separated from unincorporated ligation primers (20–25 bp) the LCR mutation assay is suited to very rapid CE separation techniques. Indeed, CE-LIF using short capillary columns (7.5-cm effective length) and fields of 400 V/cm has been used to simultaneously detect three point mutations in human mitochondrial DNA resulting in Leber's hereditary optic neuropathy (LHON) with high speed *(38)*. CE-LIF has also been utilized for the rapid separation and highly sensitivity quantitative detection of <1 μL samples of LCR products amplified from the *lacI* gene in a silicon-glass chip *(39)*.

4. Methods to Detect Unknown Mutations

CE is an emerging technology and its application to the detection of DNA mutation is a recent and rapidly developing area of research. In the following subheadings, some of the mutation-detection methods that can make use of CE are discussed. *See* **Table 1** for a summary of the application of CE to mutation detection.

4.1. Gene Expression Analysis

4.1.1. Quantitative RT-PCR

Variation in the level of specific expression of genes is important in the diagnostic assessment of disease or metabolic states, and may result from genotypic factors in some inherited conditions. Quantitative reverse-transcriptase PCR (QRT-PCR) is used for estimating the activity and expression of particular genes or alleles, by the synthesis of cDNA from mRNA followed by quantitative amplification of the cDNA by PCR *(66–74)*. The direct estimation of RNA-PCR reactions, which reflect in vivo gene expression, may be quantified by the automated on-line detection and peak-area analysis provided by CE *(67,68)*. Importantly, the direct quantification of DNA by its UV absorption in capillary provides higher accuracy and reliability compared to the earlier indirect methods, such as scanning of a stained polyacrylamide gel electrophoresis (PAGE) gel or autoradiogram. Comparisons of levels of mRNA transcript from genes that amplify with different primer pairs cannot be easily made. Amplification of target and competitor in identical reaction environments at each critical enzymatic step in a single tube provides dependable, internally standardized quantitation of low-abundance mRNA transcripts by quantitative competitive QC-RT-PCR, which coupled to CE allows rapid separation and high-sensitivity detection of products ***(69,70,72,73)***. It is expected that the greatly improved capacity of CE systems will allow accurate, high-throughput comparison of cDNAs.

4.1.2. Differential-Display Reverse Transcriptase-PCR

Differential-display reverse-transcriptase-PCR (DD RT-PCR) ***(67,68,74)*** is a technique for the identification and comparison of the relative levels of expression of genes under different tissue conditions, by analyzing mRNA fragments expressed in the tissues. Analysis of DD RT-PCR on CE systems ***(68,74)*** suggests that CE could provide comparable quality to sequencing PAGE, but with greater speed. The high sensitivity of competitive RT-PCR using CE and LIF detection was used to detect down to attogram amounts of the proto-oncogene *ets-2* gene transcript in the brain ***(67)***. This level of sensitivity provides neurobiology with a powerful analytical tool for the role of such genes in brain biology.

4.2. DNA Conformation-Based Assays

CE techniques have also been developed for DNA polymorphism scanning methodologies that detect polymorphism through alteration in the electrophoretic mobility of DNA fragments. Methods including single-strand comformational polymorphism assay SSCP ***(48–51)***, heteroduplex DNA (HPA) analysis ***(52,53)*** and sensitive methods to amplify heteroduplex polymorphism such as constant denaturant capillary electrophoresis (CDCE) ***(54,55)***, which is a modified version of denaturant gradient gel electrophoresis, are frequently employed. Thermal programmed capillary electrophoresis (TPCE) ***(26,27,29)*** in which a variable temperature is increased during a run using computer-controlled thermal ramping have typically been applied for detection of defined polymorphism in genes. These approaches should be applicable to the identification of low frequency mutations, and also applicable to genetic screening of pooled samples for detection of rare DNA variants.

4.2.1. Heteroduplex Polymorphism Assay and Denaturing CE

These techniques involve reannealing of denatured allelic (PCR-amplified) DNA fragments to give a mixture of both wild-type and novel mutant reassociated-heteroduplex dsDNAs. Mismatches within the heteroduplexes result in conformational changes, which retard their electrophoretic mobilities relative to the homoduplex. HPA depends on local conformational changes to duplex DNA, and so the sensitivity decreases with the increase in both DNA-fragment length and the GC content neighboring the mismatch. Denaturing gradient capillary electrophoresis ***(29)*** employs a thermal gradient environment in the capillary during nonisocratic CZE to potentiate the mobility shift differences of heteroduplex molecules at defined temperatures by local strand melting at the mismatch locus. Temperature-programmed CZE has been demonstrated for point mutants ranging from low, intermediate, and high stability. The thermal environment is created by computer manipulation of Joule effects within the capillary and thus lends itself to automated and highly reproducible analyses ***(26,27)***.

Constant denaturant capillary electrophoresis (CDCE), which uses cooperative melting equilibrium of distinct high and low melting DNA domains to identify SNP mutations in the lower melting DNA domain ***(54)***, is used to determine the first mutational spectrum of a mitochondrial sequence in human cells and tissues without prior phenotypic selection. The combination of high-fidelity DNA amplification with CDCE can detect mutants at a fraction of 10^{-6}. Increasing the DNA loading capacity of CDCE also allows for analysis of rare mutations in large, heterogeneous DNA populations, such as samples derived from human tissues. However, serial analyses using different constant capillary conditions are necessary to construct a database of characteristic mobilities ***(55)***. Genotype analysis of a small number of characteristic gene regions can be readily acquired for target genes, whereas differential fluorescent labeling of individual DNA fragments allows simultaneous parallel analysis of different DNAs within the same capillary.

4.2.2. Single-Strand Conformation Polymorphism

SSCP involves dissociation of the double-stranded DNA fragment, after which each of the two single-stranded fragments assumes a folded conformation determined by the specific nucleotide sequence ***(48–51)***. The sensitivity of SSCP analysis depends on whether the mutation affects the folding of the DNA, and hence the electrophoretic mobility of the ssDNA molecules. Nucleotide variants occurring within single-stranded loops may have not effected the ssDNA conformation and consequently may not be detected.

4.2.3. Thermal-Profile SSCP Analysis

Thermal-profile SSCP analysis is a rapid diagnostic tool that is particularly suited to capillary electrophoresis ***(48,49)***. Based on the observation that ssDNA assumes different characteristic mobilities (determined by ssDNA folding) at different temperatures, a database of conformation polymorphism which are characteristic of each mutation in a panel of DNA fragments representing the ten most common mutations of the human *p53* tumor suppressor gene was created. Notably, different mobilities corresponding to different conformational isomers could be detected for single strands electrophoresed under different thermal conditions. The computer control of CE apparatus and rapid dissipation of Joule heating provides unparalleled uniformity and reproducibility of the thermal environment, allowing direct comparison of each analysis. "On-line in-capillary melting of PCR strands" in which strand melting and conformation formation are rapidly achieved, and strands are electrophoretically separated before significant reannealing can occur ***(51)***, would increase the speed of SSCP analysis.

4.3. DNA Mismatch Cleavage Assays

4.3.1. Enzyme Mismatch Cleavage

Enzyme mismatch cleavage (EMC) creates a DNA fragment length polymorphism by using bacteriophage T4 endonuclease VII ***(57)*** or Cleavase ***(58)*** to cleave heteroduplex DNAs at single-nucleotide mismatches and small heteroduplex loops. The majority of the possible mismatch combinations can be rapidly identified by cleavage scanning of 1-kbp fragments of DNA. Preliminary evidence indicates that crude PCR products can be analyzed directly by EMC.

Cleavase nuclease *(58)* and CE analysis of the DNA cleavage pattern have been used to detect mutations in the human genes, and may prove to be a useful system for automated, large-scale genetic screening. Although both CGE and ESCE can fractionate native DNA to about 10–20-bp resolution for fragment sizes up to 0.5 kb, the presence of natural DNA polymorphisms between individuals will probably limit the extension of DNA-cleavage analysis to alleles of defined coding regions, where mismatches could be used for diagnosis.

Interestingly, a sensitive PCR-based RNase I protection assay for detection of mutations in large segments of DNA has been recently developed *(59)*. The desired portion of the gene is amplified by PCR using specific oligonucleotides and hybridized to a labeled RNA probe containing the wild-type sequence. The RNA/DNA hybrid is subsequently digested with RNase I at the sites of RNA-DNA mismatch. The protected RNA fragments can be separated and size fractionated on a denaturing gel CE, providing detection of single-base changes involving all four bases.

4.3.2. Mismatch Identification DNA Analysis System

Siles and colleagues *(60,61)* have developed an enzymatic amplification system named the Mismatch Identification DNA Analysis System (MIDAS) that has an associated isothermal probe amplification step to increase target DNA detection sensitivity to attomole levels. MIDAS uses DNA glycosylases to create an apyrimidinic/apurinic (AP) site at mismatches, which is then cleaved by AP endonucleases/lyases. The mismatch repair enzymes cut the probe at the point of mismatch, and cleaved fragments are thermally unstable and fall off the target allowing another full-length probe to hybridize. Cleaved fluorophore-labelled probes were analyzed in 2 min using a novel CE matrix with LIF-CE and provide definitive evidence of a specific mismatch.

4.3.3. Chemical Mismatch Cleavage

Chemical mismatch cleavage (CMC) which identifies heteroduplex molecules and cleaves the heteroduplex to size-resolvable fragments can also be rapidly analysed using CE *(62)*. Potassium permanganate and tetraethylammonium chloride have recently been developed as safe chemical cleavage reagents *(63,64)* in which all mismatched thymidine residues were modified, with the majority of these showing strong reactivity. The Single Tube Chemical Cleavage of Mismatch Method detects both thymidine and cytidine mismatches without a cleanup step in between the two reactions, without disrupting the sensitivity and efficiency of either reaction. The development of these safe chemical cleavage procedures will speed the application of CE to CMC assays.

5. Future Trends in Diagnostic CE

Particular areas for further research and development in CE technology are those areas directed to the improvement of the automation, throughput, reliability, and sensitivity of the CE analysis. Presently, the high reproducibility of CE runs allows analysis of genetic differences which compares difference profiles between runs, such as the detection of differences in gene expression in target tissues, by analysis of specific gene expression [RT-PCR] *(70)* or by differential display scanning of random genes *(74)*. These methods use analysis of multiple mutations to develop a genetic landscape of genomes and the high throughput and highly reproducible electrochromatograms obtained between CE runs commends it to this application. The development of low viscosity, replaceable sieving matrices *(18,22,23,29)* that enhance the run-to-run reproducibility of CE, combined with the improvement of the speed and efficiency of CE analysis with shorter capillary lengths *(38,88)* and parallel analysis on capillary arrays *(8,44–46)* will each markedly increase the throughput capacity of CE. As noted earlier, significant improvement in the uniformity of signal detection across arrays of capillaries *(16)* and increased sensitivity in signal detection through new instrument designs or novel developments also will result in an improvement in CE performance by an order of magnitude *(89)*.

5.1. New Mutation Assays

5.1.1. Isothermal Strand Displacement Amplification of DNA

Strand displacement amplification (SDA) *(90,91)* and NASBA *(92)* are isothermal in vitro methods for amplification of a specific DNA sequence to concatomer lengths, or for RNA amplification, and both are used for diagnosis of specific point mutations. Burns et al. *(93)* elegantly demonstrates the extremely rapid analysis of a mutation using an integrated microchip system incorporating both DNA amplification and CE separation of SDA products. Molecular beacon probes can be employed in an NASBA amplicon detection system to generate a specific fluorescent signal simultaneously with RNA amplification *(92)*. The assay for NASBA could also be adapted for microchip CE analysis, in which retardation of the mobility of probe after hybridization to the amplificon could be monitored. This would be suitable for applications ranging from one-tube analysis, to high-throughput diagnostics using capillary array electrophoresis (CAE) microchips. Invasive cleavage of oligonucleotide probes *(94)* utilizes thermally stable flap endonucleases to cleave a "flap" structure created by the hybridization of two overlapping oligonucleotides to a target DNA strand. The cleavage is sufficiently specific to discriminate single base differences at the flap junction and is used to isothermally amplify a signal cleavage product of loci in single copy genes from genomic DNA template. Rapid analysis of the cleavage product could be performed in short capillaries or using chip CE.

5.1.2. Subtractive Oligonucleotide Hybridization Analysis

Uhlén and colleagues *(95)* have described a mutational scanning of PCR products by subtractive oligonucleotide hybridization analysis (SOHA) employing surface plasmon resonance to detect quantitative changes in free oligonucleotide(s) following hybridization to a target sequence. The SOHA procedure could be easily adapted to a CE or microchip format in which the quantitative changes in subtractive removal of one or more individual oligonucleotide probes could be quantitatively estimated *(96)*.

5.2. Capillary Affinity Electrophoresis

5.2.1. Capillary Affinity Gel Electrophoresis

Capillary affinity gel electrophoresis (CAGE) is an electrophoretic assay that uses the specificity of antibodies to select defined DNA sequences or structures for DNA mutation detection ***(97–99)***, and combined with rapid analysis by CE allows for the recognition of specific DNA bases or DNA sequence. In contrast, German et al. ***(99)*** also employs a highly selective fluorophore-labeled DNA aptamer against IgE as a selective fluorescent tag for determining IgE by CE-LIF. Separations revealed two zones: free aptamer and aptamer-bound to IgE separated within 60 s. The assay was quantitative, the ratio of free aptamer and bound aptamer varied in proportion to the amount of IgE, which could be detected with a linear dynamic range of 10^5 and a limit of 46 p*M*. The assay is specific for the selected IgG and the target DNA sequences. A novel, ultrasensitive on-line assay for stable affinity complex formation has been developed by Wan and Le ***(100)*** employing laser-induced fluorescence polarization (LIFP) detection of CE separation between reactants. Fluorescence polarization is sensitive to changes in the rotational diffusion arising from molecular association, and is capable of showing formation of affinity complexes during CE separation. The affinity complexes could be easily distinguished from the unbound molecules, despite the relative increase in fluorescence polarization varying with the molecular size of the binding pairs.

5.2.2. CE Mobility Shift Assay

Similar to CAGE, specific DNA-protein interactions are employed in the capillary electrophoretic mobility shift assay (CEMSA) ***(101–104)*** which is used to identify DNA-binding proteins of interest by the retardation of electrophoretic mobility of the complex. The dissociation constant for DNA binding of specific proteins or protein regions can be readily calculated. The assay is rapid and sequence-specific binding can be completed within <2–10 min ***(103,104)***. The direct quantification of DNA during CE is a particular benefit compared to the indirect data obtained from a stained PAGE gel or autoradiogram in determining the specificity and binding constants of the protein and DNA interactions. Although this assay is easily performed in free solution CE, a capillary electrochromatography format could be developed with immobilized protein, creating specific retardation of a mobile DNA ligand, or immobilized DNA specifically retarding a mobile protein ligand. Both CAGE and CEMSA benefit in speed and utility when miniaturized to a chip format ***(105)***.

5.3. High-Throughput CE

5.3.1. Capillary Array Electrophoresis

CAE offers all the advantages of conventional CE, but additionally provides very high throughput with up to 100 samples simultaneously analysed in parallel capillaries ***(8,16,42,44–46,48)***. Array CE can be used for DNA sequence determination ***(77–80)***, or for length polymorphism of PCR-STR alleles ***(42,44–46)***, or for RFLP analysis ***(8)***. The use of CAE for analysis of human STR alleles allows processing of up to 96 multiplex STR samples in under 70 min ***(46)*** and PCR fragment sizing in a glass-wafer microchip with a 96 capillary array in less than 8 min ***(8)***. CAE could conceivably be used for any other analytical procedure applicable to CE such as SSCP and HPA analysis ***(4)***. The advantage of high throughput could benefit direct sequence determination of sets of known characteristic polymorphic genes. CAE apparatus is capable of running and analyzing up to 48 DNA sequencing samples simultaneously, with runs of approx 1 h for about 500 bases, and thus has a throughput on the order of 720 templates/d ***(69)***. As the cost of such high-throughput equipment falls, widespread availability of such rapid analysis platforms will fuel the scope for genomics, as well as for epidemiological studies ***(7)***. In addition, small CAE chips have the capacity to rapidly analyze (2–3 min) differences in the mobility of DNA fragments in parallel in multiple different samples and offers the potential for ultra-high speed, high-throughput genotyping by RFLP analysis ***(44,46)***. CAE microplates will facilitate many types of high-throughput genetic analysis because their high assay speed can provide a throughput 50–100 times greater than that of conventional gels ***(8)***. Fully automated multicapillary electrophoresis systems ***(106)***, in which a novel detection system allows the simultaneous spectral detection and analysis of all 96 capillaries without any moving parts, can process as many as 40 microtiter plates totaling up to 15,000 samples before manual reloading.

5.3.2. CE on a Microchip

Recently, the rapid electrophoretic separation of DNA restriction fragments in 50 s, ranging from 75–1632 bp, in channels in a microfabricated chip formed by two glass-glass layers was reported ***(8,10–12)***. A similar analysis of two PCR-RFLP fragments of 440 and 1075 bp took 140 s (*see* **Fig. 2**). This analysis format has significant advantages over conventional CE, being 10 times quicker and the considerable saving in both reagent and sample (submicroliter level) was coupled with a dramatic reduction in the size of the separation and detection apparatus. The quality of fragment resolution and separation is not compromised by the reduction in analysis format and ultra-high-speed analysis of DNA fragments ranging from dsDNA PCR-products ***(10,107)***, through to single base resolution of DNA-sequencing products ***(108)*** has also been demonstrated on microfabricated CAE chips.

5.3.3. Microelectro-Chromatography and Mass Spectroscopy

Capillary electrochromatography (CEC) is an emerging technology in which electroosmotic flow (EOF) is used to transport the mobile phase in a chromatographic mode, and modifications of the surface of the immobile phase provides selective interactions with analytes for chromatographic separation. The separation of analytes during CEC is a result of interactions with the immobile phase and (partially) to an electrophoretic mobility component ***(9,109)***. CEC can be applied to the separation of neutral compound mixtures and is an alternative to micellar electrokinetic chromatography ***(110,111)***. CEC has recently been developed in a chip format in which the stationary phase is immobilized on the microchannel walls, which are themselves fabricated *in situ* on the chip surface ***(9)***. Mass spectrographic analysis (MS) can be applied with both CEC/MS and CE/MS for the separation of low-weight nucleotide and nucleoside adducts and subsequent mass identification ***(112–119)***. The develop-

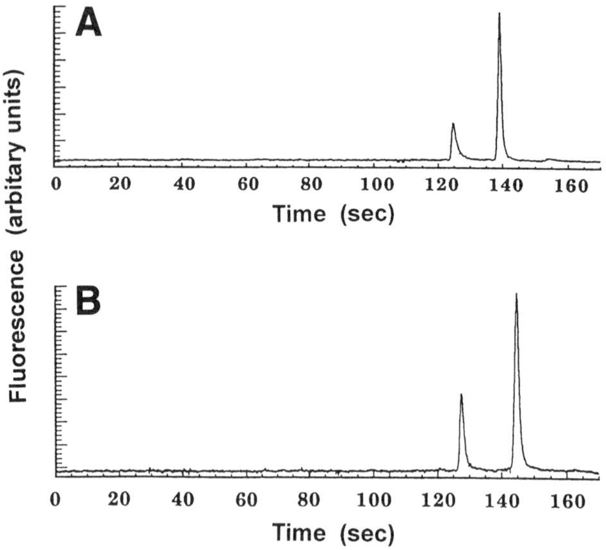

Fig. 2. The rapid electrophoretic separation of DNA restriction fragments in channels in a microfabricated chip formed by two glass-glass layers. The analysis of two PCR-RFLP fragments of 440 bp and 1075 bp took only 140 s. Reprinted from Mitchelson, K. R., Cheng, J., and Kricka, L. J. (1997) Use of capillary electrophoresis for point mutation screening. *Trends in BioTechnology* **15,** 448–458. Copyright (1997), with the permission of Elsevier Science.

ment of suitable modifications to the stationary phase promise to allow selective chromatographic separation of longer DNA molecules. These surface modifications might include base selective binding agents, and specific ligands for particular DNA sequences or structures. Incorporation of entangled polymer solutions in the mobile phase may also add to the electrophoretic mobility component of the separation mechanisms. Vouros and colleagues ***(119)*** have recently demonstrated a novel method to sequence map guanines in oligonucleotides up to 10 bases in length using a chemical apurination reaction followed by analysis using electrospray ionization ion trap mass spectrometry (ESI-MS). Development of very rapid analysis methods employing MS for low molecular weight oligonucleotides and DNA products will advance.

5.3.4. Microchip Electrophoresis of Single Cells

Random amplification of single equivalents of the human genome using the degenerate oligonucleotide primed-polymerase chain reaction (DOP-PCR) was performed in a silicon-glass chip, and immediately applied for locus-specific, multiplex PCR of the dystrophin gene exons which were then analyzed by microchip CE ***(10)***. Whole genome amplification products from DOP-PCR were suitable template for multiplex PCR, requiring the amplicon size <250 bp, but sufficient for detection of defined mutations. The successful analysis of all target multiplex PCR products powerfully demonstrates the feasibility of performing complex PCR assays using miniature microfabricated devices.

5.3.5. Integrated DNA Analysis Devices

The integration of several different apparatus for DNA mutation analysis onto single integrated microdevices has also been progressively reported ***(3–12,93,96,107, 108,120)***. This combination of devices, together on a disposable silicon chip, for thermal-cycled PCR-amplification and for CE and signal detection ***(93,96,120)*** creates a complete microanalytical device. Development of serial electrodes that provide for high "sweeping fields" separation using low-voltage supplies are suited to transportable and hand-held devices ***(121,122)***. Such pocket-sized devices have achieved PCR-amplification in 15 min and CE analysis in 2 min, to provide complete analysis in under 20 min ***(93,107)***, and will bring the analytical possibilities of CE in a easily transportable format. In the near future, real-time monitoring of PCR-amplification reactions with integrated microanalytical devices will allow direct diagnostic analysis such as quantification of gene dosage by PCR-RFLP analysis ***(41)*** and quantification of gene expression by QRT-PCR analysis to the doctors' surgery and to the scientist in the field ***(93,120)***. Recent reports ***(2,123)*** that end-labeled free-flow electrophoresis (ELFSE) can support DNA sequence analysis of several 100 bases in less than 30 min offers an attractive potential alternative to polymer solutions for DNA sequencing in capillaries and microchips, as well as to new non-electrophoretic pyrosequencing techniques ***(124)***.

6. Summary

CE fractions may also be collected and then subjected to additional analysis. Nanoliter fractions containing size or shape fractionated DNA fragments can be collected on moving affinity membranes ***(125)*** or into sample chambers ***(126)***. The exact timing of the collection steps is achieved by determining the velocity of each individual zone measured between two detection points near the end of the capillary. The DNA samples may subsequently be identified by probe hybridization, or by PCR-linked sequencing. Capillary fractions containing metabolites and derivatives of DNA and small DNA adducts can also be sampled, and then characterized directly by highly sensitive MALDI-TOF atomic analysis ***(112–118)*** and ESI-MS ***(118,119)***. The automation and integration of PCR and CE analysis (PCR-CE) on a microchip ***(3–12,96)*** will also contribute greatly to its adoption as the analysis tool of choice. Significantly, these tools will be applied for DNA sequencing ***(75,108)***, for genome mapping ***(65)*** and genotyping ***(42–46)***, for improved certainty in disease detection ***(3–6,107,120)*** and for DNA mutation analysis ***(2–12,27,58)***. Recent improvements in the design CAE arrays and associated equipment such as the radial CAE microplate and rotary confocal signal detection system ***(127)*** overcome some of the detection limitations of linear CAE and microchip devices and allow the parallel genotyping of 96 samples in about 120 s. The integration of microreactive capillary surface assays ***(128)*** and "in-capillary" analysis will also lead to further increases in the speed and sensitivity of CE-based analysis.

The recent announcement of the completion of the first draft sequence of the 90% of the entire human genome within 6 mo by Celera Genomics by sequencing random DNA fragments using several hundred ABI 3700 machines ***(129)*** illustrates the enormous efficiency realized through the automation of DNA sequencing by CAE.

Sequencing was performed at an average rate of ~6×10^9 bases/yr. The CAE machines will now be employed for a concerted resequencing of genome elements to create an extremely high-density polymorphism map of the entire genome ***(130)***. This map will be based principally on single nucleotide polymorphisms, and will catapult human medicine into a new era of closely detailed genetic trait mapping to identify the genetic basis of multi-gene diseases.

Acknowledgment

The support of Forbio Research Pty. Ltd. during the preparation of this article is much appreciated.

References

1. Slater, G. W., Kist, T. B., Ren, H., and Drouin, G. (1998) Recent developments in DNA electrophoretic separations. *Electrophoresis* **19,** 1525–1541.
2. Slater, G. W., Desruisseaux, C., and Hubert, S. J. (2001) DNA separation mechanisms during capillary electrophoresis, in *Capillary Electrophoresis of Nucleic Acids,* Vol. 1 (Mitchelson, K. R. and Cheng, J., eds.), Humana Press, Totowa, NJ, pp. 27–41.
3. Xie, W., Yang, R., Xu, J., Zhang, L., Xing, W., and Cheng, J. (2001) Microchip-based capillary electrophoresis systems, in *Capillary Electrophoresis of Nucleic Acids,* Vol. 1 (Mitchelson, K. R. and Cheng, J., eds.), Humana Press, Totowa, NJ, pp. 67–83.
4. McKenzie, S. E., Mansfield, E., Rappaport, E., Surrey, S., and Fortina, P. (1998) Parallel molecular genetic analysis. *Eur. J. Hum. Genet.* **6,** 417–429.
5. Effenhauser, C. S., Bruin, G. J, and Paulus, A. (1997) Integrated chip-based capillary electrophoresis. *Electrophoresis* **18,** 2203–2213.
6. Eggers, M. and Ehrlich, D. (1995) A review of microfabricated devices for gene-based diagnostics. *Hematol. Pathol.* **9,** 1–15.
7. Schmalzing, D., Tsao, N., Koutny, L., Chisholm, D., Srivastava, A., Adourian, A., Linton, L., McEwan, P., Matsudaira, P., and Ehrlich, D. (1999) Toward real-world sequencing by microdevice electrophoresis. *Genome Res.* **9,** 853–858.
8. Simpson, P. C., Roach, D., Woolley, A. T., Thorsen, T., Johnston, R., Sensabaugh, G. F., and Mathies, R. A. (1998) High-throughput genetic analysis using microfabricated 96-sample capillary array electrophoresis microplates. *Proc. Natl. Acad. Sci. USA* **95,** 2256–2261.
9. Regnier, F. E., He, B., Lin, S., and Busse, J. (1999) Chromatography and electrophoresis on chips: critical elements of future integrated, microfluidic analytical systems for life science. *Trends Biotechnol.* **17,** 101–106.
10. Fortina, P., Cheng, J., Kricka, L. J., Waters, L. J., Jacobson, S. C., Wilding, P., and Ramsey, J. M. (2001) DOP-PCR amplification of whole genomic DNA and microchip-based capillary electrophoresis, in *Capillary Electrophoresis of Nucleic Acids,* Vol. 2 (Mitchelson, K. R. and Cheng, J., eds.), Humana Press, Totowa, NJ, pp. 211–219.
11. Ferrance, J., Giordano, B., and Landers, J. P. (2001) Toward effective PCR-based amplification of DNA on microfabricated chips, in *Capillary Electrophoresis of Nucleic Acids,* Vol. 2 (Mitchelson, K. R. and Cheng, J., eds.), Humana Press, Totowa, NJ, pp. 191–204.
12. Schmalzing, D., Koutny, L., Adourian, A., Chisholm, D., Matsudaira, P., and Ehrlich, D. (2001) Genotyping by microdevice electrophoresis, in *Capillary Electrophoresis of Nucleic Acids*, Vol. 2 (Mitchelson, K. R. and Cheng, J., eds.), Humana Press, Totowa, NJ, pp. 163–173.

13. Cheng, J., Sheldon, E. L., Wu, L., Uribe, A., Gerrue, L. O., Carrino, J., Heller, M. J., and O'Connell, J. P. (1998). Preparation and hybridization analysis of DNA/RNA from E. coli on microfabricated bioelectronic chips. *Nat. Biotechnol.* **16,** 541–546.
14. Righetti, P. G. and Bossi, A. (1998) An isoelectrically trapped enzyme reactor operating in an electric field. *Electrophoresis* **19,** 1075–1080.
15. Lieberwirth, U., Arden-Jacob, J., Drexhage, K. H., Herten, D. P., Muller, R., Neumann, M., et al. (1998) Multiplex dye DNA sequencing in capillary gel electrophoresis by diode laser-based time-resolved fluorescence detection. *Anal. Chem.* **70,** 4771–4779.
16. Lu, S. X., and Yeung, E. S. (1999) Side-entry excitation and detection of square capillary array electrophoresis for DNA sequencing. *J. Chromatogr. A* **853,** 359–369.
17. Sumita, C., Tsuhako, M., and Baba, Y. (1999) Simultaneous analysis of genes by capillary electrophoresis with a laser-induced fluorescence detector using a stepwise field strength gradient. *Chem. Pharm. Bull. (Tokyo)* **47,** 111–113.
18. Chiari, M., Damin, F., Melis, A., and Consonni, R. (1998) Separation of oligonucleotides and DNA fragments by capillary electrophoresis in dynamically and permanently coated capillaries, using a copolymer of acrylamide and β-D-glucopyranoside as a new low viscosity matrix with high sieving capacity. *Electrophoresis* **19,** 3154–3159.
19. Chiari, M. and Cretich, M. (2001) Capillary coatings: choices for capillary electrophoresis of DNA, in *Capillary Electrophoresis of Nucleic Acids,* Vol. 1 (Mitchelson, K. R. and Cheng, J., eds.), Humana Press, Totowa, NJ, pp. 125–138.
20. Inoue, H., Tsuhako, M., and Baba, Y. (1998) Enhanced separation of DNA sequencing products by capillary electrophoresis using a stepwise gradient of electric field strength. *J. Chromatogr. A.* **802,** 179–184.
21. Chu, B., Liu, T., Wu, C., and Liang, D. (2001) DNA capillary electrophoresis using block copolymer as a new separation medium, in *Capillary Electrophoresis of Nucleic Acids,* Vol. 1 (Mitchelson, K. R. and Cheng, J., eds.), Humana Press, Totowa, NJ, pp. 225–238.
22. Magnúsdóttir, S., Viovy, J.-L., and Francois, J. (1998) High resolution capillary electrophoretic separation of oligonucleotides in low-viscosity, hydrophobically end-capped polyethylene oxide with cubic order. *Electrophoresis* **19,** 1699–1703.
23. Madabhushi, R. S. (2001) DNA sequencing in noncovalently coated capillaries using low viscosity polymer solutions, in *Capillary Electrophoresis of Nucleic Acids,* Vol. 2 (Mitchelson, K. R. and Cheng, J., eds.), Humana Press, Totowa, NJ, pp. 309–315.
24. Gelfi, C., Mauri, D., Perduca, M., Stellwagen, N. C., and Righetti, P. G. (1998) Capillary zone electrophoresis of ds-DNA in isoelectric buffers: effect of adding of competing, nonamphoteric ions. *Electrophoresis* **19,** 1704–1710.
25. Stellwagen, N.C., Gelfi, C., and Righetti, P. G. (1999) DNA-histidine complex formation in isoelectric histidine buffers. *J. Chromatogr. A* **838**, 179–189.
26. Righetti, P. G. and Gelfi, C. (1997) Capillary electrophoresis of DNA for molecular diagnostics. *Electrophoresis* **18,** 1709–1714.
27. Gelfi, C., Cremoresi, L., Ferrari, M., and Righetti, P. G. (2001) Point mutation detection by temperature-programmed capillary electrophoresis, in *Capillary Electrophoresis of Nucleic Acids,* Vol. 2 (Mitchelson, K. R. and Cheng, J., eds.), Humana Press, Totowa, NJ, pp. 73–88.
28. Liang, D., Song, L., Zhou, S., Zaitsev, V. S., and Chu, B. (1999) Poly(*N*-isopropylacrylamide)-g-poly(ethyleneoxide) for high resolution and high speed separation of DNA by capillary electrophoresis. *Electrophoresis* **20,** 2856–2863.
29. Righetti, P. G. and Gelfi, C. (1998) Analysis of clinically relevant, diagnostic DNA by capillary zone and double-gradient gel slab electrophoresis. *J. Chromatogr. A.* **806,** 97–112.

30. Mitchelson, K. R., Cheng, J., and Kricka, L. J. (1997) Use of capillary electrophoresis for point mutation screening. *Trends BioTech.* **15,** 448–458.
31. Syvänen, A.-C. (1999) From gels to chips: "minisequencing" primer extension for analysis of point mutations and single nucleotide polymorphisms. *Hum. Mutat.* **13,** 1–10.
32. Piggee, C. A. and Karger, B. L. (2001) Single-nucleotide primer extension assay by capillary electrophoresis laser-induced fluorescence, in *Capillary Electrophoresis of Nucleic Acids,* Vol. 2 (Mitchelson, K. R. and Cheng, J., eds.), Humana Press, Totowa, NJ, pp. 89–94.
33. Tomita-Mitchell, A., Muniappan, B. P., Herrero-Jimenez, P., Zarbl, H., and Thilly, W. G. (1998) Single nucleotide polymorphism spectra in newborns and centenarians: identification of genes coding for rise of mortal disease. *Gene* **223,** 381–391.
34. Matsunaga, H., Kohara, Y., Okano, K., and Kambara, H. (1996) Selecting and amplifying one fragment from a DNA fragment mixture by polymerase chain reaction with a pair of selective primers. *Electrophoresis* **17,** 1833–1840.
35. Okano, K. and Kambara, H. (1996) Fragment walking for long DNA sequencing by using a library as small as 16 primers. *Gene* **176,** 231–235.
36. Li, T., Okano, K., and Kambara, H. (2001) Selective primer sequencing from a DNA mixture by capillary electrophoresis, in *Capillary Electrophoresis of Nucleic Acids,* Vol. 2 (Mitchelson, K. R. and Cheng, J., eds.), Humana Press, Totowa, NJ, pp. 317–336.
37. Carrera, P., Righetti, P. G., Gelfi, C., and Ferrari, M. (2001) Amplification refractory mutation system analysis of point mutations by capillary electrophoresis, in *Capillary Electrophoresis of Nucleic Acids,* Vol. 2 (Mitchelson, K. R. and Cheng, J., eds.), Humana Press, Totowa, NJ, pp 95–108.
38. Muth, J., Williams, P. M., Williams, S. J., Brown, M. D., Wallace, D. C., and Karger, B. L. (1996) Fast capillary electrophoresis-laser induced fluorescence analysis of ligase chain reaction products: human mitochondrial DNA point mutations causing Leber's hereditary optic neuropathy. *Electrophoresis* **17,** 1875–1883.
39. Cheng, J., Shoffner, M. A., Mitchelson, K. R., Kricka, L. J., and Wilding, P. (1996) Analysis of ligase chain reaction products amplified in a silicon-glass chip using capillary electrophoresis. *J. Chromatogr. A* **732,** 151–158.
40. Woolley, A. T., Hadley, D., Landre, P., deMello, A. J., Mathies, R. A., and Northrup, M. A. (1996) Functional integration of PCR amplification and capillary electrophoresis in a microfabricated DNA analysis device. *Anal. Chem.* **68,** 4081–4086.
41. Butler, J. M. and Reeder, D. J. (2001) Detection of DNA polymorphisms using PCR-RFLP and capillary electrophoresis, in *Capillary Electrophoresis of Nucleic Acids,* Vol. 2 (Mitchelson, K. R. and Cheng, J., eds.), Humana Press, Totowa, NJ, pp. 49–56.
42. Zhang, N., Tan, H., and Yeung, E. S. (1999) Automated and integrated system for high-throughput DNA genotyping directly from blood. *Anal. Chem.* **71,** 1138–1145.
43. Mansfield, E. S., Wilson, R. B., and Fortina, P. (2001) Analysis of short tandem repeat markers by capillary array electrophoresis, in *Capillary Electrophoresis of Nucleic Acids,* Vol. 2 (Mitchelson, K. R. and Cheng, J., eds.), Humana Press, Totowa, NJ, pp. 151–161.
44. Mansfield, E. S., Robertson, J. M., Vainer, M., Isenberg, A. R., Frazier, R. R., Ferguson, K., Chow, S., Harris, D. W., Barker, D. L., Gill, P. D., Budowle, B., and McCord, B. R. (1998) Analysis of multiplexed short tandem repeat (STR) systems using capillary array electrophoresis. *Electrophoresis* **19,** 101–107.
45. Wang, Y., Hung, S. C., Linn, J. F., Steiner, G., Glazer, A. N., Sidransky, D., and Mathies, R. A. (1997) Microsatellite-based cancer detection using capillary array electrophoresis and energy-transfer fluorescent primers. *Electrophoresis* **18,** 1742–1749.

46. Woolley, A. T., Sensabaugh, G. F., and Mathies, R. A. (1997) High-speed DNA genotyping using microfabricated capillary array electrophoresis chips. *Anal. Chem.* **69,** 2181–2186.
47. Moeseneder, M. M., Arrieta, J. M., Muyzer, G., Winter, C., and Herndl, G. J. (1999) Optimization of terminal-restriction fragment length polymorphism analysis for complex marine bacterioplankton communities and comparison with denaturing gradient gel electrophoresis. *Appl. Environ. Microbiol.* **65,** 3518–3525.
48. Wenz, H.-M., Baumhueter, S., Ramachandra, S., and Worwood, M. (1999) A rapid automated SSCP multiplex capillary electrophoresis protocol that detects the two common mutations implicated in hereditary hemochromatosis. *Hum. Genet.* **104,** 29–35.
49. Hayashi, K., Wenz, H.-M., Inazuka, M., Tahira, T., Sasaki, T., and Atha, D. H. (2001) SSCP analysis of point mutations by capillary electrophoresis, in *Capillary Electrophoresis of Nucleic Acids,* Vol. 2 (Mitchelson, K. R. and Cheng, J., eds.), Humana Press, Totowa, NJ, pp. 109–126.
50. Arakawa, H., Tsuji, A., Maeda, M., Kamahori, M., and Kambara, H. (1997) Analysis of single-strand conformation polymorphisms by capillary electrophoresis with laser induced fluorescence detection. *J. Pharm. Biomed. Anal.* **15,** 1537–1544.
51. Kuypers, A. W., Linssen, P. C., Willems, P. M., and Mensink, E. J. (1996) On-line melting of double-stranded DNA for analysis of single-stranded DNA using capillary electrophoresis. *J. Chromatogr. Biomed. Appl.* **675,** 205–211.
52. Barbieri, A. M., Soriani, N., Ferlini, A., Michelato, A., Ferrari, M., and Carrera, P. (1996) Seven novel additional small mutations and a new alternative splicing in the human dystrophin gene detected by heteroduplex analysis and restricted RT-PCR heteroduplex analysis of illegitimate transcripts. *Eur. J. Hum. Genet.* **4,** 183–187.
53. Cheng, J., Kasuga, T., Lightly, E., Mitchelson, K. R., Watson, N. D., Martin, L., and Atkinson, D. (1994) PCR heteroduplex polymorphism analysis by entangled solution capillary electrophoresis (ESCE). *J. Chromatogr. A* **677,** 169–177.
54. Muniappan, B. P. and Thilly, W. G. (1999) Application of constant denaturant capillary electrophoresis (CDCE) to mutation detection in humans. *Genet. Anal.* **14,** 221–227.
55. Khrapko, K., Collier, H. A., Li-Sucholeiki, X.-C., André, P. C., and Thilly, W. G. (2001) High resolution analysis of point mutations by constant denaturant capillary electrophoresis (CDCE), in *Capillary Electrophoresis of Nucleic Acids,* Vol. 2 (Mitchelson, K. R. and Cheng, J., eds.), Humana Press, Totowa, NJ, pp. 57–72.
56. Tubiello, G., Carrera, P., Soriani, N., Morandi, L., and Ferrari, M. (1995) Mutational analysis of muscle and brain specific promoter regions of dystrophin gene in DMD/BMD Italian patients by denaturing gradient gel electrophoresis (DGGE). *Mol. Cell. Probes* **9,** 441–446.
57. Babon, J. J., McKenzie, M., and Cotton, R. G. (1999) Mutation detection using fluorescent enzyme mismatch cleavage with T4 endonuclease VII. *Electrophoresis* **20,** 1162–1170.
58. Andersen, P. S., Larsen, L. A., Kanters, Jr., Havndrup, O., Bundgaard, H., Brandt, N. J., Vuust, J., and Christiansen, M. (1998) Mutation detection by cleavase in combination with capillary electrophoresis analysis: Application to mutations causing hypertrophic cardiomyopathy and long-QT syndrome. *Mol. Diagnos.* **3,** 105–111.
59. Murthy, K. K., Shen, S. H., and Banville, D. (1995) A sensitive method for detection of mutations—a PCR-based RNase protection assay. *DNA Cell. Biol.* **14,** 87–94.
60. Siles, B. A., O'Neil, K. A., Tung, D. L., Bazar, L., Collier, G. B., and Lovelace, C. I. (1998) The use of dynamic size-sieving capillary and mismatch repair enzymes for mutant DNA analysis. *J. Capillary Electrophor.* **5,** 51–58.

61. Bazar, L. S., Collier, G. B., Vanek, P. G., Siles, B. A., Kow, Y. W., Doetsch, P. W., Cunningham, R. P., and Chirikjian, J. G. (1999) Mutation identification DNA analysis system (MIDAS) for detection of known mutations. *Electrophoresis* **20,** 1141–1148.
62. Ren, J. (2001) Chemical mismatch cleavage analysis by capillary electrophoresis with laser-induced fluorescence, in *Capillary Electrophoresis of Nucleic Acids,* Vol. 2 (Mitchelson, K. R. and Cheng, J., eds.), Humana Press, Totowa, NJ, pp. 231–239.
63. Roberts, E., Deeble, V. J., Woods, C. G., and Taylor, G. R. (1997) Potassium permanganate and tetraethylammonium chloride are a safe and effective substitute for osmium tetroxide in solid-phase fluorescent chemical cleavage of mismatch. *Nucleic Acids Res.* **25,** 3377–3378.
64. Lambrinakos, A., Humphrey, K. E., Babon, J. J., Ellis, T. P., and Cotton, R. G. (1999) Reactivity of potassium permanganate and tetraethylammonium chloride with mismatched-bases and a simple mutation detection protocol. *Nucleic Acids Res.* **27,** 1866–1874.
65. Valentini, A., Timperio, A. M., Cappuccio, I. and Zolla, L. (1996) Random amplified polymorphic DNA (RAPD) interpretation requires a sensitive method for the detection of amplified DNA. *Electrophoresis* **17,** 1553–1554.
66. Williams, S. J., Schwer, C., Krishnarao, A. S., Heid, C., Karger, B. L., and Williams, P. M. (1996) Quantitative competitive polymerase chain reaction: analysis of amplified products of the HIV-1 gag gene by capillary electrophoresis with laser-induced fluorescence detection. *Anal. Biochem.* **236,** 146–152.
67. Fasco, M. J. (1997) Quantitation of estrogen receptor mRNA and its alternatively spliced mRNAs in breast tumor cells and tissues. *Anal. Biochem.* **245,** 167–178.
68. Schatzmann-Turhani, D., Greber-Platzer, S., Cairns, N., and Lubec, G. (1999) Determination of the protooncogene ets-2 gene transcript in human brain at the atto-gram-level by the use of competitive RT/PCR. *Amino Acids* **16,** 13–19.
69. Martinelli, G., Testoni, N., Montefusco, V., Amabile, M., Saglio, G., Ottaviani, E., Terragna, C., Bonifazzi, F., de Vivo, A., Pane, F., Rosti, G., and Tura, S. (1998) Detection of bcr-abl transcript in chronic myelogenous leukemia patients by reverse-transcription-polymerase chain reaction and capillary electrophoresis. *Haematologica* **83,** 593–601.
70. Borson, N. D., Strausbauch, M. A., Wettstein, P. J., Oda, R. P., Johnston, S. L., and Landers, J. P. (1998) Direct quantitation of RNA transcripts by competitive single-tube RT-PCR and capillary electrophoresis. *Biotechniques* **25,** 130–137.
71. Beckmann, A., Gebhardt, F., and Brandt, B. H. (1998) Direct quantification of polymerase chain reaction fragments using field-amplified sample injection in capillary zone electrophoresis for gene dosage estimation. *J. Chromatogr. B* **710,** 75–80.
72. Williams, S. J. and Williams, P. M. (2001) Quantitation of mRNA by competitive PCR using capillary electrophoresis, in *Capillary Electrophoresis of Nucleic Acids,* Vol. 2 (Mitchelson, K. R. and Cheng, J., eds.), Humana Press, Totowa, NJ, pp. 243–252.
73. Stanta, G., Bonin, S., and Lugli, M. (2001) Quantitative RT-PCR from fixed paraffin-embedded tissues by capillary electrophoresis, in *Capillary Electrophoresis of Nucleic Acids*, Vol. 2 (Mitchelson, K. R. and Cheng, J., eds.), Humana Press, Totowa, NJ, pp. 253–258.
74. Zhao, X. and George, K. S. (2001) Differential display analysis by capillary electrophoresis, in *Capillary Electrophoresis of Nucleic Acids,* Vol. 2 (Mitchelson, K. R. and Cheng, J., eds.), Humana Press, Totowa, NJ, pp. 259–267.
75. Liu, S., Shi, Y., Ja, W. W., and Mathies, R. A. (1999) Optimization of high-speed DNA sequencing on microfabricated capillary electrophoresis channels. *Anal. Chem.* **71,** 566–573.

76. Marsh, M., Tu, O., Dolnik, V., Roach, D., Solomon, N., Bechtol, K., Smietana, P., et al. (1997). High-throughput DNA sequencing on a capillary array electrophoresis system. *J. Capillary Electrophor.* **4,** 83–89.
77. Salas-Solano, O., Carrilho, E., Kotler, L., Miller, A. W., Goetzinger, W., Sosic, Z., Karger, B. L. (1998) Routine DNA sequencing of 1000 bases in less than one hour by capillary electrophoresis with replaceable linear polyacrylamide solutions. *Anal. Chem.* **70,** 3996–4003.
78. Lindberg, P. and Roeraade, J. (2001) DNA sequencing at elevated temperature by capillary electrophoresis, in *Capillary Electrophoresis of Nucleic Acids,* Vol. 2 (Mitchelson, K. R. and Cheng, J., eds.), Humana Press, Totowa, NJ, pp. 289–308.
79. Quesada, M. A. and Menchen, S. (2001) Replaceable polymers for DNA sequencing by capillary electrophoresis, in *Capillary Electrophoresis of Nucleic Acids,* Vol. 1 (Mitchelson, K. R. and Cheng, J., eds.), Humana Press, Totowa, NJ, pp. 139–166.
80. Kheterpal, I. and Mathies, R. A. (1999) Capillary array electrophoresis DNA sequencing. *Anal. Chem.* **71,** 31A–37A.
81. Kim, Y. and Yeung, E. S. (1997) DNA sequencing with pulsed-field capillary electrophoresis in poly(ethylene oxide) matrix. *Electrophoresis* **18,** 2901–2908.
82. Xiong, Y., Park, S. R., and Swerdlow, H. (1998) Base stacking: pH-mediated on-column sample concentration for capillary DNA sequencing. *Anal. Chem.* **70,** 3605–3611.
83. Chen, N., Wu, L., Palm, A., Srichaiyo, T., and Hjerten, S. (1996) High-performance field inversion capillary electrophoresis of 0.1-23 kbp DNA fragments with low-gelling, replaceable agarose gels. *Electrophoresis* **17,** 1443–1450.
84. Palm, A. K. (2001) Capillary electrophoresis of DNA fragments with replaceable low-gelling agarose gels, in *Capillary Electrophoresis of Nucleic Acids,* Vol. 1 (Mitchelson, K. R. and Cheng, J., eds.), Humana Press, Totowa, NJ, pp. 279–290.
85. Madabhushi, R. S. (2001) DNA sequencing in noncovalently coated capillaries using low viscosity polymer solutions, in *Capillary Electrophoresis of Nucleic Acids,* Vol. 2 (Mitchelson, K. R. and Cheng, J., eds.), Humana Press, Totowa, NJ, pp. 309–315.
86. Heller, C., Magnúsdóttir, S., and Viovy, J.-L. (2001) Robust field inversion capillary electrophoretic separation of long DNA fragments, in *Capillary Electrophoresis of Nucleic Acids,* Vol. 1 (Mitchelson, K. R. and Cheng, J., eds.), Humana Press, Totowa, NJ, pp. 293–305.
87. Morris, M. D., Schwinefus, J. J., and de Carmejane, O. (2001) Pulsed-field capillary electrophoresis separation of large DNA fragments, in *Capillary Electrophoresis of Nucleic Acids,* Vol. 1 (Mitchelson, K. R. and Cheng, J., eds.), Humana Press, Totowa, NJ, pp. 307–321.
88. Klepárník, K., Müller, O. M., and Foret, F. (2001) Ultra-fast DNA separations using capillary electrophoresis, in *Capillary Electrophoresis of Nucleic Acids,* Vol. 2 (Mitchelson, K. R. and Cheng, J., eds.), Humana Press, Totowa, NJ, pp. 19–39.
89. Zhang, J., Voss, K. O., Shaw, D. F., Roos, K. P., Lewis, D. F., Yan, J., et al. (1999) A multiple-capillary electrophoresis system for small-scale DNA sequencing and analysis. *Nucleic Acids Res.* **27,** e36.
90. Lizardi, P. M., Huang, X., Zhu, Z., Bray-Ward, P., Thomas, D. C., and Ward, D. C. (1998) Mutation detection and single-molecule counting using isothermal rolling-circle amplification. *Nat. Genet.* **19,** 225–232.
91. Banér, J., Nilsson, M., Mendel-Hartvig, M., and Landegren, U. (1998) Signal amplification of padlock probes by rolling circle replication. *Nucleic Acids Res.* **26,** 5073–5078.
92. Leone, G., van Schijndel, H., van Gemen, B., Kramer, F. R., and Schoen, C. D. (1998) Molecular beacon probes combined with amplification by NASBA enable homogeneous, real-time detection of RNA. *Nucleic Acids Res.* **26,** 2150–2155.

93. Burns, M. A., Johnson, B. N., Brahmasandra, S. N., Handique, K., Webster, J. R., Krishnan, M., et al. (1998) An integrated nanoliter DNA analysis device. *Science* **282,** 484–487.
94. Lyamichev, V., Mast, A. L., Hall, J. G., Prudent, J. R., Kaiser, M. W., Takova, T., et al. (1999) Polymorphism identification and quantitative detection of genomic DNA by invasive cleavage of oligonucleotide probes. *Nat. Biotechnol.* **17,** 292–296.
95. Nilsson, P., Larsson, A., Lundeberg, J., Uhlén, M., Nygren, P. A. (1999) Mutational scanning of PCR products by subtractive oligonucleotide hybridization analysis. *Biotechniques* **26,** 308–316.
96. Woolley, A. T., Lao, K., Glazer, A. N., Mathies, R. A. (1998) Capillary electrophoresis chips with integrated electrochemical detection. *Anal. Chem.* **70,** 684–688.
97. Baba, Y. (2001) Capillary affinity gel electrophoresis, in *Capillary Electrophoresis of Nucleic Acids,* Vol. 2 (Mitchelson, K. R. and Cheng, J., eds.), Humana Press, Totowa, NJ, pp. 347–354.
98. Baba, Y., Sawa, T., Kishida, A., Akashi, M. (1998) Base-specific separation of oligodeoxynucleotides by capillary affinity gel electrophoresis. *Electrophoresis* **19,** 433–436.
99. German, I., Buchanan, D. D., and Kennedy, R. T. (1998) Aptamers as ligands in affinity probe capillary electrophoresis. *Anal. Chem.* **70,** 4540–4545.
100. Wan, Q. H. and Le, X. C. (1999) Fluorescence polarization studies of affinity interactions in capillary electrophoresis. *Anal. Chem.* **71,** 4183–4189.
101. Xian, J. (2001) Capillary DNA-protein mobility shift assay, in *Capillary Electrophoresis of Nucleic Acids,* Vol. 2 (Mitchelson, K. R. and Cheng, J., eds.), Humana Press, Totowa, NJ, pp. 355–367.
102. Stebbins, M. A., Hoyt, A. M., Sepaniak, M. J., and Hurlburt, B. K. (1996) Design and optimization of a capillary electrophoretic mobility shift assay involving trp repressor-DNA complexes. *J. Chromatogr. B* **683,** 77–84.
103. Xue, B., Gabrielsen, O. S., and Myrset, A. H. (1997) Capillary electrophoretic mobility shift assay (CEMSA) a protein-DNA complex. *J. Capillary Electrophor.* **4,** 225–231.
104. Foulds, G. J. and Etzkorn, F. A. (2001) Protein-DNA binding affinities by capillary electrophoresis, in *Capillary Electrophoresis of Nucleic Acids,* Vol. 2 (Mitchelson, K. R. and Cheng, J., eds.), Humana Press, Totowa, NJ, pp. 369–378.
105. Mangru, S. D. and Harrison, D. J. (1998) Chemiluminescence detection in integrated post-separation reactors for microchip-based capillary electrophoresis and affinity electrophoresis. *Electrophoresis* **19,** 2301–2307.
106. Behr, S., Matzig, M., Levin, A., Eickhoff, H., and Heller, C. (1999) A fully automated multicapillary electrophoresis device for DNA analysis. *Electrophoresis* **20,** 1492–1507.
107. Wilding, P., Kricka, L. J., Cheng, J., Hvichia, G., Shoffner, M. A., and Fortina, P. (1998) Integrated cell isolation and polymerase chain reaction analysis using silicon microfilter chambers. *Anal. Biochem.* **257,** 95–100.
108. Soper, S. A., Williams, D. C., Xu, Y., Lassiter, S. J., Zhang, Y., Ford, S. M., and Bruch, R. C. (1998) Sanger DNA-sequencing reactions performed in a solid-phase nanoreactor directly coupled to capillary gel electrophoresis. *Anal. Chem.* **70,** 4036–4043.
109. He, B. and Regnier, F. E. (1998) Microfabricated liquid chromatography columns based on collocated monolith support structures. *J. Pharm. Biomed. Anal.* **17,** 925–932.
110. Ding, J., Barlow, T., Dipple, A., and Vouros, P. (1998) Separation and identification of positively charged and neutral nucleoside adducts by capillary electrochromatography-microelectrospray mass spectrometry. *J. Am. Soc. Mass Spectrom.* **9,** 823–829.

111. Marzilli, L. A., Koertje, C., and Vouros, P. (2001) Capillary electrophoresis–mass spectrometric analysis of DNA adducts, in *Capillary Electrophoresis of Nucleic Acids,* Vol. 1 (Mitchelson, K. R. and Cheng, J., eds.), Humana Press, Totowa, NJ, pp. 395–406.
112. Ding, J. and Vouros, P. (1999) Advances in CE/MS. *Anal. Chem.* **71,** 378A–385A.
113. Apruzzese, W. A. and Vouros, P. (1998) Analysis of DNA adducts by capillary methods coupled to mass spectrometry: a perspective. *J. Chromatogr. A.* **794,** 97–108.
114. Nackerdien, Z. E., Siles, B. A., Nevins, S. A., and Atha, D. H. (2001) Analysis of Environment-Induced DNA Damage by Capillary Electrophoresis, in *Capillary Electrophoresis of Nucleic Acids,* Vol. 1 (Mitchelson, K. R. and Cheng, J., eds.), Humana Press, Totowa, NJ, pp. 407–417.
115. Xing, J. Z., Carnelley, T., Lee, J., Watson, W. P., Weinfeld, M., and Le, X. C. (2001) Assay for DNA damage using immunochemical recogition and capillary electrophoresis, in *Capillary Electrophoresis of Nucleic Acids,* Vol. 1 (Mitchelson, K. R. and Cheng, J., eds.), Humana Press, Totowa, NJ, pp. 419–428.
116. Saevels, J., Van Schepdael, A., and Hoogmartens, J. (2001) Integration of phosphodiesterase-induced degradation of oligonucleotides with capillary polymer-sieving electrophoresis, in *Capillary Electrophoresis of Nucleic Acids,* Vol. 1 (Mitchelson, K. R. and Cheng, J., eds.), Humana Press, Totowa, NJ, pp. 443–457.
117. Xu, G., Liebich, H. M., Lehmann, R., and Müller-Hagedorn, S. (2001) Capillary electrophoresis of urinary normal and modified nucleosides of cancer patients, in *Capillary Electrophoresis of Nucleic Acids,* Vol. 1 (Mitchelson, K. R. and Cheng, J., eds.), Humana Press, Totowa, NJ, pp. 459–474.
118. Deforce, D. L. and Van den Eeckhout, E. E. (2001) Analysis of DNA damage using capillary zone electrophoresis and electrospray mass spectrometry, in *Capillary Electrophoresis of Nucleic Acids,* Vol. 1 (Mitchelson, K. R. and Cheng, J., eds.), Humana Press, Totowa, NJ, pp. 429–441.
119. Marzilli, L. A., Barry, J. P., Sells, T., Law, S. J., Vouros, P., and Harsch, A. (1999) Oligonucleotide sequencing using guanine-specific methylation and electrospray ionization ion trap mass spectrometry. *J. Mass Spectrom.* **34,** 276–280.
120. Burns, M. A., Mastrangelo, C. H., Sammarco, T. S., Man, F. P., Webster, J. R., et al. (1996) Microfabricated structures for integrated DNA analysis. *Proc. Natl. Acad. Sci. USA* **93,** 5556–5561.
121. Ying, F., Mastrangelo, C. H., Burke, D. T., and Burns, M. A. (1998) Electrophoretic separations using sweeping fields. *Electrophoresis* **19,** 1388–1393.
122. Duke, T. A. J., Austin, R. H., Cox, E. C., and Chan, S. S. (1996) Pulsed-field electrophoresis in microlithographic arrays. *Electrophoresis* **17,** 1073–1079.
123. Ren, H., Karger, A. E., Oaks, F., Menchen, S., Slater, G. W., and Drouin, G. (1999) Separating DNA sequencing fragments without a sieving matrix. *Electrophoresis* **20,** 2501–2509.
124. Ronaghi, M., Uhlén, M., and Nyren, P. (1998) A sequencing method based on real-time pyrophosphate. *Science* **281,** 363–365.
125. Magnúsdóttir, S., Heller, C., Sergot, P., and Viovy, J.-L. (2001) Collection of capillary electrophoresis fractions on a moving membrane, in *Capillary Electrophoresis of Nucleic Acids,* Vol. 1 (Mitchelson, K. R. and Cheng, J., eds.), Humana Press, Totowa, NJ, pp. 323–331.
126. Minarik, M., Foret, F., and Karger, B. L. (2000) Fraction collection in micropreparative capillary zone electrophoresis and capillary isoelectric focusing. *Electrophoresis* **21,** 247–254.

127. Shi, Y., Simpson, P. C., Scherer, J. R., Wexler, D., Skibola, C., Smith, M. T., Mathies, R. A. (1999) Radial capillary array electrophoresis microplate and scanner for high-performance nucleic acid analysis. *Anal. Chem.* **71,** 5354–5361.
128. Kenis, P. J., Ismagilov, R. F., and Whitesides, G. M. (1999) Microfabrication inside capillaries using multiphase laminar flow patterning. *Science* **285,** 83–85.
129. http://www.pecorporation.com/press/prccorp011000.html
130. http://pubs.acs.org/hotartcl/ac/99/oct/focus.html

2

DNA Separation Mechanisms During Electrophoresis

Gary W. Slater, Claude Desruisseaux, and Sylvain J. Hubert

1. Introduction

This chapter describes the separation mechanisms used for DNA electrophoresis. The focus is on the concepts that may help the researcher understand the methodology, read the theoretical literature, analyze experimental data, identify the relevant separation regimes, and/or design optimization strategies. But first, let's look at some key definitions. Since capillary electrophoresis (CE) is a "finish line" technique, the mobility $\mu(M)$ and the velocity $v(M)$ of a molecule of size M (in bases or base pairs) in an electric field E are generally defined as:

$$\mu(M) = [v(M)]/E = L/[t(M)E]$$

in which, L is the distance migrated during the elution time $t(M)$. Clearly, this definition is valid only if $v(M)$ is constant during the run. This requires time-independent and uniform (i.e., along the capillary) conditions (e.g., field, temperature, and so on), something that is rarely checked and is rather unlikely. This definition may thus lead, in some cases, to dubious conclusions *(1)*. Successful separation of molecular sizes M_1 and M_2 requires the time spacing $t_1–t_2$ between these electrophoresis peaks to be larger than their full (time) width at half-maximum (FWHM), $w_{1,2}$. A useful measure of the resolution is thus given by the separation factor S, which gives the smallest resolvable size difference:

$$S = [(w_1 + w_2) \times (M_2 - M_1)]/[2 \times (t_1 - t_2)]$$

The FWHM is related to the processes of peak broadening, which can be both process (e.g., diffusion), or instrument related (e.g., sample injection).

1.1. Cations, Capillary Walls, and DNA Molecules

The inner wall of a fused silica capillary is negatively charged when in contact with standard buffers. The wall then attracts cations that form the so-called double-layer, a thin layer of cations of thickness $\lambda_D \cong 1–10$ nm, termed the Debye length. In the

From: *Methods in Molecular Biology, Vol. 162:*
Capillary Electrophoresis of Nucleic Acids, Vol. 1: Introduction to the Capillary Electrophoresis of Nucleic Acids
Edited by: K. R. Mitchelson and J. Cheng © Humana Press Inc., Totowa, NJ

presence of an electric field, the diffuse part of this layer moves and drags the liquid toward the cathode: this is the electroosmotic flow (EOF) *(2)*. It is often preferable to suppress the EOF for DNA applications. For example, the EOF may not be constant along the capillary, and the resulting axial flow gradient may affect the resolution of analytes. More importantly however, one must take the EOF into account in order to test theories, since the apparent mobility is then given by $\mu=\mu_{electrophoretic}+\mu_{EOF}$. The μ_{EOF} contribution can in principle be measured directly, for example by using an uncharged marker. Several buffer additives and capillary wall coating agents (covalent or dynamic) have been proposed in order to eliminate the EOF.

But what happens to DNA during free solution electrophoresis? When E=0, the collective hydrodynamic effects make the DNA coil acts like an impermeable sphere (**Fig. 1A**) with a radius-of-gyration $R_g \sim M^{1/2}$ and a friction coefficient $\xi \sim R_g \sim M^{1/2}$. However, much like the walls, the DNA is negatively charged and attracts a cloud of counter-ions in its vicinity. When $E \neq 0$, the DNA and the cations move in opposite directions. Moreover, the hydrodynamic interactions between the different parts of the DNA molecule are then screened over distances larger than λ_D. This screening kills the collective effects inside the DNA coil, and the friction coefficient now scales like $\xi \sim M$. Since the mobility $\mu(M) = Q(M)/\xi(M)$, where $Q \sim M$ is the charge of the DNA molecule, the resulting mobility is independent of the DNA size M! This is the famous (and electrophoretically unfavorable) free-draining property of DNA (**Fig. 1B**). Size-dependent mobilities are sometimes observed when the ionic strength is too weak to hinder hydrodynamic interactions (giving $\lambda_D \gtrsim R_g$), but this is an extreme case of little practical value. The current use of sieving matrices in CE is because of this microscopic phenomenon.

1.2. Resolution, Diffusion, and Band Broadening

CE being an analytical tool, it is useful to define one or several performance parameters. We have already seen the separation factor S; in our opinion, this is the key parameter. Clearly, small separation factors require large peak spacings and small peak widths. Separation mechanisms that give mobilities $\mu(M)$ with a strong molecular size dependence naturally maximize peak spacing. Peak widths, on the other hand, are related to a number of nonideal physical effects. The latter effects include various stochastic processes (such as diffusion, Joule effects, and wall-analyte interactions), that are characterized by the observation that the final (spatial) peak width increases like $\sqrt{t}$. The factors that lead to a fixed spatial peak width include factors such as the injection sample width and the detector window size. We recommend **ref.** ***2*** for more details about these effects.

However, here it is worth stressing two points that are usually underestimated: (1) The diffusion coefficient of DNA in absence of a field is irrelevant since the separation mechanism often affects diffusion *(3,4)*. (2) Axial gradients (field, temperature, and so on) always reduce the resolution and may make the identification of the main source of band broadening difficult. Although any optimization scheme must try to estimate the relative contribution of the various peak broadening mechanisms, keeping these two points in mind may help the user avoid reaching misleading conclusions.

A

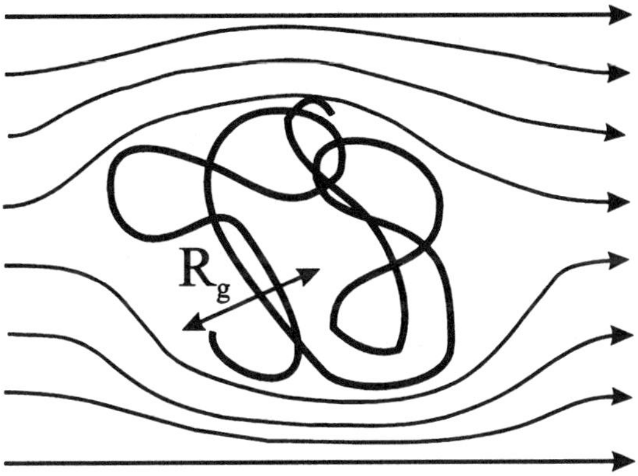

B

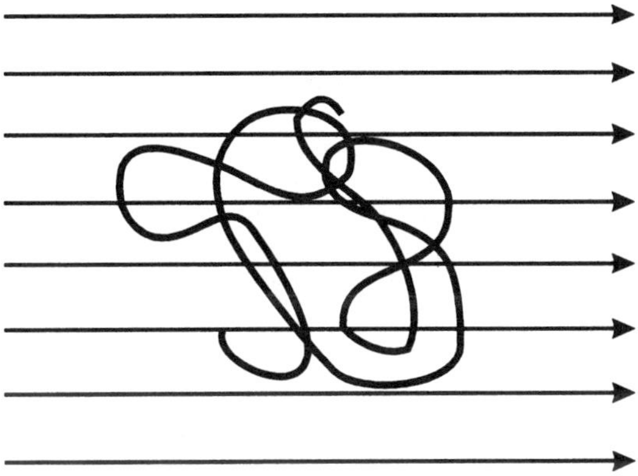

Fig. 1. A random coil DNA molecule with a radius-of-gyration R_g is moving in a fluid. (**A**) In the absence of an electric field, the hydrodynamic interactions between the different parts of the polymer make the coil move like an impermeable sphere of size R_g. (**B**) During electrophoresis, the counter-ions screen the hydrodynamic interactions and the flow penetrates the random coil.

Unfortunately, S does not always allow one to distinguish between the factors limiting the resolution. The plate height $H=\sigma^2/L$ is also useful, where σ^2 is the (spatial) variance of the peak and L is the distance the analyte migrated *(2)*. Note that for a Gaussian peak, one has the relationship $\sigma=FWHM/[8\ln 2]^{1/2}$. Many key factors make unique contributions to the value of H. For example, diffusion gives $H \sim D/v$, where D is the diffusion coefficient and v the velocity, whereas injection and wall-analyte interactions give $H \sim 1/L$ and $H \sim v$, respectively. A study of H as a function of v and L may help the user identify some of the relevant peak broadening mechanisms.

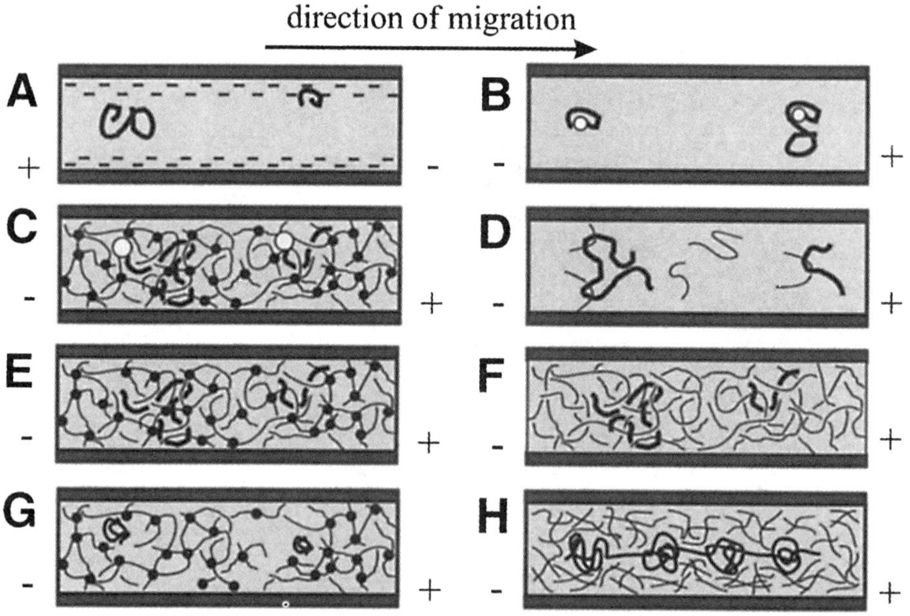

Fig. 2. A schematic view of the various DNA separation mechanisms: (**A**) the mechanism of Iki et al. *(5)* in which smaller DNA fragments are slowed by the wall double-layer. (**B**) ELFSE of DNA: The large empty sphere is a neutral label whose role is to slow down the smaller DNA fragments. (**C**) Trapping electrophoresis in a gel: The large label is sterically trapped if the DNA fragment chooses a too narrow path. (**D**) In Barron's *(13)* ultra-dilute polymer solutions, the DNA molecule drags along a few polymer molecules that it collides with and the latter then act similar to the label in ELFSE mechanism. (**E**) Reptation in a gel: A large DNA molecule must move head-first through the dense pore structure of the gel. (**F**) Reptation in a concentrated polymer solution: the situation is similar to that encountered in gels, except that the sieving matrix is not quenched. (**G**) Ogston regime: The small random coil DNA molecules migrate through the very porous gel structure as hard spheres would do (if their radius-of-gyration R_g is close to the mean pore size $\hat{a}$ of the gel, entropic trapping may occur). (**H**) The process of Ueda et al. *(37)*, in which extremely long DNA molecules move along the field direction, but local clumps form every 5 μm or so.

1.3. Separating DNA Molecules Using EOF

Although EOF is often a nuisance during CE, one can actually exploit it for the purpose of DNA separation. Such a surprising idea was recently demonstrated by Iki, Kim, and Yeung *(5)*. These authors did indeed separate DNA fragments by size, without using any sieving medium or other buffer additive! The principle behind this novel method is that small fragments can access the diffuse layer (at the fused silica-running buffer interface) more readily than larger fragments because of their larger radial diffusion coefficient and their smaller size. The excess of positive charge in this layer increases the friction felt by the negatively charged DNA moving in the opposite direction, thus lowering its electrophoretic mobility against the dominating EOF (**Fig. 2A**).

Therefore, smaller fragments elute first. It is not yet clear whether this new method will turn out to be useful.

1.4. End-Labeled Free Solution Electrophoresis (ELFSE)

The need to use a sieving medium is a CE "dogma" motivated by the free-draining properties of DNA (*see* **Subheading 1.1.**). Unfortunately, loading a viscous sieving medium in a capillary is no easy matter. In **Subheading 1.3.**, we described an example where one can get around this problem by exploiting local ionic gradients. Although many suggestions were made to directly alter the DNA free-draining properties, no data was available until the first separation of long ssDNA molecules in free solution was first reported in 1997 ***(6)***. The idea here is to label one end of the DNA fragments with a neutral object which provides extra friction but no charge (**Fig. 2B**), hence the name end-labeled free solution electrophoresis (ELFSE). Because this affects only the denominator of the ratio $\mu = Q/\xi$, the mobility μ becomes size-dependent and size separation becomes possible. In most cases, the mobility μ of the end-labeled DNA fragment is given by:

$$\mu(M)/\mu_0 = 1/(1 + \alpha/M)$$

for which, μ_0 is the free mobility of DNA and α is the label's friction coefficient (relative to the friction coefficient of one DNA monomer). Analyzing data to find α using this relationship is trivial. The label must be quite mono-disperse in order to obtain sharp peaks. More importantly however, since this equation predicts poor peak spacing when $M >> \alpha$, large labels are required even for sequencing applications (*see* **Note 1**). The only label currently known is streptavidin, which is easily attached to DNA primers; however with $\alpha = 30$, it is too small to provide truly competitive results ***(7)***. The future of this technique will depend on our ability at designing labels for specific applications.

1.5. Trapping Electrophoresis (TE)

Ulanovsky, Drouin, and Gilbert ***(8)*** attempted to separate streptavidin end-labeled DNA (S-DNA) molecules in polyacrylamide gels 7 yr before the advent of ELFSE. The idea behind the trapping electrophoresis (TE) concept is that the S-DNA molecule may become sterically trapped after its unlabeled, leading end enters a pore whose radius (a) is smaller than that (R_s) of the label (**Fig. 2C**). The electric force pulling on the S-DNA molecule then keeps it trapped in this state until a thermally activated backward "jump" makes the leading head of the DNA disengage from the narrow pore and choose a different, wider path. Because the depth of the trap is related to the electric force $QE \sim ME$ pulling on the molecule, larger DNA fragments should be more severely trapped than shorter ones. Experimentally, one does indeed observe a very abrupt (exponential) decrease of the mobility beyond a certain critical molecular size $M_{TE} \sim E^{-2/3}$. Initially, TE seemed to be a promising alternative to normal gel sieving electrophoresis. However, it was shown both experimentally ***(9)*** and theoretically ***(10)*** that the distribution of detrapping times was so wide that the resulting diffusion coefficient made it impossible to exploit the amazingly large inter-peak spacing. This is quite unfortunate, since TE actually kills the famous plateau regime that restricts the usefulness of gel electrophoresis to small DNA sizes (*see* **Subheading 1.8.**)! Pulsed fields were used initially to modulate TE, but with limited success. More recently, Griess and Serwer ***(11)*** designed a ratchet separation process based on TE and special pulsed fields. In this case, a delicate balance between the TE of labeled DNA molecules and the field-dependent mobility of unlabeled DNA molecules leads to remarkable separations where these two different types of molecules move in opposite directions ***(12)***!

1.6. Separation of DNAs in Dilute Polymer Solutions

Because DNA is free-draining, dense polymer matrices are normally used to sieve DNA molecules according to their molecular size. However, Barron et al. ***(13)*** have discovered that even ultra-dilute polymer solutions (whose concentration C can be two orders of magnitude below their entanglement threshold C^*!) can give rapid separation of dsDNA in uncoated capillaries. In essence, this surprising new and unexpected mechanism is like a stochastic ELFSE process, where the migrating DNA fragment collides with and captures free polymer coils which then act as drag-labels (**Fig. 2D**). Although the association between the two polymers is temporary, the hydrodynamic resistance of the captured polymers does reduce the mean velocity of the DNAs.

The theoretical basis of this new process is still immature. Given the average number (n) of polymer chains dragged by the DNA fragment at any given time, we can distinguish between two limits: $n>1$ (the ultra-dilute regime), and $n<1$ (the hyper-dilute regime). In the latter regime, the DNA molecule migrates freely most of the time, but sometimes drags along one polymer chain; this is expected to be less efficient since the total drag force exerted by the polymers is fairly small, whereas the large velocity fluctuations should lead to broader peaks. The $n>1$ regime was studied by Hubert et al. ***(14)***. The value of n is the product of the polymer number concentration (C) and the volume $\mathrm{v}\tau S$ scanned by the DNA during the lifetime (τ) of a DNA-polymer contact, i.e., $n=\tau \mathrm{v} CS$, where S is the collision cross-section and v is the DNA velocity. The mean velocity v is the ratio of the electric force pulling the DNA to the total friction (including the drag of the n neutral polymers). Using several assumptions about S, the lifetimes τ and the various drag forces, Hubert et al. ***(14)*** obtained:

$$\mu(\mathrm{C,M}) = \mu_0 \times \frac{1}{1 + \dfrac{\gamma \mathrm{C}}{1 + \mathrm{b/M}}}$$

for which, γ and b are constants that depend on the contour length of the sieving polymer. This relation could explain the original data of Barron et al. ***(13)*** in the limit of small polymer molecules, however many other sieving regimes cannot be described by this theory. Therefore, there is room for more theoretical development of this phenomenon, whereas video-microscopy of migrating DNA will be helpful in providing practical experimental data for the further development of a theoretical understanding of the process.

Current empirical and theoretical knowledge indicate that longer DNA molecules can only be resolved using longer sieving polymers, probably because the relaxation (or escape) times of the two molecules must be close for optimal resolution. If a mixture of both low and high molecular weight polymers is used as the sieving media, the size range of the DNA separation can be increased *(15)*. Furthermore, stiffer polymers improve the resolution of the elution peaks, because the larger radius-of-gyration of the polymer molecules results in more collisions between the DNAs and the free polymer coils.

DNA fragments up to 23 kb in size can be separated in less than 20 min, an enormous improvement in speed compared to standard gel methods. Pulsed-field capillary electrophoresis can apparently extend the range of dsDNA sizes up to several Mbp long, which can be separated in ultra-dilute polymer solutions *(16)*. The optimum separation conditions for field inversion electrophoresis are not yet fully understood, but simple pulse protocols appear to be effective. The relation between the pulse duration and the polymer escape times τ now needs to be clarified.

1.7. Ogston Sieving in a Gel

The most common CE methods for DNA separations make use of sieving matrices, either crosslinked gels, or entangled polymer solutions (**Fig. 2E,F**). The separation mechanisms are similar in both cases, but are not identical. Considering the long history of gel electrophoresis, it is not surprising that the related theories are more advanced than those relating specifically to polymer solutions. **Subheadings 1.7.–1.9.** review the main theories of gel electrophoresis, whereas in **Subheadings 1.10.** and **1.11.**, we discuss polymer solutions.

As explained in **Subheading 1.1.**, random coil DNA fragments can be described by a parameter, radius-of-gyration $R_g \sim M^{1/2}$ in free solution (**Fig. 1**). In the limit, in which the mean pore size $\hat{a}(C)$ of a gel of concentration C is larger than R_g, it is reasonable to assume that for low field intensities the DNA fragment should migrate through the gel much like an undeformed ball of radius R_g. That is, it should move along a percolating path made of pores in the sieving matrix of size $a > R_g$. The net mobility must then be related to the tortuosity of the path as well as to the DNA-gel fiber interactions (**Fig. 2G**). This is the Ogston regime. The simplest model of such sieving assumes that the ratio μ/μ_0 is equal to the fraction $f(R_g)$ of the gel volume that is made of pores of size $a \geq R_g$ *(17)*. Ogston *(18)* calculated that for a gel composed of long, noncrosslinked and randomly orientated fibers, $\ln[f(R_g)] \cong -(\pi/4) \times [(R_g + r)/\hat{a}]^2$, where r is the fiber radius and $\hat{a}(C) \sim 1/\sqrt{C}$. Putting these several ideas together, we obtain the prediction:

$$\ln(\mu/\mu_0) = -K(M)C$$

in which, $K(M) \sim (R_g + r)^2$ is the retardation factor. The Ferguson plot $\ln[\mu/\mu_0]$ vs C is often used for fundamental electrophoresis studies. Two microscopic parameters can be estimated from this plot: the mean pore size $\hat{a}$ is approximately the size R_g of the DNA coil for which $\ln[\mu/\mu_0] = -1$, while the fiber radius r is given by the extrapolated value of R_g for which $K=0$ (*see* **Fig. 3**).

Our group has recently introduced a more microscopic model of Ogston sieving that takes into account the exact gel structure *(19)*. Although our results indicate that the exponential function must be replaced by a series expansion of the form $\mu/\mu_0 = 1 - b_1C - b_2C^2 - \ldots$, we found that the latter series is also a function of the fundamental

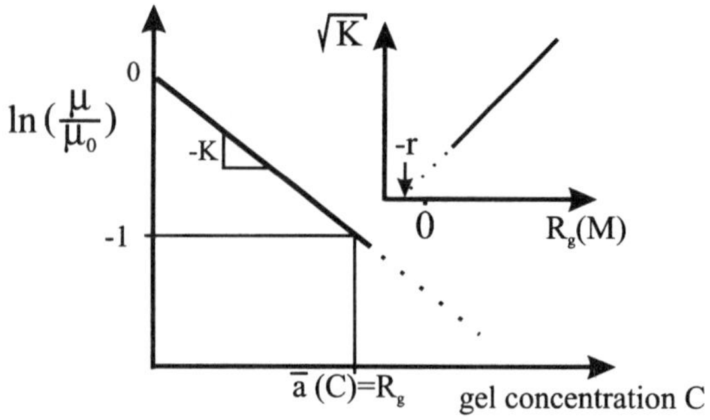

Fig. 3. Data analysis according to the Ogston model. Main figure, semi-log (Ferguson) plot of the relative mobility μ/μ_0 vs the gel concentration C for a molecule of radius R_g. According to this model, the decay should be linear (at least for low concentrations) with a slope –K. The gel giving $\mu/\mu_0 \cong e^{-1}$ is believed to have a pore size $\hat{a}=R_g$. Inset, a plot of the root of the retardation factor K vs the molecular radius R_g. Extrapolating the straight line fit to K=0 gives the gel fiber radius r.

ratio $[(R_g+r)/\hat{a}]^2$; consequently, the data analysis method suggested by the Ogston model remains useful to obtain semiquantitative information about the nature of the gel. However, a high-field Ogston model is still missing from the theoretical armory of researchers.

The Ogston regime normally provides excellent resolution. Yet, the model predicts negligible mobilities when $R_g > \hat{a}(C)$. Experimentally, the mobility of large DNAs does not decrease as fast as predicted by this model, rather the mobility even saturates for very long DNAs (*see* **Fig. 4**), which is a result that certainly contradicts the assumption that DNA molecules represent nondeformable coils (*see* **Note 2**).

1.8. Reptation in a Gel

When the radius-of-gyration $R_g(M)$ of the DNA fragment is larger than the average gel pore size $\hat{a}(C)$, the fragment must deform in order to migrate through the gel. Rigid particles cannot deform, and thus could not migrate over the macroscopic distances that DNA actually moves. Thus, a flexible DNA fragment actually finds its way through the gel, like a snake through thick grass (*see* **Fig. 2E**). As originally proposed by De Gennes for polymer melts, DNA is reptating in a tube of gel pores ***(20,21)***.

The theories related to this electrophoresis concept have evolved considerably over the last 15 yr, and they currently represent our best tool for understanding the separation of long DNA molecules, at least when the electric field is not too high.

Figure 4 presents a schematic mobility-DNA size plot with all the different regimes one might observe in a gel. Very useful is the concept of the effective mean pore size M_a, defined as the molecular size of a DNA molecule for which $R_g(M_a) = \hat{a}$. The Ogston sieving (*see* regime A, **Subheading 1.7.**) and entropic trapping (regime B, **Subheading 1.9.**) are relevant when $R_g \leq \hat{a}$ or $M \leq M_a$.

One distinguishes between two DNA reptation regimes: (1) the reptation of random coil fragments, which applies to small molecules $M_a<M<M^*(E)$; and (2) the reptation

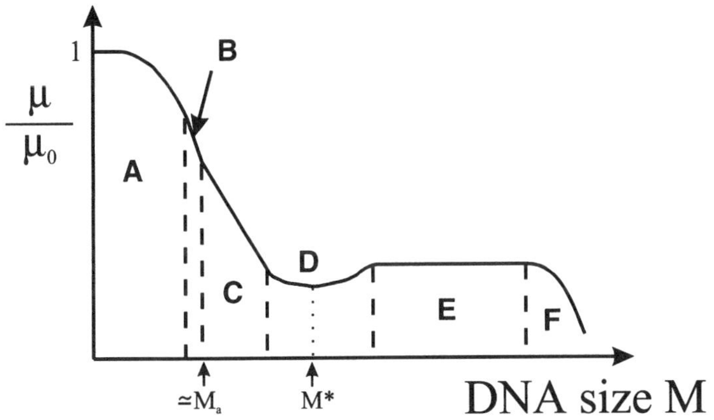

Fig. 4. A schematic diagram of μ/μ_0 vs M, showing the various separation mechanisms that one may observe when using gel electrophoresis. (**A**) Ogston sieving (*see* **Subheading 1.7.**), (**B**) entropic trapping (*see* **Subheading 1.9.**), (**C**) reptation without orientation, (**D**) band inversion, (**E**) the plateau mobility (*see* **Subheadings 1.8.** and **1.10.**), and (**F**) very large DNA fragments occasionally refuse to enter the gel.

of longer fragments (M>M*) oriented along the field direction. In these two limits, the biased reptation model predicts that the mobility of a fragment of size M in a field E should scale according to ***(22–27)***:

$$(\mu/\mu_0) \sim (1/[\min\{M, M^*(E)\}]) \qquad M > M_a$$

for which, the critical size $M^*(E) \sim E^{-\delta}$, with $2 \geq \delta > 0$. For small sizes $M < M^*$, the 1/M scaling law provides excellent separations, in agreement with experimental data. In the opposite limit $M > M^*$, size separation is impossible since the mobility plateaus at a field-dependent value $\mu \sim 1/M^* \sim E^{\delta}$. Clearly, the molecular orientation that leads to the latter effect is a major nuisance. This effect has been studied extensively, and it is due to the fact that the external field biases the direction taken by the DNA-snake as it migrates through the gel. Several pulsed-field methods can be used to circumvent this problem ***(28)***; in essence, these methods recover some size separation beyond M* by controlling the magnitude and/or the direction of the molecular orientation. The reptation model also predicts a minimum in the mobility for $M \cong M^*$ but not for $M \to \infty$ as previously proposed. Although this band inversion effect is observed, in practice it plays a minor role ***(29)***.

To go beyond this simple relation, we must compare the mean pore size â to the DNA persistence length p to see whether the gel is "tight" (p>â), or is not (p<â). In the latter case, the reptation models predicts the following results:

$$(\mu/\mu_0) \cong \begin{cases} (M_a/3M) & M < M^* \\ (\varepsilon/2) & M > M^* \end{cases}$$

Here, we used the reduced field intensity $\varepsilon = \eta \hat{a}^2 \mu_0 E / k_B T < 1$, where k_B is the Boltzmann constant, T the temperature, and η the solvent viscosity ***(26)***. Note that

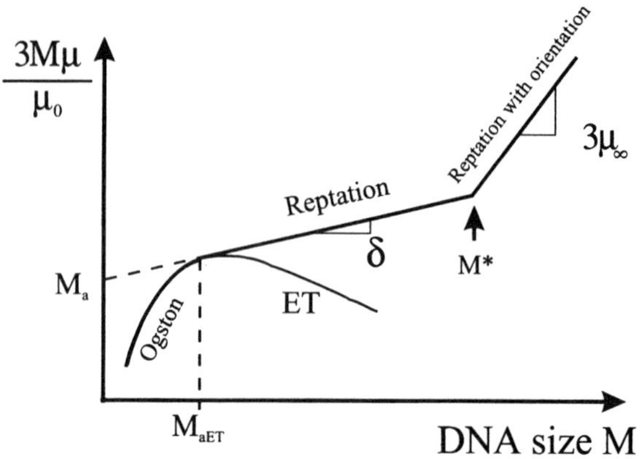

Fig. 5. The reptation plot: 3 *M* μ/μ_0 is plotted vs the DNA molecular size, M. At low field intensity, the Ogston regime is followed by the entropic trapping (ET) regime; the maximum of the curve then defines the effective entropic mean pore size M_{aET}. At high fields, the Ogston regime is followed by the two reptation regimes (without and with orientation, respectively). Note that only the ET regime has a negative slope in this type of plot, which aids identification. The molecular size M* indicates the presence of band inversion (*see* also **Fig. 4**). The extrapolation of the straight line fit for the first reptation regime gives the effective reptation mean pore size M_a. The slopes δ and $3\mu_\infty$ measure the field-driven molecular orientation in the two reptation regimes.

$M^*(E)/M_a \cong 1/\varepsilon$ and $M_a \sim \hat{a}^2$ in this situation. "Tight sieves" are discussed in **Subheading 1.10.**, as they are often more relevant to polymer solutions. An easy way to analyze data is then to plot $3M\mu/\mu_0$ vs size M (*see* **Fig. 5**), as this is the "reptation plot" ***(30)***. The Ogston model predicts a curve with a positive slope and a negative curvature, whereas the two reptation regimes both predict straight lines. In the case of M<M*, the extrapolation of the line gives the characteristic size M_a, as shown in **Fig. 5**, and sometimes some residual orientation (δ). The slope of the other line (M>M*) gives the plateau mobility $\mu_\infty = \varepsilon/2$. For discussion of the entropic trapping (ET) regime, *see* **Subheading 1.9.**

Diffusion is often (but not always) an important contributor to band broadening. In fact, the maximum performance achievable with a separation system is obtained when peak sharpness is diffusion-limited. The reptation model predicts that the electric field actually increases the rate at which diffusion broadens the peaks. This phenomenon has been overlooked for a long time, and often still is! More precisely, it predicts the three following diffusion regimes ***(31)***:

$$D \propto \begin{cases} M^{-2}E^{0} & \text{for } M<M^{**} \\ M^{-1/2}E^{1} & \text{for } M^{**}<M<M^{*} \\ M^{0}E^{3/2} & \text{for } M>M^{*} \end{cases}$$

in which, the new critical size $M^{**} \sim E^{-2/3}$. Tinland and colleagues ***(3)*** have recently confirmed these predictions. This means that the Einstein relationship between the mobility and the diffusion coefficient is invalid during gel electrophoresis. The enhanced peak dispersion is unique to the reptation process and must be taken into account when optimizing a process.

These results point thus to an interesting conclusion regarding DNA sequencing. Peak spacing start to diminish when $M \cong M^*$. However, peak broadening begins to increase before that state, because $M^{**} < M^*$. Therefore, electric field-driven thermal diffusion plays a major role in limiting our ability of sequencing DNA beyond about 1000 bases. This is the reason that pulsed fields are of little use for improving the length of sequencing runs.

1.9. ET in a Gel

DNA is not a rigid ball, since it possesses internal entropy. Together, entropy and Brownian motion of DNA molecules lead to random coil conformations in free solution (*see* **Fig. 1**). This entropy factor remains almost intact in the Ogston limit $R_g < \hat{a}$, although it is quite reduced in the reptation limit $R_g > \hat{a}$. If the electric field is sufficiently low and $R_g \cong \hat{a}$, the entropic forces actually dominate the dynamics of DNA motion in the gel. In this intermediate case between the Ogston and reptation regimes (*see* **Fig. 4**), the DNA coil tries to maximize its internal entropy by "hopping" between voids of size $a > R_g$. However, narrow channels separate these voids and a DNA fragment must lose entropy when it uses the narrow channels. In other words, high entropic energy barriers separate the large voids ***(32)***. This hopping process is called "Entropic Trapping" (ET). Computer simulations and experimental data show that in this situation $\mu(M) \sim 1/M^{1+\alpha}$, in which the exponent $\alpha \geq 0$ is a measure of the strength of the entropic effects ***(30)***. Reptation is recovered if $\alpha = 0$, while dense gels and weak fields typically give $2 > \alpha > 0$ ***(30,33)***. The best way to identify ET occurring is through the reptation plot ***(30)***, since it is the only regime with a negative slope (*see* **Fig. 5**) (*see* **Note 3**). The ET regime is irrelevant in most experimental cases, since electric fields higher than $E \cong 30$ V/cm are enough to overcome the entropic effects. In principle, ET could be exploited to design novel separation methods, but no satisfactory system has yet been built.

1.10. From Gels to Polymer Solutions

Entangled polymer solutions have great practical advantages over gels for most CE applications. In principle, entangled polymers should behave like a gel as long as the DNA residence time in a "pore" is long, compared to the lifetime of the pore itself. This subtle point has recently been studied by Cottet, Gareil, and Viovy ***(34)***. Most of the mechanisms present in gel electrophoresis also play a role in entangled polymer solutions. The main problem actually is in defining the effective mean pore size, â. Polymer solutions are made up of linear polymer molecules of molecular size M_p, radius-of-gyration R_{gp} and concentration C. The group of Viovy has shown that the so-called "blob size" used by polymer physicists is the relevant length scale for describing the sieving properties of a polymer solution. This blob size is given by $\hat{a} \cong 1.43\ R_{gp}\ (C/C^*)^{-3/4}$, whereas the entanglement concentration $C^* \sim M_p/(R_{gp})^3$ (corresponding to one polymer chain per volume $R_{gp}{}^3$) can be directly related to the intrinsic viscosity of the polymer solution *(34–36)*. Examples of blob sizes and overlap concentrations C* for various sieving polymers are tabulated in *(25,34)*.

If the polymer solution is not "tight," i.e., if $p<\hat{a}$, then reptation takes place as described in **Subheading 1.8.** for gels, with the possible exception of the effect of the lifetime of the entanglements *(34)*. Alternatively, in a "tight" sieving solution ($p>\hat{a}$; note that this situation can also happen in a gel) the "head" of the reptating DNA molecule cannot easily change direction although migrating through the "gel" because of the intrinsic stiffness of its backbone. In this case, one still has $\mu/\mu_0 \sim 1/\min\{M,M^*(E)\}$, but we now have three possible regimes when the ratio $\hat{a}/p$ decreases: $M^* \sim 1/E$, $1/E^{2/5}$, and $1/E^2$. The $M^* \sim 1/E^{2/5}$ regime has been observed by Viovy et al. *(34–36)*.

In conclusion, polymer solutions are rather similar to gels for sieving purposes, and data should be analyzed in a similar way. However, one should always be careful because the sieving matrix has its own internal dynamics. The entropic trapping regime has never been reported in polymer solutions, which is not surprising since the entropic traps would have very short lifetimes (*see* **Note 4**).

1.11. Very Concentrated Polymer Solutions

Although DNA fragments can be separated using either unentangled ($C<C^*$; *see* **Subheading 1.6.**), or entangled ($C>C^*$; *see* **Subheadings 1.7.–1.10.**) polymer solutions, the separation of 0.1–10 Mbp fragments normally requires pulsed fields *(28)*. However, Ueda et al. *(37)* have recently shown the fast (in min) separation of *Saccharomyces pombe* chromosomes (up to 5.7 Mbp) in C = 7% linear polyacrylamide solutions ($C^* \cong 0.7\%$), using DC fields! At such high polymer concentrations, the average mesh size of the sieving matrix is $\hat{a} \cong 20$ Å. This mesh size is much smaller than the mean pore size of agarose ($\hat{a} \cong 2{,}000$ Å), or of polyacrylamide ($\hat{a} \cong 200$ Å) gels, and is some 30X times smaller than the persistence length of dsDNA ($p \cong 600$ Å). Clearly, the use of high polymer concentration solutions is a new separation mechanism!

These authors also show that the DNA fragments then migrate in an "I-shape" conformation that has several globular and immobile regions of high density (like lakes connected by straits) separated by about 5 µm in their case (*see* **Fig. 2H**). Fractionation is achieved when the average end-to-end distance becomes independent of the field strength; in this case, the mobility increases like $\mu \sim E^{0.6}$. Obviously, DNA fragments with contour lengths shorter than about 5 µm do not show the same type of motion and the relevant mechanism must then be different.

This new mechanism is based on the formation of temporary "voids" in the concentrated polymer solution. The DNA fills (and probably enlarges) the unstable voids that it encounters during its migration, and the resulting dynamics provides efficient size separation. A theory of this new and exciting mechanism is yet to be derived. Interestingly, high electric fields result in a better separation of the large DNA molecules with this system, whereas low fields yield broad peaks.

2. Notes

1. When analyzing ELFSE experiments, one must make sure that EOF is negligible or that its contribution (e.g., as measured with a marker) is subtracted from the apparent mobility. Plotting $t(M)/t_0$ vs $1/M$, where $t(M)$ is the elution time of a labeled DNA fragment of

size M and t_0 is elution time of unlabeled DNA fragments, then gives a straight line with a slope α, the effective friction coefficient of the label.

2. The scaling law $R_g \sim M^{1/2}$ is valid only if the contour length $\Lambda \sim M$ of the DNA fragment is much larger than its persistence length p, we expect $R_g \sim M$ in the opposite (rigid rod) limit. The following Kratky-Porod equation gives the radius-of-gyration in the general case:

$$R^2_g = (\Lambda p/3) \times [1 - (3p/\Lambda) + 6(p/\Lambda)^2 - 6(p/\Lambda)^3 \times (1 - e^{-\Lambda/P})]$$

   However, this relation does not take into account excluded volume interactions.

3. The mean pore size M_a (which actually gives the molecular size of a DNA molecule whose radius-of-gyration $R_g(M_a)$ is equal to the mean pore size â) can be found from the reptation plot, as shown in **Fig. 5**. The transition between the Ogston and entropic trapping regimes, on the other hand, happens at a molecular size M_{aET}. It is important to realize that these two pore sizes measure different aspects of the gel randomness. In practice, $M_{aET} > M_a$. Once M_a is found, the mean pore size â can be calculated using the Kratky-Porod equation (*see* **Note 2**) with R_g= â and Λ=Mb, where b is the contour length of a DNA monomer (one then needs to know the persistence length p). Moreover, the reptation plot gives directly the reduced plateau mobility $\mu_\infty = \varepsilon/2$, as shown.
4. The references ***(2,4,20,21,25–28,36,38–43)*** are useful review articles and book chapters that we recommend for detailed examination.

References

1. Desruisseaux, C., Slater, G. W., and Drouin, G. (1998) The gel edge electric field gradients in denaturing polyacrylamide gel electrophoresis. *Electrophoresis* **19,** 627–634.
2. Grossman, P. D. (1992) Factors affecting the performance of capillary electrophoresis separations: Joule heating, electroosmosis, and zone dispersion, in *Capillary Electrophoresis Theory and Practice* (Grossman, P. D. and Colburn, J. C., eds.), Academic Press, San Diego, pp. 3–43.
3. Meistermann, L. and Tinland, B. (1998) Band broadening in gel electrophoresis of DNA: measurements of longitudinal and transverse dispersion coefficients. *Phys. Rev. E* **58,** 4801–4806.
4. Issaq, H. J. (2001) Parameters affecting capillary electrophoretic separation of DNA, in *Capillary Electrophoresis of Nucleic Acids,* Vol. 1 (Mitchelson, K. R. and Cheng, J., eds.), Humana Press, Totowa, NJ, pp. 189–199.
5. Iki, N., Kim, Y., and Yeung, E. S. (1996) Electrostatic and hydrodynamic separation of DNA fragments in capillary tubes. *Anal. Chem.* **68,** 4321–4325.
6. Heller, C., Slater, G. W., Mayer, P., Dovichi, N. J., Pinto, D., Viovy, J.-L., and Drouin, G. (1998) Free-solution electrophoresis of DNA. *J. Chromatogr. A* **806,** 113–121.
7. Ren, H., Karger, A. E., Oaks, F., Menchen, S., Slater, G. W., and Drouin, G. (1999) DNA sequencing using end-labeled free-solution electrophoresis. *Electrophoresis,* **20,** 2501–2509.
8. Ulanovsky, L., Drouin, G., and Gilbert, W. (1990) DNA trapping electrophoresis. *Nature* **343,** 190–192.
9. Desruisseaux, C., Slater, G. W., and Drouin, G. (1998) On using DNA trapping electrophoresis to increase the resolution of DNA sequencing gels. *Macromolecules* **31,** 6499–6505.
10. Slater, G. W., Desruisseaux, C., Villeneuve, C., Guo, H. L., and Drouin, G. (1995) Trapping electrophoresis of end-labeled DNA: An analytical model for mobility and diffusion. *Electrophoresis* **16,** 704–712.

11. Griess, G. A., and Serwer, P. (1998) Gel electrophoretic ratcheting for the fractionation of DNA-protein complexes. *Biophys. J.* **74,** A71.
12. Desruisseaux, C., Slater, G. W., and Kist, T. B. L. (1998) Trapping electrophoresis and ratchets: a theoretical study for DNA-protein complexes. *Biophys. J.* **75,** 1228–1236.
13. Barron, A. E., Blanch, H. W., and Soane, D. S. (1994) A transient entanglement coupling mechanism for DNA separation by capillary electrophoresis in ultra-dilute polymer solutions. *Electrophoresis* **15,** 597–615.
14. Hubert, S. J., Slater, G. W., and Viovy, J.-L. (1996) Theory of capillary electrophoresis separation of DNA using ultra-dilute polymer solutions. *Macromolecules* **29,** 1006–1009.
15. Bunz, A. P., Barron, A. E., Prausnitz, J. M., and Blanch, H. W. (1996) Capillary electrophoretic separation of DNA restriction fragments in mixtures of low- and high-molecular-weight hydroxyethylcellulose. *Ind. Eng. Chem.* **35,** 2900–2908.
16. Kim, Y. and Morris, M. D. (1995) Rapid pulsed-field capillary electrophoretic separation of megabase nucleic acids. *Anal. Chem.* **67,** 784–786.
17. Rodbard, D. and Chrambach, A. (1970) Unified theory for gel electrophoresis and gel filtration. *Proc. Nat. Acad. Sci. USA* **65,** 970–977.
18. Ogston, A. G. (1958) The spaces in a uniform random suspension of fibers. *Trans. Faraday Soc.* **54,** 1754–1757.
19. Mercier, J.-F. and Slater, G. W. (1998) An exactly solvable Ogston model of gel electrophoresis IV: Sieving through periodic three-dimensional gels. *Electrophoresis* **19,** 1560–1565.
20. de Gennes, P. G. (1979) *Scaling Concepts in Polymer Physics.* Cornell University Press, NY.
21. Doi, M. and Edwards, S. F. (1986) *The Theory of Polymer Dynamics.* Oxford University Press, NY.
22. Duke, T. A. J., Viovy, J.-L., and Semenov, A. N. (1994) Electrophoretic mobility of DNA in gels I: new biased reptation theory including fluctuations. *Biopolymers* **34,** 239–248.
23. Semenov, A. N., Duke, T. A. J., and Viovy, J.-L. (1995) Gel electrophoresis of DNA in moderate fields: the effect of fluctuations. *Phys. Rev. E* **51,** 1520–1537.
24. Heller, C., Duke, T. A. J., and Viovy, J.-L. (1994) Electrophoretic mobility of DNA in gels II: systematic study in agarose gels. *Biopolymers* **34,** 249–259.
25. Heller, C. (2001) Influence of polymer concentration and polymer composition on capillary electrophoresis of DNA, in *Capillary Electrophoresis of Nucleic Acids,* Vol. 1 (Mitchelson, K. R. and Cheng, J., eds.), Humana Press, Totowa, NJ, pp. 111–123.
26. Viovy, J.-L. (2001) Mechanisms of polyelectrolyte gel electrophoresis. Submitted for publication. This is a comprehensive review by one of the leading researcher in this field.
27. Slater, G. W. (1997) Electrophoresis theories, in *Analysis of Nucleic Acids by Capillary Electrophoresis* (Heller, C., ed.), Vieweg and Son, Wiesbaden, pp. 24–66.
28. Burmeister, M., and Ulanovsky, L., eds. (1992) *Pulsed-Field Gel Electrophoresis: Protocols, Methods and Theories,* Vol. 12. Humana Press, Totowa, NJ, pp. 1–467.
29. Noolandi, J., Rousseau, J., Slater, G. W., Turmel, C., and Lalande, M. (1987) Self-trapping and anomalous dispersion of DNA in electrophoresis. *Phys. Rev. Lett.* **58,** 2428–2431.
30. Rousseau, J., Drouin, G., and Slater, G. W. (1997) Entropic trapping of DNA during gel electrophoresis: effect of field intensity and gel concentration. *Phys. Rev. Lett.* **79,** 1945–1948.
31. Semenov, A. N. and Joanny, J.-F. (1997) Formation of hairpins and band broadening in gel electrophoresis of DNA. *Phys. Rev. E* **55,** 789–799.
32. Arvanitidou, E. and Hoagland, D. (1991) Chain-length dependence of the electrophoretic mobility in random gels. *Phys. Rev. Lett.* **67,** 1464–1466.
33. Smisek, D. L. and Hoagland, D. A. (1990) Electrophoresis of flexible macromolecules: evidence of a new mode of transport in gels. *Science* **248,** 1221–1223.

34. Cottet, H., Gareil, P., and Viovy, J.-L. (1998) The effect of blob size and network dynamics on the size-based separation of polystyrenesulfonates by capillary electrophoresis in the presence of entangled polymer solutions. *Electrophoresis* **19,** 2151–2162.
35. Mitnik, L., Salomé, L., Viovy, J.-L., and Heller, C. (1995) Systematic study of field and concentration effects in capillary electrophoresis of DNA in polymer solutions. *J. Chromatogr. A* **710,** 309–321.
36. Viovy, J.-L. and Heller, C. (1996) Principles of size-based separations in polymer solutions, in *Capillary Electrophoresis in Analytical Biotechnology* (Righetti, P. G., ed.), CRC Press, Boca Raton, pp. 477–508.
37. Ueda, M., Oana, H., Baba, Y., Doi, M., and Yoshikawa, K. (1998) Electrophoresis of long DNA molecules in linear polyacrylamide solutions. *Biophys. Chem.* **71,** 113–123.
38. Slater, G. W., Kist, T.B.L., Ren, H., and Drouin, G. (1998) Recent Developments in DNA Electrophoretic Separations. *Electrophoresis* **19,** 1525–1541.
39. Quesada, M. A. (1997) Replaceable polymers in DNA sequencing by capillary electrophoresis. *Curr. Opin. Biotech.* **8,** 82–93.
40. Quesada, M. A. and Menchen, S. (2001) Replaceable polymers for DNA sequencing by capillary electrophoresis, in *Capillary Electrophoresis of Nucleic Acids,* Vol. 1 (Mitchelson, K. R. and Cheng, J., eds.), Humana Press, Totowa, NJ, pp. 139–166.
41. Dovichi, N. J. (1997) DNA sequencing by capillary electrophoresis. *Electrophoresis* **18,** 2393–2399.
42. Dovichi, N. J. and Zhang, J.-Z. (2001) DNA sequencing by capillary array electrophoresis, in *Capillary Electrophoresis of Nucleic Acids,* Vol. 1 (Mitchelson, K. R. and Cheng, J., eds.), Humana Press, Totowa, NJ, pp. 85–94.
43. Barron, A. E. and Blanch, H. W. (1995) DNA separations by slab gel and capillary electrophoresis: theory and practice. *Sep. Purif. Methods* **24,** 1–118.

3

Purification Methods for Preparing Polymerase Chain Reaction Products for Capillary Electrophoresis Analysis

Joseph M. Devaney and Michael A. Marino

1. Introduction

The analysis of polymerase chain reaction (PCR) products by capillary electrophoresis (CE) is often compromised by the presence of a high concentration of salt. Typical PCR reactions contain 50 m*M* of KCl and 1.5 m*M* of $MgCl_2$, which cause irreproducible electrokinetic injections and migration times as a result of the chloride ions competing with the DNA for migration into the capillary ***(1,2)***. An alternative mode of injection, pressure injection, eliminates the problems induced by salt however it significantly reduces the resolution of analytes ***(3–11)***. A high salt concentration also induces localized heating within the column; hence, PCR products must be desalted prior to CE analysis (*see* **Note 1**). In addition, the PCR mixtures contain unused dNTPs, primers, and primer dimers. These moieties inhibit the injection of the desired DNA fragment as well as hinder any additional analysis, such as sequencing. In a previous study, Devaney et al. ***(12)*** evaluated commercial cleanup systems for their effectiveness in preparing PCR products for CE analysis based on the resolution, the salt removal, the DNA recovery, the processing time, and the cost of procedures. In the following discussion, several protocols for PCR cleanup for CE analysis are outlined, including membrane dialysis and high performance liquid chromatography (HPLC) ***(13)***.

2. Materials

2.1. PCR Amplification

The HUMTHO1 locus alleles ***(8,10)*** are the PCR products used to demonstrate the membrane dialysis cleanup procedure. The alleles 8 and 10 are 191 and 199 bp, respectively (approx 116,350 and 131,750 Dalton) when generated using the ABI PRISM STR protocol (Applied Biosystems, Foster City, CA). The 50-µL PCR mixture contains:

From: *Methods in Molecular Biology, Vol. 162:*
Capillary Electrophoresis of Nucleic Acids, Vol. 1: Introduction to the Capillary Electrophoresis of Nucleic Acids
Edited by: K. R. Mitchelson and J. Cheng © Humana Press Inc., Totowa, NJ

1. 10–20 ng of DNA template.
2. 0.5 μ*M* HUMTHO1 primer pair.
3. 200 μ*M* each of dNTP.
4. 10 m*M* Tris-HCl, pH 8.3.
5. 50 m*M* KCl.
6. 1.5 m*M* $MgCl_2$.
7. 0.1% Triton X-100.
8. 2.5 U of Ampli*Taq*, DNA Polymerase (Perkin-Elmer, Norwalk, CT).

2.2. Membrane Dialysis

1. Membrane filter VS-0.025 μm (Millipore Corporation, Bedford, MA).
2. Deionized water.
3. Medium size weighing dish.
4. Timer.

2.3. High Performance Liquid Chromatography

The ion-pair reversed-phase-high performance liquid chromatography (IPRP-HPLC) system consists of the following:

1. WAVE® DNA Fragment Analysis system with gradient pump and UV detector (Transgenomic, Inc., San Jose, CA).
2. DNAsep® column 4.6 mm × 50 mm (Transgenomic, Inc.).
3. Variable volume sample loop.
4. HPLC grade acetonitrile (ACN) (EM Science, Gibbstown, NJ).
5. Triethylammonium acetate (TEAA) (Applied Biosystems, Foster City, CA).
6. Glacial acetic acid (Mallinkrodt Specialty Chemicals, Paris, KY).
7. 0.6-mL Eppendorf tubes.

2.4. CE (Genotyping)

1. Fused silica capillaries (Polymicro Technologies, Phoenix, AZ) with an internal diameter of 75 μm and an effective length determined by the CE system.
2. 5% polymer DNA Fragment Analysis Buffer (Applied Biosystems).

2.5. CE (Sequencing)

All sequence analyses used the ABI Prism Sequencing Kit and the 310 Genetic Analyzer (Perkin-Elmer Applied Biosystems Division, Foster City, CA). The cycle sequencing of the HPLC purified alleles follows the manufacturer's instructions using the forward primer of the THO1 locus. The following are used for CE sequencing:

1. 1X Analysis buffer with EDTA.
2. Fused silica capillaries 50 μm × 47 cm (effective column length 30 cm).
3. High Resolution Polymer POP 6 (PE-Applied Biosystems Division).
4. Thermal suppression reagent (TSP).

3. Methods

3.1. PCR Amplification

The PCR amplification parameters for the HUMTHO1 locus are: 4 min at 95°C, 10X (94°C 30 s, 64°C 30 s, 70°C 45 s), and 20X (90°C 30 s, 64°C 30 s, 70°C 45 s).

3.2. Membrane Dialysis

1. Fill the weighing dish with deionized water.
2. Float the membrane filter (MF-Millipore, Milton, MA) on deionized water with the matte side down (shiny side up) (*see* **Note 2**).
3. Pipet 5–25 μL of the sample on top of the center of the membrane filter.
4. Allow the sample to float for 15 min for ion diffusion to occur (*see* **Note 3**).
5. Transfer the sample by pipeting from the membrane into a clean vial.

3.3. CE Analysis of Crude and Membrane Dialysis PCR Product (Genotyping)

1. PCR products (purified and unpurified) are loaded onto the CE system.
2. Perform the CE analysis at a column temperature of 30°C.
3. Electrokinetically inject the sample at –7 kV for 7 s.
4. Run the sample at a voltage of –11 kV for empirically determined periods.
5. Detect the DNA using UV absorption at 260 nm.
6. **Figure 1** represents the CE analysis of the HUMTHO1 *(8,10)* PCR products. Both electropherograms contain the same two alleles of 191 and 199 bp in length. The unpurified PCR product shown in panel A represents the primers, but no PCR-amplified alleles are seen. The desalted PCR sample is shown in panel B, in which both of the alleles are detected after 17 min run time.

3.4. High Performance Liquid Chromatography

The parameters used for the IPRP-HPLC separations and purification of the DNA samples are as follows (*see* **Note 4**):

1. Flow rate of 0.9 mL/min.
2. The temperature of column is maintained at 55°C.
3. The elution gradient mixture consists of two eluants (eluant A = 0.1 *M* TEAA, eluant B = 0.1 *M* TEAA/25% ACN). Eluant B is increased from 39–89% over a period of 7 min. To prepare the 25% acetonitrile solution in 0.1 *M* TEAA: 50 mL of a 2 *M* TEAA stock solution is added to 250 mL of acetonitrile in a 1000-mL volumetric flask and the final volume is made to be 1000 mL with deionized water. The pH of the mobile phase is adjusted to pH 7.0 with using glacial acidic acid.
4. Variable wavelength detector is set at 260 nm (range - 0.001 absorbance units).
5. Inject 25 μL of the PCR product (*see* **Note 5**).
6. Calculate the exit time of the fragment by multiplying the flow rate and the volume of the tubing from the detector to the waste outlet (*see* **Note 6**). Once the DNA fragment is detected start the timer and collect the appropriate fraction into a clean 0.6-mL Eppendorf tube when the exit time is achieved.
7. A typical volume of 75–110 μL of TEAA/ACN containing the DNA allele should be collected (*see* **Note 7**).

3.5. CE Sequencing of Purified Product

Samples of 5 μL of the HPLC purified DNA are cycle sequenced using with the ABI Prism Sequencing Kits and Thermal Cycler 9600 (Perkin-Elmer) following the manufacturer's instructions *(14)*. The thermal cycling conditions used for these alleles are: 96°C for 15 s, 50°C for 5 s, 60°C for 120 s, for 25 cycles.

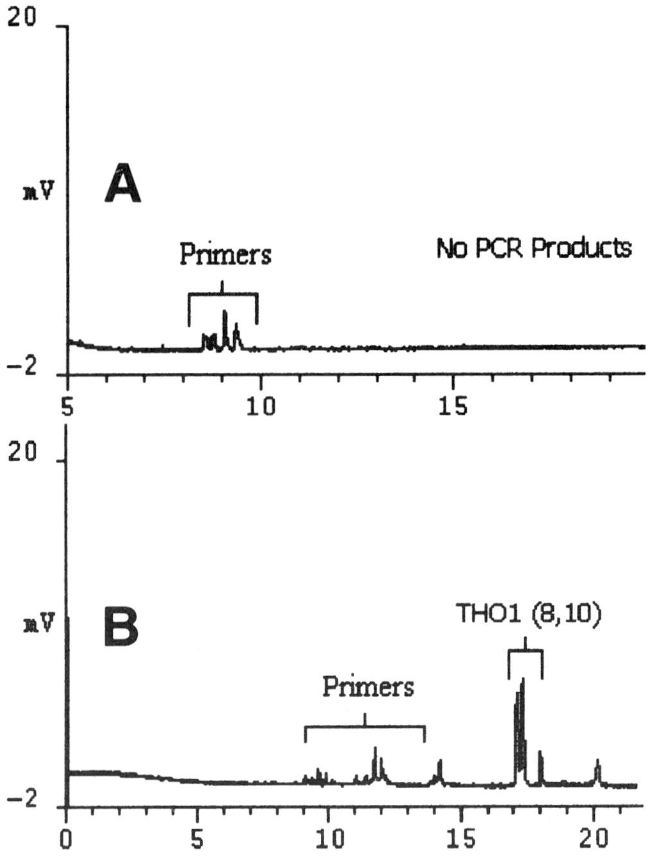

Fig. 1. CE analysis of HumTHO1 alleles 8,10. CE analysis of unpurified and purified HUMTHO1 PCR products which consisted of alleles, 8 and 10 (191 and 199 bp, respectively). Each analysis is performed at 30°C at a run voltage of –11 kV and detected by absorbance at a wavelength of 260 nm. The capillary had an effective length of 40 cm × 75 μm id and contained the 5% (w/w) DNA Fragment Analysis Buffer (Applied Biosystems) as a seiving matrix. The PCR products are: **(A)** unpurified products and **(B)** products purified using a MF-Millipore® Membrane Filter (Millipore, Bedford, MA).

1. The samples are electrokinetically injected at –2 kV (42.5 V/cm) for 30 s.
2. The samples are run at a voltage of –319 V/cm.
3. The column system temperature is 50°C.
4. Data are collected and analyzed using a Power Macintosh PC 6100/66 with the ABI 310 Genetic Analyzer data collection software version 2.1.2 and Sequence Navigator version 1.0 (Perkin-Elmer-Applied Biosystems Division, Foster City, CA). Using the HPLC system, the crude PCR mixtures are separated and captured in a single analysis. **Figure 2** shows that the alleles 9 and 10 of the THO1 locus can be resolved using HPLC and sequenced using the CE system.

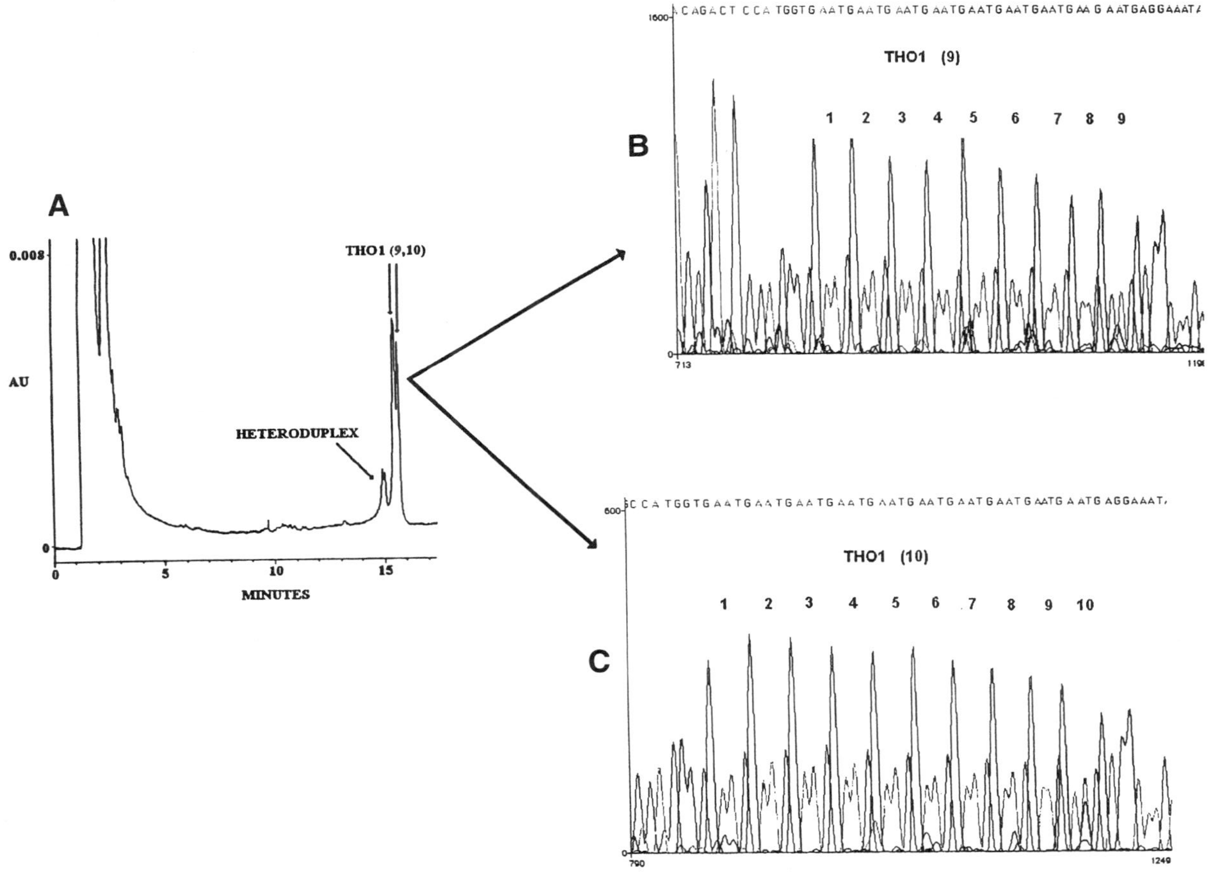

Fig. 2. HPLC peak capture and CE sequencing of the THO1 9,10 alleles. The separation and isolation of the THO1 9 and 10 alleles using the HPLC with resolution of 0.657 and UV detection at 260 nm is shown in panel **(A)**. Panels **(B)** and **(C)** are the sequence in the forward direction with the repeat of AATG of each individual allele.

4. Notes

1. Problems of salt concentration are also overcome by dilution of sample with deionized water. This technique reduces the DNA concentration and necessitates the sensitivity of fluorescence for detection of the alleles.
2. During the membrane dialysis procedure, the salt diffuses through the membrane into the deionized water bath. The water also diffuses into the sample, therefore the sample will become diluted the longer it remains on the membrane. Do not exceed 15-min dialysis. A briefer dialysis time may be sufficient for some samples.
3. The membrane filters will move although floating on the water, therefore only use one membrane filter per weighing dish to prevent sample mixup.
4. For consistent sequencing reactions, the IPRP-HPLC isolated alleles should be quantitated using UV absorbance readings.
5. Additional 0.005-id tubing (approx 40 cm) is placed between the injection valve and the separator column within the heated body of the instrument. This extra tubing adds to the flow path and allows time for the eluant and sample to heat up to the operating temperature before reaching the column.
6. The distance between the end of the column and the detector cell is minimized to reduce diffusion and band broadening.
7. If the fraction collection volume is greater than 120 μL, spin dry the sample and resuspend with deionized water to 50 μL.

Acknowledgments

The authors wish to acknowledge the Armed Forces Institute of Pathology–Center for Medical and Molecular Genetics (Rockville, MD) where the experimental protocols were developed. We also wish to recognize Transgenomic Inc. (San Jose, CA) for the continued support of this research.

References

1. Williams, P. E., Marino, M. A., Del Rio, S. A., Turni, L., Devaney, J. M. (1994) A practival approach to DNA restriction fragment and PCR product analysis by capillary electrophoresis. *J. Chromatogr. A* **680,** 525–540.
2. Anon. (1992) Capillary gel electrophoresis of PCR products: Easy sample preparation method. J & W Scientific, *CE Application Note #2.*
3. Carson, S., Cohen, A. S., Belenkii, A., Ruiz-Martinez, M. C., Berka, J., and Karger, B. L. (1993) DNA sequencing by capillary electrophoresis: Use of a two-laser-two-window intensified diode array detection system. *Anal. Chem.* **65,** 3219–3226.
4. Chen, D., harke, H. R., and Dovichi, N. J. (1992) Two-label peak-height encoded DNA sequencing by capillary gel electrophoresis: Three examples. *Nucleic Acids Res.* **20,** 4873–4880.
5. Guttman, A. (1994) Separation of DNA by capillary electrophoresis, in *Handbook of Capillary Electrophoresis* (Landers, J. P., ed.), CRC Press, Ann Arbor, MI, p. 137.
6. Karger, A. E., Harris, J. L. and Gesteland, R. F. (1991) Multiwavelength fluorescence detection for DNA sequencing using capillary electrophoresis. *Nucleic Acids Res.* **19,** 4955–4962.
7. Lu, H., Arriaga, E., Chen, D. Y., and Dovichi, N. J. (1994) High-speed and high-accuracy DNA seequencing by capillary gel electrophoresis in a simple, low cost instrument two-color peak-height encoded sequencing at 40°C. *J. Chromatogr. A* **680,** 497–501.

8. McCord, B. R., McClure, D. L., and Jung, J. M. (1993) Capillary electrophoresis of polymerase chain reaction-amplififed DNA using fluorescence detection with an intercalating dye. *J. Chromatogr. A* **652,** 75–82.
9. Ruiz-Matrinez, M. C., Berka, J., Belenii, A., Foret, F., Miller, A. W., and Karger, B. L. (1993) DNA sequencing by capillary electrophoresis with replaceable linear polyacrylamide and laser-induced fluorescence detection. *Anal. Chem.* **65,** 2851–2858.
10. Srinivasan, K., Girard, J. E., Wiliams, P., Roby, R. K., Weedn, V. M., Morris, A. C., Kline, M. C., and Reeder, D. J. (1993) Electrophoretic separations of polymerase chain reaction-amplified DNA fragments in DNAtyping using a capillary electrophoresis-laser indecued fluorescece system. *J. Chromatogr. A* **652,** 83–91.
11. Wang, Y., Ju, J., Carpenter, B. A., Athrton, J. M., Sensabaugh, G. F., and Mathies, R. A. (1995) Rapid sizing of short tandem repeat alleles using capillary array electrophoresis and energy-transfer fluorescent primers. *Anal. Chem.* **67,** 1197–1203.
12. Devaney, J. M., Marino, M. A., Williams, P. E., Weaver, K. R., Del Rios, S. A., Turner, K. A., and Belgrader, P. (1995) The evaluation of fast purification methods for preparing polymerase chain reaction (PCR) products for capillary electrophoresis analysis. *Appl. Theor. Electrophor.* **5,** 129–132.
13. Marino, M. A., Devaney, J. M., Smith, J. K., and Girard, J. E. (1998) Sequencing using capillary electrophoresis of short tandem repeat alleles separated and purified by high-performance liquid chromatography. *Electrophoresis* **19,** 108–118.
14. Kamahori, M. and Kambara, H. (2001) Capillary array electrophoresis analyser, in *Capillary Electrophoresis of Nucleic Acids*, Vol. 2 (Mitchelson, K. R. and Cheng, J., eds.), Humana Press, Totowa, NJ pp. 271–287.

4

Quantitative Measurements

Bruce McCord

1. Introduction

The fundamental role of any separation process is the isolation and quantitation of the individual components of a mixture. In the analysis of DNA by capillary electrophoresis (CE), the migration time and the sample quantity are the basic parameters obtained from any electropherogram. Migration time is used to estimate the DNA fragment size, and the quantity of product gives the analyst polymerase chain reaction (PCR) reaction yields, the efficiency of an extraction process, or the level of expression of a particular gene. Chemical processes such as PCR amplification or restriction enzyme digestion can vary in their yield, making measurement of the product quantity a critical parameter. The need for exact quantitative measurements is increasing with the use of methods such as quantitative PCR, in which an analyte and an internal standard of similar sequence are amplified together and compared. Many different applications, such as mutational analysis, and the detection of length polymorphisms, require the ability to reliably detect the size of a DNA fragment. DNA typing can require the routine determinations of differences in fragment size of two bases or less.

A number of procedures have been used over the years to estimate DNA product quantity. These include spectrophotometry, gel electrophoresis, and hybridization techniques. Visualization of the products is performed using UV absorbance, fluorescence, chemiluminescence, or radioactive labeling. CE is perhaps the newest of these techniques. CE methods are relatively fast and highly informative compared to other methods of quantitation and are easily automated.

Although it is difficult to quantitate DNA extracted from cell nuclei using CE becasue of its large and indeterminate fragment lengths, CE has proven to be particularly useful for quantitative analysis of DNA fragments of 40,000 bp, or smaller. DNA in this form is commonly produced as a result of enzyme digestion and PCR amplification. Quantitative analysis of such samples is particularly important if further processing such as sequencing will be performed.

From: *Methods in Molecular Biology, Vol. 162:*
Capillary Electrophoresis of Nucleic Acids, Vol. 1: Introduction to the Capillary Electrophoresis of Nucleic Acids
Edited by: K. R. Mitchelson and J. Cheng © Humana Press Inc., Totowa, NJ

In addition to the quantity of the product, CE can also be used to evaluate the product size and quality. Cycle sequencing requires the DNA template be clean and free from extraneous material such as primers and enzyme that can interfere with or inhibit the subsequent sequencing reaction. Quantitation of the product is important in order to optimize reaction conditions. CE has been shown to be superior to other methods for accessing the quality of the PCR products prior to sequencing as it permits the analyst to detect any remaining primers and to quantitate the resultant product. The accuracy of on-column analysis in CE is superior to what is commonly available in agarose yield gels in which product quantity is estimated using the spot size of different standards applied to the gel *(1)*. This chapter reviews several of the factors and conditions that can affect the performance of CE and which must be considered when undertaking the quantitative measurement of DNA. For this purpose, the chapter is a broad discussion of those factors, and specific practical points are illustrated with experimental data. The methods described therefore are not exhaustive, but are a brief guide to the performance of those experiments.

2. Methods

2.1. Preparation of Buffers

1. DNA fragments are separated using a sieving buffer containing hydroxyethyl cellulose (HEC), although other water-soluble polymers give similar results. For these buffers, concentration and molecular weight of the polymer is a critical parameter for successful application of the method.
2. HEC (Aldrich 30,863-3, Milwaukee, WI), or equivalent (approx 100 cp for a 2% solution at 25°C) is used in all preparations.
3. The buffer is prepared by dissolving 1% of the polymer in a solution of 100 m*M* Tris-base, 100 m*M* boric acid and 2 m*M* EDTA, pH adjusted to 8.6 with CsOH. The polymer must be added slowly with stirring to the solution to avoid clumping. The buffer mixture is then left to stir overnight. This solution is stable for 6 mo or more.
4. Prior to analysis, 25 mL of the buffer is filtered through a 5-µm Milex-SV syringe filter to remove undissolved solids and YO-PRO-1 dye (Molecular Probes, Eugene, OR) is added to a concentration of 500 ng/mL. The buffer solution must be kept dark at this point as YO-PRO and similar dyes are light sensitive.
5. For the analysis of denatured DNA, the concentration of the HEC polymer is 2–3% and urea is added to the buffer for a final concentration of 7 *M*. Bashkin *(2)* recommends deionizing the solution of HEC and urea prior to the addition of buffer salts to improve the stability of the analysis. This is done by stirring with 0.5 g of Amberlight mixed bed ion exchange resin (Sigma, St Louis, MO) in approx 32 mL of solution. The mixture is centrifuged to pellet the resin and the polymer containing supernatant is collected.

2.2. Capillary Electrophoresis

1. CE is carried out using 50 µm J&W DB-17 capillaries using a CE equipped with a 488-nm argon-ion laser.
2. The capillaries are conditioned by repeatedly rinsing with methanol and buffer until efficient and reproducible separations are obtained. Prior to each run, the capillary is rinsed for 1 min with methanol and 2 min with buffer. Capillaries are stored overnight in distilled water.

3. Native DNA samples are diluted 1:40 in deionized water prior to analysis. If excess salts are present, the sample may be dialyzed by pipeting it onto a 0.1 μm Millipore VSWP (Milipore, Bedford, MA) filter floating on a dish of distilled water for 20 min *(3)*.
4. Internal standards for native DNA separations were obtained from Bioventures (Murfreesboro, TN), or Gensura (DelMar, CA).
5. Denatured, dye labeled DNA samples are amplified via the PCR and prepared by diluting 2 μL of sample and 2 μL of internal standard in 24 μL of formamide *(4)*.
6. Samples are heated to 95°C for 3 min, and snap cooled prior to analysis.
7. The 3% HEC sieving buffer used for these separations is viscous and requires CE systems capable of high-pressure capillary filling.
8. Dye labeled primers and ROX labeled internal standards are obtained through Perkin-Elmer (Norwalk, CT), or Promega (Madison, WI).

3. Analytical Methods for Quantitative Measurement

1. In general, there is a tradeoff in applying CE methods for the analysis of DNA fragments in terms of speed and separation power. Often, analytical methods for product quantitation are used in a routine setting. In these situations, the method should be rapid (*see* **Note 1**). Due to the high efficiency of capillary gel electrophoresis, a useful procedure is to use the shortest possible capillary and adjust the field strength to obtain the desired resolution. Typically, the size and quantity of a single product is estimated, which may contain a mutant, or wild-type allele. It should be noted that the temperature, the concentration, and the molecular weight of the polymer constituents of the sieving buffer may also affect the chromatographic efficiency.
2. **Figure 1** illustrates the relationship between field strength, resolution, and migration time using two sets of PCR-amplified DNA fragments at various field strengths using a 27-cm long capillary column. Depending on the desired resolution and fragment length, the entire analysis can take from 5 to 70 min.
3. For example, the rapid assay of a single PCR product can be performed in less than 5 min using an electric field strength of 300 V/cm across the capillary. The same column can then be used at lower field strengths to affect a more complete separation of mixtures.
4. Note that proper field strength is particularly important for the resolution of larger DNA fragments, presumably due to orientation with the field. Electrophoresis at a field below 75 V/cm results in the increase in migration times to an unacceptable level.

3.1. Buffers and Capillaries

1. Highly efficient DNA separations can be obtained in buffered solutions of water-soluble polymers, and a number of different polymer systems have been used for DNA separations (*see* **Note 2**).
2. Among the most popular systems include linear polyacrylamide, poly(dimethyl acrylamide), poly(ethylene oxide), and modified celluloses. The development of effective buffer systems is covered elsewhere in this book in depth *(5–7)*.

3.1.1. Contamination of Capillaries

1. A critical factor in choosing a sieving buffer is its interaction with the capillary. The lifetime of capillaries is of paramount importance in the routine analysis of DNA and much research has been carried out to develop buffers and capillary coatings that are stable and reusable for many runs *(8)*.

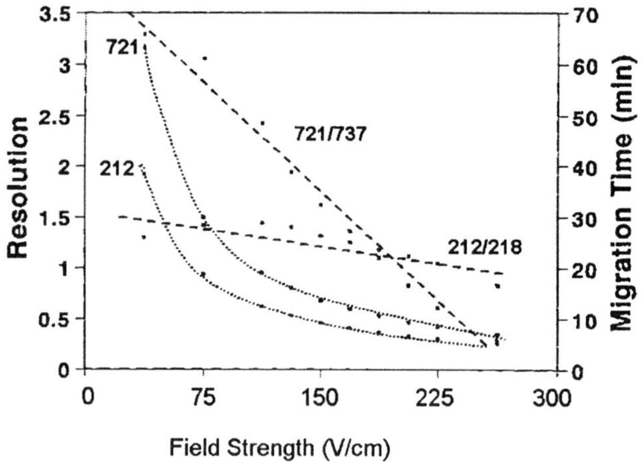

Fig. 1. The relationship between field strength, resolution (dashed lines), and migration time (dotted lines) for the analysis of fragment-pairs, 212/218-bp fragments and 721/737-bp fragments *(47)*. Samples are run in nondenaturing conditions. A capillary (27-cm effective length, DB-17) is used with a sieving buffer of 1% HEC, 500 ng/mL YO-PRO-1, 100 m*M* Tris-borate, pH 8.6, 1 m*M* EDTA. Laser-induced fluorescence detection at 488-nm excitation, 520-nm emission. Reprinted from **ref.** ***47***.

2. The largest problem for stability of the capillary seems to be contamination of the capillary walls. This process can affect both the migration time and the separation efficiency by creating active sites on the capillary and by inducing the electroosmotic flow (EOF).
3. To avoid wall contamination, capillaries can be washed with various solutions prior to the buffer fill in order to remove accumulated adsorbates. The type of wash buffer used depends upon whether the capillary is coated or not.
4. Uncoated capillaries can be washed with basic solutions to clean and etch the silica walls of the capillary, followed by HCl to protonate active sites on the capillary *(9)*.
5. Methanol has been shown to be an effective media for rinsing coated capillaries ***(10)***. Phenyl methyl coated capillaries have been shown to last for months when maintained in this fashion.
6. Alternatively, capillaries can be filled with polymer solutions that dynamically coat the walls and minimize adsorption ***(8,11)***. Such systems provide capillary lifetimes of 100 runs or more before it is necessary to replace the capillary.

3.2. Sample Injection

1. The selection of the proper injection mode to be used in capillary electrophoresis is dependent upon the intended application. There are currently two methods, hydrodynamic injection or electrokinetic injection.
2. In general, hydrodynamic (pressure) injections are more reproducible because they are not dependent on the sample matrix as long as the viscosity remains constant. The volume of sample, V, injected using a pressure injection may be calculated by the following equation:

$$V = \pi P r^4 t / 8 \eta l \quad (1)$$

Here P is the pressure, r is the capillary radius, t is the injection time, η is the viscosity, and l is the capillary length ***(12)***.

3. Hydrodynamic injections should be considered the preferred method for quantitative analysis. However, the relatively poor resolution and sensitivity of the method limits its application.
4. The alternative injection mode, electrokinetic injection, utilizes an applied voltage to induce sample ions to migrate into the capillary orifice. The quantity, Q of DNA injected onto the capillary is described by the following equation:

$$Q_{DNA} = E\pi r^2 [DNA](\mu_{ep} + \mu_{eof}) \quad (2)$$

Where, E is the field strength, r is the capillary radius, μ_{ep} is the electrophoretic flow, and μ_{eof} is the electroosmotic flow. Electrokinetic injections are often preferred because they produce highly efficient separations due to sample stacking.

5. However, these injections must be used with caution. The quantity of material injected Q_{DNA} can vary greatly as it is a function of not only the field-induced mobility of the sample DNA, but is also influenced by the ionic strength of the sample matrix.
6. Chloride and other buffer ions with higher charge to mass ratios than DNA will be selectively injected into the capillary. Changes in ionic strength from one sample to the next will influence the injection field strength ***(13)***. These variable factors can result in poor run-to-run reproducibility when electrokinetic injection is used.
7. In a study utilizing a HEC polymer as the sieving media and a coated capillary, the relative standard deviation (RSD) of peak area for 10 repetitive injections of a 100-bp DNA fragment is measured at 28% ***(14)***. By adding an internal standard to the sample, peak area reproducibility is improved to 8% RSD. This value is equivalent to that obtained with hydrodynamic injection.

3.3. Buffer Depletion

1. Maintaining the integrity of the sample and buffer are also necessary in order to obtain reproducible results.
2. The buffers used in DNA analysis are highly viscous and can adhere to the exterior of the capillary during the buffer fill step. In the subsequent injection step, this material can contaminate the sample, changing its ionic strength, and reducing the quantity of material injected. Thus, it is important to rinse the capillary in water prior to injection of the sample.
3. Another problem is the potential for buffer depletion ***(15)***. Repetitive analysis of samples from a limited volume of reservoir buffer can result in gradual changes in the pH of the solution.
4. Additionally additives such as intercalating dyes can become depleted as they migrate out of the buffer during the electrophoresis step. This problem can be avoided by periodic replenishment of the buffer, and by using three buffer reservoirs, two for electrical contacts and one to fill the capillary ***(16)***.
5. **Figure 2** illustrates the effect of buffer age on the estimated size (number of repetitive 16-bp segments) of a PCR-amplified DNA fragment.

3.4. Detection of DNA Using Dyes

1. Early work in DNA analysis by CE-utilized ultraviolet absorption (UV) as the method of detection. Extensive sample preparation and preconcentration are required prior to analysis because of the low sensitivity and poor selectivity of the UV detection procedure between different nucleic acid species ***(17–19)***.

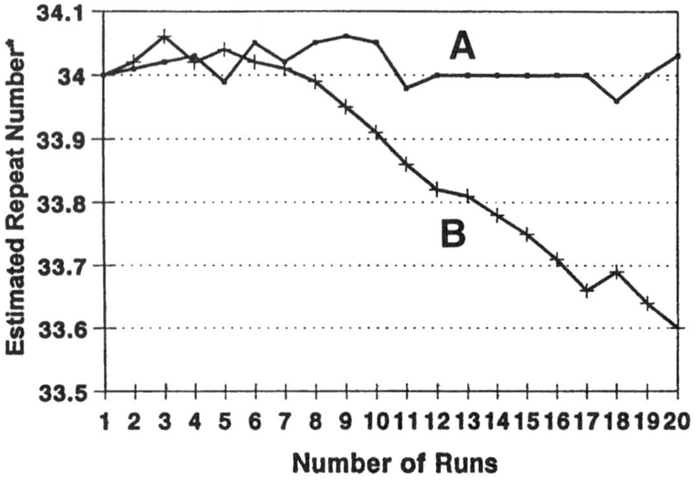

Fig. 2. The effect of buffer depletion on the size estimation of a DNA fragment containing a 16 base sequence repeated 34 times *(42)*. The fragment is approx 690 bp in size and each tenth of a repeat number is 1.6 bases. Experimental conditions are as shown in **Fig. 1**. Electric field strength is 167 V/cm. Reprinted from **ref.** ***42***.

2. The introduction of laser-induced fluorescence (LIF) solved the problem of sensitivity by focusing the light beam directly onto the capillary window *(20–22)*. Detection enhancements of 400X-fold or more have been achieved using LIF when compared to the sensitivity of UV detection *(20)*.
3. Fluorescence detection of dsDNA is primarily achieved through the use of intercalating dyes. These dyes bind to the DNA molecule by inserting themselves into or along the DNA helix, affecting the configuration of the aromatic rings of the dye and enhancing the fluorescence signal *(23)*.
4. Intercalating dyes are especially important in performing size analysis of native DNA as they have a strong effect on the conformation of the DNA molecule in solution. The relationship between migration time and native DNA fragment length is fairly linear to approx 500 bp, however, differences in sequence can create situations in which the elution order of peaks does not correspond with the ascending order of fragment sizes *(24,25)*.
5. Thus, the separation between any two fragments in an analysis should not be used to determine overall separation efficiency of a method. This error is commonly made in reports in the literature. The correct calculation of separation efficiency should utilize the slope of the relationship between DNA size and elution order.
6. It is also interesting to note that intercalating dyes minimize effects of DNA structure on migration rate, resulting in better estimates of fragment lengths *(10,23,26)*.
7. Changes in concentration or type of dye used can alter the selectivity of the procedure. Another advantage of using intercalating dyes is their low background florescence when free in solution and uncomplexed to DNA. The uncomplexed dyes can be added directly to the CE buffer without appreciably increasing the background signal.
8. Monomeric intercalating dyes such as ethidium bromide, thiazole orange, and oxazole yellow have proven to be the most useful, providing precise and reproducible estimates of DNA size and quantity *(27,28)*.

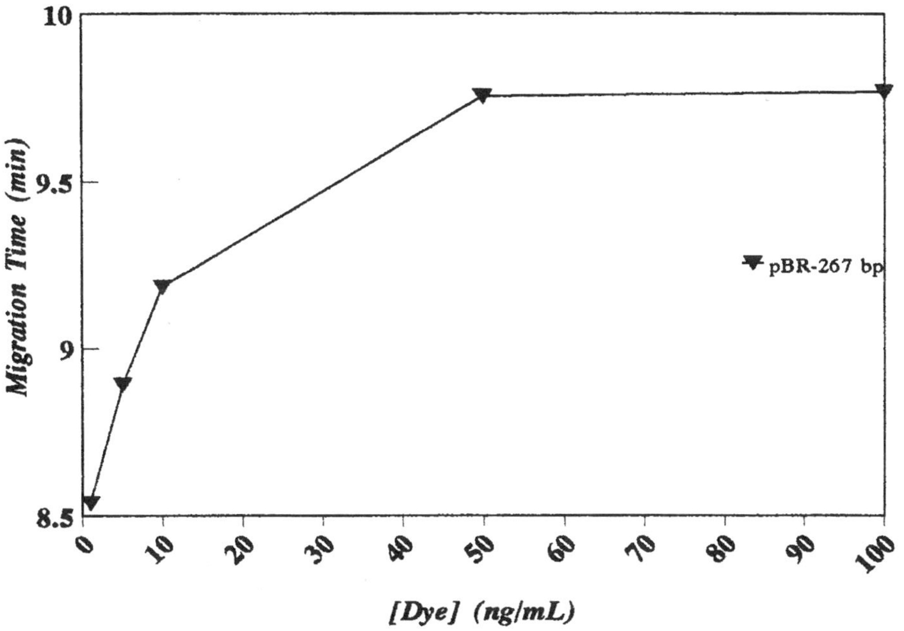

Fig. 3. The effect of intercalator dye concentration on DNA migration times for the 267-bp fragment of a *Hae*III digest of pBR322 *(48)*. Electrophoresis conditions are as in **Fig. 1**, however, the electric field strength is 185 V/cm.

9. Dimeric dyes can also be used, however both intermolecular and intramolecular crosslinking can cause problems for quantitation using these dyes unless special steps are taken *(23,29)*.

3.5. Dyes and DNA Fragment Size

1. The concentration of the intercalating dye in the CE buffer is important because the dye binds to the DNA in multiple locations along the entire length of the molecule. This process affects the mobility and fluorescence intensity of the DNA/dye complex.
2. Because these dyes are positively charged, they alter the migration of the DNA through the capillary by changing the net charge on the DNA fragment. If the DNA fragments in solution are not saturated with dye, both the fluorescence intensity and elution times of the fragments can vary from run to run.
3. **Figure 3** shows that the concentration of dye in the buffer should be optimized by examining a plot of DNA mobility vs the dye concentration, and the analysis should take place at concentrations where the mobility has become stable.
4. Additionally, the fluorescent intensity is a function of the length of the molecule *(30)*. The larger the DNA fragment, the greater its fluorescent intensity. Thus, in quantitative analysis, a 243-bp fragment cannot be directly compared with a 200-bp internal standard because they will differ in intensity due to the length differences.
5. The estimated quantity can be calculated by multiplying peak areas of the internal standard by a factor, which relates intensity to fragment length.

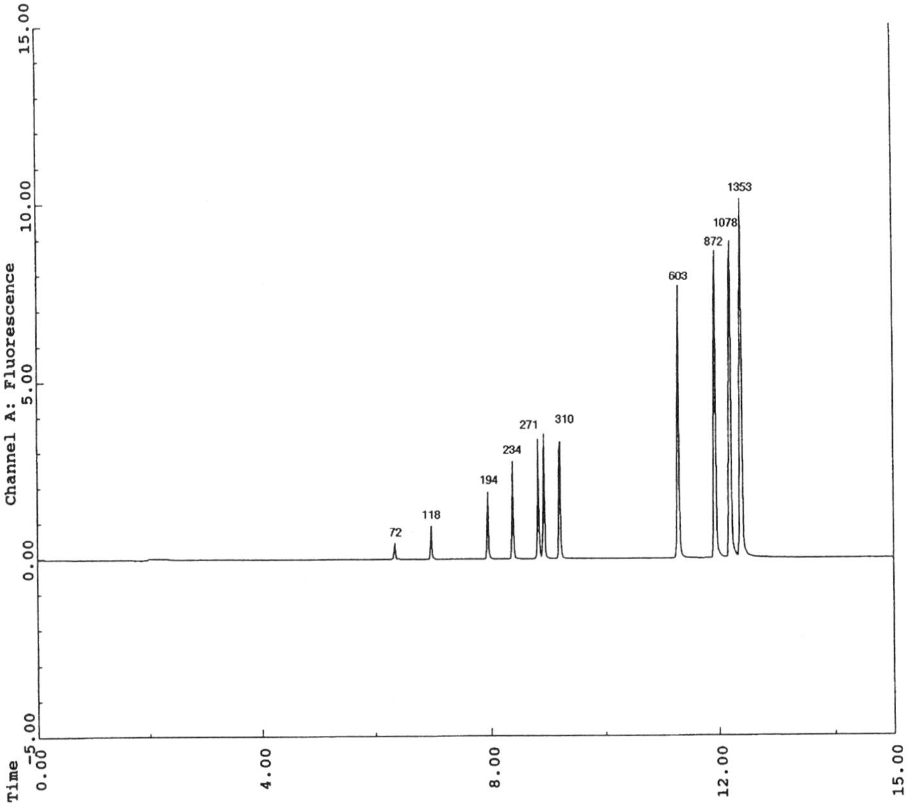

Fig. 4. The analysis of a *Hae*III digest of ΦX174 DNA using LIF and an intercalating dye. Note that the concentration of each fragment in the digest is the same. Conditions are as shown in **Fig. 1**, and an electric field of 185 V/cm.

$$Q_{243} = K\ Q_{200} \quad (4)$$

This factor, $K = (F_{243}/F_{200})$ can be estimated by examining the fluorescence intensity of the restriction digest of a plasmid fragment length on intensity.

6. This relationship is illustrated in **Fig. 4**, in which a restriction enzyme has cut a circular piece of DNA only at specific sites where the GGCC motif occurs. Thus, each fragment in the digested sample has a unique length and all fragments have the same concentration. As can be seen, the intensity of each peak gradually increases with size even though concentration of each peak is the same. By plotting the peak intensity vs the size, the relationship between intensity, molecular weight, and concentration can be determined.

3.6. Applications of CE Analysis

1. There are a number of important applications in the analysis of native DNA, which have been explored using sieving buffers and intercalating dyes. Many of these, such as quantitation of gene alleles ***(31,32)*** and oligonucleotides ***(33–35)***, will be covered in later

chapters of this book. The quantitation of levels of cellular RNA by analysis of cDNA products using the accurate detection equipment on CE apparatus are also discussed in later chapters ***(36,37)***.

2. CE can be particularly useful in situations where standard yield gels lack precision. CE systems give quantitative estimates as a result of on-column detection, whereas slab gels require the user to compare band density against known standards, an indirect and tedious process.
3. Fine detail such as peaks near the shoulder of a main peak can be missed in slab-gel analysis unless densitometric analysis is used. Another particular advantage of the CE system is that numerical data can be electronically stored for later retrieval.
4. One example of a particularly useful application for CE is in the analysis of amplified products prior to DNA sequencing ***(31,38)***. For high-quality results in cycle sequencing reactions, it is important to know the quality and quantity of the product DNA before the reaction is started so that an optimal quantity of reagents can be used. In addition, residual primers from the prior amplification must be removed to eliminate their interference with the subsequent sequencing reaction.
5. **Figure 5** illustrates the CE-LIF analysis of an amplicon (HV2) from the control region of mitochondrial DNA. In the figure, several important features of the CE method are illustrated. The quantity of the sample is estimated by comparison of the intensity of the product peak with that of the 200-bp internal standard. The method also detects a nonspecific PCR by-product at 16 min and reveals the presence of primer peaks.
6. The analyst can use this information to refine reaction conditions and extraction protocols to develop cleaner sample analysis. Because of the wide separation between the peaks in this analysis, the method can be further refined to provide a separation of all three peaks in under three min ***(39)***.
7. In other situations where poor results are obtained, the CE system can also be used access the quality of the primer pair ***(40)***.

3.6.1. Estimating DNA Size

1. Internal standards can be used in a similar fashion to determine DNA size. Sizing applications such as the DNA typing of short tandem repeats require precise measurements ***(28,31,32,41)***.
2. These genetic loci, used in human identification and linkage studies, contain repeated segments of DNA 2–7 bases long. In these applications using these loci, allele sizes must be determined to a precision of 1–2 bases. This precision is difficult to achieve in a capillary system because variables such as temperature, sample and buffer ionic strength, solution viscosity, osmotic flow, and electric field can all influence DNA mobility.
3. Fortunately, although these effects alter the migration time of peaks from one run to the next, the migration rate of fragments within runs remains stable. Thus, judicious choice of internal standards permits the comparison of one run to the next (*see* **Note 3**).
4. Additionally, a linear relationship exists between peak migration time and DNA size for fragments up to approx 500 bp. Therefore, the analyst can ensure better precision by bracketing the fragments of interest between two internal standards and interpolating the their size based upon this relationship.
5. **Figure 6** illustrates an allelic ladder consisting of the seven most common alleles of the HUMTHO1 system. The DNA fragments are sized by comparing their migration time with that of the two internal standards. The size is then calculated by assuming a linear relation between the two peaks and interpolating.

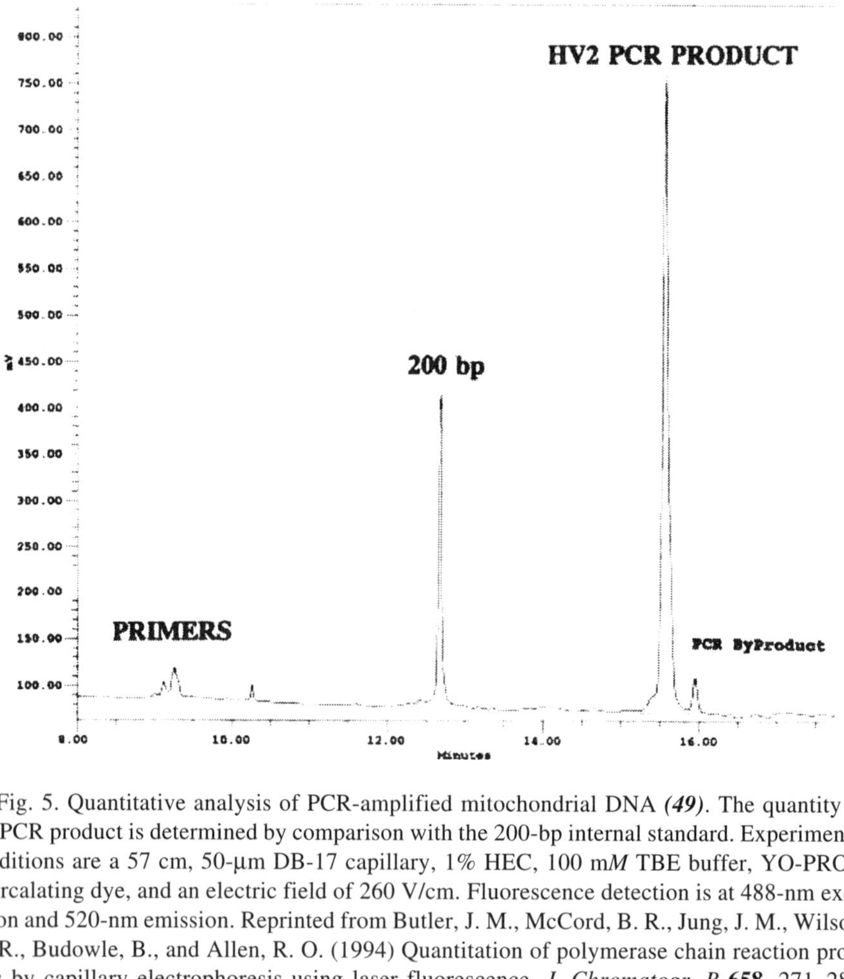

Fig. 5. Quantitative analysis of PCR-amplified mitochondrial DNA *(49)*. The quantity of the PCR product is determined by comparison with the 200-bp internal standard. Experimental conditions are a 57 cm, 50-μm DB-17 capillary, 1% HEC, 100 m*M* TBE buffer, YO-PRO-1 intercalating dye, and an electric field of 260 V/cm. Fluorescence detection is at 488-nm excitation and 520-nm emission. Reprinted from Butler, J. M., McCord, B. R., Jung, J. M., Wilson, M. R., Budowle, B., and Allen, R. O. (1994) Quantitation of polymerase chain reaction products by capillary electrophoresis using laser fluorescence. *J. Chromatogr. B* **658**, 271–280, with permission from Elsevier Science.

6. Absolute determination of DNA fragment length is not possible with this procedure since sequence variations between the internal standards and the HUMTHO1 alleles will affect the relative migration time between the standard and the allele. Thus, estimation of DNA size using this procedure is a precise, but not necessarily an accurate technique, because the calculated size may not correspond with the true, sequenced size of the fragment.
7. Thus, the process of measuring allele size by migration time is a two-step procedure. First, the size of an individual fragment is determined based upon its relationship with the internal standards. Then the calculated size of the allele is compared to that of the external standard (the allelic ladder) to determine its identity.
8. This two-step process is necessary to correct for the fact that long term variations in temperature and ionic strength may influence the migration of the internal standard and analyte in different ways.

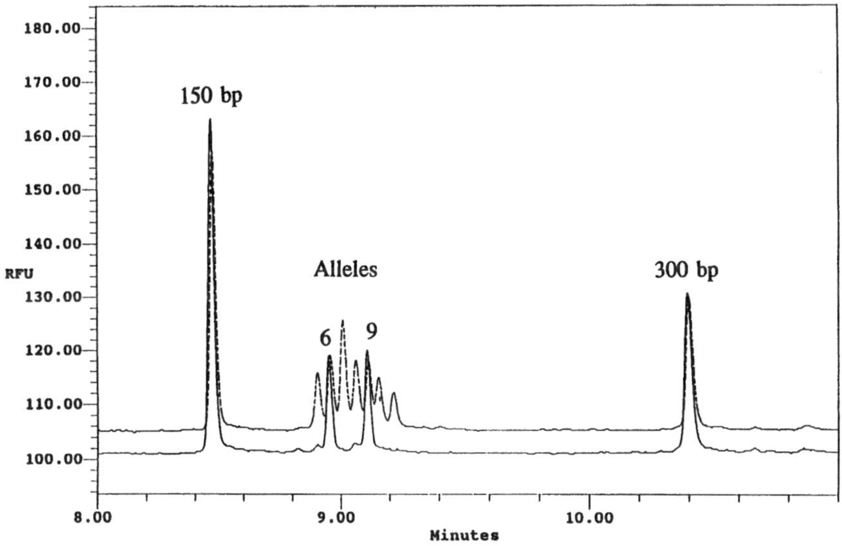

Fig. 6. Sizing of two PCR-amplified alleles from the HUMTHO1 genetic locus. DNA standards of 150 and 300 bp are used to bracket the two alleles of interest (lower trace) and the size of each peak is determined through interpolation *(48)*. The results are compared with the sizes calculated using the external standard (upper trace) which consists of all of the common alleles in the locus. Precise size estimates are obtained with this method. Electrophoresis conditions are as described for **Fig. 4**. Reprinted from **ref.** ***38***.

9. A similar procedure can be used for accurate sizing of alleles larger than 500 bp using nonlinear curve fitting *(42)*.

3.6.2. Fluorescent Derivatization

1. Fluorescent dye molecules may also be covalently bound to the DNA fragments in a process known as fluorescent derivatization *(43)*.
2. The easiest way to do this is to label one or both primers and attach the tagged primer to the fragment of interest via the PCR process. Using this labeling process for detection is not as efficient as using intercalating dyes as only a single dye molecule is added to each fragment.
3. However, the trade off for this loss in intensity is a reduction in the background signal as the covalent labeling process eliminates the problem of residual fluorescence of unincorporated intercalating dyes. Upon completion of the PCR, all of the target DNA molecules are labeled with a fluorophore.
4. Covalent derivitization is particularly well suited for use with ssDNA and it allows multiple dyes to be used. These dyes absorb light at similar wavelengths but they have different emission profiles.
5. A multichannel analyzer can then identify the specific PCR product by detecting the emission maxima of each of the different bound dyes *(44)*. Product size and quantity can be estimated by comparing the results with an internal standard labeled with a different fluorescent dye.

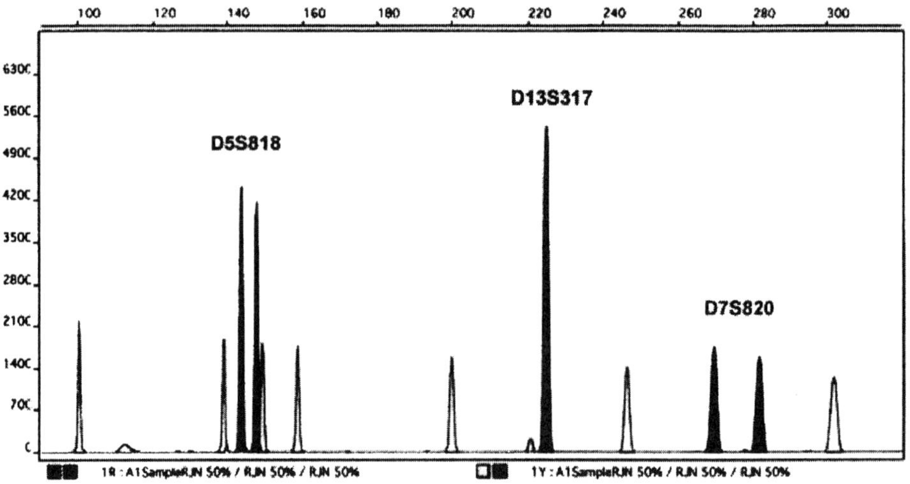

Fig. 7. Illustration of the analysis of a PCR multiplex (black) consisting of three genetic loci coanalyzed with an internal size standard. Fragment size and quantity are determined by comparing peak areas and migration times with those of the known internal standard peaks. Conditions *(31)* are 3% HEC in a buffer solution consisting of 100 m*M* Tris-borate, 2 m*M* EDTA, 7 *M* urea. Dye excitation is at 488 nm, and emissions are detected at multiple wavelengths. The figure is provided by courtesy of Tim Nock, Ohio University.

6. **Figure 7** illustrates this process. Three separate genetic loci, D5S818, D13S317, and D7S820 are coamplified using primers labeled with the fluorescent dye NED and analyzed via CE using a multichannel fluorescence detector. An internal standard consisting of a series of DNA fragments labeled with a ROX dye is added to the sample.
7. The size of each individual allele in the sample is determined by comparing its migration time to a regression line calculated from the size/migration relationship of the red internal standard peaks. Using this system for sizing yields a standard deviation of the estimated allele size of 0.2 bases (0.07% RSD), or better *(4,11)*.

3.6.3. Complementary Fluorescent Probes

1. A third form of labeling uses complementary fluorescent probes that hybridize to specific sequences of DNA *(45)*. This technique can be utilized to identify specific fragments in a restriction digest.
2. A similar procedure has also been developed for the quantitation of RNA using hybridization *(46)*. Extracted RNA from HIV positive patients is probed using a 26-bp DNA sequence labeled with a fluorescent dye and then digested with RNAase enzyme. The probe blocks digestion of the target RNA and an intercalating dye is then used to enhance the detection of the resulting fluorescent complex *(46)*.

4. Notes

1. There are a wide variety of applications for quantitative analysis of DNA, many of which are covered in subsequent chapters of this book *(31–38)*. The key to successful quantita-

tive measurements using CE is control and understanding of the chemistry and mechanics of the analytical process.

2. Further improvements will also require mitigation of sample matrix effects. However, for many of the DNA separations now being utilized, high reproducibility of peak areas and migration times has already been achieved.
3. The success that has been achieved thus far is due in large part to the increasingly sophisticated application of internal and external standards. Future improvements will require better control of separation and injection as internal standards only compensate for instabilities in these processes. Quantitative measurements are becoming widespread and the range of applications of CE in DNA analysis as illustrated in this book, and throughout the scientific literature, are a testament to this.

Acknowledgments

John Butler, Alice Isenberg, Janet Doyle, and the staff of the FBI Academy in Quantico, VA provided substantial contributions to this paper, and Tim Nock, Ohio University for **Fig. 7**. This work was supported by contributions from the National Institute of Justice and the University of Ohio.

References

1. Butler, J. M., McCord, B. R., Jung, J. M., Wilson, M. R., Budowle, B., and Allen, R. O. (1994) Quantitation of polymerase chain reaction products by capillary electrophoresis using laser fluorescence. *J. Chromatogr. B* **658,** 271–280.
2. Bashkin, J., Marsh, M., Barker, D., and Johnston, R. (1996) DNA sequencing by capillary electrophoresis with a hydroxyethyl cellulose sieving buffer. *Appl. Theor. Electrophoresis* **6,** 23–28.
3. Marusyk, R. and Sergent, A. (1980) A simple method for dialysis of small volume samples. *Anal. Biochem.* **105,** 403–404.
4. Isenberg, A. R., Allen, R. O., Keys, K. M., Smerick, J. B., Budowle, B., and McCord, B. R. (1998) Analysis of two multiplexed short tandem repeat systems using capillary electrophoresis with multiwavelength fluorescence detection. *Electrophoresis* **19,** 94–100.
5. Wehr, T., Zhu, M., and Mao, D. T. (2001) Sieving matrix selection, in *Capillary Electrophoresis of Nucleic Acids,* Vol. 1 (Mitchelson, K. R. and Cheng, J., eds.), Humana Press, Totowa, NJ, pp. 167–187.
6. Heller, C. (2001) Influence of polymer concentration and polymer composition on capillary electrophoresis of DNA, in *Capillary Electrophoresis of Nucleic Acids,* Vol. 1 (Mitchelson, K. R. and Cheng, J., eds.), Humana Press, Totowa, NJ, pp. 111–123.
7. Quesada, M. A. and Menchen, S. (2001) Replaceable polymers for DNA sequencing by capillary electrophoresis, in *Capillary Electrophoresis of Nucleic Acids,* Vol. 1 (Mitchelson, K. R. and Cheng, J., eds.), Humana Press, Totowa, NJ, pp. 139–166.
8. Chiari, M. and Cretich, M. (2001) Capillary coatings: choices for capillary electrophoresis of DNA, in *Capillary Electrophoresis of Nucleic Acids,* Vol. 1 (Mitchelson, K. R. and Cheng, J., eds.), Humana Press, Totowa, NJ, pp. 125–138.
9. Siles, B. A., Anderson, D. E., Buchanan, N. S., and Warder, M. F. (1997) The characterization of composite agarose/hydroxyethylcellulose matrices for the separation of DNA fragments using capillary electrophoresis. *Electrophoresis* **18,** 1980–1989.
10. McCord, B. R., McClure, D. M., and Jung, J. M. (1993) Capillary electrophoresis of PCR-amplified DNA using fluorescence detection with an intercalating dye. *J. Chromatogr.* **652,** 75–82.

11. Lazaruk, K., Walsh, P. S., Oaks, F., Gilbert, D., Rosenblum, B. B., Menchen, S., et al. (1998) Genotyping of forensic short tandem repeat (STR) systems based on sizing precision in a capillary electrophoresis instrument. *Electrophoresis* **19,** 86–93.
12. Ermakov, S. V., Zhukov, M. Y., Capelli, L., and Righetti, P. G. (1994) Quantitative studies of different injection systems in capillary electrophoresis. *Electrophoresis* **15,** 1158–1166.
13. Salas-Solano, O., Ruiz-Martinez, M. C., Carrilho, E., Kotler, L., and Karger, B.L. (1998) A sample purification method for rugged and high-performance DNA sequencing by capillary electrophoresis using replaceable polymer solutions. B: Quantitative determination of the role of sample matrix components on sequencing analysis. *Anal. Chem.* **70,** 1528–1535.
14. Butler, J. M., McCord, B. R., Jung, J. M., Wilson, M. R., Budowle, B., and Allen, R. O. (1994) Quantitation of polymerase chain reaction products by capillary electrophoresis using laser fluorescence. *J. Chromatogr. B* **658,** 271–280.
15. Corstjens, H., Billiet, H. A. H., Frank, J., and Luyben, K. C. A. M. (1996) Variation of the pH of the background electrolyte due to electrode reactions in capillary electrophoresis: Theoretical approach and *in situ* measurement. *Electrophoresis* **17,** 137–143.
16. Isenberg, A. R., McCord, B. R., Koons, B. W., Budowle, B., and Allen, R. O. (1996) DNA typing of a polymerase chain reaction amplified D1S80/amelogenin multiplex using capillary electrophoresis and a mixed entangled polymer matrix. *Electrophoresis* **17,** 1505–1511.
17. Nathakarnkitkool, S., Oefner, P. J., Bartsch, G., Chin, M. A., and Bonn, G. K. (1992) High-resolution capillary electrophoretic analysis of DNA in free solution. *Electrophoresis* **13,** 18–31.
18. Marino, M. A., Turni, L. A., Del Rio, S. A., and Williams, P. E. (1994) Molecular size determinations of DNA restriction fragments and polymerase chain reaction products using capillary gel electrophoresis. *J. Chromatogr. A* **676,** 185–189.
19. McCord, B. R., Jung, J. M., and Holleran, E. A. (1993) High resolution capillary electrophoresis of forensic DNA using a non-gel sieving buffer. *J. Liq. Chromatogr.* **16,** 1963–1981.
20. Schwartz, H. E. and Ulfelder, K. J. (1992) Capillary electrophoresis with laser-induced fluorescence detection of PCR fragments using thiazole orange. *Anal.Chem.* **64,** 1737–1740.
21. Clark, B. K. and Sepaniak, M. J. (1993) Evaluation of on-column labeling with intercalating dyes for fluorescence detection of DNA fragments separated by capillary electrophoresis. *J. Microcol. Sep.* **5,** 275–282.
22. Schwartz, H. E., Ulfelder, K. J., Chen, F.-T., and Pentoney, S. L. (1994) The utility of laser-induced fluorescence detection in applications of capillary electrophoresis. *J. Capillary Electrophor.* **1,** 36–54.
23. Zhu, H., Clark, S. M., Benson, S. C., Rye, H. S., Glazer, A. N., and Mathies, R.A. (1994) High-sensitivity capillary electrophoresis of double-stranded DNA fragments using monomeric and dimeric fluorescent intercalating dyes. *Anal. Chem.* **66,** 1941–1948.
24. Berka, J., Pariat, Y. F., Muller, O., Hebenbrock, K., Heiger, D. N., Foret, F., and Karger, B. L. (1995) Sequence dependent migration behavior of double-stranded DNA in capillary electrophoresis. *Electrophoresis* **16,** 377–388.
25. Lee, L. G., Chen, C.-H., and Chiu, L. A. (1986) Thiazole orange: A new dye for reticulocyte analysis. *Cytometry* **7,** 508–517.
26. Berka, J., Pariat, Y. F., Muller, O., Hebenbrock, K., Heiger, D. N., Foret, F., and Karger, B. L. (1995) Sequence- dependent migration behavior of double-stranded DNA in capillary electrophoresis. *Electrophoresis* **16,** 377–388.
27. Lu, W., Han, D. S., Yuan, J., and Andrieu, J. M. (1994) Multi-target PCR analysis by capillary electrophoresis and laser- induced fluorescence. *Nature* **368,** 269–271.

28. Butler, J. M., McCord, B. R., Jung, J. M., Lee, J. A., Budowle, B., and Allen, R. O. (1995) Application of dual internal standards for precise sizing of PCR products by capillary electrophoresis. *Electrophoresis* **16,** 974–980.
29. Carlsson, C., Jonsson, M., and Akerman, B. (1995) Double bands in DNA gel electrophoresis caused by bis-intercalating dyes. *Nucleic Acids Res.* **23,** 2413–2420.
30. Personett, D. A., Chouinard, M., Sugaya, K., and McKinney, M. (1996) Simplified RT/PCR quantitation of gene transcripts in cultured neuroblastoma (SN49) and microglial (BV-2) cells using capillary electrophoresis and laser-induced fluorescence. *J. Neurosci. Meth.* **65,** 77–91.
31. Butler, J. M. and Reeder, D. J. (2001) Detection of DNA polymorphisms using PCR-RFLP and capillary electrophoresis, in *Capillary Electrophoresis of Nucleic Acids,* Vol. 2 (Mitchelson, K. R. and Cheng, J., eds.), Humana Press, Totowa, NJ, pp. 47–56.
32. Khrapko, K., Collier, H. A., Li-Sucholeiki, X.-C., André, P. C., and Thilly, W. G. (2001) High resolution analysis of point mutations by constant denaturant capillary electrophoresis (CDCE), in *Capillary Electrophoresis of Nucleic Acids,* Vol. 2 (Mitchelson, K. R. and Cheng, J., eds.), Humana Press, Totowa, NJ, pp. 57–72.
33. Pearce, M. and Watson, N. (2001) Quality control of nucleotides and primers for PCR, in *Capillary Electrophoresis of Nucleic Acids,* Vol. 1 (Mitchelson, K. R. and Cheng, J., eds.), Humana Press, Totowa, NJ, pp. 347–352.
34. DeDionisio, L. A. (2001) Analysis of modified oligonucleotides with capillary gel electrophoresis, in *Capillary Electrophoresis of Nucleic Acids,* Vol. 1 (Mitchelson, K. R. and Cheng, J., eds.), Humana Press, Totowa, NJ, pp. 353–370.
35. Srivatsa, G. S., Pourmand, R., and Winters, S. (2001) Use of capillary electrophoresis for concentration analysis of phosphorothioate oligonucleotides, in *Capillary Electrophoresis of Nucleic Acids,* Vol. 1 (Mitchelson, K. R. and Cheng, J., eds.), Humana Press, Totowa, NJ, pp. 371–376.
36. Williams, S. J. and Williams, P. M. (2001) Quantitation of mRNA by competitive PCR using capillary electrophoresis, in *Capillary Electrophoresis of Nucleic Acids,* Vol. 2 (Mitchelson, K. R. and Cheng, J., eds.), Humana Press, Totowa, NJ, pp. 243–252.
37. Zhao, X. and George, K. S. (2001) Differential display analysis by capillary electrophoresis, in *Capillary Electrophoresis of Nucleic Acids,* Vol. 2 (Mitchelson, K. R. and Cheng, J., eds.), Humana Press, Totowa, NJ, pp. 259–267.
38. Butler, J. M. (1995) *Sizing and Quantitation of Polymerase Chain Reaction Products by Capillary Electrophoresis for use in DNA Typing*. Thesis Dissertation, University of Virginia, pp. 1–254.
39. Wilson, M. R., Polanskey, D., Butler, J. M., DiZinno, J. A., Replogle, J., and Budowle, B. (1995) Extraction, PCR amplification, and sequencing of mitochondrial DNA from human hair shafts. *BioTechniques* **18,** 662–669.
40. van Dinther, F. H. M., Bally, R. W., and van Sommeren, T. P. G. (1998) Evaluation of three commercially available capillary electrophoresis kits for single stranded DNA oligonuleotide analysis. *J. Chromatogr. A* **817,** 273–279.
41. Mansfield, E. S., Robertson, J. M., Vainer, M., Isenberg, A. R., Frazier, R. R., Ferguson, Chow, S., Harris, D. W., Barker, D. L., Gill, P. D., Budowle, B., and McCord, B. R. (1998) Analysis of multiplexed short tandem repeat (STR) systems using capillary array electrophoresis. *Electrophoresis* **19,** 101–107.
42. Isenberg, A. R., McCord, B. R., Koons, B. W., Budowle, B., and Allen, R.O. (1996) DNA typing of a polymerase chain reaction amplified D1S80/amelogenin multiplex using capillary electrophoresis and a mixed entangled polymer matrix. *Electrophoresis* **17,** 1505–1511.

43. Theisen, P., McCollum, C., Upadhya, K., Jacobson, K., Huynh, V., and Angius, A. (1992) Fluorescent dye phosphoramidite labelling of oligonucletides. *Tetrahedron Letts.* **33,** 5033–5036.
44. Wang, Y., Ju, J., Carpenter, B., Atherton, J. M., Sensabaugh, G. F., and Mathies, R. A. (1995) High-speed, high-throughput TH01 allelic sizing using energy transfer fluorescent primers and capillary array electrophoresis. *Anal. Chem.* **67,** 1197–1203.
45. Chen, J. W., Cohen, A. S., and Karger, B. L. (1991) Identification of DNA molecules by pre-column hybridization using capillary electrophoresis. *J. Chromatogr.* **559,** 295–305.
46. Kolesar, J. M., Allen, P. G., and Doran, C. M. (1997) Direct quantification of HIV-1 RNA by capillary electrophoresis with laser-induced fluorescence. *J. Chromatogr. B* **697,** 189–194.
47. Isenberg, A. (1998) *Separation of PCR-amplified DNA by Capillary Electrophoresis: A Study of the Effects of Polymer Characteristics and Separation Parameters*. Thesis, University of Virginia, pp. 1–171.
48. Butler, J. M., McCord, B. R., Jung, J. M., Lee, J.A., Budowle, B., and Allen, R. O. (1995) Application of dual internal standards for precise sizing of PCR products by capillary electrophoresis. *Electrophoresis* **16,** 974–980.
49. Butler, J. M., McCord, B. R., Jung, J. M., Wilson, M. R., Budowle, B., and Allen, R. O. (1994) Quantitation of polymerase chain reaction products by capillary electrophoresis using laser fluorescence. *J. Chromatogr. B* **658,** 271–280.

5

Microchip-Based Capillary Electrophoresis Systems

Wenzhang Xie, Rong Yang, Junquan Xu, Liang Zhang, Wanli Xing, and Jing Cheng

1. Introduction

Microchip-based capillary electrophoresis (CE) systems have enjoyed impressive advancement in the past few years *(1–3)*. Up until now, CE array chips have been fabricated based upon the well-understood behavior of a single channel chip system. Different materials (e.g., silicon *[4]*, glass *[5]*, and plastics *[6]*) have been used. A variety of different fabrication processes have also been developed to accommodate the complicated requirements and materials used for making such a device. Microchip-based CE systems have demonstrated uses in diverse applications such as the separation of amino acids *(7)*, analysis of blood serum cortisol *(8)*, examination of polymerase chain reaction (PCR) amplicons *(9)*, and analysis of metal ion complexes *(10)*. In this chapter, aspects related to chip-based CEs will be reviewed.

2. Methods

2.1. Technologies for Fabricating CE Chips

The micromachining technologies employed for fabricating the CE chips can be very different depending on the materials used. Photolithographic processing techniques are by far the most commonly used methods for producing microchannels in the surface of a planar silicon or glass substrate. One advantage of using these materials is that their electrophoretic and chromatographic properties and surface derivatization chemistries are extensively studied in many cases. Another advantage is that many established microfabrication processes can be easily modified and applied. Other types of fabrication methods are used for machining plastic substrates, including injection-molding, casting, imprinting, laser ablation, and stamping processes. The use of plastic as a substrate has two advantages, plastic is less expensive and easier to manipulate than glass or silicon-based substrates, and the device may be easily disposed of after use.

From: *Methods in Molecular Biology, Vol. 162:*
Capillary Electrophoresis of Nucleic Acids, Vol. 1: Introduction to the Capillary Electrophoresis of Nucleic Acids
Edited by: K. R. Mitchelson and J. Cheng © Humana Press Inc., Totowa, NJ

2.1.1. Photolithographic Fabrication of Glass CE Chips

1. Several elements are equally important in the process of fabricating a glass-based CE chip: a protective etch mask of high quality, the appropriate glass type and the composition of etchant. Wet-etching is the most widely used method for fabricating CE channels and reservoirs in glass, silica, or silicon substrates. Among the glass substrates, soda lime microscope slides (cat. no. 12-550C, Fisher Scientific) have been used from time to time in the past, owing mainly to its high etching rates (**5**). These slides have the merit that to obtain the required shallow etches, less aggressive etchants can be used along with the use of hard-baked photoresist ***(11)***.
2. However, for deep etching of microchannels, a more resistant sacrificial etch mask has to be used. Typical examples include the use of Cr/Au and amorphous silicon as sacrificial etch masks. Among these etch masks, photoresist/Au/Cr etch mask can be etched in various types of HF-based aggressive etchants such as HF/HNO_3.
3. When Borofloat glass is used with the photoresist/Au/Cr etch mask, channels with 35-μm etch depth can be obtained in approx 5 min in 49% HF.
4. Amorphous silicon is another type of etch mask that can be deposited on the substrates through plasma enhanced chemical vapor deposition ***(12)***. Masks such as amorphous silicon are sensitive to aggressive etchants, such as HF/HNO_3. Used as an etch mask, amorphous silicon proved itself possessing the best resistance to HF etching as well as the least defects. Channels with smooth side walls to a depth of 70 μm are achieved when amorphous silicon is utilized as the etch mask. Generally when used as an etch mask, high-quality amorphous silicon can withstand etchant for twice as long as Cr/Au, making it possible to etch twice as deep.
5. Among all sorts of glass substrate investigated, Schott Borofloat glass is found having the best etching quality and simplest processing ***(12)***. Borofloat is a borosilicate glass produced using the float process. This type of glass is uniform in thickness and composition and has a smooth and flat surface requiring no mechanical polishing (**Fig. 1**). Another great advantage of this material is that its background fluorescence is several times lower than that of microscope slide glass, making it the ideal candidate for experiments requiring high sensitivity of detection.
6. To make a complete piece of CE device, a planar cover glass with holes connected to the lower electrophoresis channels has to be used to seal the channels. To drill the holes on the cover glass a diamond-tipped drill bit is ideal for the rapid drilling of individual holes, requiring approx 15 s to drill for each hole. In contrast, a multitipped ultrasonic drill bit is suitable for the production of many chips with consistent hole patterns (requiring approx 15 min drilling for each chip).

2.1.2. Bonding of Glass

1. Bonding of the etched and drilled glass of the same size and type can be done using three different approaches.
2. One is to sandwich the glass substrates with two polished graphite blocks and then place them in an evacuated furnace with a stainless steel weight on top ***(12)***.
3. A second method is to have a thin layer of silicon dioxide deposited on one side of the cover glass and then allow the top and bottom glass substrates be anodically bonded. The yield of chips made this way can be much higher than the thermal fusing method.
4. The third approach is to bond the quartz glass substrates with hydrofluoric acid. The advantages of this method includes low thermal damage to the glass, low residual stress (i.e., bonding is carried out at room temperature), and simplicity of operation ***(13)***.

Fig. 1. Scanning electron micrograph of the intersection of two 25-μm deep channels etched in Borofloat using a 1700 Å thick amorphous silicon etch mask. Striations on the side walls of the capillary are evident near the intersection. Reprinted with permission from **ref.** ***12***.

2.2. Plastic Microfabrication

1. The methods developed for fabricating microchannels and reservoirs in polymeric substrates include, laser abalation, injection molding, compression molding, casting, and X-ray lithography.
2. Injection molding and casting methods are in general called replication method. Unlike laser etching method, the replication methods can generate capillary channels or reservoirs with smooth surfaces, which are demanded for the performance of CE in such plastic devices.

2.2.1. Injection Molding

1. Injection molding of a CE chip usually involves a multistep fabrication process.
2. First, a silicon master has to be wet-etched through the standard photolithographic process to obtain a "negative" image of the capillary channels. The fabricated silicon master is formed with protruded features, the height and width of these features corresponding to the specified dimensions of the capillary separation channel.
3. Second, a "positive" metal mold is formed through electroplating against the silicon master. From this metal mother (master), daughter molds can be made with the same features as that of the silicon master.
4. These daughter metal molds may then be mounted on a mold insert. The insert can be used through the injection molding or compression molding processes to produce hundreds of thousands of identical molded plastic chips using polymeric materials such as acrylic copolymer resin.

5. The preparation of a sealed plastic CE chip is achieved by first drilling a few millimeter-diameter “through holes” on the molded piece followed by thermal lamination of a thick sheet of Mylar coated with a thermally activated adhesive at raised temperature ***(6)***.

2.2.2. Casting Plastic Molds

1. Using casting process to fabricate polymeric chip is similar to the injection molding process.
2. The materials used include poly(dimethyl siloxane) (PDMS) and poly(methyl methacrylate) (PMMA) ***(14,15)***.
3. The polymer chip is cast against the wet-etched silicon master. Once formed, the PDMS replica can be peeled off easily. After punching the buffer reservoirs through, the cast chip can be placed on a slab of PDMS to form a sealed CE device.

2.2.3. Photoablation

1. Apart from replication method, photoablation can also be used to machine plastic chips ***(16)***. During the machining process, the laser pulses in the UV region can be transmitted through the mask to hit the selected areas on the plastic substrate. When the illuminated areas absorb the laser energy, the chemical bond within the long-chain polymer molecules are broken, and the photoablation generated debris such as gas, polymer molecules, and small particulate matter is ejected, leaving the desired channels and reservoirs in the plastic chip.
2. There are a variety of polymer materials that are suitable for photoablation, including polycarbonate, PMMA, polystyrene, nitrocellulose, poly(ethylene terephthalate), and Teflon.
3. A PMMA CE chip has also been fabricated using soft X-ray lithography, and a transfer or Kapton® mask ***(17)***. The main advantage of machining PMMA with soft X-rays is that narrow and deep channels (i.e., high aspect ratio) can be fabricated in the substrate. Several components are machined in PMMA including the injector, separation channel, and integrated fiber optic fluorescence detector.

2.3. Separation Methods

1. One distinct advantage of microfabricated chips is that they can be utilized as platforms for multipurpose liquid sample handling and analysis. As a result, a variety of separation methods have been developed for use with microchips.
2. The methods implemented on chips include free-solution CE, capillary gel electrophoresis, micellar electrokinetic chromatography, isotachophoresis, isoelectric focusing, open-channel electrochromatography, and free-flow electrophoresis.

2.3.1. Free-Solution CE

1. Free-solution CE was the earliest CE transferred into a microchip manifold. Representative pioneer reports were made jointly by the groups of Harrison and Manz ***(18)***. They performed the free-solution capillary electrophoretic separation of six γ-fluorescein isothiocyanate labeled amino acids in approx 15 s using a glass CE chip. Many developments have been reported since the initial report.
2. In a recent study conducted by Ramsey’s group ***(19)***, free-solution CE has been performed on a microchip CE device. It is the first time that single chromophore molecules were separated and then counted using confocal microscopy.
3. Another study conducted by the same group recently demonstrated the separation of rhodamine B and dichlorofluorescein in 0.8 ms ***(20)***. In this study, a 200-μm long and 26-μm

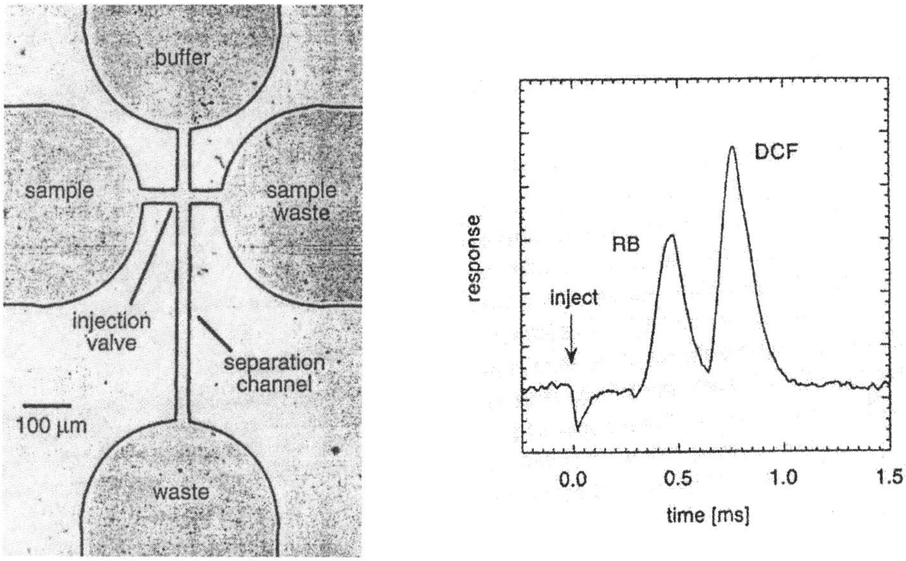

Fig. 2. *Left*: photograph of microchip used for high-speed electrophoretic separations. The area of the photograph is 0.8 × 1.2 mm, and the injection valve and separation channels are shown. The wide channels are 440 μm wide, and the narrow channels are 26 μm wide. *Right*: high-speed electropherogram of rhodamine B and dichlorofluorescein resolved in 0.8 ms using a separation field strength of 53 kV cm^{-1} and a separation length of 200 μm. The start time is marked with an arrow at 0 ms. Reprinted with permission from **ref. *20***.

wide separation channel was fabricated that achieves a 100-fold time decrease in electrophoretic separation (**Fig. 2**). This high-speed microchip electrophoresis s is also shown to have improved efficiency, which is made possible mainly by reducing the injected sample plug width and reducing the Joule heating to increase the plate height. This system undoubtedly will become a valuable tool for the monitoring of ms time-scale kinetics for chemical and biochemical reactions that are commonly required in ultra-high throughput drug screening processes.

2.3.2. Capillary Gel Electrophoresis

1. Gel electrophoresis is the most common approach for separating biopolymers such as nucleic acids and proteins. Effenhauser et al. *(21)* were the pioneers in transferring this process to the microchip, and in 1994 reported the use of microchannels machined on planar glass substrates and filled with noncrosslinked 10% T polyacrylamide for the size separation of phosphorothioate oligonucleotides ranging from 10–25 bases. The separation was completed in 45 s with a plate height of 200 nm.
2. Because of its wide application and huge commercial value, gel electrophoresis has been exploited for quite a few years on microchip platforms and research has been undertaken to advance this technique and to broaden its applications. More review on this technique is presented in the later section.

2.3.3. Micellar Electrokinetic Chromatography

1. Micellar electrokinetic chromatography (MEKC) was initially developed by Terabe et al. *(22)*. In MEKC, surfactants with concentrations larger than their critical micelle points are added to the separation buffer facilitating the separation of uncharged solutes based upon differential partitioning. The separation of charged molecules is determined by electrophoresis, electrostatic interactions, solute complexation with the surfactant, and also partitioning between two phases.
2. Microchip-based MEKC was first reported by Moore et al. *(23)* and used to separate three coumarin dyes in a separation buffer of 50 m*M* SDS and 10% methanol. Excellent reproducibility was demonstrated at field strength lower than 400 V/cm.
3. In another study, the separation of eight of the most common biogenic amines was achieved in 75 s by running the MEKC in a glass chip and also the biogenic amines from soy sauce samples were identified by the same approach *(24)*.
4. A quite different separation format was reported by von Heeren et al. *(25)*, where a glass CE chip with cyclic channels was fabricated for achieving separation of six fluorescein isothiocyanate-labeled amino acids in a few seconds. The MEKC separation efficiency in this study was found comparable to that obtained by chip-based gel electrophoresis.
5. Overall, chip-based MEKC is found to increase the separation efficiency and to achieve decreases in separation time by several 10-fold orders of magnitude, when compared to conventional MEKC performed in a fused silica capillary.
6. Additionally, the efficient heat dissipation in glass, silica, or silicon chips enables the application of very high field (up to 2 kV/cm), which permits separations in time periods ranging from milliseconds to seconds.

2.3.4. Isotachophoresis

1. The isotachophoresis method is also able to be transferred onto a glass-glass microchip, and in a recent report is used to separate two herbicides *(26)*. The herbicides paraquat and diquat could be detected by Raman spectroscopy at concentrations as low as 2.3×10^{-7} *M*.
2. The Raman microprobe is directly coupled to the microchip without any interfacing. The Raman spectra are generated by a 532-nm NdY-VO4 laser of 2 W and collected at an 8 cm^{-1} resolution with use of a holographic transmissive spectrography and a cooled charge coupled device (CCD).

2.3.5. Isoelectric Focusing

1. Capillary isoelectric focusing reportedly can be performed in the separation channel with a width of 200 µm, a depth of 10 µm, and a length of 7 cm etched on a glass microchip *(27)*. Compared to the chemical and hydrodynamic driven mobilization of analytes, electroosmotic flow driven analyte mobilization (occurring simultaneously with focusing) is the most suitable isoelectric focusing format for use with the chip format.
2. Micromachined chips are readily useable for isoelectric focusing because of their high speed, compatibility with electroosmotic flow mobilization, and their low instrumentation requirements. Using chip-based capillary isoelectric focusing, a mixture of Cy5-labeled peptides could be focused in less than 30 s with plate heights of 0.4 µm.

2.3.6. Open-Channel Electrochromatography

1. Microchip capillary platform has been adapted for techniques other than CE, such as for a chromatographic technique called open-channel electrochromatography (OCEC) *(28,29)*.

2. In this application, electroosmotic pumping is used to move the mobile phase through a serpentine-shaped microchannel. In the first study reported by Jacobson et al. ***(28)***, the interface of the microchannel is chemically modified with octadecylsilane, which acts as the stationary phase. The separation of three neutral coumarin dyes could be demonstrated in a mobile phase containing 10 m*M* sodium tetraborate and 25% acetonitrile.
3. In a further study, solvent programming has improved the efficiency for OCEC ***(29)***. The voltages applied across the chip are computer-controlled. The voltages are applied to the terminals of the chip to adjust the isocratic and gradient elution conditions. It is found that important elements for achieving enhanced selectivity and to reduce assay time include factors such as, the slope of linear solvent gradients, different start times, different duration times, and different start percentages of organic modifier. A complete run including fast reconditioning took only 60 s to accomplish.

2.3.7. Free-Flow Electrophoresis

1. In the separation of biopolymers or cells, two-dimensional methods are inherently more powerful than one-dimensional methods since the resulting peak capacity is increased. FFE is one of these two-dimensional methods.
2. FFE is different from other two-dimensional separation methods such as two-dimensional gel electrophoresis where a supporting medium (e.g., gel), high salt concentrations, or even organic solvents are used.
3. FFE has very gentle separation conditions and therefore it is especially suitable for separations involving cells and easily denatured molecules such as proteins.
4. Raymond et al. ***(30)*** has performed FFE initially in a micromachined silicon device for continuous separation of rhodamine-B isothiocyanate labeled amino acids in 20 min with an applied voltage of 40 V.
5. Raymond et al. ***(31)*** later further separated high molecular weight compounds, such as human serum albumin, bradykinin, and ribonuclease A, in a 25-µL vol FFE microstructure. They also could achieve continuous separation of more complicated sample mixtures such as tryptic digests of mellitin and cytochrome *c*. The samples and the concentration of each component was detected using laser-induced fluorescence (LIF) detection.
6. In another report, dielectrophoretic free-flow fractionation is used to separate human leukemia (HL-60) cells from peripheral blood mononuclear cells in a planar glass substrate plated with gold interdigitated electrodes ***(32)***. With this technique, an appropriate AC voltage signal is applied to the microelectrodes, the cells are levitated and suspended in the separation chamber with different equilibrium heights due to the dielectrophoretic forces generated by the applied AC signal. The separation of cells was eventually achieved based on the balance of several applied forces, including dielectrophoretic, gravitational, and hydrodynamic lift forces.
7. Two recent reports published at the same time, describe a method which is very similar in principle to that of FFE, but differs in that microfabricated fractal arrays are adapted to replace the free zones created in FFE, and can achieve much higher efficiency and separation resolution ***(33,34)***. Although both papers report works at an early stage, it is predicted that the method will achieve a calculated separation of DNA molecules with sizes ranging from 100–20,000 bp. Once realized, this technique may facilitate a simple setup and automation of DNA separation. Also, the cost of each device might be less than $1, and so it could be disposed of after a single use.

2.3.8. Pulse-Field Electrophoresis

1. Austin's group *(35)* has transferred pulsed-field electrophoresis (PFE) onto microfabricated silicon devices, in which arrays etched on the silicon are utilized as the separation matrix rather than the gels used in conventional PFE. The study indicates that the motions of biopolymers in the microarray matrix is more uniform compared to the motion in gels.
2. Less dispersion in displacement is anticipated, which may lead therefore to reduced band broadening and improved resolution. The analyte separation can be greatly increased as megabase DNA will not be trapped in the array, as happens in a gel.
3. Also, because the long DNA molecules will not be arrested, much higher separation voltage can be applied. Excellent heat dissipation through the silicon substrate further supports the application of higher fields without significant Joule heating occurring.

3. Methods

3.1. Nucleic Acid Analyses

1. The analyses of nucleic acids can be divided into two main categories. One is to achieve fragment sizing, which in most cases is related to the detection of DNA mutations. The other category is for DNA sequence analysis.
2. For the first category, polymer solution gel capillary electrophoresis has been used as the main separation media. One of the earliest separation cases, the rapid sizing of PCR-amplified *HLA-DQα* alleles as well as the spiked DNA marker with size ranging from 72 to 353 bp was obtained in approx 2 min in a glass CE device *(36)*. Hydroxyethylcellulose (HEC) was used to form the entangled free-solution sieving matrix in this study.
3. Both glass chips and plastic chips have been fabricated and used for fragment sizing *(6,15)* and for the detection of single DNA molecules *(15)*.
4. Using an injection-molded acrylic CE chip and HEC as the sieving matrix, all fragments in the DNA marker with size ranging from 72 to 1353 bp was baseline resolved in 2.5 min *(6)*. The standard deviation for run-to-run is less than 1% and for chip-to-chip is between 2–3%.
5. A PDMS-molded CE chip together with hydroxypropylcellulose (HPC) as the sieving media has been used *(15)* in the separation of the same DNA marker as that used in **refs.** ***6*** and ***36***. In this study of a single DNA molecule, detection efficiency larger than 50% is obtained *(15)*.
6. A fused silica CE chip has been fabricated and used for fast-DNA profiling *(37,38)* with a replaceable denaturing polyacrylamide matrix. Baseline resolved separation of single-locus short tandem repeats amplicons is achieved with this system, with the complete separation of four amplicons representing loci of the *CSF1PO*, *TPOX*, *THO1*, and *vWA* genes achieved in less than 2 min. This represents a 10–100-fold increase in the speed of separation compared to the conventional capillary or slab gel electrophoresis systems.
7. A glass, capillary array electrophoresis (CAE) chip filled with HEC solution is also used for high-speed DNA genotyping *(39)*. Twelve DNA samples, with the largest fragment size of 622 bp, could be separated in parallel in less than 3 min (**Fig. 3**). For all the lanes, detection is achieved using a laser-excited confocal fluorescence scanner, which achieves a temporal resolution of 0.3 s.
8. Recently, ribosomal RNA samples are reportedly separated in an injection-molded plastic microchannel with a cross-section of 100 × 40 μm and with an effective length of

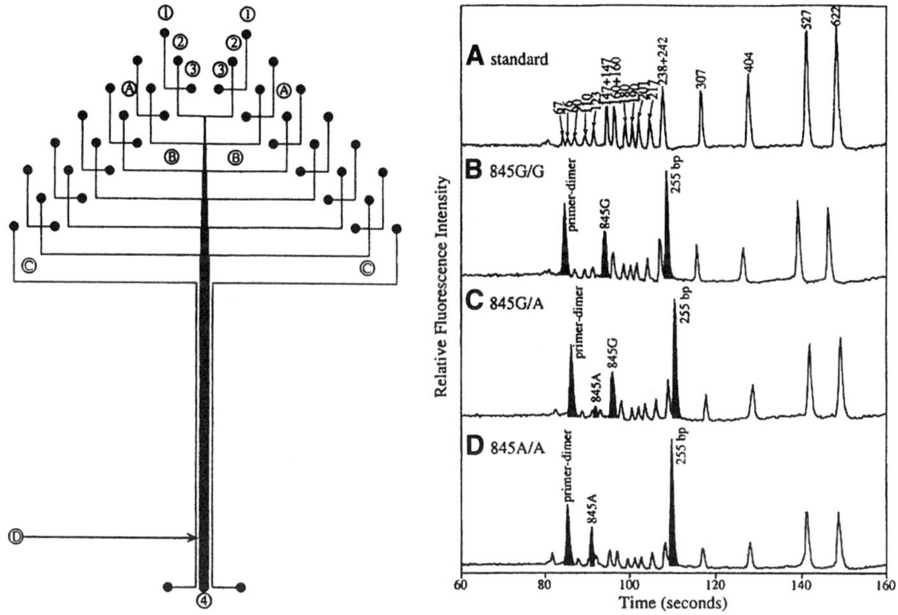

Fig. 3. *Left*: Mask design used to photolithographically pattern the 12-channel CAE chips. Key: **(A)** Injection channels, 8 mm long. **(B)** Separation channels, 60 mm long. **(C)** Optical alignment channels. **(D)** Detection region, ~45 mm from the junctions of injection and separation channels. (1) Injection waste reservoirs. (2) Ground reservoirs. (3) Injection reservoirs. (4) High-voltage anode reservoir. The actual chip size was 50 mm × 75 mm. *Right*: Representative electropherograms of the three different HLA-H nucleotide 845 genotypes **(B–D)** generated from the microfabricated CAE chip, along with the pBR322 *Mspl* standard ladder **(A)**. The HLA-H peaks are shaded. Reprinted with permission from **ref.** ***39***.

1 cm *(40)*. The sieving matrix employed is hydroxypropyl methyl cellulose (HPMC). The detection of RNA was more sensitive than an equivalent to the RNA content from a single cell, and is achieved using a fluorescent microscope equipped with a photometer.

3.2. DNA Sequence Analyses

1. Ultra high-speed DNA sequence analysis has been achieved on a glass CE chip, where a denaturing 9% T and 0% C polyacrylamide solution is used as the separation media *(41)*. A four-color detection scheme is used for the readout, and reads of approx 200 bases are obtained in 10 min in a channel with an effective separation length of 3.5 cm.
2. Following the optimization of both the electrophoretic channel design and the electrophoresis methods, a much higher readout in four-color DNA sequencing is obtained, and up to 500 bases can be read in less than 20 min *(42)* (**Fig. 4**).
3. If a 96-channel array chip is used for the purpose of fast-DNA sequence analysis, one would expect to see a significant increase in the throughput rate, compared to the conventional DNA sequencers.

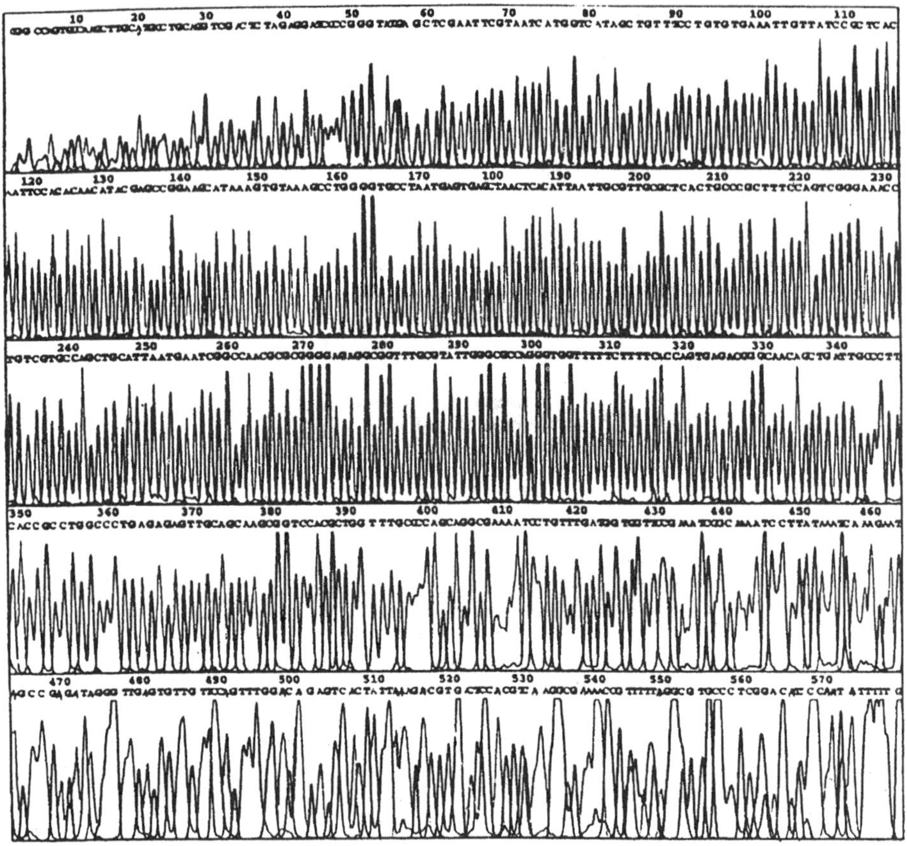

Fig. 4. Analyzed four-color M13 DNA sequencing traces from a CE chip. Separation was performed on a 7-cm long channel with a 100-µm twin-T injector using 4% LPA as the separation medium at 40°C. Separation was performed with a voltage of 160 V/cm, and the detector was 6.5 cm from the injector. Only 0.2 µg of DNA template was employed per reaction, and 1 µL of the final reaction solution (33%) was loaded on the chip. This run was complete in under 20 min. Reprinted with permission from **ref.** ***42***.

3.3. CE-Based Laboratory-On-A-Chip Systems

1. The ultimate goal of developing microchip-based devices is to be able to build a so-called "laboratory-on-a-chip" system. Generally speaking, a "laboratory-on-a-chip" system would be classically composed of the three steps usually seen with all biological assays: the sample preparation step, some form of chemical reaction(s), and sample detection ***(43)***.
2. The sample preparation step(s) generally include the cell isolation, cell lysis, and DNA or RNA extraction.
3. The chemical reaction step(s) are usually represented by various enzymatic reactions such as PCR amplification reactions, or proteinase K digestion, or chemical labeling reactions, and so on.

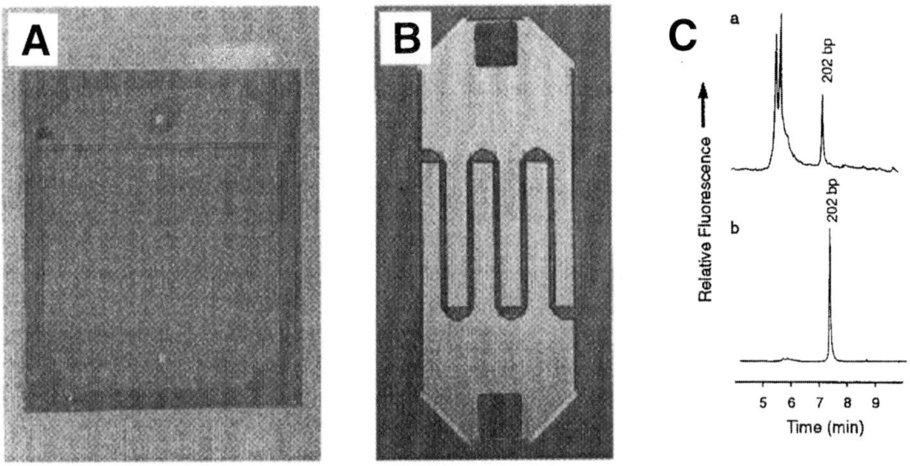

Fig. 5. Filtration-PCR chip designs and results of direct PCR for *dystrophin* gene. **(A)** Integrated filter-PCR chip based on linear weir-type filter in the PCR chamber. **(B)** Integrated filter-PCR chip based on coiled weir-type filter in the PCR chamber. **(C)** Electropherograms of 202-bp amplification product (exon 6, *dystrophin* gene) from direct PCR of DNA in filtered white blood cells using a filter-PCR chip (a) and a positive control (b). A 0.5-µL sample was removed from the chip, diluted with 99.5-µL water, and then injected (5 kV/5 s) onto a 100-mm × 27-cm long DB-1-coated capillary in a P/ACE 5050 (run 8 kV/10 min). Reprinted with permission from **ref. *44***.

4. The detection of nucleic acids is usually achieved through two approaches. One approach is a "separation-based detection" such as represented by CE or HPLC techniques. Only the efforts devoted for the construction of CE-based laboratory-on-a-chip systems are reviewed below.
5. The other detection method is represented by "hybridization-based" approaches.

3.4. Integration of Process Steps

1. The integration of processes requires the combining together of the sample preparation, chemical reaction, and CE-based detection processes on a microdevice. At present, only a partial integration of the processes has been achieved.
2. The group at University of Pennsylvania led by Wilding and Kricka ***(44)*** reported the first example of the integration of both sample preparation and chemical reaction in a single chip. Here a silicon-glass chip with a microfilter and integrated reactor is used for the separation of nucleated white blood cells from red blood cells by microfiltration, followed directly by a PCR amplification of the DNA released from the isolated white blood cells (**Fig. 5**).
3. Cheng et al. ***(45)***, using a single bioelectronic chip, made the isolation of *Escherichia coli* cells from human blood possible through use of dielectrophoretic separation processes followed by the electronic lysis of the isolated cells (**Fig. 6**). Cultured cervical carcinoma cells were similarly isolated from normal human blood cells by this group using the same device in a modified manner ***(46)***.

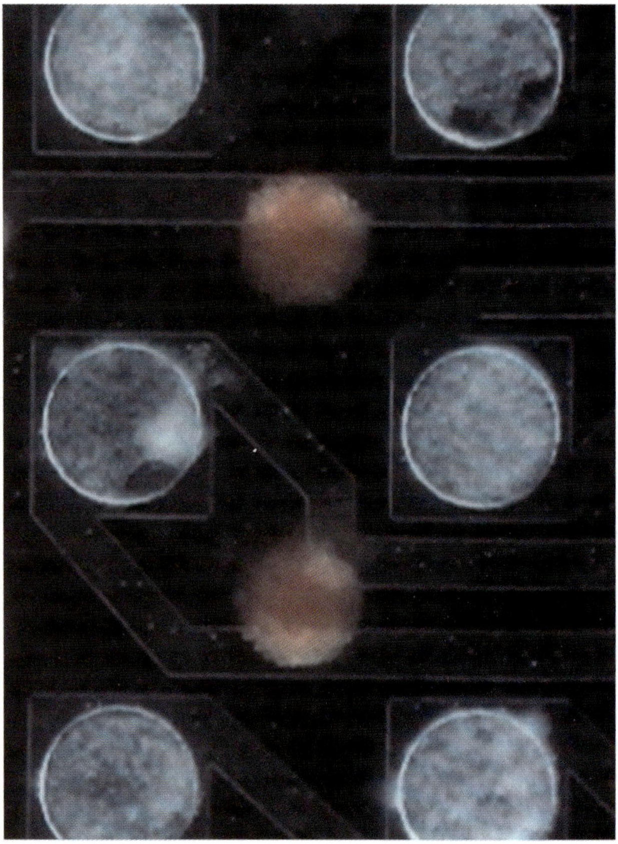

Fig. 6. Dielectrophoretic separation of *E. coli* from blood cells. The mixture is separated according to charge, bacteria localizing in zones under electric field maxima (white) and erythrocytes localizing in zones under filed minima (red). Reprinted with permission from **ref. *45***.

4. Li and Harrison ***(47)*** can achieve cell transportation in microchannels by electrophoretic pumping and/or electroosmotic pumping, which following chemical lysis using SDS buffer allows DNA liberation.
5. The partial integration of enzymatic reaction and capillary electrophoretic separation has been made on a single glass chip by Jacobson and Ramsey ***(48)***. Both the plasmid pBR322 DNA and the restriction enzyme *Hinf*I are electrophoretically pumped independently into a 0.7-nL reaction chamber in which the digestion can occur. The digested DNA fragments are then sized electrophoretically in the capillary device etched on the same chip. The entire process was completed in 5 min.
6. In a joint study ***(49,50)***, DNA was amplified randomly in a silicon-glass chip using DOP-PCR, followed by specific multiplex PCR amplification of the *dystrophin* gene in a second chip. The *dystrophin* gene amplicons are then separated in a glass microchip by CE (**Fig. 7**).

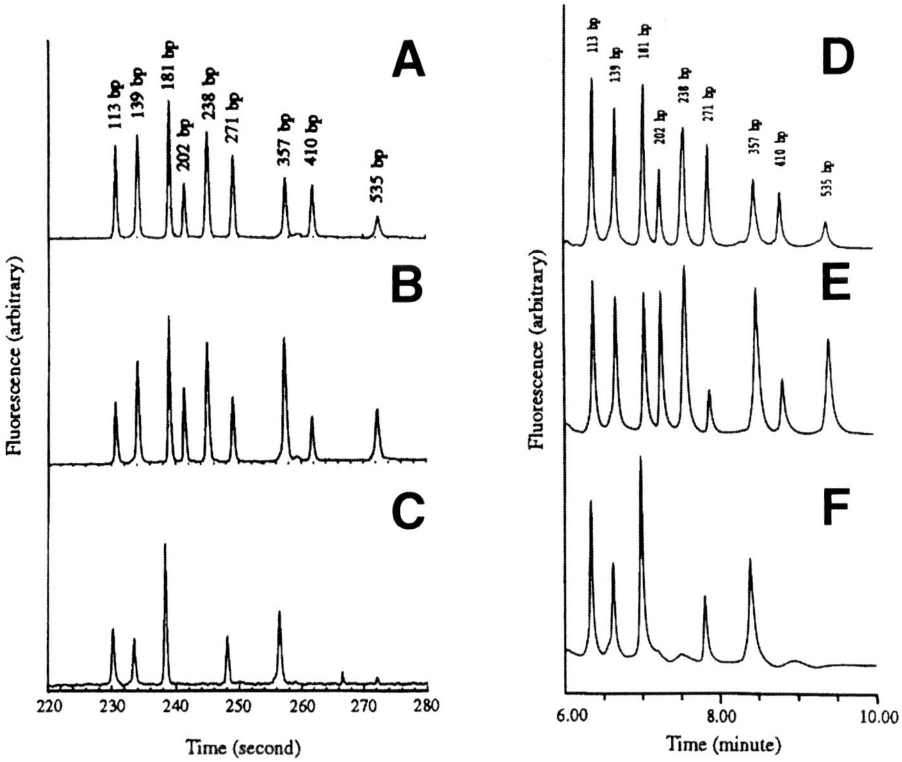

Fig. 7. Comparison of chip CE electropherograms with conventional CE electropherograms. (**A**) Control, PCR amplicons generated in a GeneAmp test tube using normal human genomic template DNA; (**B**) PCR amplicons generated in a silicon-glass chip using the same human template DNA as used in (A); (**C**) the PCR amplicons generated in a silicon-glass chip with template DNA from an affected patient. **(D–F)** Conventional CE electropherograms of the PCR amplicons used in (A–C), respectively. Reprinted with permission from **ref. *49***.

7. The functional integration of PCR amplification followed by CE fragment sizing was achieved by the coupling of the silicon PCR chip with the glass CE chip ***(51)***. Using an in-chip polysilicon heater, the rapid PCR amplification of a β-*globin* target cloned in an M13 plasmid was complete in 15 min, and the CE separation took approx 2 min to size the amplicons which followed.
8. One single glass chip has been fabricated by the Ramsey group ***(52,53)*** to perform combined steps of cell lysis, multiplex PCR, and CE separation. The PCR thermal cycling required several hours, however the size separation of amplicons took about 3 min using either HEC, or poly(dimethyl acrylamide) as the sieving gels.
9. In another recent study Burns et al. ***(54)*** reported great progress has been made in the fabrication and utilization of a microfabricated silicon chip with an integrated metering capability, a thermal pump, an isothermal reactor, and a CE structure with an integrated photodetector. The amplification of DNA is achieved through "strand displacement

amplification"–an exponential isothermal amplification technique. The amplicon was separated in the microchannel and detected by the integrated photodetector. The total performance of the device including the metering of the reactants, the isothermal DNA amplification and the CE-based separation took approx 20 min to complete. This milestone-breaking work has shown that it may be possible in the near future to make a completely integrated and self-contained, portable device for many different applications for rapid analysis following sample collection. Likely applications include "point-of-care analysis" in the hospital situation, scene-of-crime identification purposes, on-site agricultural testing and environmental monitoring, and even for outer-space exploitation.

Acknowledgments

This work is funded in part by National Natural Science Foundation (contract nos. 39880035 and 39825108), National High Technology Program (863) and the 100-Talent Project Fund from Tsinghua University.

References

1. Colyer, C. L., Tang, T., Chiem, N., and Harrison, D. J. (1997) Clinical potential of microchip capillary electrophoresis systems. *Electrophoresis* **18,** 1733–1741.
2. Effenhauser, C. S., Bruin, G. J. M., and Paulus, A. (1997) Integrated chip-based capillary electrophoresis. *Electrophoresis* **18,** 2203–2213.
3. Kopp, M. U., Crabtree, H. J., and Manz, A. (1997) Developments in technology and applications of microsystems. *Curr. Opinion Chem. Biol.* **1,** 410–419.
4. Webster, J. R. and Mastrangelo, C. H. (1997) Large-volume integrated capillary electrophoresis stage fabricated using micromachining of plastics on silicon substrates. *Transducers '97*, pp. 503–506.
5. Wooley, A. T. and Mathies, R. A. (1994) Ultra-high-speed DNA fragment separations using microfabricated capillary array electrophoresis chips. *Proc. Natl. Acad. Sci. USA* **91,** 11,348–11,352.
6. McCormick, R. M., Nelson, R. J., Alonso-Amigo, M. G., Benvegnu, D. J., and Hooper, H. H. (1997) Microchannel electrophoretic separations of DNA in injection-molded plastic substrates. *Anal. Chem.* **69,** 2626–2630.
7. Jacobson, S. C., Hergenroder, R, Moore, A. W., and Ramsey, J. M. (1994) Precolumn reactions with electrophoretic analysis integrated on a microchip. *Anal. Chem.* **66,** 4127–4132.
8. Koutny, L. B., Schmalzing, D., Taylor, T. A., and Fuchs, M. (1996) Microchip electrophoretic immunoassay for serum cortisol. *Anal. Chem.* **68,** 18–22.
9. Cheng, J., Waters, L. C., Fortina, P., Hvichia, G. E., Jacobson, S. C., Ramsey, J. M., Kricka, L. J., and Wilding, P. (1998) Degenerate oligonucleotide primed-polymerase chain reaction and capillary electrophoretic analysis of human DNA on microchip-based devices. *Anal. Biochem.* **257,** 101–105.
10. Jacobson, S. C., Moore, A. W., and Ramesey, J. M. (1995) Fused quartz substrates for microchip electrophoresis. *Anal. Chem.* **67,** 2059–2063.
11. Fan, Z. H. and Harrison, D. J. (1994) Micromachining of capillary electrophoresis injectors and separators on glass chips and evaluation of flow at capillary electrophoresis. *Anal. Chem.* **66,** 177–184.
12. Simpson, P. C., Wooley, A. T., and Mathies, R. A. (1998) Microfabrication technology for the production of capillary array electrophoresis chips. *Biomed. Microdevices* **1,** 7–26.
13. Nishimoto, T., Nakanishi, H., Abe, H., Arai, A., Nakamura, R., Yotsunoto, A., and Shoji, S. Microfabricated chips for capillary electrophoresis on quartz glass substrates

using a bonding with Hydrofluoric acid, in *Micro Total Analysis Systems '98*, (Harrison, D. J. and van den Berg, A., eds.), Kluwer Academic Publishers, pp. 311–314.
14. Effenhauser, C. S., Bruin, G. J. M., Paulus, A., and Ehrat, M. (1997) Integrated capillary electrophoresis on flexible silicone microdevices: analysis of DNA restriction fragments and detection of single DNA molecules on microchips. *Anal. Chem.* **69,** 3451–3457.
15. Martynova, L., Locascio, L. E., Gaitan, M., Kramer, G. W., Christensen, R. G., and MacCrehan, W. A. (1997) Fabrication of plastic microfluid channels by imprinting methods. *Anal. Chem.* **69,** 4783–4789.
16. Roberts, M. A., Rossier, J. S., Bercier, P., and Girault, H. (1997) UV laser machined polymer substrates for the development of microdiagnostic systems. *Anal. Chem.* **69,** 2035–2042.
17. Ford, S. M., Davies, J., Kar, B., Qi, S. D., McWhoeter, S., Soper, S. A., and Malek, C. K. (1999) Micromachining in plastics using X-ray lithography for the fabrication of microelectrophoresis devices. *J. Biomech. Eng.* **121,** 13–21.
18. Harrison, D. J., Fluri, K., Seiler, K., Fan, Z., Effenhauser, C. S., and Manz, A. (1993) Micromachining a miniaturized capillary electrophoresis-based chemical analysis system on a chip. *Science* **261,** 895–896.
19. Fister, J. C. III, Jacobson, S. C., Davis, L. M., and Ramsey, J. M. (1998) Counting single chromophore molecules for ultrasensitive analysis and separations on microchip devices. *Anal. Chem.* **70,** 431–437.
20. Jacobson, S. C., Culbertson, C. T., Daler, J. E., and Ramsey, J. M. (1998) Microchip structures for submillisecond electrophoresis. *Anal. Chem.* **70,** 3476–3480.
21. Effenhauser, C. S., Paulus, A., Manz, A., and Widmer, H. M. (1994) High-speed separation of antisense ofligonucleotides on a micromachined capillary electrophoresis device. *Anal. Chem.* **66,** 2949–2953.
22. Terabe, S., Otsuka, K., Ichikawa, K., Tsuchiya, A., and Ando, T. (1984) Electrokinetic separations with micellar solutions and open-tubular capillary. *Anal. Chem.* **56,** 111–113.
23. Moore, A. W., Jacobson, S. C., and Ramesey, J. M. (1995) Microchip separations of neutral species via micellar electrokinetic capillary chromatography. *Anal. Chem.* **67,** 4184–4189.
24. Rodriguez, I., Lee, H. K., and Li, S. F. (1999) Microchannel electrophoretic separation of biogenic amines by micellar electrokinetic chromatography. *Electrophoresis* **20,** 118–126.
25. von Heeren, F., Verpoorte, E., Manz, A., and Thormann, W (1996) Micellar electrokinetic chromatography separations and analyses of biological samples on a cyclic planar microstructure. *Anal. Chem.* **68,** 2044–2053.
26. Walker, P. A. III, Morris, M. D., Burns, M. A., and Johnson, B. N. (1998) Isotachophoretic separation on a microchip. Normal Raman spectroscopy detection. *Anal. Chem.* **70,** 3766–3769.
27. Hofmann, O., Che, D., Cruickshank, K. A., and Muller, U. R. (1999) Adaptation of capillary isoelectric focusing to microchannels on a glass chip. *Anal. Chem.* **71,** 678–686.
28. Jacobson, S. C., Hergenroder, R., Koutny, L. B. and Ramsey, J. M. (1994) Open-channel electrochromatography on a microchip. *Anal. Chem.* **66,** 2369–2373.
29. Kutter, J. P., Jacobson, S. C., and Ramsey, J. M. (1998) Solvent-programmed microchip open-channel electrochromatography. *Anal. Chem.* **70,** 3291–3297.
30. Raymond, D. E., Manz, A., and Widmer, H. M. (1994) Continuous sample pretreatment using a free-flow electrophoresis device integrated onto a silicon chip. *Anal. Chem.* **66,** 2858–2865.
31. Raymond, D. E., Manz, A., and Widmer, H. M. (1996) Continuous separation of high molecular weight compounds using a microliter volume free-flow electrophoresis microstructure. *Anal. Chem.* **68,** 2515–2522.
32. Huang, Y., Wang, X.-B., Becker, F. F., and Gascoyne, P. R. C. (1997) Introducing dielectrophoresis as a new force field for field-flow fractionation. *Biophys. J.* **73,** 1118–1129.

33. Ertas, D. (1998) Lateral separation of macromolecules and polyelectrolytes in microlithographic arrays. *Phys. Rev. Lett.* **80,** 1548–1551.
34. Duke, T. A. J., and Austin R. H. (1998) Microfabricated sieve for the continuous sorting of macromolecules. *Phys. Rev. Lett.* **80,** 1552–1555.
35. Duke, T. A. J., Austin, R. H., Cox, E. C., and Chan, S. S. (1996) Pulsed-field electrophoresis in microlithographic arrays. *Electrophoresis* **17,** 1073–1079.
36. Woolley, A. T., and Mathies, R. A. (1994) Ultra-high-speed DNA fragment separations using microfabricated capillary array electrophoresis chips. *Proc. Natl. Acad. Sci. USA* **91,** 11,348–11,352.
37. Schmalzing, D., Koutny, L., Adourian, A., Belgrader, P., Matsudaira, P., and Ehrlich, D. (1997) DNA typing in thirty seconds with a microfabricated device. *Proc. Natl. Acad. Sci. USA* **94,** 10,273–10,278.
38. Schmalzing, D., Koutny, L., Adourian, A., Chisholm, D., Matsudaira, P., and Ehrlich, D. (2001) Genotyping by microdevice electrophoresis, in *Capillary Electrophoresis of Nucleic Acids,* Vol. 2 (Mitchelson, K. R., and Cheng, J., eds.), Humana Press, Totowa, NJ, pp. 163–173.
39. Woolley, A. T., Sensabaugh, G. F., and Mathies, R. A. (1997) High-speed DNA genotyping using microfabricated capillary array electrophoresis chips. *Anal. Chem.* **69,** 2181–2186.
40. Ogura, M., Agata, Y., Watanabe, K., McCormick, R. M., Hamaguchi, Y, Aso, Y., and Mitsuhashi, M. (1998) RNA chip: quality assessment of RNA by microchannel linear gel electrophoresis in injection-molded plastic chips. *Clin. Chem.* **44,** 2249–2255.
41. Woolley, A. T., and Mathies, R. A. (1995) Ultra-high-speed DNA sequencing by using capillary electrophoresis chips. *Anal. Chem.* **67,** 3676–3680.
42. Liu, S., Shi, Y., Ja, W. W., and Mathies, R. A. (1999) Optimization of high-speed DNA sequencing on microfabricated capillary electrophoresis channels. *Anal. Chem.* **71,** 566–673.
43. Cheng, J., Fortina, P., Sorrey, S., Kricka, L. J., and Wilding, P. (1996) Microchip-based devices for molecular diagnosis of genetic diseases. *Molecular Diagnosis* **1,** 183–200.
44. Wilding, P., Kricka, L. J., Cheng, J., Hvichia, G. E., Shoffner, M. A., and Fortina, P. (1998) Integrated cell isolation and PCR analysis using silicon microfilter-chambers. *Anal. Biochem.* **257,** 95–100.
45. Cheng, J., Sheldon, E. L., Wu, L., Heller, M. J., and O'Connell, J. P. (1998) Isolation of cultured cervical carcinoma cells mixed with peripheral blood cells on a bioelectronic chip. *Anal. Chem.* **70,** 2321–2326.
46. Cheng, J., Sheldon, E. L., Wu, L., Uribe, A., Gerrue, L. O., Heller, M. J., and O'Connell, J. P. (1998) Preparation and hybridization analysis of DNA/RNA from *E. coli* on microfabricated bioelectronic chips. *Nature Biotechnol.* **16,** 541–546.
47. Li, P. C. H., and Harrison, D. J. (1997) Transport, manipulation, and reaction of biological cells on-chip using electrokinetic effects. *Anal. Chem.* **69,** 1564–1568.
48. Jacobson, S. C., and Ramsey, J. M. (1996) Integrated microdevice for DNA restriction fragment analysis. *Anal. Chem.* **68,** 720–723.
49. Cheng, J., Waters, L. C., Fortina, P., Hvichia, G. E., Jacobson, S. C., Ramsey, J. M., Kricka, L. J., and Wilding, P. (1998) Degenerate oligonucleotide primed-polymerase chain reaction and capillary electrophoretic analysis of human DNA on microchip-based devices. *Anal. Biochem.* **257,** 101–105.
50. Fortina, P., Cheng, J., Kricka, L. J., Waters, L. J., Jacobson, S. C., Wilding, P., and Ramsey, J. M. (2001) DOP-PCR amplification of whole genomic DNA and microchip-based capillary electrophoresis, in *Capillary Electrophoresis of Nucleic Acids,* Vol. 2 (Mitchelson, K. R. and Cheng, J., eds.), Humana Press, Totowa, NJ, pp. 211–219.

51. Woolley, A. T., Hadley, D., Landre, P., deMello, A. J., Mathies, R. A., and Northrup, A. (1996) Functional Integration of PCR amplification and capillary electrophoresis in a microfabricated DNA analysis device. *Anal. Chem.* **68,** 4081–4086.
52. Waters, L. C., Jacobson, S. C., Kroutchinina, N., Khandurina, J., Foote, R. S., and Ramsey, J. M. (1998) Microchip device for cell lysis, multiplex PCR amplification, and electrophoretic sizing. *Anal. Chem.* **70,** 158–162.
53. Waters, L. C., Jacobson, S. C., Kroutchinina, N., Khandurina, J., Foote, R. S., and Ramsey, J. M. (1998) Multiplex sample PCR amplification and electrophoretic analysis on a microchip. *Anal. Chem.* **70,** 5172–5176.
54. Burns, M. A., Johnson, B. N., Brahmasandra, S. N., Handique, K., Webster, J. R., Krishnan, M., Sammarco, T. S., Man, P. M., Jones, D., Heldsinger, D., Mastrangelo, C. H., and Burke, D. T. (1998) An integrated nanoliter DNA analysis device. *Science* **282,** 484–487.

6

DNA Sequencing by Capillary Array Electrophoresis

Norman J. Dovichi and JianZhong Zhang

1. Introduction

On May 9, 1998, PE Biosystems announced that they would introduce a new instrument based on "breakthrough DNA analysis technology" *(1)*. Over 700 orders for the new sequencer were received in the first 5 mo of production, which make this instrument one of the most successful new products in the history of analytical instrumentation. This new instrument is based on capillary array electrophoresis (CAE) with a sheath-flow cuvet detector. This detection technology provides nearly 100% duty cycle, where every capillary is monitored simultaneously without the need for a scanning detector system. By use of this instrument, the *Drosophila melanogaster* sequence will be completed in less than 5 mo *(2)*, and the sequence of the human genome will be substantially complete by the time that this book is in print.

DNA sequencing is used to identify genes and mutations, to confirm successful site directed mutagenesis, and to analyze phage-display libraries. The vast majority of DNA sequence is determined by use of the chain-termination method developed by Sanger's group in 1977 *(3)*. In the original method, a DNA sequencing ladder is synthesized, radioactively labeled, separated by polyacrylamide gel electrophoresis, detected by autoradiography, and interpreted by a skilled technician. This technology, although cumbersome and manually intensive, has remained the method of choice for DNA sequencing. Several advances have been made to improve the technology by replacing the use of radioactive labels, autoradiographic detection, and manual data interpretation with fluorescent labels, laser-induced fluorescence detection, and computer-based automatic data analysis *(4–6)*. Several instruments based on these principles have been commercialized.

1.1. Automated Sequencing

The commercialization of automated DNA sequencers is an important event in biotechnology. A group of visionary biologists realized that this new technology would allow the completion of very large DNA sequencing projects, and the Human Genome

From: *Methods in Molecular Biology, Vol. 162:*
Capillary Electrophoresis of Nucleic Acids, Vol. 1: Introduction to the Capillary Electrophoresis of Nucleic Acids
Edited by: K. R. Mitchelson and J. Cheng © Humana Press Inc., Totowa, NJ

Project was initiated in 1988, two years after the first report of an automated DNA sequencer *(7)*. The National Institutes of Health of the United States has provided the lion's share of public funding for the Human Genome Project, which has grown into an international effort of global proportions. The Welcome Foundation in Great Britain has been notable in their support of the Human Genome Project. In the last year, Celera has become an important private-sector company active in very large-scale sequencing efforts.

The original goals of the Human Genome Project envisioned that the entire genome would be sequenced by the year 2005. Improved DNA sequencing technology has become available that dramatically speeds the sequencing effort. It is estimated that 90% of the human genome will be sequenced in the year 2000 *(8)*. The advances in DNA sequencing technology include the development of large-scale shot-gun sequencing *(9)*, highly automated DNA sequencers, and powerful data processing algorithms. This chapter focuses on the latest generation of highly automated DNA sequencer, which is based on CAE.

1.2. Sequencing in Slab Gels

Conventional DNA separations are performed using slab gel electrophoresis that are coated with a ~200-μm thick layer of crosslinked polyacrylamide. The polymer is allowed to polymerize between two glass plates. The relatively thick, crosslinked polymer is not ideal for DNA separations. From a practical point of view, the thick, rigid polymer must be scraped from the glass plates after every separation. Once the tedious scraping step is completed, the plates must be washed and reassembled. The acrylamide mixture is then poured between the plates and allowed to polymerize. DNA is manually loaded into discrete wells at the top of the gel and the plates are inserted into the sequencing instrument. This procedure is very expensive in large-scale sequencing centers, where the personnel cost associated with handling the sequencing plates dominates the cost of DNA sequencing.

On a more fundamental point of view, the thick gels used in classic sequencing are not able to rapidly generate DNA sequence. DNA sequencing speed is related to the electric field applied to separate the fragments. To first approximation, the sequencing speed increases linearly with the applied electric field. However, application of a voltage across a material with high conductivity results in the generation of heat; this effect is called Joule heating. Because of their thickness, classic slab gels cannot efficiently radiate heat, and Joule heating limits the maximum applied electric field to rather modest levels, which results in rather slow sequencing speeds.

2. Capillary Electrophoresis (CE)

2.1. Sequencing by CE

CE has replaced conventional electrophoresis and has dramatically increased the speed of DNA sequencing. These capillaries are typically 50 μm in inner diameter, are about 30 cm in length, and are made from high-purity fused silica. The small inner diameter of the fused silica capillaries result in excellent thermal properties, which reduces Joule heating to negligible levels and allows the use of extremely high electric fields for very rapid separations of DNA sequencing fragments. However, the phenomenon of biased reptation with orientation is observed at high electric field, and this degrades the separation of longer sequencing fragments ***(10)***. In practice, very high electric fields are not useful in DNA sequencing applications. Instead, the most important property of the fused silica capillaries is that they are highly flexible and easily incorporated into an automated instrument. It is the ease of automation, rather than fundamental thermal properties, that have resulted in the successful development of advanced sequencing instruments.

2.2. Capillary Array Electrophoresis

Whereas the performance of CE is superior to that of conventional slab-gel electrophoresis, a single-capillary instrument does not offer significant advantages compared with a multilane slab-gel system. Instead, it is necessary to operate an array of capillaries to obtain throughput that is comparable with conventional electrophoresis systems.

The first report of CAE for DNA sequencing appeared in 1990 ***(11)***. The detector of that instrument scanned across the capillaries, recording the fluorescence signal sequentially from each capillary. Other groups have also used this approach of scanning a detector across the capillary array for DNA detection. However, the scanning of capillaries for DNA detection suffers from a limited duty cycle period: although one capillary is being interrogated by the detector, DNA is also migrating from the other unscanned capillaries, so that most DNA is not being detected. Limited duty cycle can be a severe restraint, for example, each capillary in a 96-capillary array is probed for 1% of the time, and 99% of the DNA is lost without detection.

2.3. Continuous Capillary Monitoring

The more successful capillary array sequencers rely on continuous monitoring of each capillary. Two instruments are of particular interest, one from the laboratory of H. Kambara at Hitachi and the other from this group ***(12,13)***. Both instruments rely on a sheath-flow cuvet to simultaneously monitor fluorescence from a linear array of capillaries. This technology allows one laser beam to simultaneously illuminate samples migrating from all capillaries. An optical system images the fluorescence from each capillary onto a charged coupled device (CCD) camera or an array of photodiodes. In this way, fluorescence from each capillary is monitored continuously, and the instrument's duty-cycle approaches 100%.

2.3.1. Single Capillary Sheath-Flow Detection

The sheath-flow detector was originally developed for use in flow cytometry, where a dilute suspension of fluorescently labeled cells is pumped into the center of a flowing sheath stream under laminar flow conditions ***(14)***. The cellular suspension flows through the center of the flow chamber and intersects a focused laser beam. If the sheath buffer is similar to the buffer in which the cells are suspended, then there is little refractive index difference between the two streams, and no light is scattered at their interface. By use of a flow chamber with flat windows, the fluorescence signal from each cell may be measured with very low background signal from scattered laser light. Flow cytometry operates at high flow rates in order to process large numbers of cells, so that the cellular population may be characterized with high precision.

Dick Keller's laboratory at Los Alamos Scientific Laboratory pioneered the use of the sheath-flow cuvet as a high-sensitivity detector for fluorescent dyes ***(15,16)***. In favorable cases, single fluorescent molecules have been detected as they passed through the focused laser beam ***(17)***. My laboratory has pioneered the use of the cuvet as a detector in CE ***(18–20)***. The outstanding detection performance of the cuvet, coupled with the very small probe volume produced at the intersection of the laser beam and sample stream, make the detector well suited for application in CE. However, the volumetric flow rate employed in CE is many orders of magnitude lower than that employed in flow cytometry, and care is required to optimize the performance of the detector.

In these instruments, the separation capillary is inserted into a square flow chamber, illustrated in **Fig. 1**. We typically use a 150-μm od fused silica capillary for electrophoresis. The flow chamber is typically 165 μm square and 1.5 cm long. The windows on the cuvet are usually 1 mm thick. Sheath buffer is pumped at a low flow rate from the top of the cuvet to surround the sample stream as it migrates from the capillary. A low-power laser beam is focused into the cuvet, typically forming a 50-μm spot about 150 μm downstream from the capillary exit (*see* **Fig. 2**). Since fluorescence is detected in the flow chamber and not in the capillary, this detector is a rare example of a postcolumn detector in CE. Fluorescence is collected with a high-efficiency microscope objective, spectrally filtered to reduce scattered laser light, imaged onto a mask to further block scattered laser light, and detected with a high-efficiency photomultiplier tube. The flow chamber is held in a stainless-steel fixture, which is held at ground potential and which completes the electrophoresis circuit.

2.3.2. Sensitivity of Sheath-Flow Cuvet Detection

The sheath-flow cuvet has been used as a detector for the CE analysis of zeptomole amounts of fluorescently labeled amino acids, peptides, proteins, monosaccharides, oligosaccharides, and oligonucleotides ***(21–26)***. The detection limits are routinely in the yoctomole range. We have also detected single molecules of β-phycoerythrin after CE separation with this detector ***(27)***.

In 1990 this group ***(28)*** reported the first application of CE with sheath-flow detection for the separation of DNA sequencing fragments, and the first four-color sequencing application was reported in the following year ***(29)***. The sheath-flow cuvet has proven to be quite reliable and able to generate long sequencing read-lengths when combined with high-temperature electrophoresis separation of sequencing fragments ***(30,31)***.

A reviewer of one of our proposals naively felt that the sheath-flow cuvet could not function as a detector in DNA sequencing applications. Because there is no electroosmotic flow from the gel-filled capillary to carry analyte into the cuvet, the reviewer assumed that the analyte would remain trapped at the capillary outlet and would not migrate into the flowing sheath stream. Of course, diffusion rapidly carries analyte into the sheath stream, where it is efficiently swept downstream to the fluorescence excitation volume. Fluorescence must be resolved into spectral bands for DNA sequencing applications. We have reported the use of both a rotating filter wheel and a dichroic filter assembly for this purpose.

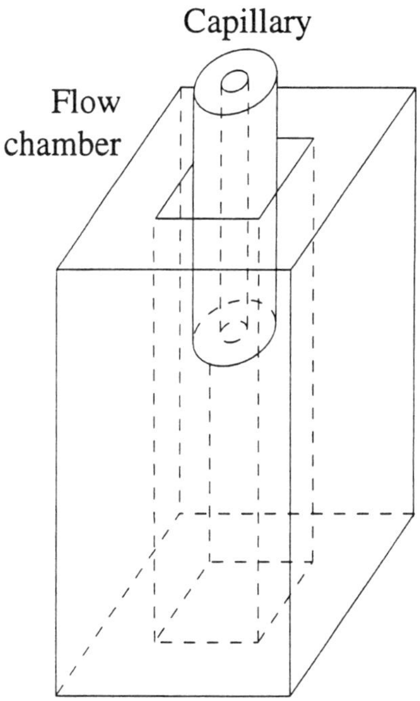

Fig. 1. Single capillary sheath-flow cuvet. A fused silica capillary is placed inside of a square quartz flow chamber.

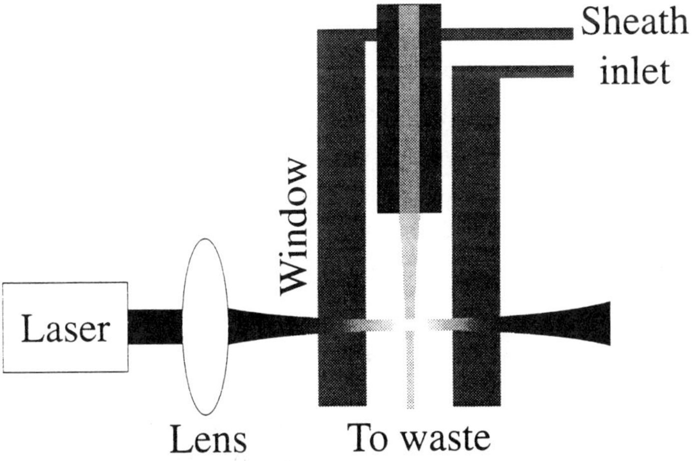

Fig. 2. Single capillary sheath-flow cuvet instrument. Sheath fluid draws the analyte into a thin stream in the center of the flow chamber. A laser beam is focused beneath the capillary tip on the sample stream. A lens collects the fluorescence emission signal, which is then spectrally filtered, then imaged onto an aperture and detected with a photomultiplier tube.

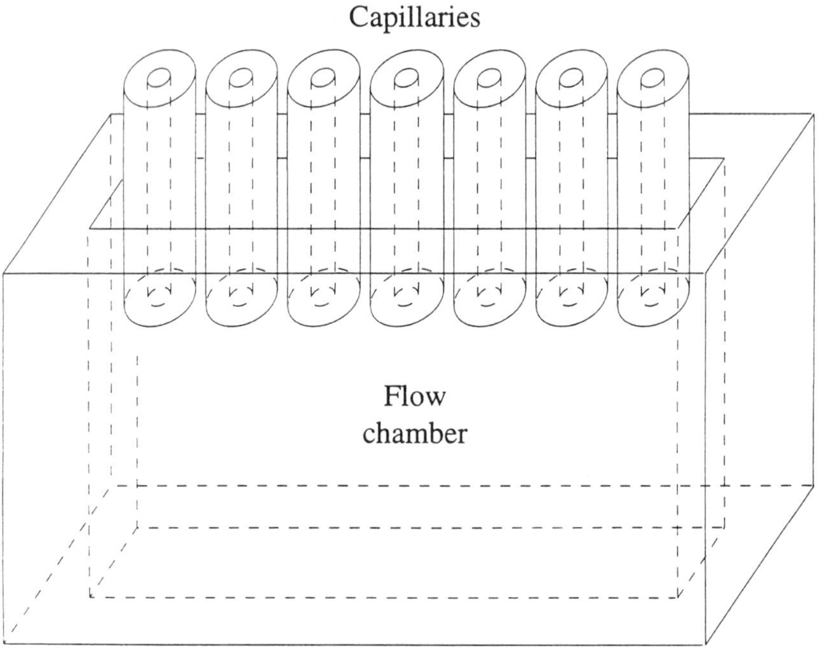

Fig. 3. A linear-array capillary, sheath-flow cuvet. A linear array of fused silica capillaries is placed inside of a rectangular glass flow chamber.

2.3.3. Capillary Array Sheath-Flow Detection

As part of his PhD thesis, JianZhong Zhang developed a capillary array DNA sequencer based on a sheath-flow cuvet *(32)*. Simultaneously and independently, H. Kambara's group developed a similar instrument *(33,34)*. In these instruments, a linear array of capillaries is inserted into a rectangular sheath-flow cuvet, illustrated in **Fig. 3**. Sheath fluid is pumped through the interstitial space between the capillaries and entrains the DNA sequencing fragments as discrete streams, with one stream per capillary. A laser beam is focused into the cuvet and skims beneath the capillary tips, as shown in **Fig. 4**, exciting fluorescence from all of the capillaries, simultaneously. When observed from the front, the detector generates a set of fluorescent spots, with one spot beneath each capillary and with the spots separated by the outer diameter of the capillary. **Reference *35*** presents a color photograph of a cuvet in operation. A high numerical aperture lens is used to collect the fluorescence and to image each fluorescent spot onto a discrete photo-detector. In our original systems, we used a set of fiber-optic coupled avalanche photodiodes, which were operated in the photon counting mode. More recently, we have used a CCD camera to image the fluorescence. The laser beam traverses the set of sample streams that are migrating from each capillary. Fortunately, there is a negligible loss of laser power in traversing the cuvet. The analyte concentration is quite low and the optical path-length across each sample stream is

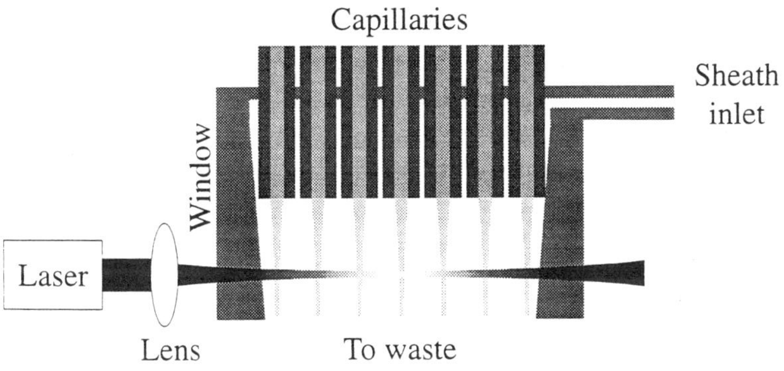

Fig. 4. Capillary linear-array sheath-flow cuvet instrument. Sheath fluid draws the analyte into thin streams in the center of the flow chamber, with a single stream produced downstream from each capillary. A laser beam is focused beneath the capillary tips on the sample streams. A lens collects the fluorescence emission signal, which is then spectrally filtered and detected with either an array of photodiodes, or with a CCD camera.

small. As a result, the loss of laser power owing to absorbance of the beam by the sample is much less than 1%, even for the highest concentration primer peak that migrates early in the separation.

2.4. Capillary Spacing

If the capillaries are evenly spaced within the cuvet, then the sample streams will also be evenly spaced. If the capillaries are not evenly spaced, the hydrodynamic flow of the sheath fluid between the capillaries will not be uniform. Instead, the sheath-flow rate will be higher in the region where the capillaries are spaced wider than normal and the sheath-flow rate will be lower in the regions where the capillaries are closer together. These differences in the rate of sheath flow, cause the sample streams to be drawn together where the capillaries are close together. When illuminated by the laser beam, the fluorescent spots will not be uniformly spaced, and alignment of the detectors with the images of the spots will be difficult and irreproducible. We find that the capillary spacing needs to be within ~5 μm to ensure even spacing of the fluorescence spots.

There are two methods of generating evenly spaced capillaries. In the first case, the cuvet is designed so that the capillaries are in intimate contact. As a result, their center-to-center spacing is determined by the outer diameter of the capillary, which is typically held to within 1 μm across a 100-m long reel of capillary. However, the reel-to-reel dimensions of the capillary can vary by more than 10 μm. To avoid the custom manufacture of a cuvet for each reel of capillary, we developed a wedge-shaped cuvet, wherein the walls of the cuvet are tapered. The capillaries are forced into contact as they are inserted into the tapered cuvet.

In the second approach to ensuring uniform capillary spacing, a set of fingers can be micromachined in the inner wall of the cuvet. These features are raised with respect to the cuvet walls and act to hold the capillaries on uniform centers.

3. Commercial CAE Instruments

The CAE instrument has been commercialized by PE Applied Biosystems as the model 3700 DNA Sequencer ***(1,34,36)***. The instrument operates with 96 capillaries in its sheath-flow detector for convenient mating with a 96-well microtiter plate. There are also eight spare capillaries in the instrument. A robot automatically and rapidly transfers samples from a 96-well microtiter plate to an injection block, where samples are loaded onto the capillaries. Sample loading, sieving matrix replacement, electrophoresis, data collection, and base-calling are all automated, so that the instrument can run 24 h without intervention.

The commercial instrument uses a concave spectrograph to image fluorescence onto a CCD camera, which records the fluorescence spectrum generated by each capillary. This spectral data is monitored continuously, which ensures nearly a 100% duty cycle for detection. The data is then processed to obtain sequence data. Unlike slab-gel data, there is no need for lane-finding algorithms to process the data; the capillary format ensures that each sample is imaged to a specified location on the camera.

Although the instrument is specified as producing 550 bases of sequence data at 98.5% accuracy, the performance of this instrument has exceeded expectation ***(2)***. Celera has announced that they have generated over 500 million bases of DNA sequence from the *Drosophila* genome in 1 million electropherograms. It is expected that this performance will improve as the sequencing software is optimized.

4. Conclusions

The human genome project is the most ambitious and important effort in the history of biology. It will provide the complete genetic blueprint for human life, and will provide important insights into human health and development. High-throughput DNA sequencers based on CAE with a sheath-flow detector will do the lion's share of the sequencing effort. The highly flexible, automated, and efficient CE technology with the high sensitivity and high duty-cycle sheath-flow cuvet is making the human genome project a reality. This technology is also one of the best selling products in the history of analytical instrumentation, with sales of over 700 instruments in the first few months of production. Combined, the sales of these instruments have a value of $200 million ***(37)***.

There will be many more applications of this high-throughput technology beyond the initial sequencing of the human genome. Comparative genomics will require that the DNA sequence from many individuals be obtained and analyzed. Human disease diagnosis and prognosis will become very important in the near future, particularly as genetic markers are found that are associated with cancer stage and prognosis; these markers will be routinely sequenced in clinical laboratories. The genome from other vertebrate organisms will be sequenced to gain insight into evolution. The genomes of agriculturally important plants and animals will be sequenced to provide better control and manipulation of economically important traits. Plant and animal breeding programs will require that large numbers of organisms be genotyped.

We are at the threshold of a new era in the biological sciences. This era is based on genomic information that was unimaginable two decades ago. The development of novel analytical technology has made the genome project possible and is decreasing dramatically the cost of the sequencing effort. This is a very exciting time in the biological sciences!

Acknowledgments

This work was supported by the Canadian Genetic Diseases Network, the Natural Sciences and Engineering Research Council, and SCIEX.

References

1. http://www.pecorporation.com/press/prc5448.html
2. http://www.pecorporation.com/press/prccorp0728.html
3. Sanger, F., Nicklen, S., and Coulson, A. R. (1977) DNA sequencing with chain-terminating inhibitors. *Proc. Nat. Acad. Sci. USA* **74,** 5463–5467.
4. Smith, L. M., Sanders, J. Z., Kaiser, R. J., Hughes, P., Dodd, C., Connell, C. R., Heiner, C., Kent, S. B. H., and Hood, L. E. (1986) Fluorescence detection in automated DNA sequence analysis. *Nature* **321,** 674–679.
5. Prober, J. M., Trainor, G. L., Dam, R. J., Hobbs, F. W., Robertson, C. W., Zagursky, R. J., Cocuzza, A. J., Jensen, M. A., and Baumeister, K. (1987) A system for rapid DNA sequencing with fluorescent chain-terminating dideoxynucleotides. *Science* **238,** 336–341.
6. Ansorge, W., Sproat, B. S., Stegemann, J., and Schwager, C. (1986) A non-radioactive automated method for DNA sequence determination. *J. Biochem. Biophys. Meth.* 13**,** 315-317.
7. http://www.nhgri.nih.gov/HGP
8. http://www.nhgri.nih.gov:80/NEWS/Finish_sequencing_early/Intent_to_finish_ sequencing_early.html
9. Venter, J. C., Adams, M. D., Sutton, G. G., Kerlavage, A. R., Smith, H. O., and Hunkapiller, M. (1998) Shotgun sequencing of the human genome. *Science* **280,** 1540–1542.
10. Viovy J.-L. and Duke, T. (1993) DNA electrophoresis in polymer solutions: Ogston sieving, reptation and constraint release. *Electrophoresis* **14,** 322–329.
11. Zagursky, R. J. and McCormick, R. M. (1990) DNA sequencing separations in capillary gels on a modified commercial DNA sequencing instrument. *BioTechniques* **9,** 74–79.
12. Dovichi, N. J. and Zhang, J. Z. (1995) Multiple capillary biochemical analyzer. US Patent 5,439,578; 8 August 1995.
13. Takahashi, S. and Kambara, H. (1996) DNA detector and DNA detection method. US Patent 5,529,679; 25 June 1996.
14. Steinkamp, J. A. (1984) Flow cytometry. *Rev. Sci. Instrum.* **55,** 1375.
15. Dovichi, N. J., Martin, J. C., Jett, J. H., Trkula, M., and Keller R. A. (1984) Laser-induced fluorescence of flowing samples as an approach to single-molecule detection in liquids. *Anal. Chem.* **56,** 348–354.
16. Dovichi, N. J., Martin, J. C., Jett, J. H., and Keller R. A. (1983) Attogram detection limit for aqueous dye samples by laser-induced fluorescence. *Science* **219,** 845–847.
17. Hguyen, D. C., Keller, R. A., Jett, J. A., and Martin, J. C. (1987) Detection of single molecules of phycoerythrin in hydrodynamically focused flows by laser-induced fluorescence. *Anal. Chem.* **59,** 2158–2161.
18. Cheng, Y. F. and Dovichi, N. J. (1988) Subattomole amino acid analysis by capillary zone electrophoresis and laser-induced fluorescence. *Science* **242,** 562–564.
19. Cheng, Y. F., Wu, S., Chen, D. Y., and Dovichi, N. J. (1990) Interaction of capillary zone electrophoresis with a sheath flow cuvette detector. *Anal. Chem.* **62,** 496–503.

20. Wu, S. and Dovichi, N. J. (1989) High-sensitivity fluorescence detector for fluorescein isothiocyanate derivatives of amino acids separated by capillary zone electrophoresis. *J. Chromatogr.* **480,** 141–155.
21. Zhao, J. Y., Chen, D. Y., and Dovichi, N. J. (1992) Low-cost laser-induced fluorescence detector for micellar capillary zone electrophoresis: detection at the zeptomol level of tetramethylrhodamine thiocarbamyl amino acid derivatives. *J. Chromatogr.* **608,** 117–120.
22. Zhao, J. Y., Waldron, K. C., Miller, J., Zhang, J. Z., Harke, H. R., and Dovichi, N. J. (1992) Attachment of a single fluorescent label to peptides for determination by capillary zone electrophoresis. *J. Chromatogr.* **608,** 239–242.
23. Pinto, D. M., Arriaga, E. A., Craig, D., Angelova, J., Sharma, N., Ahmadzadeh, H., Dovichi, N. J., and Boulet, C. A. (1997) Picomolar assay of native proteins by capillary electrophoresis - precolumn labeling, sub-micellar separation, and laser induced fluorescence detection. *Anal. Chem.* **69,** 3015–3021.
24. Zhang, Y., Arriaga, E., Diedrich, P., Hindsgaul, O., and Dovichi, N. J. (1995) Nanomolar determination of aminated sugars by capillary electrophoresis. *J. Chromatogr.* **716,** 221–229.
25. Le, X.-C., Scaman, C., Zhang, Y., Zhang, J. Z., Dovichi, N. J., Hindsgaul, O., and Palcic, M. M. (1995) Analysis by capillary electrophoresis-laser-induced fluorescence detection of oligosaccharides produced from enzyme reactions. *J. Chromatogr.* **716,** 215–220.
26. Figeys, D., Arriaga, E., Renborg, A., and Dovichi, N. J. (1994) Use of the fluorescent intercalating dyes POPO-3, YOYO-3 and YOYO-1 for ultrasensitive detection of double stranded DNA separated by capillary electrophoresis with hydroxypropylmethyl cellulose and non-cross-linked polyacrylamide. *J. Chromatogr.* **669,** 205–216.
27. Chen, D. Y. and Dovichi, N. J. (1996) Single-molecule detection in capillary electrophoresis: molecular shot noise as a fundamental limit to chemical analysis. *Anal. Chem.* **68,** 690–696.
28. Swerdlow, H., Wu, S., Harke, H., and Dovichi, N. J. (1990) Capillary gel electrophoresis for DNA sequencing—laser-induced fluorescence detection with the sheath flow cuvette. *J. Chromatogr.* **516,** 61–67.
29. Swerdlow, H., Zhang, J. Z., Chen, D. Y., Harke, H. R., Grey, R., Wu, S., Fuller, C., and Dovichi, N. J. (1991) Three DNA sequencing methods using capillary gel electrophoresis and laser-induced fluorescence. *Anal. Chem.* **63,** 2835–2841.
30. Zhang, J. Z., Fang, Y., Hou, J. Y., Ren, H. J., Jiang, R., Roos, P., and Dovichi, N. J. (1995) Use of non-cross-linked polyacrylamide for four-color DNA sequencing by capillary electrophoresis separation of fragments up to 640 bases in length in two hours. *Anal. Chem.* **67,** 4589–4593.
31. Yan, J. Y., Best, N., Zhang, J. Z., Ren, H. J., Jiang, R., Hou, J., and Dovichi, N. J. (1996) The limiting mobility of DNA sequencing fragments for both cross-linked and noncross-linked polymers in capillary electrophoresis: DNA sequencing at 1200 V cm^{-1}. *Electrophoresis* **17,** 1037–1045.
32. Zhang, J. Z. (1994) DNA sequencing by single and multiple capillary gel electrophoresis. *PhD thesis*, Department of Chemistry, University of Alberta.
33. Kambara, H. and Takahashi, S. (1993) Multiple-sheath flow capillary array DNA analyzer. *Nature* **361**, 565–566.
34. Kamahori, M. and Kambara, H. (2001) Capillary array electrophoresis analyser, in *Capillary Electrophoresis of Nucleic Acids,* Vol. 2 (Mitchelson, K. R. and Cheng, J., eds.), Humana Press, Totowa, NJ, pp. 271–287.
35. Anonymous comment. *Science* 1998, **280,** 995.
36. http://www2.perkin-elmer.com/ga/3700/features.html
37. http://www.pecorporation.com/press/dlw042999.html

7

Determining Dye-Induced DNA Mobility Shifts for DNA Sequencing Fragments by Capillary Electrophoresis

John Bashkin

1. Introduction

Over the past decade, fluorescent end-labeling of DNA fragments has evolved into the preferred method of DNA detection for a wide variety of applications, including DNA sequencing and polymerase chain reaction (PCR) fragment analysis. One of the advantages inherent in fluorescent detection methods is the ability to perform multicolor analyses. Unfortunately, labeling DNA fragments with different fluorescent tags may impart disparate relative electrophoretic mobilities to the labeled fragments. In these instances, mobility-shift corrections must be applied to the electrophoretic data. These corrections may lead to increased error in the estimation of DNA fragment sizes and reduced confidence in DNA sequence information.

Previously, we published a systematic study of the relationship between dye structure and the mobility shifts for primer-labeled DNA sequencing fragments observed during capillary electrophoresis (CE) *(1)*. Based on these and other results *(3–7)*, the dye-primer sequencing reagents built on the fluorescence energy-transfer concept (DYEnamic ET® Kit, Amersham Pharmacia Biotech) have developed to the point that mobility shifts are no longer significant. Other dye-primer labeling reagents have also been developed that exhibit minimal shifts *(8)*. The software used to analyze DNA sequencing data has also developed to the point that no *a priori* knowledge of mobility shifts is required for proper corrections to be made, as long as the shifts are less than ~1.5 bases. On the MegaBACE 1000® DNA Analysis System (Amersham Pharmacia Biotech), for instance, adaptive methods are used within the Sequence Analyzer® software that correct mobility shifts "on the fly" without the need for data look-up tables or other input parameters.

Dye-terminator DNA sequencing reagents have proven more difficult to optimize with regard to mobility shifts than dye-labeled sequencing primers. Synthetic constraints and the interaction of the labeled terminators with DNA polymerases place limitations on terminator reagent design not found with primer reagents. Furthermore,

From: *Methods in Molecular Biology, Vol. 162:*
Capillary Electrophoresis of Nucleic Acids, Vol. 1: Introduction to the Capillary Electrophoresis of Nucleic Acids
Edited by: K. R. Mitchelson and J. Cheng © Humana Press Inc., Totowa, NJ

continuous improvements in DNA sequencing dyes require that mobility shifts continue to be evaluated, and minimizing such shifts remains one of the primary criteria by which potential dye sets are judged. Here, we review briefly our previous results for dye-labeled primer DNA sequencing fragments and illustrate in detail the methodology employed to quantify the mobility-shift behavior for dye-terminator labeled fragments (*see* **Note 1**).

2. Methods

1. All DNA samples comprise of M13mp18, and dye-terminator samples are sequenced using the standard protocols provided with either the DYEnamic ET® sequencing reagents for the MegaBACE 1000® (Amersham Pharmacia) or the Big Dye Terminator® sequencing kits (Perkin-Elmer Biosystems).
2. All samples are then run on a MegaBACE 1000® system using linear polyacrylamide sieving matrix (Amersham Pharmacia). The data are processed through background subtraction, spectral separation, and peak height normalization using the Sequence Analyzer® (version 1.0, build 116) software available on the MegaBACE 1000® system.
3. The low sampling rate of the MegaBACE 1000® made it necessary to fit each peak in the sequence traces to a Gaussian shape to determine the precise peak center positions. Consequently, the spectrally separated data are exported to ASCII text format files. Commercially available peak analysis software (PeakFit, Jandel Scientific) is then used to perform the Gaussian peak fits and provide accurate retention times for each peak in the electrophoretic traces. In all examples, sequence base numbers refer to the standard M13mp18 numbering scheme.

3. Results

3.1. Mobility Shifts for Dye Primer Sequencing

1. The mobility shifts for dye-labeled primers are easily measured by generating ddT-termination sequencing ladders and coinjecting the samples in the same capillary. Using the spectral discrimination capabilities of the CE device, the signals from the two sequencing ladders can be distinguished and the mobility shift between peaks of the same DNA fragment size measured. This method has been well described in two previous references ***(1,2)***.
2. In general, there are three types of relative mobility profiles as a function of increasing DNA sequencing fragment size: a constant offset, an exponential decline in shift, and a mobility crossover, in which the order of electrophoretic retention is swapped between two dyes as the fragment sizes get larger. **Figures 1–3** show examples of these three types

---

Fig. 1. *(opposite page)* Electrophoretic traces for Cy5 and Cy5.5T5 primer-labeled sequencing fragments, showing the relative sequencing fragment mobilities early in the run and in the middle of the run. Samples: M13mp18 ddT-termination reactions. Electrophoresis: 185 V/cm. *Blue trace*: Cy5; *red trace*: Cy5.5T5. Adapted from *Nucleic Acids Research* **26**, 2797–2802, with the permission of Oxford University Press.

Fig. 2. *(opposite page)* Electrophoretic traces for Cy5T7 and Cy5.5T5 primer-labeled sequencing fragments, showing the relative sequencing fragment mobilities early in the run and in the middle of the run. Samples: M13mp18 ddT-termination reactions. Electrophoresis: 185 V/cm. *Blue trace*: Cy5T7; *red trace*: Cy5.5T5. Adapted from *Nucleic Acids Research* **26**, 2797–2802, with the permission of Oxford University Press.

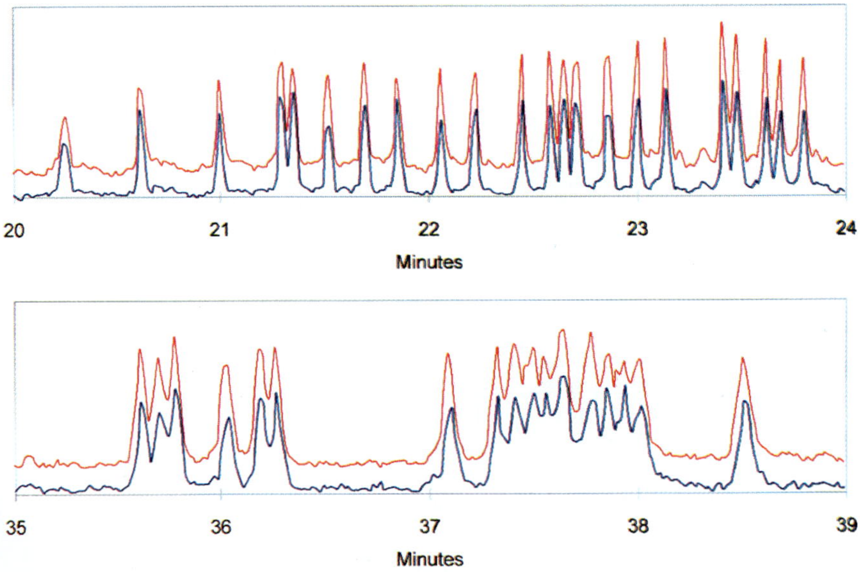
20
21
22
23
24
Minutes
35
36
37
38
39
Minutes

Fig. 1

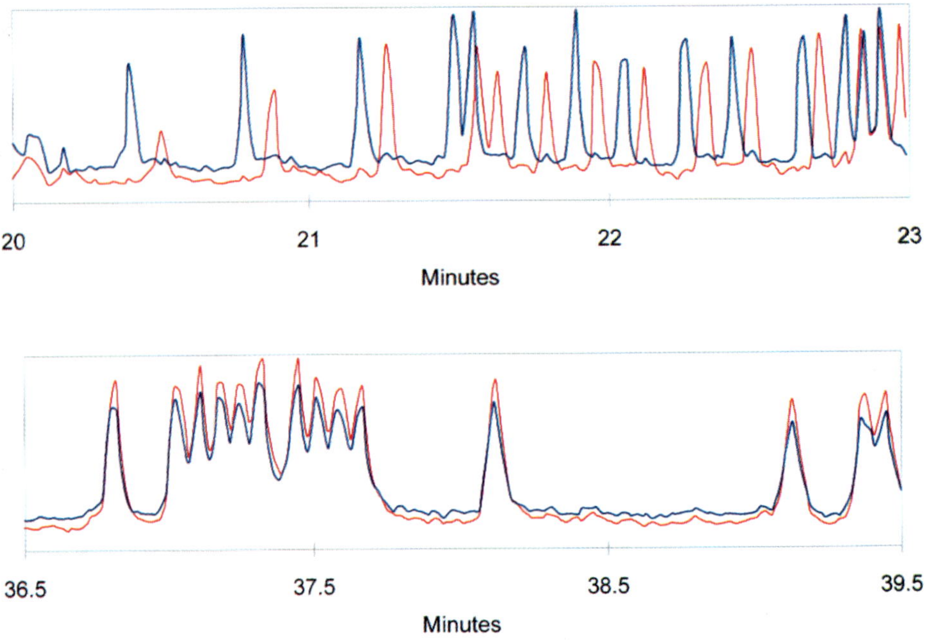
20
21
22
23
Minutes
36.5
37.5
38.5
39.5
Minutes

Fig. 2

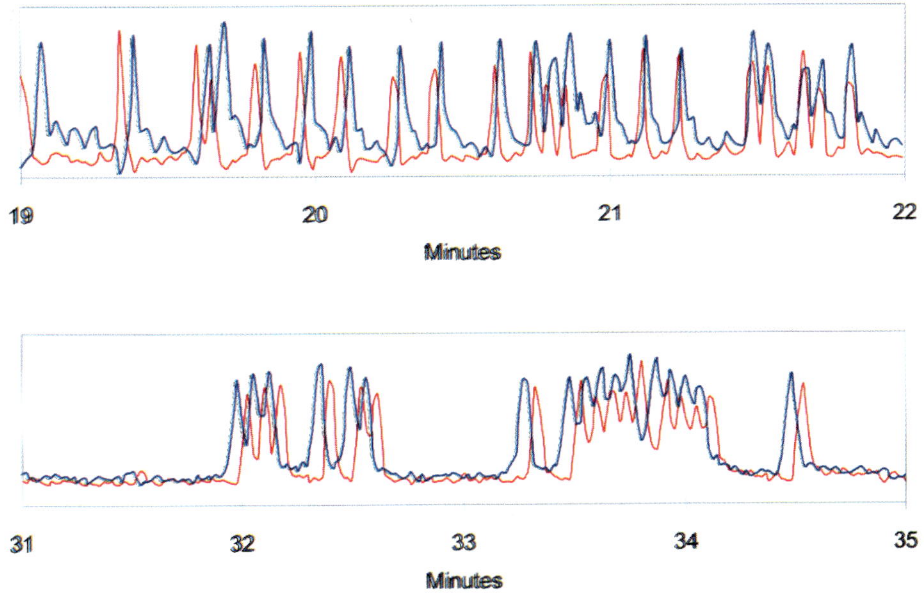

Fig. 3. Electrophoretic traces for Sq5T4 and Cy5.5T12 primer-labeled sequencing fragments, showing the relative sequencing fragment mobilities early in the run and in the middle of the run. Samples: M13mp18 ddT-termination reactions. Electrophoresis: 185 V/cm. *Blue trace*: Sq5T4; *red trace*: Cy5.5T12. Adapted from from *Nucleic Acids Research* **26**, 2797–2802, with the permission of Oxford University Press.

of behavior for CY5 and CY5.5 dye derivatives used as sequencing primer labels ***(1)***. The ddT traces for dye pairs are shown early in the M13 sequencing run, where mobility shifts are generally largest, and at around 300 bases.

3. In our study of the relationship between dye structure and electrophoretic mobility, we found several important factors that contribute to the dye-induced mobility shifts ***(1)***. These included the overall charge on the dye, placement of charge within the dye structure, interactions between the dye and the DNA (electrostatic and hydrophobic), and the linkage between the dye and the primer.
4. These results were subsequently translated into a revised set of energy-transfer (ET) primers for the MegaBACE 1000®. The initial set of ET primers commercialized by Amersham Pharmacia consisted of the family F10F, F10REG, F10T, F10R, where fluorescein, denoted F, was the donor fluor for each ET construct. The acceptor dyes were fluorescein, rhodamine 6G, TAMRA, and ROX, respectively. The only nonrhodamine acceptor in this group was fluorescein and not surprisingly, the sequencing extension products labeled with this primer exhibited a substantial mobility shift relative to the others.
5. Replacement of the fluorescein acceptor on F10F with rhodamine 110 (F10R110) yielded a primer family with structurally similar rhodamines in all acceptor positions and exhibiting almost no relative electrophoretic mobility shifts. This is the ET reagent set currently used in the primer-labeled sequencing kit for the MegaBACE 1000® (*see* **Note 2**).

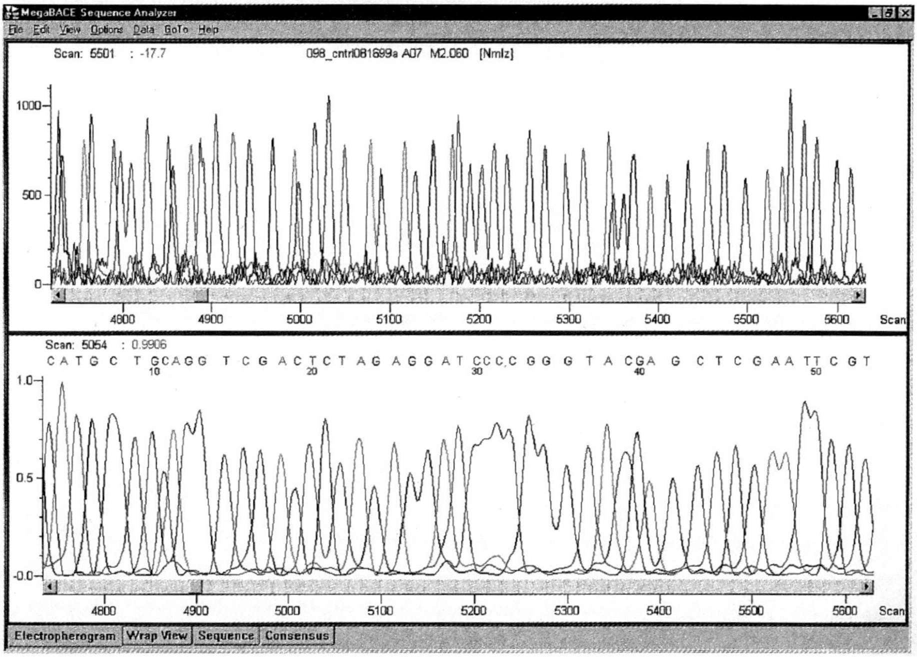

Fig. 4. M13mp18 sequencing data obtained with the second generation ET primers F10R110, F10R6G, F10TMRA, F10ROX (Amersham Pharmacia). *Upper panel* shows the data processed through background subtraction, spectral separation, and peak height normalization. *Lower panel* shows the fully processed data, including adaptive mobility shift corrections. Electrophoresis: 140 V/cm.

6. **Figure 4** (upper panel) shows the beginning portion of an M13 sequencing trace generated with this second generation ET primer set, processed through background subtraction, spectral separation, and peak height normalization. The residual shift of approx +0.2 bp in the ddA and –0.2 bp for the ddC peaks is observable in this early portion of the sequence data, but these small offsets are easily corrected in software using the adaptive methods now available.
7. **Figure 4** (lower panel) also shows the same data region fully processed by the Sequence Analyzer software.

3.2. Mobility Shifts for Dye Terminator Sequencing

1. For dye-terminator sequencing the situation is more complicated than that for dye-primer sequencing. Unlike dye-primer sequencing, a thorough exploration of the relationship between dye-terminator structure and mobility has not yet been done and the rules that govern electrophoretic mobility behavior are not as well understood (*see* **Note 3**).
2. The same dyes that exhibit no relative mobility shifts when used as primer labels may still exhibit substantial shifts when used as terminator labels. In addition, the dyes induce different relative electrophoretic mobility profiles when attached to the different terminators.

3. Unlike dye-labeled sequencing primers, the fluor-terminator construct is intimately involved in the enzymology of the sequencing reaction. This can severely limit the fluor-terminator combinations that can be used effectively, and the incorporation efficiencies of these constructs by the DNA polymerase enzymes tends to be a dominant criterion for selecting dye-terminator combinations rather than electrophoretic mobility behavior.
4. Interactions between the dyes and the DNA sequence fragments may also give rise to sequence-specific compression artifacts in the data. These issues can mandate a particular choice of dye structure or fluor-terminator combination independent of mobility-shift considerations.

3.3. DYEnamic ET Terminators

1. In addition, the ET terminator dyes may be chosen in order to match the fluorescence emission spectra to those for dye-primer reagents to make all reagents compatible with a particular analysis system.
2. The DYEnamic ET® terminators consist of a fluorescein donor and rhodamine 110, rhodamine 6G, tetramethylrhodamine, and rhodamine X acceptors to match the spectra of the DYEnamic ET® primers (although the fluor-nucleotide labeling order is different). The result of all these factors is that reagent development groups tend to have far less latitude in the design of dye-terminator reagents than for dye-primer reagents.
3. The necessity of coupling the dye-terminator development to the enzymology of the DNA polymerases used in DNA sequencing kits makes the process much more difficult and lengthy than for dye-primer reagents.
4. Because the dye-terminator moieties will exhibit different electrophoretic mobilities depending on which dye is attached to which terminator, one cannot determine the mobility shift induced by a given dye by running two ddT tracts in parallel.
5. Rather, the mobility shifts must be determined indirectly based on complete four-color sequencing traces. Typically, one dye terminator is chosen as a reference and the relative electrophoretic mobilities of the remaining three dyes are expressed relative to the reference. It is also possible to express the electrophoretic mobilities of all four terminators relative to an ideal reference (e.g., a straight line), but using one of the terminators as a reference permits that terminator to act as a check on the mobility model, since it should yield a zero shift relative to itself.
6. To determine the extent of mobility shifts between the DYEnamic ET® dye terminators, we processed the data for a M13mp18 sequencing trace through background subtraction, spectral separation, and peak height normalization, then saved the data to ASCII text format. The results are shown in the upper panel of **Fig. 5**, in which the relative terminator mobility shifts early in the sequence are readily observable, particularly when these data are compared to the ET primer data in **Fig. 4**. The trace data in ASCII format are then processed with peak analysis software to precisely determine the pixel locations of each peak center.
7. As discussed earlier, the low data density obtained from the MegaBACE 1000® requires that each peak be fit to a Gaussian shape, in order to accurately determine the peak retention times. These peak location data were then used to plot peak retention times vs fragment size for the ddC termination reaction (*see* **Fig. 6**). This is our reference curve against which the other dye-terminator mobilities are compared and is used to predict the retention times for the sequencing fragment sizes represented by the remaining dye terminators.
8. In order to calculate the predicted retention times for ddA, ddG, and ddT-termination fragments, the peak position vs scan line curve for the ddC data is empirically fit to a

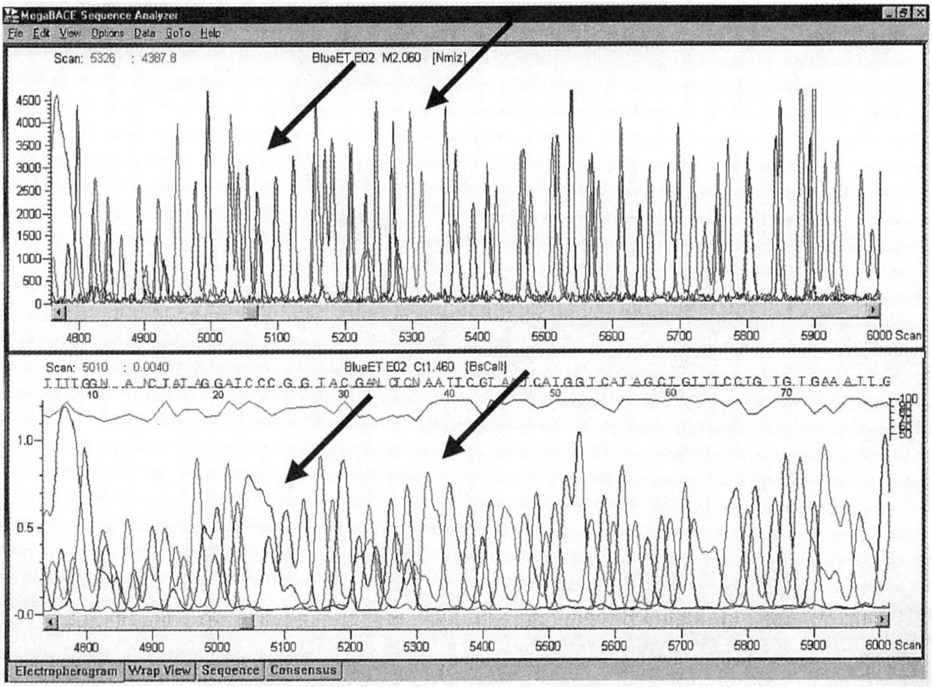

Fig. 5. Early section of M13mp18 sequencing data obtained with ET dye terminators (Amersham Pharmacia). *Upper panel* shows the data processed through background subtraction, spectral separation, and peak height normalization. *Lower panel* shows the data fully processed for base-calling, including adaptive mobility-shift corrections, using the Cimarron Dye Terminator algorithm. Right arrows indicate a doublet A sequence at M13 base number 52. Left arrows indicate one peak mobility-shift error that remains present in the fully processed data. Electrophoresis: 140 V/cm.

high-order polynomial. No single polynomial curve gives a satisfactory fit over the entire fragment size range, so the data are partitioned into four segments and each segment fitted separately.

9. In this case, the segments ranged from M13 base number 0–50, 51–200, 201–400, and 401–600. Using this composite polynomial fit, the predicted peak positions for each peak from the remaining three termination reactions were then calculated. The difference between the calculated values and the observed peak positions represent the mobility shift for that dye-terminator.
10. Next, a plot of peak-spacing per base vs fragment size is generated from the ddC data (*see* **Fig. 7**). This curve is also fit to a high order polynomial and used to calculate the reference peak-spacing per base for all other base positions.
11. Finally, the calculated mobility shifts for each fragment was divided by the reference peak spacing at that position to yield the mobility shifts in terms of fractions of a base pair shift (*see* **Fig. 8**).

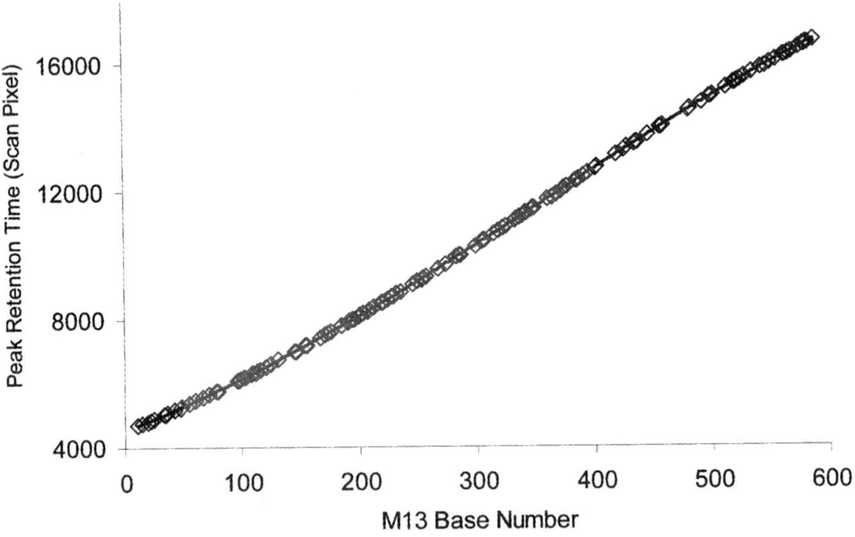

Fig. 6. Retention time (scan pixel) vs fragment size (M13 base number) for ddC termination fragments. Polynomial fits to the sectioned data are shown. R^2 values > 9.999E-01 for all fits.

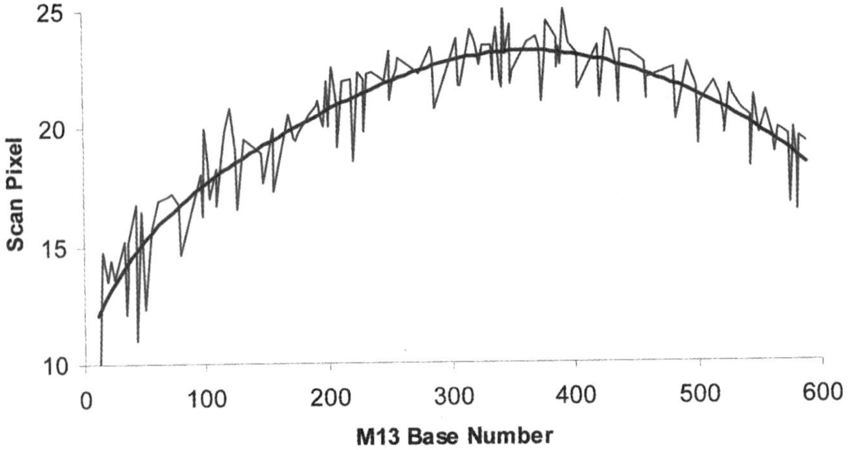

Fig. 7. Plot of peak spacing per base (scan pixel) vs fragment size (M13 base number) for the ddC termination fragments. Polynomial fit is also shown.

3.4. Overview of ET Terminators

1. **Figure 9** shows the final results for all ET terminators. Mobility shifts relative to ddC are expressed in fractions of base pairs vs fragment size (M13 base number), and a positive shift indicates that the terminator has a slower electrophoretic mobility than ddC.

T	T Center	T Fit	T Difference	Fit Peak Spacing/Base	T Ratio Difference/Peak Spacing
9	4690.81	4670.34	20.47	11.87	1.72
13	4741.16	4728.40	12.76	12.28	1.04
19	4828.39	4809.68	18.71	12.85	1.46
24	4894.80	4881.07	13.73	13.30	1.03
26	4923.80	4910.26	13.54	13.47	1.01
33	5033.38	5013.14	20.25	14.03	1.44
41	5159.75	5136.82	22.92	14.61	1.57
49	5272.67	5262.08	10.59	15.14	0.70
54	5352.73	5338.22	14.51	15.44	0.94
55	5366.96	5354.74	12.22	15.50	0.79
58	5416.11	5404.29	11.82	15.67	0.75
61	5466.53	5453.92	12.60	15.84	0.80
64	5513.68	5503.75	9.93	16.00	0.62
67	5568.07	5553.84	14.23	16.16	0.88
70	5615.33	5604.29	11.05	16.31	0.68
74	5686.05	5672.17	13.89	16.51	0.84
76	5722.25	5706.40	15.85	16.60	0.95
77	5739.51	5723.59	15.92	16.65	0.96
78	5755.91	5740.83	15.08	16.70	0.90
81	5803.95	5792.87	11.08	16.83	0.66
83	5845.07	5827.83	17.24	16.92	1.02
85	5882.61	5863.00	19.61	17.01	1.15
90	5972.90	5951.83	21.06	17.22	1.22
91	5989.96	5969.75	20.20	17.26	1.17
93	6024.35	6005.74	18.62	17.35	1.07
94	6041.10	6023.80	17.29	17.39	0.99
96	6072.78	6060.07	12.71	17.47	0.73
101	6165.25	6151.52	13.74	17.66	0.78

Fig. 8. Example data for the mobility shift calculations of ddT fragments out to base number 100. T: base number for ddT M13 fragments; T Center: ddT peak retention time (scan pixel) as determined from Gaussian peak fits; T Fit: predicted ddT peak retention time (scan pixel) as determined from the polynomial fit to the ddC data in **Fig. 6**. T Difference: T Center – T Fit values; Fit Peak Spacing/Base: Peak spacing per base for the ddT fragment size, based on the polynomial fit to the ddC data in **Fig. 7**. T Ratio Difference/Peak Spacing: the ratio of T Difference to the Peak Spacing/Base, yielding the final mobility shift for that ddT fragment relative to ddC is given in fractions of a base separation.

2. The ddC data show scatter about zero shift with a standard deviation of 0.10 bases, indicating the accumulation of errors through the fitting processes is acceptable.
3. The ddA fragments exhibit a consistent offset of –0.25 bases throughout the sequencing trace, whereas ddG fragments show a negligibly positive shift.
4. The ddT fragments show the most marked mobility shift, approaching +2 bases very early in the sequence and decaying to a steady +0.5 base shift after 150 bases.
5. Examination of the trace data in **Fig. 5** (upper panel) reaffirms that these values are reasonable. **Figure 5** (lower panel) shows the same data file after full data processing, including mobility-shift corrections, by the MegaBACE 1000® Sequence Analyzer software (Cimarron terminator base-caller).
6. Comparison of the upper and lower panels of **Fig. 5** illustrates the limitations of the adaptive mobility correction software, at least in the version currently available to the author. In the lower panel of **Fig. 5** the mobility shifts are shown to be correctly accounted for beyond the doublet A at base number 52, but prior to that position a number of mobility-shift errors in peak spacing and peak order are still present.

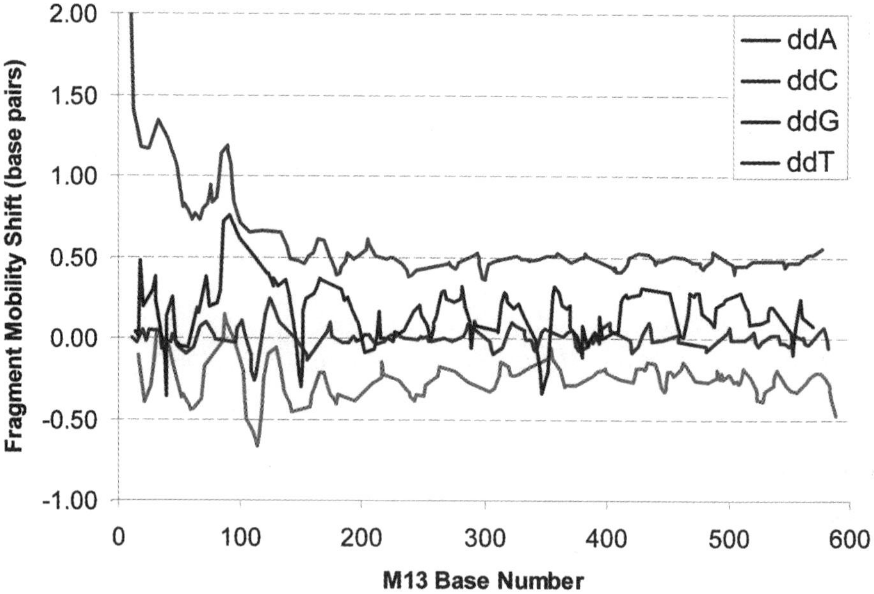

Fig. 9. Plot of fragment mobility shift (base pairs) vs fragment size (M13 base number) for all ET dye terminators, relative to ddC.

7. Comparison of the lower panels from **Figs. 4** and **5** also show the extent to which the larger mobility shifts associated with the ET terminators, independent of residual free dye-terminator "blobs," affect the data quality early in the sequence trace relative to the ET primer data. These errors will limit how close to the end of a sequencing primer useful data can be obtained with the ET dye terminators or, at least, affect the accuracy of the sequence base-calling in what is generally the vector-containing region of a typical insert clone template. Such errors can in turn adversely impact vector sequence recognition algorithms and data quality score assignments for sequence data. These tools are commonly used to automatically process data files in large genome sequencing projects.
8. We have performed the same mobility-shift analysis on the Big Dye® terminator reagents available from Perkin-Elmer Biosystems. The upper panel of **Fig. 10** again shows the early section of a M13mp18 sequencing trace for which the data have been processed through background subtraction, spectral separation, and peak height normalization. It is apparent that these dye-terminators also show substantial relative electrophoretic mobility shifts (*see* **Note 4**).
9. In the lower panel of **Fig. 10**, the fully processed and base-called data are shown (Cimarron Terminator algorithm). The large mobility shifts in the data contribute to truncation of the early part of the sequence by the base-calling algorithm and residual peak spacing variations cause numerous errors in the initial called sequence. The mobility shifts are not properly accounted for until after the doublet A at M13, base number 59, indicated by the arrow in **Fig. 10**.
10. The final mobility shift results are shown in **Fig. 11**, with ddC again used as the reference. While the ddA terminator exhibits no shift relative to ddC, ddG, and ddT show shifts of –2 and –1 base, respectively, early in the sequence before settling in to 0 and –0.5 base

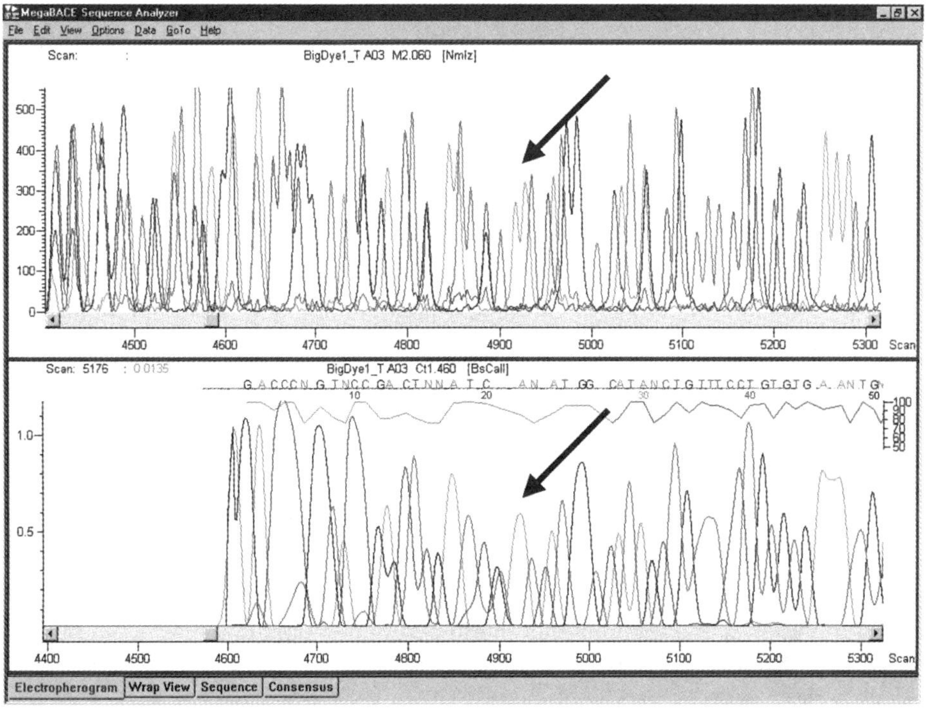

Fig. 10. Early section of M13mp18 sequencing data obtained with Big Dye® dye terminators (Perkin-Elmer Biosystems). *Upper panel* shows the data processed through background subtraction, spectral separation, and peak height normalization. *Lower panel* shows the data fully processed for base-calling, including adaptive mobility-shift corrections, using the Cimarron Dye Terminator algorithm. The arrow indicates the doublet A at M13 base number 59. Electrophoresis: 140 V/cm.

shifts after 100–150 bases. The ddG and ddT terminators exhibit a nearly constant offset relative to each other of ~0.5 base.

4. Notes

1. As CE becomes a prominent technique for high-throughput DNA analysis, the continued development of fluorescent nucleotide reagents will be an important aspect of research efforts. The high-resolution separations possible with CE place demands on reagent purity and stability not encountered with slab-gel electrophoresis systems. The minimization of dye-induced mobility shifts will be a significant criterion in the reagent development process, particularly for applications such as forensics, where any data corrections or manipulations weaken the evidentiary integrity of the results.
2. The relationship between dye structure and the electrophoretic mobility of labeled DNA fragments is reasonably well understood and controlled for primer labeling techniques.
3. More work needs to be done to foster a similar understanding of dye-terminator behavior, where the interactions of the terminators with DNA polymerases put constraints on the

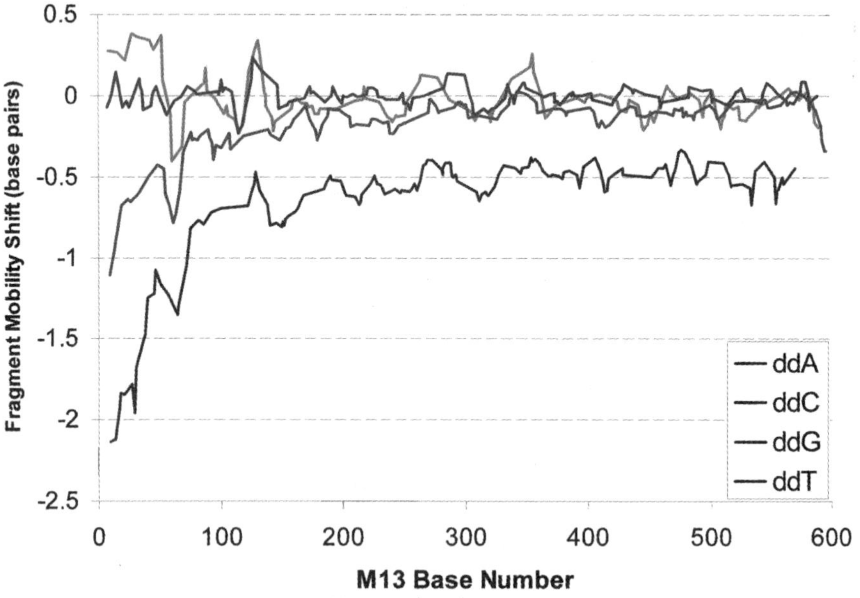

Fig. 11. Plot of fragment mobility shift (base pairs) vs fragment size (M13 base number) for all Big Dye® dye terminators, relative to ddC.

structure of the dye-nucleotide conjugates, and sequence-specific dye-DNA interactions can lead to dye-induced sequencing compressions.

4. The current DNA sequencing reagents available for the MegaBACE 1000® exhibit modest mobility shifts, and whereas dye-terminators perform less well than dye-primers in this regard, "on the fly" adaptive mobility corrections are able to handle the corrective manipulations, except for the earliest few bases in a sequence trace. Either further improvements to the reagents or the data analysis software developments will undoubtedly overcome this issue in due course.

Acknowledgments

The author would like to thank Mojgan Amjadi of Molecular Dynamics/Amersham Pharmacia for providing the raw data used for these analyses.

References

1. Tu, O., Knott, T., Marsh, M., Bechtol, K., Harris, D., Barker, D., and Bashkin, J. (1998) The influence of fluorescent dye structure on the electrophoretic mobility of end-labeled DNA. *Nucleic Acids Res.* **26,** 2797–2802.
2. Tan, H. and Yeung, E. S. (1997) Characterization of dye-induced mobility shifts affecting DNA sequencing in poly(ethylene oxide) sieving matrix. *Electrophoresis* **18,** 2893–2900.
3. Ju, J., Ruan, C., Fuller, C. W., Glazer, A. N., and Mathies, R. A. (1995) Fluorescence energy transfer dye-labeled primers for DNA sequencing and analysis. *Proc. Natl. Acad. Sci. USA* **92,** 4347–4351.

4. Ju, J., Kheterpal, I., Scherer, J. R., Ruan, C., Fuller, C. W., Glazer, A. N., and Mathies, R. A. (1995) Design and synthesis of fluorescence energy transfer dye-labeled primers and their application for DNA sequencing and analysis. *Anal. Biochem.* **231,** 131–140.
5. Ju, J., Glazer, A. N., and Mathies, R. A. (1996) Cassette labeling for facile construction of energy transfer fluorescent primers. *Nucleic Acids Res.* **24,** 1144–1148.
6. Hung, S. C., Ju, J., Mathies, R. A., and Glazer, A. N. (1996) Cyanine dyes with high absorption cross section as donor chromophores in energy transfer primers. *Anal. Biochem.* **243,** 15–27.
7. Hung, S. C., Mathies, R. A., and Glazer, A. N. (1997) Optimization of spectroscopic and electrophoretic properties of energy transfer primers. *Anal. Biochem.* **252,** 78–88.
8. Metzker, M. L., Lu, J., and Gibbs, R. A. (1996) Electrophoretically uniform fluorescent dyes for automated DNA sequencing. *Science* **271,** 1420–1422.

II

Capillary Matrix: Considerations for Capillary Electrophoresis

8

Influence of Polymer Concentration and Polymer Composition on Capillary Electrophoresis of DNA

Christoph Heller

1. Introduction

1.1. The Separation Matrix

In traditional slab-gel electrophoresis, the gel has two functions: it serves as an anticonvective medium, as well as a sieving matrix that provides separation. Because of the low convection, CE permits a gel-free separation at high voltages, and thus yields unprecedented resolution and speed. However, the situation is different for biopolymers such as RNA, dsDNA, ssDNA, or SDS-denatured proteins, which have a constant charge: size ratio. In this case, no separation occurs in free solution and some sort of sieving matrix still has to be used (however, the anticonvective property of the matrix is not needed any more).

Early attempts to apply CE to the size-separation of biomolecules, especially for the separation of small molecules, were based on gel-filled capillaries (e.g., agarose or crosslinked polyacrylamide) *(1,2)*. The gels are prepared in the same manner as slab-gels, by adding the catalysts to the monomer solution which is then pumped into the capillary, where the polymerisation takes place. However, gel-filled capillaries have several disadvantages: First, the filling of the capillary has to be done with great caution in order to avoid the introduction of air bubbles. The shrinkage of the gel during polymerisation can also be a source of bubbles *(3)*. Gels, and in particular polyacrylamide, can be degraded by hydrolysis, especially at the alkaline pH normally used to separate biopolymers. However, the main problem is at the entrance of the gel, as this zone receives all particles and impurities during injection, and is therefore very prone to clogging. It has also been observed that during repeated use, bubbles can form at the sample-injection end of the capillary *(4,5)*. Drying out of the gel at the ends can also occur.

From: *Methods in Molecular Biology, Vol. 162:*
Capillary Electrophoresis of Nucleic Acids, Vol. 1: Introduction to the Capillary Electrophoresis of Nucleic Acids
Edited by: K. R. Mitchelson and J. Cheng © Humana Press Inc., Totowa, NJ

A solution to these problems is to use a polymer solution instead, which could be replaced after each run. For separating DNA in CE, a number of different polymers have been used so far ***(6)***. Most of the polymers used are modified polysaccharides: agarose and its derivatives aimed at modifying the gelling temperature ***(1,7)***, various cellulose derivatives such as methyl cellulose (MC) ***(8–10)***, hydroxyethyl cellulose (HEC) ***(11,12)***, hydroxypropyl cellulose (HPC) ***(9,13)***, hydroxypropyl methyl cellulose (HPMC) ***(9)***, and glucomannan ***(13)***. From the synthetic polymers, linear polyacrylamide ***(15–20)*** and alkyl-substituted derivatives such as poly(*N*,*N*-dimethyl acrylamide (pDMA) ***(21–23)***, poly(*N*-acryloylaminoethoxy ethanol) (pAAEE) ***(23,24)***, poly(acryloylamino propanol) (pAAP) ***(25)***, and poly([acryloylaminoethoxy]ethyl-glucopyranoside) (pAEG) ***(23)*** as well as poly(ethylene glycol) (PEG) ***(26)***, poly(ethylene oxide) (PEO) ***(27–29)*** or polyvinylpyrrolidone (PVP) ***(30)*** have been used.

1.2. Polymer Solutions

From a physical point of view, three regimes of polymer solutions can be distinguished: dilute, semi-dilute, and concentrated solutions. In dilute solutions, the polymer chains are hydrodynamically isolated from each other and we can regard the properties of a single chain (***6,31,32***).

When the concentration of the polymer solution is increased, the transition from the dilute to a new regime, called “semi-dilute” occurs: the polymer chains become entangled, forming a transient network of obstacles. In a systematic study, it could be shown that for small dsDNA such entangled polymer solutions give superior separations over dilute solutions ***(13)***. Therefore, for example in DNA sequencing, we most likely will need the properties of an entangled network. This entanglement takes place above the overlap threshold, or entanglement threshold, c*. For a given polymer, this threshold can be experimentally determined by measuring the viscosity of the polymer solution at different concentrations and finding the point of departure from linearity on the viscosity vs concentration plot ***(11)***. It can also be estimated theoretically by using the following definition, derived from polymer physics ***(33,34)***:

$$c^* \approx 3\ M_w/4\ \pi\ N_A\ R_g^{\ 3} \qquad (1)$$

with R_g and M_w being the radius of gyration and the molecular weight of the polymer, respectively, and N_A being the Avogadro number.

Equation 1 is the geometrical definition of the entanglement threshold, i.e., c* is the concentration at which the polymer chains (modeled as coils) touch each other. Therefore, in order to determine this threshold, we need to know the radius of gyration of a particular polymer. There are several ways to measure the radius of gyration experimentally, but if M_w is known, a simple way to determine it is to measure the intrinsic viscosity in the dilute regime ***(35)***:

$$[\eta] = (\eta - \eta_s)/\eta_s c \equiv 6.2\ R_g^{\ 3}\ N_A/M_w \qquad (2)$$

in which “c” is the concentration of the polymer and η_s is the solvent viscosity. The value [η] (also called limiting viscosity number) is a measure of the capacity of a polymer to enhance the viscosity of a solution and it increases with the molecular weight. It has been measured for a number of polymers and it was found that for many of them it fits to the following empirical expression (Mark-Houwink-Sakurada equation) ***(36)***.

$$[\eta] \equiv K\,M_w^{\,a} \qquad (3)$$

where "a" and "K" are characteristic constants for a given polymer-solvent system. The Mark-Houwink coefficients "K" and "a" have been determined for many polymer-solvent systems (e.g., ***36***). For flexible molecules, the value of "a" varies between 0.5 (polymer in an ideal solvent) and 0.8 for so-called "swollen" chains.

Combining **Eqs. 1–3**, we get a simple expression for the overlap threshold:

$$c^* \equiv 1.5\,[\eta]^{-1} \equiv (1.5/K)\,M_w^{\,-a} \qquad (4)$$

For electrophoretic separations, we might want to know the "pore size" of such a network. To estimate this value, we can regard a polymer chain segment between two entanglement points as an independent subunit, which can undergo a so-called "random walk." The volume, which this chain segment can take, is called a "blob" and is described by the "blob size," ξ_b. Viovy and Duke ***(34)*** argued that this "blob size" could be used as a "pore size" of the network. It can be calculated as ***(33)***:

$$\xi_b = 1.43\,R_g\,(c/c^*)^{-(1+a)/3a} \qquad (c > c^*) \qquad (5)$$

Following Cottet et al. ***(37)***, **Eq. 5** can be rewritten to:

$$\xi_b \equiv 1.43\,(K/6^{3/2}\Phi)^{-1/3a}\,c^{-(a+1)/3a}\,(1.5/2\pi\,N_A)^{(a+1)/3a} \qquad (c > c^*) \qquad (6)$$

(ξ_b is in centimeters and concentration in g/mL), where Φ is the modified Flory viscosity constant, taking into account excluded volume effects by: $\Phi = \Phi_0\,(1 - 2\,\beta + 2.86\,\beta^2)$ with $\beta = (2a - 1)/3$ and Φ_0 being 2.5×10^{23} mol^{-1} for polyelectrolytes ***(38)***. We have to point out here that the equations mentioned above are based on scaling laws and the prefactors cannot be given exactly. **Equation 6** means that the "pore size" of an entangled polymer solution does not depend on the degree of polymerization but only on its concentration, "c" and the nature of the polymer (through "K" and "a").

1.3. Resolution

For comparing different separation techniques or protocols, we need a measure for the quality of the separation. For this purpose, we can use the resolution, which is defined by the ratio of the distance between two bands or peaks to the average peak width:

$$R_s = \Delta X /\ 1/2(W_1 + W_2) \qquad (7)$$

Both ΔX and W are given in units of length, but may also be regarded as the corresponding quantities in time units. The peak width can be measured at the baseline or at half height, W_h. From this definition, it is clear that the resolution depends on two factors: The interband spacing is given by the length of the migration path and the relative velocity difference between the two species (which in turn is determined by the migration mechanism): $\Delta X = L\,\Delta v/v$. The second factor is the band width, which can be influenced by the migration mechanism through enhanced diffusion ***(39,40)***, as well as by other independent dispersion effects (Joule heating, adsorption, and so on).

For the special case of DNA separation, it might be more convenient to take the reciprocal value of R_s, and normalize it to the DNA size ***(41,42)***. When using an array of closely spaced length standards, the peak width doesn't change considerably from one peak to the next, and we can use the width of a single (e.g., the first) peak:

$$RSL \approx W_h/(\Delta X/\Delta M) \qquad (8)$$

This "resolution length" has the advantage that it directly gives the smallest difference in base pairs (or bases) that can be resolved.

2. Materials

2.1. Device

1. For CE, a number of devices are commercially available. Although DNA can be detected by UV absorption, fluorescence detection is preferable, due to its higher sensitivity or even absolutely necessary for applications such as DNA sequencing.
2. Important features to consider are the temperature regulation, the kind of capillaries (diameter, length) which can be accommodated and the ability to inject viscous solutions.

2.2. Buffer

1. As electrophoresis buffer (background electrolyte), we recommend 0.5X TBE (45 m*M* Tris-base, 45 m*M* boric acid, 1 m*M* EDTA, pH 8.2) or 50 m*M* TAPS (*N*-Tris[hydroxymethyl]methyl-3-aminopropane-sulfonic acid), adjusted to pH 8.0 with NaOH (*see* **Note 1**).
2. For fluorescence detection without covalent label, intercalating dyes such as TOPRO-1, (Quinolinium, 4-[(3-methyl-2(3H)-benzothiazolylidene)methyl]-1-[3-(trimethylammonio) propyl]-, diiodide), SYBR Green I, SYBR Gold, and so on (all from Molecular Probes) are added to a final concentration of 0.1 μ*M* (*see* **Note 2**).
3. For denaturing conditions, urea is added to a final concentration of 7 *M*.

2.3. Capillaries

1. A number of suppliers offer capillaries convenient for CE. Although DNA can be separated in presence of electroosmotic flow (EOF) ***(12)***, we recommend to use conditions where the osmotic flow is suppressed. This can be achieved by an inner coating of the capillary.
2. Again, a number of coated capillaries are commercially available or can be produced according to numerous published protocols (for review, *see* **refs. *43*** and ***44***) (*see* **Note 3**).
3. A simpler way to achieve such conditions, is to use a separating polymer with self-coating abilities. So far, PEO, PVP, and pDMA have been identified as useful "self-coating" polymeric matrices.

2.4. Polymers

1. Polymers, such as linear polyacrylamide, PEO, HPC, HEC, dextran, and so on, are available in a wide range of molecular weight from different suppliers (e.g., Polysciences, Sigma-Aldrich) (*see* **Note 4**).
2. Some polymers, such as long chain PAA or pDMA are not commercially available, but can be synthesized from the monomer by radical polymerization. The chain length can be influenced in the following way: increasing the amount of initiator (in this case ammo-

nium persulphate [APS], in combination with *N,N,N',N'*-tetramethyl-1,2-diaminoethane [TEMED]) or increasing the temperature leads to a higher rate of nucleation and decreases the polymer length (*see* **Note 5**). A more precise method to control the polymer length is to stop the chain growth by using radical traps ("chain transfer reagents"), such as isopropanol.

3. We recommend to prepare pDMA according to a protocol by Chiari et al. ***(23)***: *N,N*-dimethyl acrylamide (Polysciences or Fluka) is dissolved in water to a final concentration of 10% w/w and isopropanol is added at various amounts (0–5%). After degassing (with vacuum and ultrasound), 1 µL TEMED per mL of solution and 1 µL/mL of a 40% ammonium persulfate solution are added (*see* **Note 6**).
4. After careful mixture, the solution is allowed to react for 1 h at 25°C or 50°C, followed by incubation at 25°C overnight.
5. The reaction product is extensively dialyzed against water using 25,000 mol wt cutoff (MWCO) dialysis membranes (Roth or Spectrum) and lyophilized.

3. Methods

3.1. Choice of Polymer

1. The result of **Eq. 5** means, in other words, that two solutions of the same type of polymer and with the same concentration but different molecular weight will have the same "pore size" as long as they are entangled. However, their viscosity is dependent on the molecular weight, which has important consequences for the choice of the appropriate polymer for CE.
2. The "pore size" determines the resolution power of a separation matrix. For a useful evaluation of different polymer preparations, we therefore have to compare the viscosities of different matrices having the same mesh size. Instead of a plot of mesh size vs concentration, we would like to compare the mesh size of different polymer solutions at a given viscosity. For this goal, we calculated the mesh sizes for a number of different polymer networks for all concentrations above c* with **Eq. 6**. Having measured the viscosity of a number of different polymer solutions (**Fig. 1**) we obtained a plot of measured viscosity vs estimated pore size (**Fig. 2**). This diagram allows us to choose the matrix with the lowest viscosity at a given pore size (in the entangled regime).
3. For all pore sizes (in the range of 1–100 nm), dextran (molecular mass 500 kDa) and two pDMA preparations (216 and 1150 kDa) show by far the lowest viscosity. For example, a network consisting of dextran 500 kDa, having a mesh size of 10 nm (probably required for oligonucleotide separation) has a viscosity of less than 10 mm^2/s, whereas matrices made of HEC (140 kDa), PAA (5 mio Dalton), or PEO (8 mio Dalton), which have been proposed for DNA sequencing ***(27,45–48)*** show a much higher viscosity (about three orders of magnitude higher; outside the range of **Fig. 2**) at the same mesh size.
4. Polymer networks most probably won't separate better than gels, but are chosen for their convenience in CE. To achieve highest separation (e.g., for DNA sequencing) we would want to have similar properties as in a gel, so we need to choose a polymer which is far above its entanglement threshold at the concentration needed, e.g., 5–10 times or more. At the same time, we want to keep the viscosity low, so we will choose the smallest possible molecular weight (*see* **Note 2**).

3.2. Separation of DNA

1. For separation of dsDNA, we recommend to use pDMA, because of its unique properties such as its self-coating ability and its low viscosity ***(49–51)***. However, the principles of separation shown below should also be valid for other polymers.

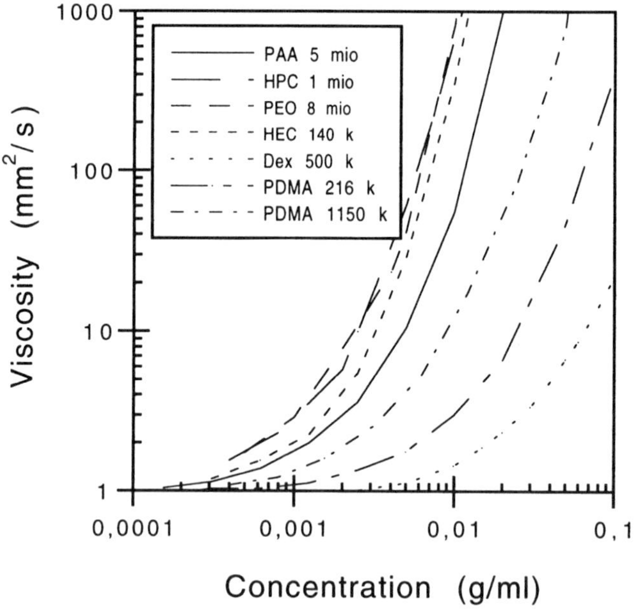

Fig. 1. Kinematic viscosity (in mm^2/s or centiStokes) of different polymer solutions in dependence on the concentration.

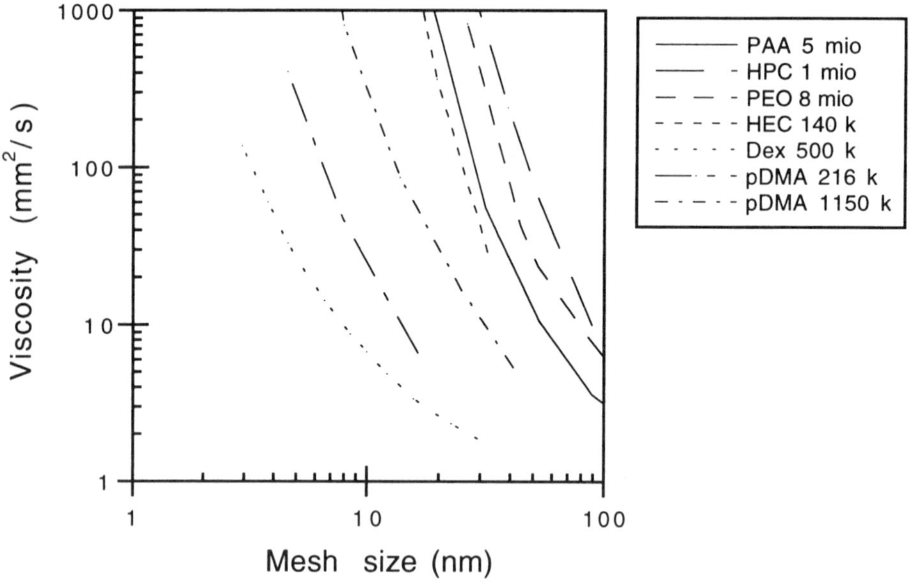

Fig. 2. Kinematic viscosity (in mm^2/s or centiStokes) of different entangled polymer solutions ($c > c^*$) in dependence of the "mesh size" of the network formed by the polymer chains.

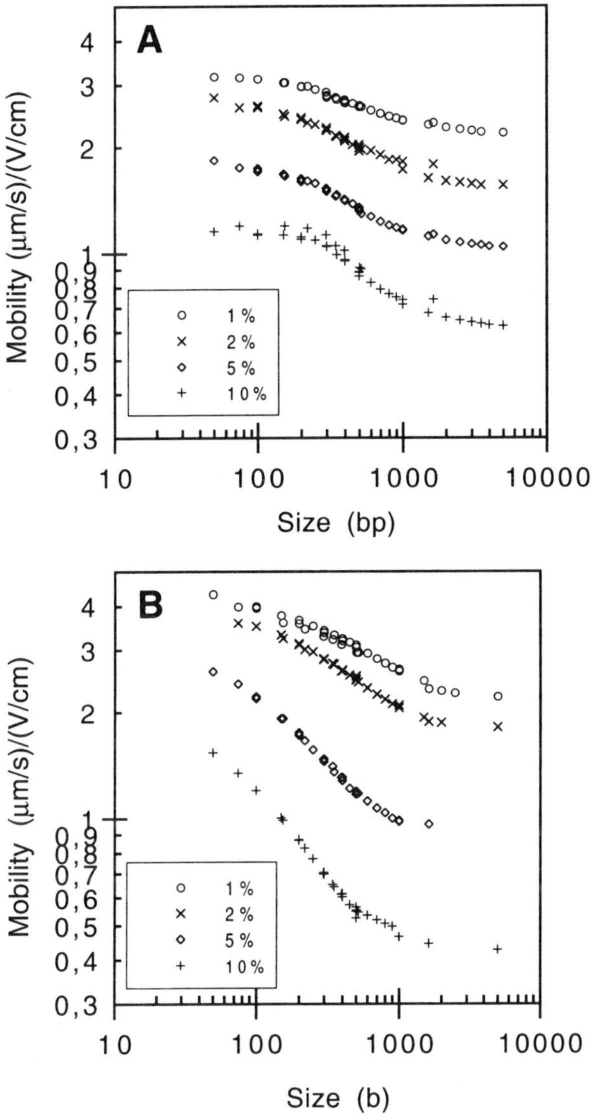

Fig. 3. Dependence of the electrophoretic mobility of linear DNA fragments on the size, separated in pDMA (216 kDa) of different concentrations. **(A)** dsDNA, **(B)** ssDNA. Electrophoresis conditions: 0.5X TBE at 210 V/cm in a 47-cm long (36 cm to the detector) fused silica (id = 100 µm) capillary. A: Separation at 25°C, B: separation in 4 *M* urea and 50°C. Reproduced from **ref. *53***.

3.2.1 Influence of Polymer Concentration

1. The separation of DNA under different conditions can be viewed best on double logarithmic mobility vs size plots. **Figure 3** shows the dependence of electrophoretic mobility of

dsDNA and ssDNA on pDMA concentration at an electric field strength of 210 V/cm. Such plots usually show a sigmoidal behavior, indicating the regions of good separation (steep slope) and those size ranges where DNA can hardly be separated.

2. With increasing polymer concentration, separation gets better in the medium size range but at the cost of lower mobility and therefore increased run times. In other words, low concentrations are useful for fast screenings at low resolution, whereas higher concentrations allow to "zoom" into certain regions in the medium size range at high resolution. In the case of dsDNA, a concentration of 2–5% pDMA is often the best compromise and offers good separation over a wide range of sizes.
3. In the case of ssDNA (denaturing conditions), the situation is slightly different. The onset of orientation ("critical size") is shifted to smaller sizes with increasing polymer concentration. This has important consequences for different applications. If short molecules such as oligonucleotides are to be separated, high polymer concentrations are used to achieve high resolution. For separating DNA fragments over a large size range (e.g., for long reads in DNA sequencing), we should lower the polymer concentration. This corresponds to the situation in slab gels, where the resolution of larger DNA could be improved by lowering the gel concentration to 4%T ***(52)***.

3.2.2. Influence of Polymer Length

1. As long as the polymer solutions are entangled ($c > c^*$), the polymer length does not have much effect on the separation (i.e., the peak distance) of DNA molecules, just as expected from theory. In fact, as can be seen from **Fig. 4**, in the case of native DNA, the resolution does not change considerably with polymer chain length.
2. However, for ssDNA, the polymer chain length should be chosen according to the polymer concentration used to ensure a sufficient entanglement. For example, for oligonucleotide separation where high polymer concentrations are used, we can afford to use rather short chains and the polymer solution will still be entangled.
3. For separating DNA fragments over a large size range (e.g., for long reads in DNA sequencing), we should lower the polymer concentration. In this case, we need to use rather long polymer chains, to keep the solution well entangled. This behavior is reflected in **Fig. 4**. At high polymer concentration, both long chains and short chains give an equally good resolution for short ssDNA.
4. For separating a larger size range, however, it is better to use longer polymer chains and—in contrast to dsDNA—to use lower polymer concentrations. Beside the lower viscosity, this leads to a more uniform resolution in function of the DNA size. This is especially important for DNA sequencing, where long reads are demanded.
5. In other words, for long read lengths we would prefer to use low polymer concentrations, which give a more uniform peak spacing and a more uniform resolution length (RSL) vs size. However, we then have to choose a long chain polymer preparation with a strong entanglement in order to avoid possible negative effects (e.g., larger peak width) due to the dynamic nature of the network. Generally, we can say that for DNA sequencing, we need highly entangled polymer solutions, which can only be achieved by rather long polymer chains. On the other hand, we would like to keep the viscosity low; therefore, we need to find a compromise between these two contrary requirements.

3.2.3. Different Polymers

1. The principles outlined above should also be valid for other entangled polymer solutions. In fact, a number of different polymers have been used so far as separation matrices

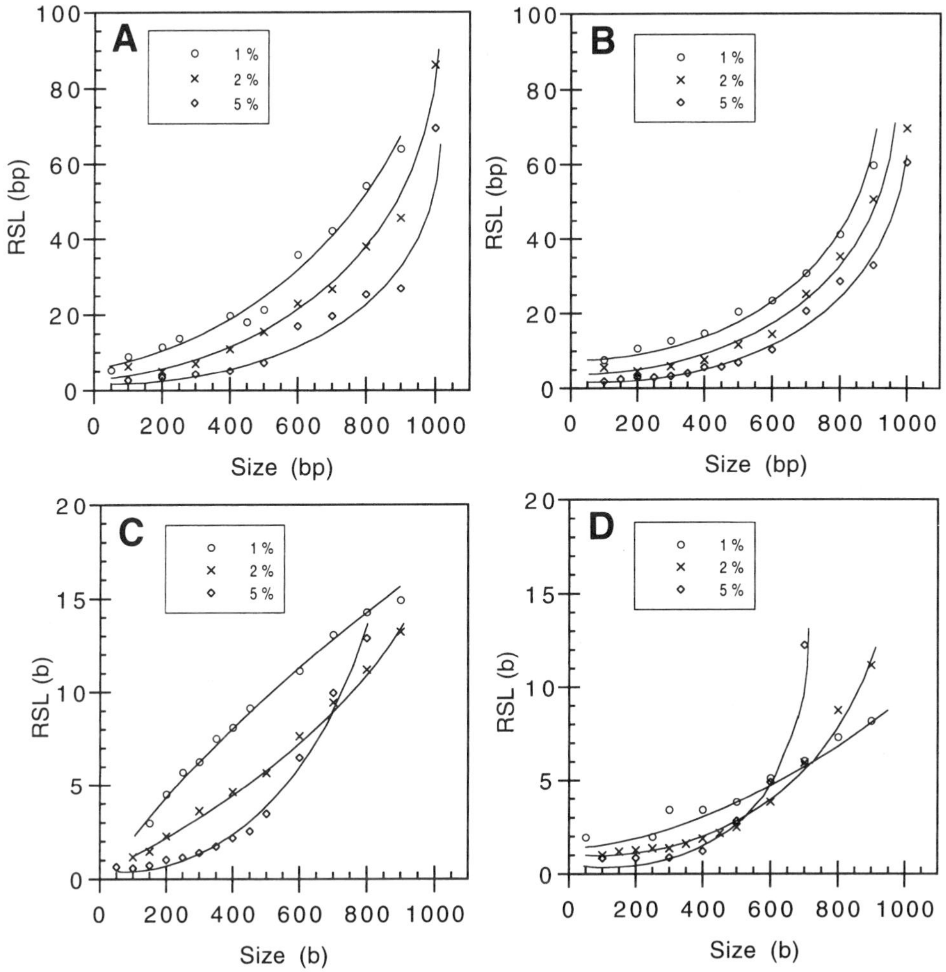

Fig. 4. Resolution length (i.e., the smallest difference in DNA size in b or bp that can be resolved) vs size for DNA separated in pDMA of different molecular mass and at different concentrations. **(A)** dsDNA in short chain pDMA; **(B)** dsDNA in long chain pDMA; **(C)** ssDNA in short chain pDMA, and **(D)** ssDNA in long chain pDMA. Conditions are as described in **Fig. 3**. Lines are guide to the eyes. Note the different scale on the *y*-axis in A, B and C, D. Reproduced from **ref. *54***.

for dsDNA as well as for ssDNA. However, when choosing a polymer, two major aspects should be considered:

a. Viscosity: Depending of the physical properties, different polymers have a different intrinsic viscosity and also form networks with different mesh sizes. We have tested a number of polymers for these parameters. From **Fig. 2**, we can identify for example dextran and pDMA as polymers which offer small pore size at low viscosity.

b. Coating: Most polymeric separation matrices require an inner coating of the capillary to prevent EOF. However, a few polymers show self-coating abilities. They adhere to the inner capillary wall and provide a so-called "dynamic coating." So far, two polymers, i.e., PVP and pDMA have been identified as polymers with good coating abilities and good separation properties. The use of self-coating capillaries allows the use of cheaper "naked" fused silica capillaries. Such capillaries can be washed under harsh conditions and easily regenerated. Therefore, their lifetime can be longer than that of coated capillaries.

4. Notes

1. We recommend to prepare stock solutions for the buffers, i.e., 5X TBE or 500 m*M* TAPS. For TBE, higher concentrations than 5X are not recommended due to possible precipitation over prolonged storage times.
2. Most polymer solutions and electrophoresis buffers are UV-transparent, allowing DNA to be detected by absorption at 260 nm. However, with laser-induced fluorescent detection, the sensitivity is 100–1000× higher. Best results are obtained with covalently labeled DNA, as background fluorescence from buffers, and so on, is minimized. The peak heights show a uniform distribution.

   If covalent labeling is to be avoided, DNA binding dyes such as SYBR Green, SYBR Gold, and so on, can be used. They are usually put at low concentration (about 0.1 μm) into the polymer solution as well as the electrode buffer. However, as the amount of bound dye increases with DNA size, signal (peak height) increases accordingly. This can cause difficulties if the dynamic range of the detector is limited and a large range of DNA sizes is to be separated.
3. When using noncoated capillaries, the capillary window can be easily burnt by using a flame or with a heating coil. However, for coated capillaries, lower temperatures are necessary as the heat can destroy the inner coating. The use of sulfuric acid or nitric acid at 80–90°C is recommended.
4. For the synthesis of polymers, all reagents should be of the highest quality and purity. Poor quality dimethyl acrylamide and acrylamide can contain:
   a. Acrylic acid, the hydrolysis product of acrylamide (the aminocarbonyl group is converted into a carboxyl group under basic conditions). Acrylic acid will polymerize with acrylamide and therefore change the properties of the polymer solution. The degradation reaction of acrylamide is catalyzed by light, storage in the dark is therefore recommended. The hydrolysis reaction can also occur in the polymerized chain (i.e. in the gel or polymer solution), leading to a negatively charged matrix. Poly(dimethyl acrylamide) (pDMA) is much more hydrolitically stable than polyacrylamide.
   b. Linear polyacrylamide or poly(dimethyl acrylamide), caused by catalytic contaminants in the dry monomer. It will affect the polymerization and the effective acrylamide concentration, leading to a loss of reproducibility.
   c. Metal ions as contaminants, which can inhibit or accelerate the polymerization or affect the mobility of the DNA.
5. APS is very hygroscopic and decomposes almost immediately when dissolved in water. The result is a loss of activity. As this compound affects the rate of polymerization, it is important to prepare it fresh daily in order to achieve reproducible results. Alternatively, aliquots of a stock solution can be frozen and discarded after use. TEMED is very reactive and subject to oxidation. The oxidized form is yellow and less reactive. As TEMED

is also hygroscopic, it will accumulate water, which again accelerates the oxidative decomposition. Therefore, only water-free TEMED, greater than 99% pure, should be used.

6. When working with pDMA as separation matrix and under denaturing conditions (i.e., in presence of urea), 50 m*M* TAPS as background electrolyte is recommended. For unknown reasons, the use of Tris buffers in combination with urea and pDMA leads to bad results (peak broadening).

References

1. Motsch, S. R., Kleemiß, M.-H., and Schomburg, G. (1991) Production and application of capillaries filled with agarose gel for electrophoresis. *J. High Res. Chromatogr.* **14,** 629–632.
2. Cohen, A. S., Najarian, D. R., Paulus, A., Guttman, A., Smith, J. A., and Karger, B. L. (1988) Rapid separation and purification of oligonucleotides by high-performance capillary gel electrophoresis. *Proc. Natl. Acad. Sci. USA* **85,** 9660–9663.
3. Dolnik, V., Cobb, K. A., and Novotny, M. (1991) Preparation of polyacrylamide gel-filled capillaries for capillary electrophoresis. *J. Microcol. Sep.* **3,** 155–159.
4. Swerdlow, H., Dew-Jager, K. E., Grey, R., Dovichi, N. J., and Gesteland, R. (1992) Stability of capillary gels for automated sequencing of DNA. *Electrophoresis* **13,** 475–483.
5. Dubrow, R. S. (1992) Stability of capillary gels for automated sequencing of DNA, in *Capillary Electrophoresis* (Grossman, P. D. and Colburn, J. C., eds.), Academic Press, San Diego, pp. 133–156.
6. Slater, G. W., Desruisseaux, C., and Hubert, S. J. (2001) DNA separation mechanisms during electrophoresis, in *Capillary Electrophoresis of Nucleic Acids*, Vol. 1 (Mitchelson, K. R. and Cheng, J., eds.), Humana Press, Totowa, NJ, pp. 27–41.
7. Kleparnik, K., Mala, Z., Doskar, J., Rosypal, S., and Bocek, P. (1995) An improvement of restriction analysis of bacteriophage DNA using capillary electrophoresis in agarose solution. *Electrophoresis* **16,** 366–376.
8. Strege, M. and Lagu, A. (1991) Separation of DNA restriction fragments by capillary electrophoresis using coated fused silica capillaries. *Anal. Chem.* **63,** 1233–1236.
9. Baba, Y., Ishimaru, N., Samata, K., and Tsuhako, M. (1993) High-resolution separation of DNA restriction fragments by capillary electrophoresis in cellulose derivative solutions. *J. Chromatogr. A.* **653,** 329–335.
10. McGregor, D. A. and Yeung, E. S. (1993) Optimization of capillary electrophoretic separation of DNA fragments based on polymer filled capillaries. *J. Chromatogr. A* **652,** 67–73.
11. Grossman, P. D. and Soane, D. S. (1991) Capillary electrophoresis of DNA in entangled polymer solutions. *J. Chromatogr.* **559,** 257–266.
12. Barron, A. E., Soane, D. S., and Blanch, H. W. (1993) Capillary electrophoresis of DNA in uncross-linked polymer solutions. *J. Chromatogr. A.* **652,** 3–16.
13. Mitnik, L., Salomé, L., Viovy, J. L., and Heller, C. (1995) Systematic study of field and concentration effects in capillary electrophoresis of DNA in polymer solutions. *J. Chromatogr.* **710**, 309–321.
14. Izumi, T., Yamaguchi, M., Yoneda, K., Isobe, T., Okuyama, T., and Shinoda, T. (1993) Use of glucomannan for the separation of DNA fragments by capillary electrophoresis. *J. Chromatogr.* **652,** 41–46.
15. Heiger, D. N., Cohen, A. S., and Karger, B. L. (1990) Separation of DNA restriction fragments by high performance capillary electrophoresis with low and zero crosslinked polyacrylamide using continuous and pulsed electric fields. *J. Chromatogr.* **516,** 33–48.
16. Guttman, A. and Cooke, N. (1991) Capillary gel affinity electrophoresis of DNA fragments. *Anal. Chem.* **63,** 2038–2042.

17. Chiari, M., Nesi, M., Fazio, M., and Righetti, P. G. (1992) Capillary electrophoresis of macromolecules in 'syrupy' solutions: Facts and misfacts. *Electrophoresis* **13,** 690–697.
18. Paulus, A. and Hüsken, D. (1993) DNA digest analysis with capillary electrophoresis. *Electrophoresis* **14,** 27–35.
19. Pariat, Y. F., Berka, J., Heiger, D. N., Schmitt, T., Vilenchik, M., Cohen, A. S., Foret, F., and Karger, B. L. (1993) Separation of DNA fragments by capillary electrophoresis using replaceable linear polyacrylamide matrices. *J. Chromatogr.* **652,** 57–66.
20. Heller, C. and Viovy, J.-L. (1994) Electrophoretic separation of oligonucleotides in replenishable polyacrylamide-filled capillaries. *Appl. Theor. Electrophoresis* **4,** 39–41.
21. Madabhushi, R. S., Menchen, S. M., Efcavitch, J. W., and Grossman, P. D., (1996) *United States Patent.* 5,552,028.
22. Rosenblum, B. B., Oaks, F., Menchen, S., and Johnson, B. (1997) Improved single-strand DNA sizing accuracy in capillary electrophoresis. *Nucleic Acids Res.* **25,** 3925–3929.
23. Chiari, M., Riva, S., Gelain, A., Vitale, A., and Turati, E. (1997) Separations of DNA fragments by capillary electrophoresis in N-substituted polyacrylamides. *J. Chromatogr. A* **787,** 347–355.
24. Chiari, M., Nesi, M., and Righetti, P. G. (1994) Capillary zone electrophoresis of DNA fragments in a novel polymer network: poly(*N*-acryloylaminoethoxyethanol). *Electrophoresis* **15,** 616–622.
25. Gelfi, C., Simo-Alfonso, E., Sebastiano, R., Citterio, A., and Righetti, P. G. (1996) Novel acrylamido monomers with higher hydrophilicity and improved hydrolitic stability: III. *Electrophoresis* **17,** 738–743.
26. Menchen, S., Johnson, B., Winnik, M. A., and Xu, B. (1996) Flowable networks as DNA sequencing media in capillary columns. *Electrophoresis* **17,** 1451–1459.
27. Fung, E. N. and Yeung, E. S. (1995) High-speed DNA sequencing by using mixed poly(ethylene oxide) solutions in uncoated capillary columns. *Anal. Chem.* **67,** 1913–1919.
28. Kim, Y. and Yeung, E. S. (2001) Capillary electrophoresis of DNA fragments using poly(ethylene oxide) as a sieving material, in *Capillary Electrophoresis of Nucleic Acids*, Vol. 1 (Mitchelson, K. R. and Cheng, J., eds.), Humana Press, Totowa, NJ, pp. 215–223.
29. Madabhushi, R. S., Vainer, M., Dolnik, V., Enad, S., Barker, D. L., Harris, D. W., and Mansfield, E. S. (1997) Versatile low-viscosity sieving matrices for nondenaturing DNA separations using capillary array electrophoresis. *Electrophoresis* **18,** 104–111.
30. Gao, Q. and Yeung, E. S. (1998) A matrix for DNA separation: Genotyping and sequencing using poly(vinylpyrrolidone) solution in uncoated capillaries. *Anal. Chem.* **70,** 1382–1388.
31. Viovy, J.-L. and Heller, C. (1996) Principles of size-based separations in polymer solutions, in *Capillary Electrophoresis: An Analytical Tool in Biotechnology Series* (Righetti, P. G., ed.) pp. 477–508.
32. Heller, C. (1997) The separation matrix, in *Analysis of Nucleic Acids by Capillary Electrophoresis* (Heller, C., ed.) Vieweg Verlagsgesellschaft, Wiesbaden, pp. 3–23.
33. Broseta, D., Leibler, L., Lapp, A., and Strazielle, C. (1986) Universal properties of semidilute polymer solutions: A comparison between experiments and theory. *Europhys. Lett.* **2**, 733–737.
34. Viovy, J. L. and Duke, T. (1993) DNA electrophoresis in polymer solutions: Ogston sieving, reptation and constraint release. *Electrophoresis* **14,** 322–329.
35. DeGennes, P. G. (1979) *Scaling Concepts in Polymer Physics*, Cornell University Press, Ithaca, N.Y.
36. Kurata, M. and Tsunashima, Y. (1989) Viscosity-molecular weight relationships and unperturbed dimensions of linear chain molecules, in *Polymer Handbook* (Brandrup, J. and Immergut, E. H., eds.), Wiley, New York, pp. VII/1–VII/46.

37. Cottet, H., Gareil, P., and Viovy, J. L. (1998) The effect of blob size and network dynamics on the size-based separation of polystyrenesulfonates by capillary electrophoresis in the presence of entangled polymer solutions. *Electrophoresis* **19,** 2151–2162.
38. Thielking, H. and Kulicke, M. W. (1996) On-line coupling of field-flow fractionation and multiple laser light scattering for the characterization of macromolecules in aqueous solution as illustrated by sulfonated polystyrene samples. *Anal. Chem.* **68,** 1169–1173.
39. Slater, G. W. (1993) Theory of band broadening for DNA gel electrophoresis and sequencing. *Electrophoresis* **14,** 1–7.
40. Tinland, B. (1996) Dispersion coefficients of lambda DNA in agarose gels during electrophoresis by fringe recovery after photobleaching. *Electrophoresis* **17,** 1519–1523.
41. Heller, C. (1997) Update on improvements in DNA separation, in *Analysis of Nucleic Acids by Capillary Electrophoresis* (Heller, C., ed.) Vieweg, Weinheim, pp. 297–308.
42. Heller, C., Slater, G. W., Mayer, P., Dovichi, N., Pinto, D., Viovy, J.-L., and Drouin, G. (1998) Free-Solution Electrophoresis of DNA. *J. Chromatogr. A* **806,** 113–121.
43. Chiari, M. and Gelain, A. (1997) Developments in capillary coating and DNA separation matrices, in *Analysis of Nucleic Acids by Capillary Electrophoresis* (Heller, C., ed.) Vieweg Verlagsgesellschaft, Wiesbaden, pp. 135–173
44. Chiari, M. and Cretich, M. (2001) Capillary coatings: Choices for capillary electrophoresis of DNA, in *Capillary Electrophoresis of Nucleic Acids*, Vol. 1 (Mitchelson, K. R. and Cheng, J., eds.), Humana Press, Totowa, NJ, pp. 125–138.
45. Ruiz-Martinez, M. C., Berka, J., Belenkii, A., Foret, F., Miller, A. W., and Karger, B. L. (1993) DNA sequencing by capillary electrophoresis with replaceable linear polyacrylamide and laser-induced fluorescence detection. *Anal. Chem.* **65,** 2851–2858.
46. Wu, C., Quesada, M. A., Schneider, D. K., Farinato, R., Studier, F. W., and Chu, B. (1996) Polyacrylamide solutions for DNA sequencing by capillary electrophoresis: Mesh sizes, separation and dispersion. *Electrophoresis* **17,** 1103–1109.
47. Bashkin, J., Marsh, M., Barker, D., and Johnston, R. (1996) DNA sequencing by capillary electrophoresis with a hydroxyethylcellulose sieving buffer. *Appl. Theor. Electrophoresis* **6,** 23–28.
48. Tan, H. and Yeung, E. S. (1997) Integrated on-line system for DNA sequencing by capillary electrophoresis: From template to called bases. *Anal. Chem.* **69,** 664–674.
49. Madabhushi, R. S. (1998) Separation of 4-color DNA sequencing extension products in noncovalently coated capillaries using low viscosity polymer solutions. *Electrophoresis* **19,** 224–230.
50. Madabhushi, R. S. (2001) DNA sequencing in noncovalently coated capillaries using low viscosity polymer solutions, in *Capillary Electrophoresis of Nucleic Acids*, Vol. 2 (Mitchelson, K. R. and Cheng, J., eds.), Humana Press, Totowa, NJ, pp. 309–315.
51. Heller, C. (1998) Finding a universal low viscosity polymer for DNA separation II. *Electrophoresis* **19,** 3114–3127.
52. Ansorge, W. and Barker, R. (1984) System for DNA sequencing with resolution of up to 600 base pairs. *J. Biochem. Biophys. Methods* **9,** 33–47.
53. Heller, C. (1999) Separation of double-stranded and single-stranded DNA in polymer solutions: I. Mobility and separation mechanism. *Electrophoresis* **20,** 1962–1977.
54. Heller, C. (1999) Separation of double-stranded and single-stranded DNA in polymer solutions: II. Separation, peak width and resolution. *Electrophoresis* **20,** 1978–1986.

9

Capillary Coatings

Choices for Capillary Electrophoresis of DNA

Marcella Chiari and Marina Cretich

1. Introduction

1.1. Electroosmotic Flow

One of the most distinguishing features of capillary electrophoresis (CE) is the electroosmotic flow (EOF). This electrokinetic phenomenon originates from the presence on the capillary surface of weakly acid hydroxyl groups attached to the silicon atoms. Different authors have reported an overall pKa value of 5.3 *(1)* or 6.3 *(2)* for surface silanols. In practice, when the capillary surface is bathed in any buffer solution with a pK higher than 2, immobilized negative charges are formed on the wall. The presence of this layer of immobilized charges causes adsorption of positive counter ions from the buffer solution. As a result, an electric double layer is generated at the silica-buffer solution interface consisting of an immobile layer of cationic species, tightly held to the ionized silanols, and a diffuse layer of hydrated ions loosely bound to the surface. **Figure 1** shows a schematic representation of the electrical double layer structure at the surface-solution interface. Two planes can be envisioned in the double layer: the so-called (inner) Helmholtz plane (IHP) located in the center of the compact layer, and the outer Helmholtz plane (OHP) that delimits the compact portion of the double layer. In the diffuse part of the double layer, the ion concentration diminishes until it reaches that of the bulk solution. When an electric field is tangentially applied to the surface, the diffuse ions of the double layer migrate towards the cathode with their surrounding water molecules generating a flow of liquid in the capillary. This overall solution transport phenomenon is called EOF and has dramatic influences on the time the analytes reside inside the capillary *(3)*. As a consequence of the forces exerted by the electric field on the ions of the diffuse layer, a plane of shear is formed between the diffuse and the mobile layer regions and the potential generated at this shear plane is termed zeta potential (ζ). In a typical buffer for DNA electrophoresis,

From: *Methods in Molecular Biology, Vol. 162:*
Capillary Electrophoresis of Nucleic Acids, Vol. 1: Introduction to the Capillary Electrophoresis of Nucleic Acids
Edited by: K. R. Mitchelson and J. Cheng © Humana Press Inc., Totowa, NJ

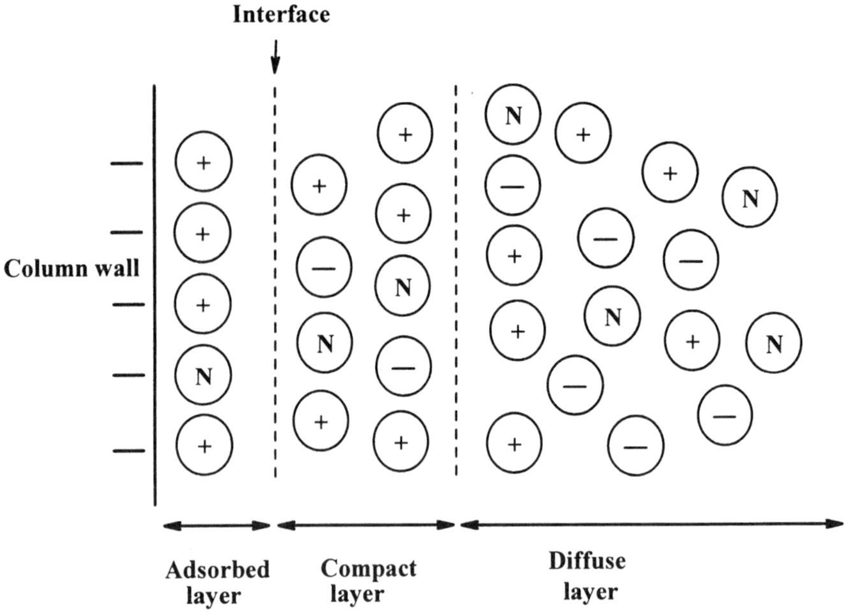

Fig. 1. Schematic representation of the electric double layer that occurs on the interface of a silica capillary filled with buffer.

the double layer is about 3 nm thick. A velocity gradient develops owing to the differences in the magnitude of frictional and electric forces within the double layer. The flow velocity that is zero across the wall increases in the region of the double layer reaching a maximum value at a very small distance from the charged surface. The bulk of the solution migrates with this maximum velocity. This means that the flow stream that originates at the wall of the capillary has an almost flat velocity distribution across the capillary diameter, deviating only within a few nanometers from the capillary surface. Unlike the parabolic hydrodynamic flow, EOF is almost plug-like and does not necessarily cause a deterioration of the separation.

1.2. Factors Influencing EOF

The electroosmotic velocity υeo is defined by the Helmholtz-Smoluchowski equation *(4)*, *see* **Eq. 1**:

$$\upsilon_{eo} = -(E\varepsilon/\eta_0)\zeta \tag{1}$$

According to this equation, the electroosmotic velocity (mm/s) depends on parameters such as the zeta potential (ξ), the applied electric potential (E), the dielectric constant (ε), and the buffer viscosity (η_0). In turn, these parameters depend on other physical and chemical properties of the solid and the liquid phases. The ξ potential can be expressed as follows, in **Eq. 2**:

$$\zeta = (4\pi\delta e)/\varepsilon \quad (2)$$

In which, ε is the dielectric constant of the buffer, e is the total excess charge in solution per unit area, and δ is the double layer thickness. The magnitude of the EOF depends on the pH of the running buffer, since the pH influences the fraction of silanols, which are in their ionized form. The zeta potential can have a magnitude of 127 mV in a fused capillary, resulting in an EOF velocity of 4 mm/s. At pH <2.0, the silica surface is neutral because the dissociation of silanol groups is suppressed. However, at pH >2.0, the surface silanols are deprotonated according to the following equilibrium:

$$SiOH + H_2O — SiO^- + H_3O^+$$

Figure 2 shows the dependence of the EOF on pH for fused silica capillaries from different scientific suppliers *(5)*. The four curves have a similar sigmoidal profile with an inflection point at pH 5.0–6.0. However, at pH values higher than 4.0, the EOF values vary significantly in the four capillaries. These differences result from the previous history of each capillary batch. EOF is also markedly influenced by the buffer ionic strength, since ξ is inversely related to the square root of the buffer concentration. Organic solvents are also able to modify EOF due to their influence on buffer dielectric constant and viscosity. The addition of methanol or isopropanol to the running buffer has been reported to decrease the EOF *(1)*. The effect of temperature on EOF depends on its effect on viscosity of the matrix buffer. The higher the temperature, the lower the resulting viscosity of the matrix, with a concomitant effect of an increase in the flow velocity.

1.3. Suppression of EOF During CE

In protein analysis, a variety of CE electrolyte buffers and additives can be used to optimize separation of analytes. In contrast, in the analysis of DNA there is less freedom in the choice of the buffers whose characteristics are imposed by the nature of DNA molecules themselves. For DNA separation, 1X TBE buffer (89 m*M* Tris-HCl, pH 8.3, 89 m*M* boric acid, 2.5 m*M* EDTA) is normally used. More recently the use of low conductive, zwitterionic buffers at a pH close to their isoelectric point have also been proposed to reduce dispersion effects caused by Joule heating. In any case, the pH of the background electrolyte must be around 8.0 to ensure a high negative charge on DNA molecules and to prevent degradation that occurs rapidly at low pH. Under the experimental conditions typically used for DNA separations (300 V/cm), the electroosmotic mobility within a fused silica capillary is 1–3 times greater then the electrophoretic mobility of DNA molecules, and the EOF pulls all the species past the detector regardless of their mobility. However, the presence of a rapid flow of buffer decreases the apparent mobility of DNA molecules that migrate against EOF. This does not necessarily cause a deterioration of separation efficiency *(6)*. The use of electroosmosis as a "pump" for CE of DNA has been advocated in the so-called countermigration technique *(7)* in which DNA molecules are dragged past the detector by the EOF, in the order of decreasing size. However, the difficulty in exerting a proper control on the phenomenon represents a serious drawback of EOF.

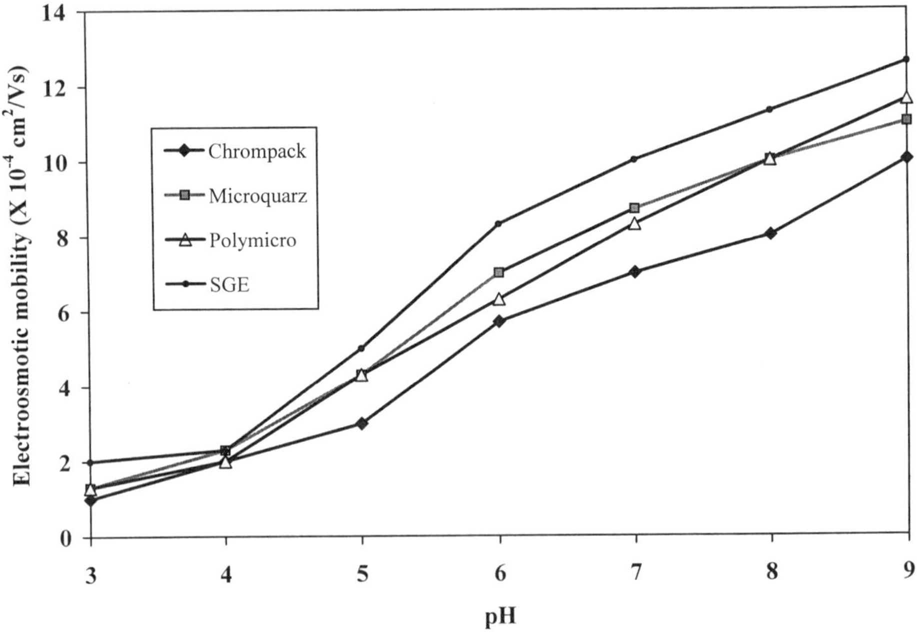

Fig. 2. The dependence of EOF on the pH of the buffer used in fused silica capillaries from different suppliers. Reproduced from *Journal of Chromatography A,* Vol. 652. Kohr, J., and Engelhardt, H. Characterization of quartz capillaries for capillary electrophoresis. Copyright (1993), pp. 309–316, reprinted with the permission of Elsevier Science.

1.3.1. Eddy Migration

Negative charges immobilized on the capillary wall cause adsorption of cationic analytes, which leads to a variation of the EOF in time through the length of the capillary. Proteins contained in DNA samples as contaminants, or cationic fluorescent DNA labels can adsorb to the wall and locally neutralize or even reverse the wall charge causing an "eddy migration." This effect, which seriously compromises the separation efficiency, is more evident with slow separands, e.g., long DNA sequencing fragments, which spend a relatively long time in eddy migration segments. For all these reasons, most of the DNA separations reported so far in CE have been carried out in capillaries with reduced or suppressed EOF. This chapter describes some of the strategies used to control EOF and reduce wall-analyte adsorption that are compatible with DNA separations.

1.4. Suppression of EOF in CE of DNA

Only a few of the methods that have been developed to control EOF and to prevent adsorption of proteins in CE can be used in DNA separation. In general, all the methods to control surface properties that require the use of a low pH, high ionic strength or addition of cationic additives or organic solvent to the background electrolyte (BGE) are not suitable for DNA. Separation of DNA in CE requires reduction of EOF to a negligible value by means of methods that act on the chemistry of the surface rather than on the composition of the BGE. In addition, EOF suppression must be efficient under critical conditions such as high pH, high temperature, and in the presence of highly concentrated urea. EOF suppression can be accomplished, by masking the charged sites on the wall through adsorption of neutral linear polymers that provide a viscous layer on the capillary surface or by chemical derivatization of surface silanols with anchoring groups for covalent immobilization of a viscous polymeric layer.

2. Materials

2.1. Preparation of Dynamic Coatings

1. Poly(ethylene oxide) (PEO) of mol wt 8,000,000.
2. Poly(vinyl alcohol) (PVA), hydrolysis grade 99%, average mol wt 2,000,000.

2.2. Activation of Silica with Allyl Groups

1. Methacryloxypropyltrimethoxysilane (MAPT).
2. 99% Thionyl chloride, distilled prior to use.
3. Vinyl magnesium bromide (1 *M* solution in tetrahydrofuran).
4. Tetrahydrofuran and toluene dried over molecular sieve.
5. Triethoxysilane (TES).
6. 2,5-di-tert-butylhydroquinone.
7. Hexachloroplatinic acid.
8. Allylmethacrylate.

2.3. Polymer Coating

1. Distilled water and all buffers filtered through 0.2-μm Nylon-66 membrane filters.
2. Acrylamide, ammonium.
3. Peroxydisulphate and *N,N,N',N'*-tetramethylenediamine (TEMED).

3. Methods

3.1. Methods to Measure Electroosmotic Flow

3.1.1. Neutral Marker Injection

The simplest method to measure EOF is by UV detection of a neutral marker that, upon application of an electric field, is dragged past to the detector with the electrolyte *(8)*. The electroosmotic mobility can be calculated from the following **Eq. 3**:

$$\mu_{EOF} = (l_d l_t)/(t_m V) \quad (3)$$

in which l_d is the length of the capillary from inlet to detector, l_t is the total length, t_m is the migration time of the neutral marker, and V is the applied voltage. Although this method is very simple, it has stringent requirements for the properties of the marker, such as detectability, true neutrality, and inertness toward the capillary wall to provide a correct EOF estimation (*see* **Note 1**).

3.1.2. Gravimetric Method

Altria et al. *(9)* proposed a method to weigh the amount of electrolyte that leaves the capillary upon application of an electric field for a known amount of time. The gravimetric method is not influenced by the analyte-wall interaction, but it requires highly precise measurements of very small quantities of liquid that can be affected by solvent evaporation.

3.1.3. Current-Monitoring Method

The rate of EOF can be measured by recording the current-time profile during capillary zone electrophoresis (CZE) operation in a discontinuous buffer system ***(10)***. The procedure involves filling the capillary tube and the cathodic reservoir with 20 m*M* sodium phosphate buffer, pH 7.0, whereas the anodic reservoir is filled with the same electrolyte diluted with water (19:1) so that its concentration is about 5%. The EOF is directed toward the cathode when the polarity of the power supply is positive. Under these conditions, EOF causes the electrolyte at lower concentration (C') to enter from the anodic reservoir into the capillary tube, thus displacing an equal volume of electrolyte at higher concentration. As a consequence, the total resistance of the fluid in the capillary changes and the current, I, drops continuously until the entire capillary is filled with electrolyte of concentration C'. A resistor (metal film 10.0 kΩ, 2W) is inserted between the cathodic reservoir and the ground. In this way, 1 μA current change would produce a 10-mV potential drop across the resistor. A chart recorder is connected across the resistor to record I as a function of time during CZE operation. The EOF μ_{eo} is given by $L/\Delta t$, where L is the distance between the reservoirs, and Δt is the time required for complete filling of the capillary column.

3.1.4. Pressure Driven Method

Practical problems arise when very small electroosmotic mobilities have to be determined, since a long electrophoresis time is required for the neutral marker to migrate past the detector. Several groups have developed procedures for accelerating EOF rate determination by pressure driven mobilization of the neutral marker ***(11,12)***. The idea behind these type of methods is to measure the neutral marker displacement only in a short fraction Δl of the effective length l, over a short time interval t_v. Sandoval and Chen ***(11)*** suggested the injection of a neutral marker before and after a timed voltage application. The plug of electrolyte, sandwiched between the two sample plugs, is pressure-driven out of the capillary under controlled conditions. EOF is deduced from the distance between the two marker plugs that is proportional to the distance migrated by electroosmotic mobility during the time the voltage is applied, according to the following **Eq. 4**:

$$\mu_{eo} = Ll/Vt_v(1 - [t_1/t_2]) \tag{4}$$

in which L is the total capillary length, l is the effective length, V is the applied voltage, t_v is the period during which the voltage is applied, and t_1 and t_2 are the migration times of the two neutral marker plugs.

3.2. Dynamic Coatings

Several linear polymers strongly interact with the capillary wall and dynamically coat its surface. The amount and the conformation of the adsorbed polymer and the coating thickness depend on physical variables such as the molecular mass of the poly-

Table 1
EOF Values of Some Polymer Matrix Solutions Used for CE

Polymer	EOF[a] (X 10^{-4} cm^2/V s)
Pure buffer	6.29
Poly(dimethyl acrylamide) (98 kDa)	0.20
Vinylpyrrolidone (40 kDa)	0.87
Vinylpyrrolidone (360 kDa)	0.47
Poly(ethylene glycol) (35 kDa)	1.44
Polyacrylamide (150 kDa)	2.43
Poly(*N*-isopropyl acrylamide) (50 kDa)	0.34

[a]The EOF is measured using 0.01% (w/v polymer in 100 m*M* glycineglycine buffer, pH 8.0, at 30°C. Reprinted from **ref. *13*** with the permission of Wiley-VCH.

mer and thermodynamic interaction parameters between polymer-solvent and the polymer-adsorbing surface. **Table 1** shows the EOF values measured in capillaries filled with dilute solutions of different polymers ***(13,14)***. The EOF, which may be taken as an indicator of the polymer adsorption, is higher for poly(ethylene glycol) and polyacrylamide than for poly(dimethyl acrylamide) or poly(*N*-isopropyl acrylamide). This is not surprising, as hydrophilic polymers interact more with the polar components of the aqueous buffers than with the surface. In a dynamic coating made up of hydrophilic polymers, the adsorption/desorption equilibrium constant can be unfavorable, and a small amount of the polymer must be included in the buffer to replace the coating that is eroded during electrophoresis (*see* **Notes 2** and **3**).

3.3. Capillary Coating Procedures

3.3.1. PEO Coating (15)

1. Prior to the first run of the day, the capillary is rinsed with 0.1 *M* NaOH and 0.1 *M* sodium dodecylsulphate solution, followed by an extensive rinse with water.
2. A 1 *M* HCl solution is forced to flow into the capillary by a syringe connected to the capillary by means of a tube with an internal diameter of 0.38 µm.
3. The capillary is then filled with a solution containing PEO and finally with the running buffer. The procedure must be repeated before each run and each step requires 5–15 min.
4. The polymer solution containing 0.2% w/v of PEO, M_r 8,000,000 is prepared by adding polymer powder into water during vigorous stirring in an Erlenmeyer flask that is heated at 95°C in a water bath by using a magnetic stirrer and hot plate. After complete dissolution, HCl is added up to a final concentration of 0.1 *M*.

3.3.2. PVA Coating (16)

1. The capillary to be coated is connected with a cut of shrinking tube to a 3 m cut of capillary of the same internal diameter that acts as a flow resistance.
2. A 10% w/w water solution of PVA (mol wt 50,000, 99% hydrolyzed) is forced through the capillary under a nitrogen pressure of 2.0 MPa using the device depicted in **Fig. 3**.
3. The coating solution is discharged from the capillary at a nitrogen pressure of 0.15 MPa.

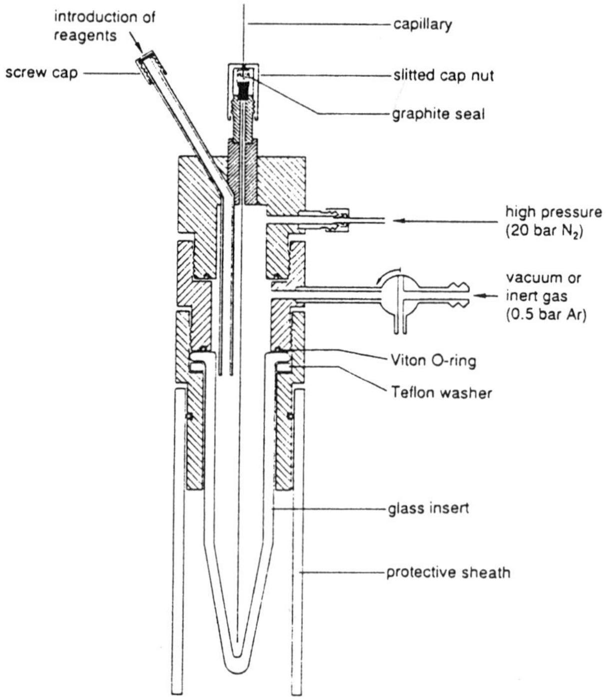

Fig. 3. Schematic representation of a device used for the introduction of derivatization reagents into capillary tubing under an inert atmosphere. Reprinted from **ref.** *27*.

4. The immobilization of PVA chains is obtained by heating the capillary at 140°C under a flow of nitrogen for several hours. This procedure is based on the well-known property of a PVA layer to become pseudocrystalline upon a thermal treatment, which leads to the production of a water-insoluble permanent coating. The heat treatment increases the mobility of polymer chains, which become more strongly associated by hydrogen bonding and form microcrystalline regions. This type of coating completely suppresses the EOF and is stable over a wide pH interval.

3.4. Covalent Coatings

1. Covalent coatings provide alternative means of controlling the EOF that do not require the addition of polymeric additives to the buffer. The most popular covalent CE phases for DNA analysis are composed of linear neutral polymers, bonded at multiple sites to the silica wall. These coatings are prepared by chemically treating the silica wall with an appropriate modifier bearing functional group that can be subsequently incorporated into a polymer.

   The whole process requires three steps:

   a. Chemical pretreatment of the native fused silica surface.
   b. Derivatization of the surface by organo-silanization to attach an appropriate R functionality to the silica.
   c. Incorporation of the R functionality into a polymer.

3.4.1. Capillary Pretreatment for Covalent Coating

1. The capillary is treated with 1.0 *M* NaOH for 15 min followed by 15 min of washing with deionized water (*see* **Note 4**).
2. Residual water is eliminated by connecting the capillary to a gas chromatography oven at 120°C for 45 min under nitrogen pressure (*see* **Note 5**).

3.4.2. Organosilanization

1. Fused-silica capillaries are typically reacted with organosilanes to yield an Si-R functionality attached to the support through an Si-O-Si (siloxane) linkage according to the following scheme:

$$|\text{Si-OH} + R_{4-n}\text{SiX}_n \rightarrow |\text{Si-OX}_{n-1}R_{4-n}$$

   in which, n = 1–3, R is an alkyl group, X is an easily hydrolyzable group such as halide, amine, alkoxy or acyloxy and the vertical line represents the surface of the support.
2. Organosilane compounds typically used as intermediates in the production of permanent coatings are:
   a. 3-methacryloxypropyltrimethoxysilane (MAPT),
   b. 3-glycidoxypropyltrimethoxysilane (GPT).
3. MAPT provides double bonds that copolymerize with acrylamide to form linear or crosslinked polymer layers ***(17)***, whereas GPT bears epoxy groups that are prone to react with different nucleophiles, such as hydroxyl groups or amines.

3.4.2.1. MAPT Organosilanization

1. A 50% solution of MAPT in tetrahydrofurane is pushed through the capillary under a nitrogen pressure for 20 min, and allowed to sit in the capillary for 12 h. After this treatment the capillary is extensively flushed with tetrahydrofuran and water.
2. Alternatively the capillary is filled with a solution of 15 µL of MAPT in 0.5 mL of 50% ethanol (pH 3.6). After 1 h of reaction at room temperature, the silane solution is withdrawn and the capillary washed with water.

3.5. Other Procedures to Graft Double Bonds to the Capillary Surface

1. One of the problems with covalently bonded phases is the limited stability of the Si-O-Si linkage. The formation of Si-C bonds provides an alternative way to graft an organic functionality onto a silica support. The great hydrolytic stability of this phase has been demonstrated.

3.5.1. The Grignard Approach

1. Cobb et al. ***(18)*** applied Grignard chemistry to attach the olefinic moiety to the capillary surface according to the following scheme (*see* **Note 6**):
   The procedure requires two steps:

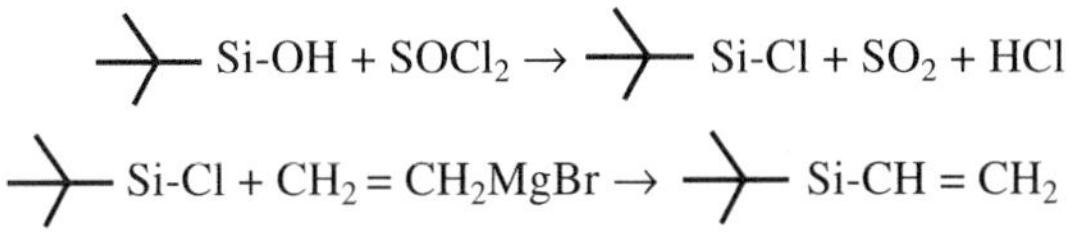

3.5.1.1. Surface Chlorination

1. Oxygen is removed from a 2-mL vial of thionyl chloride placed in a pressurizing chamber, by flushing it with nitrogen for 15 min.

2. One end of the capillary is placed into chamber to allow nitrogen to flow through the capillary for several min. Thionyl chloride is then flowed in the capillary by applying a nitrogen pressure into the sealed pressurizing chamber.
3. After the capillary is filled with thionyl chloride, one end is sealed by using a small propane torch and the other end quickly connected to a vacuum apparatus while keeping the capillary at 60°C by immersing it in a heating bath for about 20 min.
4. The end of the capillary that was connected to vacuum is sealed and the sealed capillary placed in a heating bath at 70°C for 12 h.

3.5.1.2. The Grignard Reaction

1. A 10% w/v solution of vinyl magnesium bromide in dry tetrahydrofuran (THF) is prepared under a nitrogen purge (*see* **Note 7**). One end of the chlorinated capillary is opened although immersed in a THF solution and quickly placed into the vinyl magnesium bromide/THF solution.
2. The solution is sucked into the capillary by connecting the second end of the capillary to a vacuum line. After several minutes, the capillary end that was immersed in the solution is sealed and the capillary placed in a 50°C heating bath while maintaining the capillary connected to the vacuum (60 mTorr or less) for 30 min.
3. Finally, the capillary is sealed and placed into a 70°C heating bath for 12 h, and then washed with THF for several minutes followed by a rinse with distilled water.

3.5.2. Hydrosilylation Approach

1. The general process requires two steps ***(19,20)***. First step, an intermediate hydride-modified substrate is produced by reaction of TES with the silica substrate in the presence of water and an acid catalyst. Secondly, the methacryloxypropyl group is covalently linked to the hydride-modified substrate by hydrosilation according to the following scheme:

$$1\ |SiOH + X_3SiH \xrightarrow[\text{dioxane}]{H_2O,\ HCl} |\ SiO\text{-}\underset{OX}{\overset{OX}{\underset{|}{\overset{|}{Si}}}}\text{-}H \qquad X = \text{–}OCH_2CH_3$$

$$2\ |\ SiO\text{-}\underset{OX}{\overset{OX}{\underset{|}{\overset{|}{Si}}}}\text{-}H + H_2C = CHR \xrightarrow{\text{Pt cat}} |\ SiO\text{-}Si\text{-}CH_2\text{-}CH_2\text{-}R$$

3.5.2.1. Preparation of the Hydride Intermediate

1. Prepare a TES solution by mixing 8.5 mL of dioxane, 435 µL of 2.3 *M* HCl, and 1.1 mL of 1.0 *M* TES/dioxane solution with continuous stirring.
2. Flush a preconditioned capillary with dioxane, then treat it with TES solution at 100±1°C for 90 min.
3. Rinse the tube with THF for 1 h, 75% v/v THF/water for 2 h and then THF for 4 h at room temperature. Finally, dry the capillary under a stream of nitrogen at 110±1°C for 24 h.

3.5.2.2. Methacryloxypropylsilica by Hydrosilylation of Allylmethacrylate

1. The hydride capillary is rinsed with dry toluene and then treated with a solution prepared by dissolving 375 mg of 2,5-di-tert-butyl hydroquinone in 75 mL of 25% v/v of

allylmethacrylate in dry toluene to which 50 μL of Speier's catalyst was added (*see* **Note 8**).
2. The mixture is heated at 70±5°C for 1 h before being introduced into the capillary.
3. The solution is forced into the capillary by applying a nitrogen pressure on a pressurizing chamber at 100±1°C for 60 h.
4. The capillary is rinsed at room temperature with toluene for 2 h followed by THF (2 h) and, finally, dried with nitrogen at 110±1°C for over 20 h.

3.5.3. Polymer Grafting Procedure

1. Acrylamide (10 mg) is first weighted in a airtight vial provided with a silicone septum and dissolved in 1 mL of water (*see* **Note 9**).
2. After degassing the acrylamide solution by bubbling nitrogen for 30 min, the vial is closed and solutions are added by means of a syringe. First, 2.5 μL of TEMED (10% v/v) is added to the monomer solution, followed by the addition of 2.5 μL of 10% w/v ammonium peroxydisulfate (APS), (*see* **Note 10**).
3. Taking the previously silylated capillary, (by one of the methods reported in **Subheading 3.4.2.**), one end is quickly dipped into the acrylamide solution by insertion into the vial through the septum. A nitrogen pressure of 1.5 bar is applied into the vial to force the polymerizing solution to flow into the capillary for 1 min (*see* **Note 11**).
4. Next, both ends of the capillary are sealed by puncturing into a piece of silicone rubber and the reaction allowed to proceed to completion overnight inside the sealed capillary at room temperature. Polymer that is not bonded to the capillary wall is washed away with water (*see* **Notes 12** and **13**).

4. Notes

1. Benzyl alcohol, mesityloxide, and dimethylsulfoxide are most commonly used neutral markers.
2. The addition of a small amount (0.1–0.05% w/v) of polymer to the running buffer does not represent a problem since the separation of DNA in CE is typically performed in a solution of linear polymer. When hydroxyethyl cellulose, PEO, ***(21)*** poly(dimethyl acrylamide) ***(13,14)***, or polyvinylpyrrolidone ***(22)*** are used as the sieving matrices they interact with the capillary wall to form temporary wall-coating, allowing for several tens of successful separation in bare capillaries.
3. The main advantage of adsorbed coating is the simplicity of preparation and the fact that the surface can be repeatedly regenerated, by using strong basic solution and thus extending the capillary lifetime and run-to-run reproducibility. However, in most of the cases capillary regeneration is not a straightforward procedure since some of the sample constituents and polymer segments bind irreversibly to the wall. In addition, the stability of the adsorbed coating is strongly influenced by the temperature and the composition of the buffer. For instance, operation conditions required by DNA sequencing are incompatible with the adsorption of most hydrophilic polymers. The amount of urea contained in the buffer and the high temperature used in this application hinder the formation hydrogen bonds between the polymers and the silica surface.
4. Preconditioning of the capillary wall with alkaline solutions is required to maximize the number of Si-OH binding sites on the surface. On silica surfaces, part of the hydroxyl groups are condensed to form siloxane bridges. In an alkali solution, condensed silanols that can be hydrolyzed reach up to a concentration of 4.8 hydroxyl groups/nm^2. The chemical status and the quality of the silica surface may be quite different from one batch

of capillaries to another. Therefore, an accurate pretreatment of the capillary wall eliminates the differences observed in the capillaries purchased from different sources.

5. It is advisable before performing organo-silanization reaction on silica to quantitatively remove any water physically adsorbed onto the surface, otherwise the modifier may be hydrolyzed and condensed yielding a physiosorbed organo-silicon polymer. Water can be removed from the capillary surface by heating the capillary under vacuum at 120–140°C. Note that the temperature is critical, because warming the capillary above 150°C can lead to water-cleavage of the contiguous silanols, thus diminishing surface binding sites. Overheating effectively limits the removal of water molecules from the silica surface.
6. The Grignard procedure was further modified by Dolnik et al. ***(23)*** to allow preparation of a 100 m of continuous capillary.
7. If the solution of THF becomes cloudy upon addition of vinyl magnesium bromide the procedure must be repeated with fresh reagents to make a clear solution.
8. Speier's catalyst is a 0.10 *M* solution of hexachoroplatinic acid in 2-propanol. Anhydrous conditions are very important during the preparation of this solution. The desired amount of hexacloroplatinic acid must by weighted under nitrogen atmosphere and dissolved in 2-propanol that has been dried over a molecular sieve.
9. Different authors have reported the instability of polyacrylamide coating over time, because of its unsatisfactory hydrolytic stability toward alkaline solutions. Hydrolysis of polyacrylamide causes reoccurrence of EOF. A strong effort to replace acrylamide with more stable acrylamide derivatives led to introduction of acryloylaminoethoxy ethanol ***(20)***, acryloaminoethylethoxy glucopyranose ***(24)***, and acryloylamino propanol ***(25,26)***. These *N*-substituted acrylamides can be used in the production of more stable polymeric coatings.
10. The catalyst stock solutions should be freshly prepared.
11. Insertion of the acrylamide solution containing the catalysts in the capillary must be done quickly after addition of the catalysts to prevent initiation of polymerization outside the capillary resulting in a viscosity increase of the solution and in a poor covalent attachment of the polymer.
12. Due to the high viscosity of the unreacted polyacrylamide solution, its removal from 50-μm id capillaries may take 5–10 min or even longer even when a pressure of 5–7 atmospheres is applied to the pressurizing chamber.
13. Special precautions are required for the removal of the external polyimide coating from silica tubing to obtain an optical window in the coated capillaries. It is not possible to remove the external coating by burning as is usually done in uncoated capillaries, as this would damage the polymer coating inside the capillary. A simple method to obtain an optical window involves the rotation of a 50–80-cm length capillary tube on a flat surface although slightly pressing it with a very sharp single-edge razor blade positioned at a 45° angle with respect to the capillary axis. Short back and forth motions at the same angle over a 1–2-mm section results in multiple spiral cuts on the coating until a clear window is visible.

References

1. Schwer, C. and Kenndler, E. (1991) Electrophoresis in fused-silica capillaries: the influence of organic solvents on the electroosmotic velocity and the z potential. *Anal. Chem.* **63,** 1801–1807.
2. Bello, M. S., Capelli, L., and Righetti, P. G. (1994) Dependence of electroosmotic mobility on the applied electric field and its reproducibility in capillary electrophoresis. *J. Chromatogr. A* **684,** 311–322.

3. Pretorious, V., Hopkins, B. J., and Schieke, J. D. (1974) Electroosmosis: A new concept for high-speed liquid chromatography. *J. Chromatogr.* **99,** 23–30.
4. Bier, M. (1959) *Electrophoresis: Theory, Methods and Applications.* Academic Press Inc., New York.
5. Kohr, J. and Engelhardt, H. (1993) Characterization of quartz capillaries for capillary electrophoresis. *J. Chromatogr. A* **652,** 309–316.
6. Jorgenson, J. W. and Lukacs, K. D. (1981) High resolution separations based on electrophoresis and electroosmosis. *J. Chromatogr.* **218,** 209–216.
7. Chin, A. M. and Colburn, J. C. (1989) Counter-migration capillary electrophoresis (CMCE) in DNA restriction fragment analysis. *Am. Biotech. Lab. News Edn.* **7,** 16.
8. Tsuda, T., Nakagawa, G., Sato, M., and Yagi, K. (1983) Separation of nucleotides by high-voltage capillary electrophoresis. *J. Appl. Biochem.* **5,** 330–336.
9. Altria, K. D. and Simpson, C. F. (1986) Measurement of electroosmotic flows in high-voltage capillary zone electrophoresis. *Anal. Proc.* **23,** 453–454.
10. Huang, X., Gordon, M. J., and Zare, R. N. (1988) Current-monitoring method for measuring the electroosmotic flow rate in capillary electrophoresis. *Anal. Chem.* **60,** 1837–1838.
11. Sandoval, J. E. and Chen, S.-M. (1996) Method for the accelerated measurement of electroosmosis in chemically-modified tubes for capillary electrophoresis. *Anal. Chem.* **68,** 2771–2775.
12. Williams, B. and Vigh, G. (1996) Fast, accurate mobility determination method for capillary electrophoresis. *Anal. Chem.* **68,** 1174–1180.
13. Madabhushi, R. S. (1998) Separation of 4-color DNA sequencing extension products in noncovalently coated capillaries using low viscosity polymer solutions. *Electrophoresis* **19,** 224–230.
14. Madabhushi, R. S. (2001) DNA sequencing in noncovalently coated capillaries using low viscosity polymer solutions, in *Capillary Electrophoresis of Nucleic Acids,,* Vol. 2 (Mitchelson, K. R. and Cheng, J., eds.), Humana Press, Totowa, NJ, pp. 309–315.
15. Iki, N. and Yeung, E. S. (1996) Non-bonded poly(ethylene oxide) polymer-coated column for protein separation by capillary electrophoresis. *J. Chromatogr. A* **731,** 273–282.
16. Gilges, M., Kleemiss, M. H., and Schomburg, G. (1994) Capillary zone electrophoresis separation of basic and acidic proteins using poly(vinylalcohol) coatings in fused silica capillaries. *Anal. Chem.* **66,** 2038–2046.
17. Hjertén, S. (1985) High performance electrophoresis: Elimination of electroendoosmosis and solute adsorption. *J. Chromatogr.* **347,** 191–198.
18. Cobb, K. A., Dolnik, V., and Novotny, M. (1990) Electrophoretic separation of proteins in capillaries with hydrolytically stable surface structure. *Anal. Chem.* **62,** 2478–2483.
19. Sandoval, J. E. and Pesek, J. J. (1991) Hydrolytically stable bonded chromatographic phases prepared through hydrosilation of olefins on a hydride-modified silica intermediate. *Anal. Chem.* **63,** 2634–2641.
20. Chiari, M., Nesi, M., Sandoval, J. E., and Pesek, J. J. (1995) Capillary electrophoretic separation of proteins using stable, hydrophilic poly(acryloylaminoethoxyethanol)-coated columns. *J. Chromatogr.* **717,** 1–13.
21. Fung, E. N. and Yeung, E. S. (1995) High-speed DNA sequencing by using mixed poly(ethylene oxide) solutions in uncoated capillary columns. *Anal. Chem.* **67,** 1913–1919.
22. Gao, Q. F. and Yeung, E. S. (1998) A matrix for DNA separation: genotyping and sequencing using poly(vinylpyrrolidone) solution in uncoated capillaries. *Anal. Chem.* **70,** 1382–1388.

23. Dolník, V., Xu, D., Yadav, A., Bashkin, J., Marsh, M., Tu, O., Mansfield, O., et al. (1998) Wall coating for DNA sequencing and fragment analysis by capillary electrophoresis. *J. Microcol. Sep.* **10,** 175–184.
24. Chiari, M., Dell'Orto, N., and Gelain, A. (1996) Synthesis and characterization of capillary columns coated with glycoside-bearing polymer. *Anal. Chem.* **68,** 2731–2736.
25. Simò-Alfonso, E., Gelfi, C., Sebastiano, R., Citterio, A., and Righetti, P. G. (1996) Novel acrylamido monomers with higher hydrolytic stability. 1. Synthetic route and product characterization. *Electrophoresis* **17,** 723–731.
26. Simò-Alfonso, E., Gelfi, C., Sebastiano, R., Citterio, A., and Righetti, P. G. (1996) Novel acrylamido monomers with higher hydrolytic stability. 2. Properties of N-acryloyaminoisopropanol. *Electrophoresis* **17,** 732–737.
27. Gulges, M., Husmann, H., Kleemiss, M. H., Motsch, S. K., and Schomburg, G. (1992) CZE spearations of basic proteins at low pH in fused silica capillaries with surfaces modified by silane derivatization and/or absorption of polar biopolymers. *J. High Res. Chromatogr.* **15,** 452–457.

10

Replaceable Polymers for DNA Sequencing by Capillary Electrophoresis

Mark A. Quesada and Steve Menchen

1. Introduction

In this chapter, we examine important parameters that affect the performance of replaceable polymers currently used for DNA sequencing by capillary electrophoresis (DNASCE). Background on physical models used to describe polymer solutions and DNA migration in semi-dilute polymer solution will be given to provide a framework for interpretation of experimental data and provide estimates of optimized polymer and electrophoresis parameters. Discussion of parameters affecting DNASCE resolution will be followed by comparison of experimental conditions for four linear polymers currently used in DNASCE *(1–10)*. Recent developments in performance optimization and inner-wall coatings will also be discussed.

Automated, multiple capillary systems are being developed to provide high-throughput sequencing of genomes for numerous organisms *(11,12)* with additional application in genetic analysis, medical diagnosis, and forensics *(12–16)*. Capillary electrophoresis (CE) offers enhanced separation speed over slab-gel methods as well as simple electrokinetic loading and the stacking of sample *(13)*. These features facilitate the design and automation of large-scale parallel systems. Early DNA separations with CE relied on polyacrylamide gels *(17)*, analogous to standard slab-gel methods used today. The routine preparation of homogeneous gels in capillaries however is not a simple task, because of gel shrinkage and the formation of bubbles inside the capillary; the slow onset of these factors during repeated use of the same gel is why the long-term usage of each gel is low *(18)*. Replaceable, noncrosslinked (linear) polyacrylamide solutions were introduced with a sieving effect comparable to crosslinked systems *(19)*, which made renewal of the sieving media possible without need for capillary replacement.

Automated, high-throughput, high-resolution DNA sequencers based on CE are becoming available today because of the high performance of many different polymer solution formulations. Ideally, DNASCE should exhibit a long read length

From: *Methods in Molecular Biology, Vol. 162:*
Capillary Electrophoresis of Nucleic Acids, Vol. 1: Introduction to the Capillary Electrophoresis of Nucleic Acids
Edited by: K. R. Mitchelson and J. Cheng © Humana Press Inc., Totowa, NJ

(500–1000 bases) in a short amount of time (10–60 min), with repeated use (50–1000×) of the same capillary. The capillary would be renewed between runs by flushing with polymer for only a few minutes (≤10 min). As we shall see, many practitioners in the field attain specifications that are reproducible, and similar to those described above.

Separation studies of double-stranded DNA (dsDNA) with CE will not be considered since this chapter concerns itself only with DNA sequencing. There are inherent differences between dsDNA and single-stranded DNA (ssDNA): the persistence length of dsDNA is 450 angstroms (Å) ***(20)***, yet is only 30 Å for ssDNA ***(21)***–and the frictional and charge differences are substantial ***(22)***. Unless otherwise stated, all references to DNASCE should be assumed to proceed at constant electric field using replaceable polymers.

2. Polymer Solutions and DNA Migration

A useful perspective on DNA migration in semi-dilute polymer solutions is obtained by examining the different ways DNA migrates electrophoretically in gel media. These systems are characterized by pores formed from a network of polymers that are covalently bound, whereas replaceable polymer solutions have pores defined by a network of entangled polymers. **Figure 1B** illustrates the various ways DNA can migrate in a gel based on the relative values of DNA size, electrophoresis voltage and pore size. Note that most of the DNA sequencing is conducted in the so-called reptation regime shown in **Fig. 1A**-iii and -iv and in **Fig. 1B**-iii and -iv. The arrival time of the DNA bases at the detector in a typical, high-performance DNASCE experiment ***(1–10)*** is proportional to the size of the DNA (*see inset*, **Fig. 1C**). A consistent body of experimental data ***(1–10,23–25)*** indicates that the mobility decays as the inverse of the molecular weight, and then levels off at a finite value depending on the electric field. These properties have been described as the signature of a "biased reptation-like behavior" ***(24,26)*** (*see* **Fig. 1**). The biased reptation model (BRM) ***(26–31)*** explains why the mobility of long DNA saturates at a finite value independent of molecular weight instead of an exponential decrease as predicted in an Ogston-type model (*see* **Fig.1A**-I and **Fig. 1B**-I). The BRM model correctly predicts that the mobility should be minimum for an intermediate molecular size (band-inversion) in constant-field electrophoresis ***(32)***. Although other predictions have been contradicted by experiments, such as a quadratic dependence of the mobility upon the field ***(33,34)***, the essential migration features of ssDNA in replaceable polymer matrices are captured by such reptation theories. In what follows, we will use a simple BRM formula to provide a convenient framework for interpretation of experimental data as BRM predictions provide

---

Fig. 1. *(opposite page)* A schematic description of the electrophoretic properties of DNA in sieving gels. **(A)** A schematic plot of DNA mobility vs the molecular size of the DNA. **(B)** A schematic drawing showing molecular conformations characterizing the various migration regimes. Dots correspond to a two-dimensional representation of gel pores, dark lines correspond to DNA, and dashed lines correspond to DNA trajectories. These regimes may not all be present for any particular choice of experimental conditions. Other parameters illustrated include:

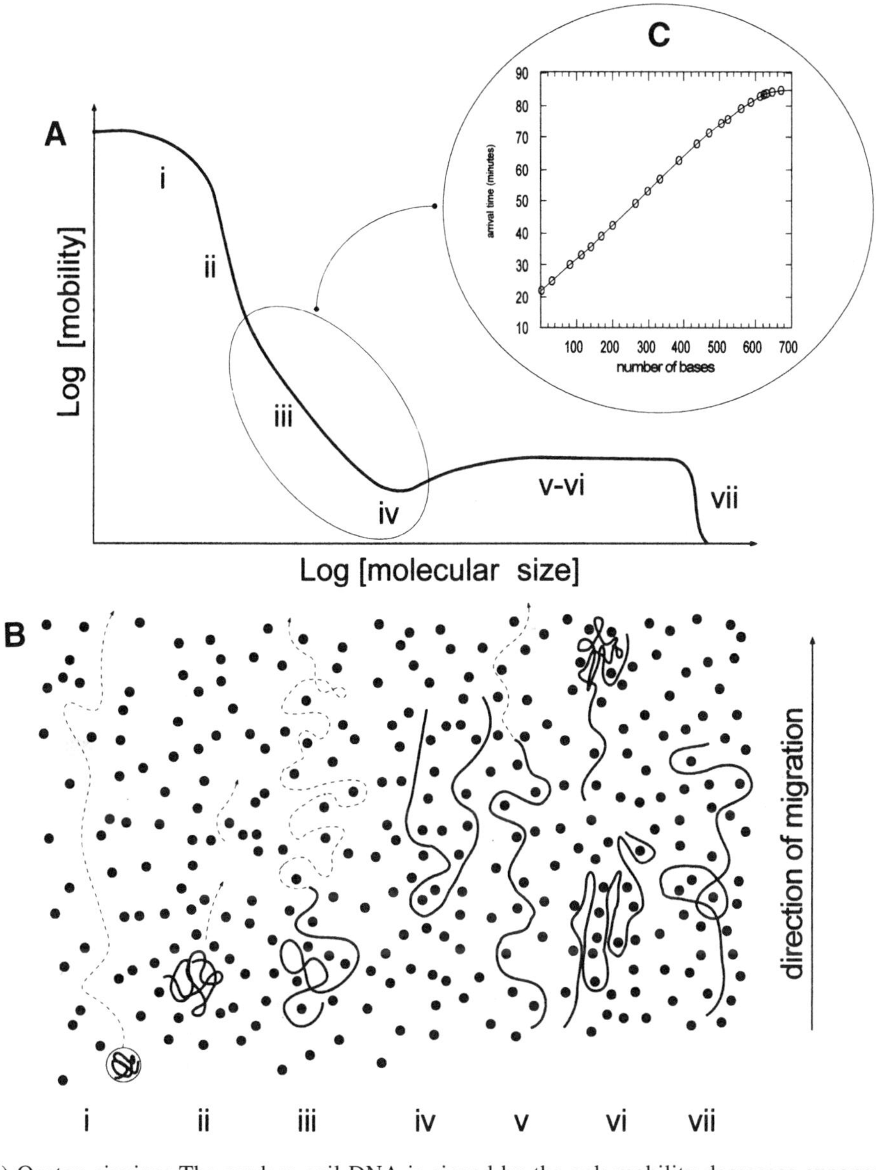

(i) Ogston sieving: The random coil DNA is sieved by the gel; mobility decreases exponentially with molecular size. (ii) Entropic trapping: The random coil DNA jumps between the larger pores. (iii) Near-equilibrium reptation: The random coil DNA migrates head-first through the gel; there is an approximate inverse relationship between mobility and size. (iv) Reptation trapping: The DNA molecule becomes trapped in U-shaped conformations and is slowed by the tug-of-war between the two arms. (v) Oriented reptation: The DNA is aligned in the direction of the electric field and reptates head-first. (vi) Geometration: For high field intensities, migration may be characterized by "hernias" and "bunching" instabilities. (vii) Complete trapping: Very large DNA molecules do not migrate through gels, possibly because they form knots around gel fibers. **(C)** Plot of the arrival time of DNA bases vs the DNA size for a typical DNASCE experiment. Reproduced from **ref. *50*.**

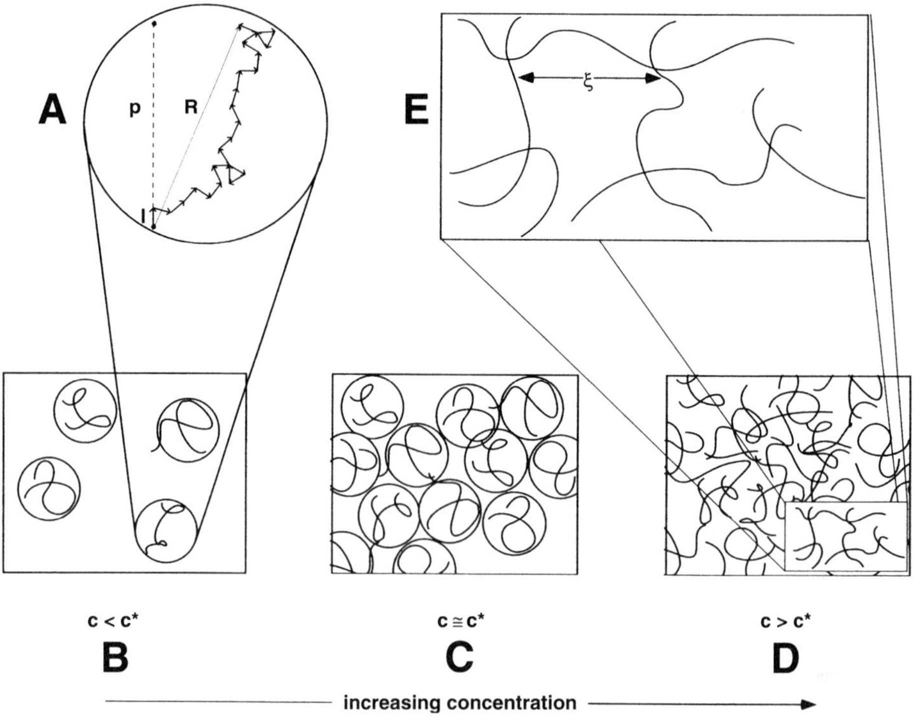

Fig. 2. Three polymer concentration regimes are used to distinguish dilute polymer solutions **(B)** from entangled semi-dilute polymer solutions **(D)** and the overlap concentration (c^*) where the polymer coils just begin to be densely packed **(C)**. C^* occurs when polymer concentration is increased to the point where each volume of the order R_G is occupied by exactly one polymer *(36,42,43)*. A single polymer chain is shown in a particular conformation in **(A)**; the average projection of the end-to-end vector, R, on the first bond of the chain, l, gives the persistence length, p. **(E)** is a schematic illustration of the average entangled mesh size, ξ, associated with semi-dilute polymer solutions (also known as the correlation length *[36,42]*) through which single-stranded DNA migrates under the influence of an electric field during DNASCE.

remarkably good qualitative, if not completely quantitative, results. First, we will review polymer solution properties to develop important attributes of semi-dilute polymer solutions in which DNASCE separations take place.

3. Sieving Media and Polymer Solution Properties

3.1. Single Polymer Chain Properties

A simple description of a single long-chain polymer in solution is the "freely jointed chain" model proposed by Kuhn in 1936 *(35)*. The chain is assumed to consist of *n* bonds of equal length *l*, joined in linear succession, where the directions of individual bonds can assume all values with equal probability (*see* **Fig. 2A**). The configuration of the polymer in this case is the same as a random walk on a three-dimensional lattice (also referred to as the random flight model). The average distance R between the two ends may be calculated with these assumptions and used as a measure of polymer size:

$$R \cong n^{1/2}\, l \qquad (1)$$

This holds even when considering the effect of short-range interactions which prevent neighboring bonds from folding back on themselves. The radius of gyration R_g is a more convenient way of expressing the size of a polymer because it can be experimentally measured. The radius of gyration represents the average distance between polymer segments and the center of mass of the polymer, and is related to R by:

$$R_g^{\,2} \cong R^2/6 \qquad (2)$$

Still, the model does permit a long chain to loop back on itself resulting in widely separated segments occupying the same volume. When the steric hindrance from this "excluded volume effect" is accounted for, however:

$$R_g \cong n^{\nu}\, l \qquad (3)$$

in which ν is approx 0.588, a value obtained from extensive investigation and numerical simulations ***(36)***. **Eq. 3** provides reasonable estimates of polymer size for large polymers. For example, we may compare the radius of gyration obtained from light scattering studies ***(37)*** of two dilute polyacrylamide polymers (2.33×10^6 and 21.3×10^6 Dalton) dissolved in 1 *M* NaCl solutions with values obtained from equation three. In this case, the measured values of 86.4 nm and 246 nm compare well with the corresponding calculated values of 69 nm and 255 nm respectively, assuming the vinyl bond length is 0.154 nm ***(38)***.

3.2. Solvent Effect

The above equations assume that the polymer is dissolved in a "good solvent" which implies the polymer is easily dissolved because of its' high affinity with the solvent and that the polymer configuration will be greatly extended. In a solvent that poorly dissolves the polymer, the configuration will be more shrunken or collapsed. To account for these solvent effects, adaptation of Flory models ***(36,39)*** provides a useful relationship:

$$R_g \cong n^{\nu'}\, l \qquad (3\text{-}1)$$

where

$$\nu' \cong \nu\,(1 - \theta/T) \qquad (4)$$

in which ν is again the excluded volume exponent 0.588 and θ is the so-called "theta temperature." Each specific polymer-solvent combination has a different theta value ***(40)***. In general, θ is small in good solvents and large in poor solvents. Strongly hydrophobic polymers dissolved in aqueous systems are one example of a polymer-solvent system with poor theta values ***(41)***.

3.3. Chain Stiffness

The extension of a single polymer chain in solution is greatly affected by its stiffness: the stiffer the chain, the more readily it is extended. Two often-used measures of chain stiffness are a polymer's persistence length, p, or its characteristic ratio, C_∞. The persistence length is a measure of the length over which the overall chain "persists" in the same direction as the first bond (**Fig. 2A**). In the limit of a fully rigid rod, the persistence length is equivalent to the polymer's overall length. The characteristic ratio of a large polymer is the value more often found in tables (*see* **ref.** ***38***) and represents the ratio of its measured end-to-end length to that predicted by the free-chain model, $(R^2_{msrd})/nl^2$. The simple relationship between p and C_∞, $(2p/l = C_\infty + 1)$, provides physical insight connecting the characteristic ratio with chain stiffness ***(20)***: namely, the larger the persistence length, p, or the characteristic ratio, C_∞, the stiffer the polymer chain. Persistence length values for certain polymers used in DNASCE are listed in the last column of **Table 1**.

4. Different Polymer Concentration Regimes

4.1. From Dilute to Semi-Dilute Polymer Concentration

It is instructive to consider three polymer solution topologies to characterize the entangled polymer media used for separating and sequencing ssDNA ***(42)***. **Figure 2** illustrates the three regions. **Figure 2B** shows a "dilute" polymer solution regime where the polymer coils are separate. **Figure 2D** shows a more concentrated, "semi-dilute" regime where the polymer coils overlap. **Figure 2C** illustrates an intermediate region between the two regimes where the concentration is such that the coils just begin to touch and become enmeshed ***(42)***. This concentration at which the coils just begin to touch is referred to as the overlap concentration, c^*. The enmeshed semi-dilute region of **Fig. 2D** is characterized as a dynamic network of entangled polymers with an average mesh size ξ (**Fig. 2E**). The mesh size reflects the average distance between polymer-polymer entanglements. It is important to note that experiments with successful polymer systems ***(1–7,10)*** used for DNASCE have all employed "semi-dilute" polymer concentrations—the concentration range characterized by polymer entanglement (*see* **Tables 1** and **2**).

4.2. Estimation and Measurements of the Overlap Concentration (c^*)

4.2.1. Predicting the Onset of Semi-Dilute Polymer Concentration for DNA Sequencing

To determine if polymers in a candidate solution are entangled, it is necessary to measure or estimate the overlap concentration. It is also informative to estimate c^* when we compare the conditions that workers use for high performance DNASCE (*see* **Table 2**). The overlap concentration (c^*) occurs when the concentration is increased to the point where each volume of the size $4\pi R^3_G/3$ is occupied by exactly one polymer ***(36,42,43)*** (*see* **Fig. 2C**). This definition of c^* provides an estimate that lies near the high end of its range:

$$c^* \cong M_{wt}/(N_A \cdot R^3_G) \tag{5}$$

in which M_{wt} is the molecular weight and N_A is Avogadro's number ***(43)***. This definition however, requires that the molecular weight and radius of gyration be determined directly. An easier approximation of c^* ***(2,22,34,44,45)***, allowing comparison of different polymer-solvent systems, involves the well-known dependence of intrinsic viscosity on molecular weight known as the Mark-Houwink-Sakurada (MHS) equation:

$$\eta = K \cdot M^{a}_{wt} \tag{6}$$

in which, η is the intrinsic viscosity, and K and a are constants for a specific polymer in a specific buffer system. The values of K and a are relatively insensitive to temperature when "a" exceeds 0.70 and may be used in a 10°C range on either side of the temperature at which the constants were determined *(38)*. In the polymers studied in this chapter, all had values of a > 0.70. *See* **refs. *13*** and ***38*** for a definition of intrinsic viscosity.

Using a hard sphere model of intrinsic viscosity ***(22,42,44)***:

$$\eta = 2.5 \cdot R^{3}_{G} \cdot N_{A} / M_{wt} \tag{7}$$

with **Eqs. 5** and **6** we may relate c^* to molecular weight as:

$$^{mhs}c^* \cong (2.5/K) \cdot (M_{wt})^{-a} \tag{8}$$

in which, $^{mhs}c^*$ refers to c^* obtained with MHS constants. The utility of this approach is that large compilations of MHS constants for different polymer solutions are available *(38)*.

4.3. Mesh Size Property of Semidilute Polymer Systems

4.3.1. Concentration Dependence, Estimation of Mesh Size for Different Polymer Systems

Figure 3 illustrates some important features of the physical mesh properties of entangled polymer solutions. Recall that the mesh size reflects the average distance between polymer-polymer entanglements. Experiments that are sensitive to concentration inhomogeneities and density fluctuations in semi-dilute polymer solutions (e.g., light scattering, small angle X-ray scattering, and neutron scattering) can yield useful spatial and temporal correlation information reflecting the polymer-polymer entanglements. Small-angle neutron scattering was used to measure the mesh size, ξ, of linear polyacrylamide at different concentrations above c^* (*see* **Fig. 2**) ***(46)***.

The measured dependence (*see* **Fig. 3A**) is:

$$\xi = 2.09 \cdot C^{-0.76} \tag{9}$$

This confirms the –3/4 dependence predicted from "scaling" arguments in polymer physics (ξ is the mesh size, in angstroms, and C is the polyacrylamide concentration in gm/mL) ***(42)***. The same measured mesh size was obtained for polyacrylamide solutions prepared in either water or in sequencing buffer (*see* **Fig. 3A**). This underscores the near universal $C^{-.75}$ dependence of mesh size in a variety of semi-dilute polymer solutions ***(42,47)***. Note that the prefactor in **Eq. 9** has a magnitude similar to the segment length of polyacrylamide (2.19 Å) ***(38,46)***. In general, a given polymer type (e.g., poly[dimethyl acrylamide]) used to prepare a semi-dilute polymer solution will obey the mesh-size-concentration dependence of **Eq. 9**, i.e., it will have the form, $\xi = \text{constant} \cdot C^{-0.75}$.

Table 1
A Compilation of Some Useful Physical Properties of Successful Polymer Systems Used in DNASCE

Polymer	Structure & Monomer M_{wt}	Mark-Houwink Sakurada constants & temperature *(a)*	Comparison of MHS predictions of c^* with measurements *(b)*: $^{mhs}C^*_{M_{wt}}$	$^{Measured}C^*_{M_{wt}}$	Method	$^{mhs}C^*$ prediction @ $M_{Wt} = 10^6$ or, $n = 14{,}000$ *(c)*	$b_{estimate}$ *(d)*	"chain" "stiffness" p *(e)*
HEC Hydroxyethyl cellulose	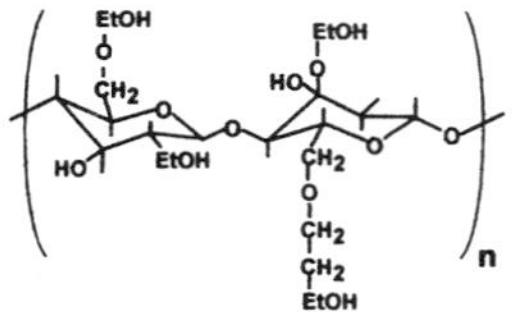 ~ 272	$K = 9.53 \cdot 10^{-3} \frac{ml}{gm}$ $a = 0.87$ $T = 25°C$	0.66% (w/v)	~ 0.35% (w/v)	Limiting viscosity [34]	0.16% (w/v) @ $M_{wt} = 10^6$ 0.05% (w/v) @ n=14000, $M_{wt} = 3.8\ 10^6$ D	5.99 Å (valid @ 25 °C in water)	8.3 nm
			$M_{wt} = 191800$					
LPA Polyacrylamide	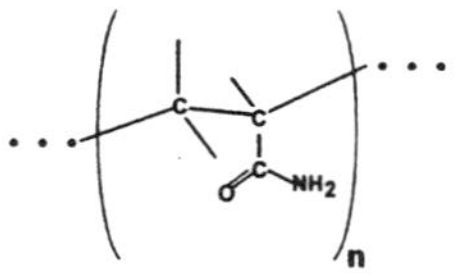 71	$K = 6.31 \cdot 10^{-3} \frac{ml}{gm}$ $a = 0.80$ $T = 30°C$	0.32% (w/v)	~ 0.60% (w/v)	Light scattering msrmnts of R_G & M_{wt} [43]	0.63% (w/v) @ $M_{wt} = 10^6$ 0.63% (w/v) @ n=14000, M_{wt} = 994,000D	2.09 Å (msrd -) 2.19 Å (valid @ 30 °C in water)	1.73 nm 30 °C in water
			$M_{wt} = 2.33 \cdot 10^6$					
PDMA Poly N,N - Dimethyl Acrylamide	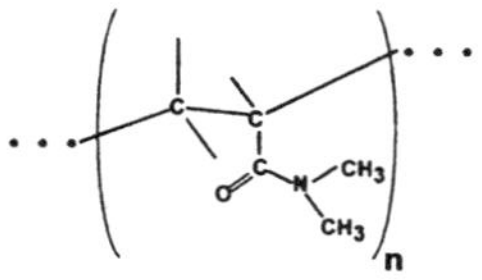 99	$K = 23.2 \cdot 10^{-3} \frac{ml}{gm}$ $a = 0.81$ $T = 25°C$	0.96% (w/v)	~ 0.50% (w/v)	Limiting viscosity [8]	0.15% (w/v) @ $M_{wt} = 10^6$ 0.114% (w/v) @ n=14000, $M_{wt} = 1.386\ 10^6$ D	2.17 Å (valid @ 25 °C in water)	1.1 nm 25 °C in water
			$M_{wt} = 100000$					
PEO Poly(ethyleneoxide)	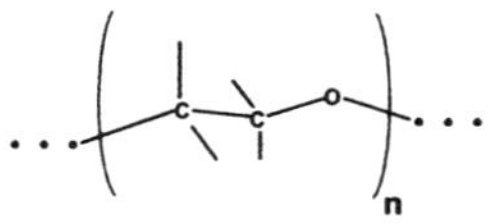 44	$K = 12.5 \cdot 10^{-3} \frac{ml}{gm}$ $a = 0.78$ $T = 30°C$	5.7% (w/v)	~ 6% (w/v)	Limiting viscosity [8]	0.42% (w/v) @ $M_{wt} = 10^6$ 0.61% (w/v) @ n=14000, M_{wt} = 616,000 D	(data only available in acetone)	0.65 nm 40 °C in .35M K_2SO_4
			$M_{wt} = 35000$					

[a] T is the temperature at which the measurements of the MHS constants, "K" and "a" were made. The values of "K" and "a" are relatively insensitive to temperature when "a" exceeds 0.70 and may be used in a 10°C range on either side of the temperature at which the constants were determined. All measurements in column three were made in water.

[b] For example, ${}^{mhs}C^*_{191,800}$ refers to the estimated value of the overlap concentration of **Eq. 8** with the HEC MHS constants and a molecular weight of 191,800. The ${}^{measured}C^*_{191,800}$ refers to the measured value of the overlap concentration according to the limiting viscosity method in **ref. *49***.

[c] These entries refer to the estimated value of the overlap concentration of **Eq. 8** with MHS constants and a molecular weight of 10^6 or $n = 14{,}000$ repeated monomer units for the four polymers listed.

[d] This column represents the associated polymers' segment length, b (Å). R_{of}/n was used as an estimate of the segment length following Kurata's nomenclature ***(38)***.

[e] These values represent the persistence length, p, taken from Kurata's Tables ***(38)*** except for HEC. The HEC persistence length was taken from **ref. *48***. The "chain stiffness" was appended to emphasize how persistence length reflects that property.

Table 2

A Compilation of Different Experimental Conditions Used for High Performance DNASCE with Replaceable Polymers

Polymer	Ref	M_{wt}	mhsC^*_{Mwt} (w/v) *(a)*	Concentration used (w/v) *(b)*	Electric Field (V/cm)	Sieving length (cm)	Internal diameter (μm)	Buffer & Denaturants *(c)*	Temp. (C°)	Cap. Coating *(d)*	Sample Load Conditions *(e)*	$R_{apparent} \cong 0.6$ (bases/min) *(f)*	LOR (bases) *(g)*
HEC	6	100,000	1.17%	2.0%	190	41	75	- 1xTBE - 6M Urea - 10% formamide	26	LPA type A	~15 sec @ 190 V/cm	~510 bases / ~62 min	500 - 600
LPA Molecular Dynamics	89	7,000,000	0.13%	3.0%	155 (fast) 132 (slow)	65	75	- Tris - Taps - EDTA - 7 MUrea	44	LPA type B	~20 sec @ 155 V/cm	~ 650 - 750 bases - time not provided	—
LPA	46	1,000,000	0.63%	8.0%	100	12	100	- 1xTBE - 3.5M Urea - 30% formamide	Rm.	LPA type A	~20 sec @ 154 V/cm	~375 bases / ~84 min	—
LPA	8	50,000 + 13,000,000	0.08% with 13,000,000	0.5% with 50,000 & 2% with 13,000,000	200	30	75	- 50 mM Tris - 50 mM Taps - 2 mM EDTA - 7M Urea	60	PVA	~10 sec @ 25 V/cm	~700 bases / ~40 min	—
LPA Beckman high throughput mode	90	7,000,000	0.13%	3.0%	155	50	100	- Tris - Taps - EDTA - 7 M Urea	42	PVA	~60 sec @ 38 V/cm	~600 bases / ~110 min	600 run duration limited in this mode
LPA Beckman long-read mode	90	7,000,000	0.13%	3.0%	132	50	100	"	35.	PVA	~60 sec @ 38 V/cm	~730 bases / ~165 min	860
DMPA Applied Biosystem	*	800,000	0.18%	6.5%	105	50	50	- 100 mM TAPS - 100 mM Tris - 5% Pyrollidone - 8 M Urea	50	Self Coat	~40 sec @ 50 V/cm	~550 bases / ~170 min	600
PEO	5	600,000 + 8,000,000	0.6% with 600,000 & 0.08% with 8,000,000	1.4% 600,000 + 1.5% 8,000,000	267	35	75	- 1xTBE - 3.5M Urea	Rm.	Self Coat	~12 sec @ 133 V/cm	—	—

Note: $R_{apparent}$ was determined by the authors' observation of where adjacent DNA peaks (of the same base type) in an electropherogram consistently appeared having a resolution, R, equal to 0.6. Gaussian peaks representing resolution equal to 0.6, 1.0, and 1.5 are shown to the right. The relationship, $R = \Delta t/\overline{w}$, was recast as $R = \Delta t/(1.6986*\text{fwhm})$ assuming the Gaussian's average base width, $\overline{w}$, is measured when the peak has diminished to $1/e^2$ of it's original value ($R \cong \Delta t/4\sigma$). Fwhm corresponds to the full width half maximum of the Gaussian, and Δt, the peak spacing.

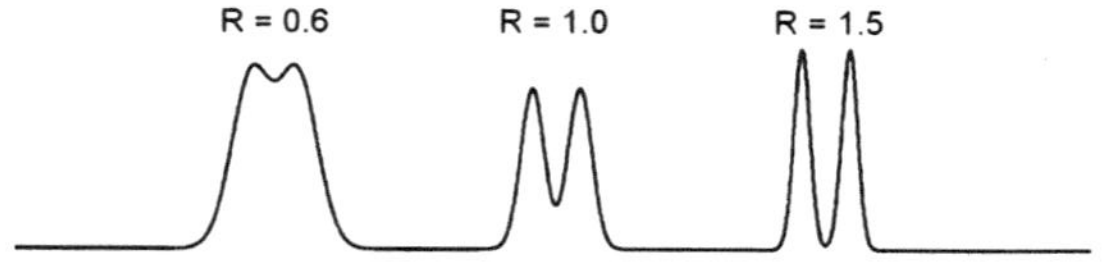

[a] ${}^{mhs}C^*_{Mwt}$ refers to the estimated value of the overlap concentration according to **Eq. 8** for the polymer and molecular weight listed in columns 1 and 3 and the appropriate Mark-Houwink-Sakurada constants (*see* **Table 1**).

[b] The polymer concentration values actually used for the DNASCE experiment. Concentration units are in % g of polymer per mL of buffered solution.

[c] All buffered and denaturing solutions had a pH near 8.

[d] The capillary used for each DNASCE experiment had one of the following coatings of the capillary inner wall. “LPA type A” refers to a linear polyacrylamide coating made according to Hjertén et al. ***(74)***, “LPA type B” refers to a linear polyacrylamide coating applied according to Cobb et al. ***(73)***, and “PVA” refers to a poly(vinyl alcohol) coating. “Self coat” refers to the inherent self coating properties of PDMA ***(2)*** and PEO ***(5)*** on silica surfaces that suppresses electroosmotic flow (*see* **Subheading 6.4.**).

[e] These entries are the electrokinetic sample injection conditions used in the DNASCE experiment, and refer to the amount of time (s) that the capillary end was immersed in the DNA sample using the indicated electric field strength (V/cm).

[f] A simple criterion is applied homogeneously to compare the “performance” between different experiments. The criterion is the largest DNA fragments for which an apparent resolution $R_{apparent}$ of 0.6 is consistently observed, is quoted in the numerator of (bases/min). The denominator referred to the time after sample injection, at which the criterion is applied. The meaning of apparent resolution is described in the bottom 2 panels. The sequencing primers, -21-M13 forward primer (ABI-Perkin-Elmer; Sequitherm-Epicentre) and M13(-40) primer (USB-Amersham) used for sequencing of m13mp18 DNA affects the length of DNA that is actually sequenced and quoted, hence all read lengths have been referenced to total physical length of DNA (i.e., counting from the 5'-end of the primer used for generating each DNA sample).

[g] Grossman ***(10)*** has defined a measureable, length-of-read parameter (LOR) that was related to the last base in a sequence for which the peak spacing is still greater than or equal to the average peak width. Those workers using this method of analysis are listed in the column. The symbol, *, in the reference column refers to information provided by one of the authors, Steve Menchen.

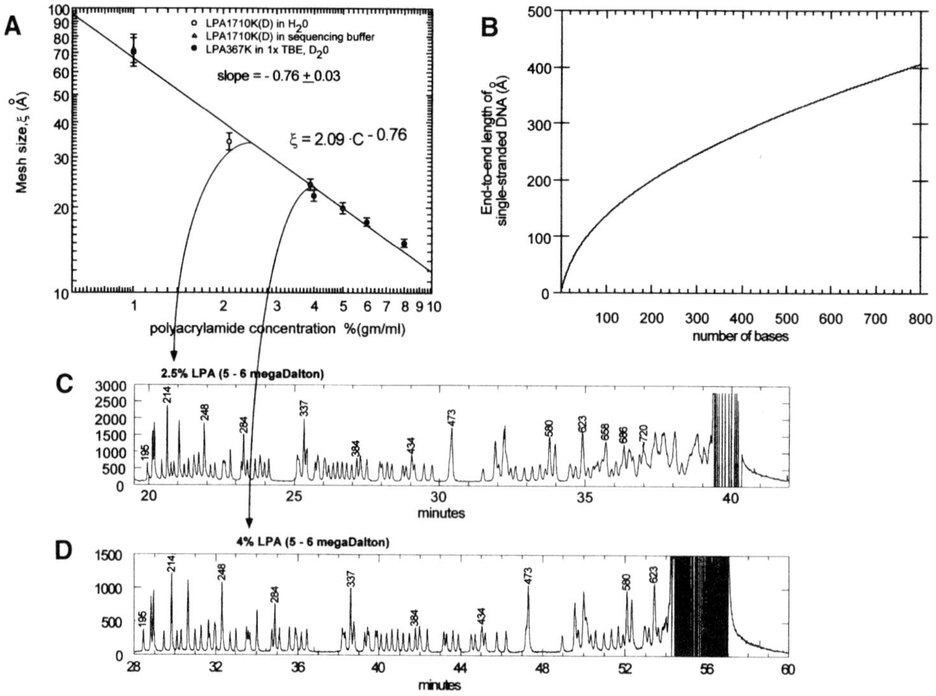

Fig. 3. Illustration of some physical mesh properties of entangled polymer solutions. **(A)** The measured values of mesh size (in angstroms, Å), obtained from small angle neutron scattering studies *(46)*, are plotted against polyacrylamide concentration (in gm/mL). The abscissa has been relabeled in % g/mL for easier comparison with experimental data. LPA1710K (D) was deuterated polyacrylamide with a molecular weight of 1.71×10^6 dissolved in water (open circles) or sequencing buffer (open triangles); LPA367K was a polyacrylamide with a molecular weight of 367×10^3 dissolved in a 1x TBE-D_2O solution (filled circles). The mesh size was independent of molecular weight but was proportional to $C^{-.75}$, the scaling law prediction for entangled polymer solutions in an athermal solvent *(42)* (C is in g/mL, not % g/mL). The factor 2.09 is similar to the segment length of polyacrylamide (2.19 Å) *(38)*. **(B)** The end-to-end length of ssDNA (in angstroms, Å) is plotted against DNA size according to the Porod-Kratky stiff chain model *(20)* assuming a persistence length of 30 Å *(21)* and a segment length of 3.4 Å per base. This plot is simply intended to give the reader a sense of the magnitude of dimensions at work between ssDNA and polymer solutions. No relationship between reptating ssDNA and the stiff chain model is implied, nor intended. **(C)** DNASCE electropherogram of "T" dye-labeled DNA (pGEM) run in a 2.5% (w/v) polyacrylamide buffered solution is shown near the unresolved large DNA fragments (high frequency vertical lines at end of trace). The number of DNA bases contained in a peak relative to a common "priming" site are indicated by the numbers above each peak. **(D)** Another DNASCE electropherogram using the same DNA sample was run in a 4.0% (w/v) polyacrylamide under identical conditions. A comparison of doublet and triplet T separations in the region between 195 and 384 bases in (C) and (D) shows that highly entangled, concentrated solutions provide better separation of the longer fragments *(6,34,46)*. Less concentrated solutions, however, could still separate longer fragments near the end of a run, albeit with poor resolution.

It is also observed that different molecular weights of the same polymer yield the same mesh size as long as each polymer is at the same concentration and well above c^*. This does not imply that polymer molecular weight plays no significant role in DNASCE, in fact, there is ample evidence to suggest that it does play a role. Carrilho et al. ***(7)*** and Wu et al. ***(46)*** have shown a pronounced reduction in peak broadening by use of increased polymer molecular weight. Consistent with the mesh size vs concentration dependence, we also observe separation of higher DNA fragments as the mesh-size was enlarged by reducing polyacrylamide concentration from 4–2.5% (*see* **Fig. 3C,D**). As shown in **Fig. 3**, highly entangled solutions (**Fig. 3D**) provide better resolution of DNA fragments than more dilute solutions (**Fig. 3C**), over a narrower range of fragment sizes ***(7,34,46)***.

We might estimate similar mesh-size—concentration relationships for other polymer systems using:

$$\xi = b \cdot C^{-0.75} \tag{10}$$

in which "b" is the segment length ***(42)*** determined from Kurata's Tables ***(38,46)***. Here we rescaled $R_{of}/M^{1/2}$ to R_{of}/n from Kurata's Tables as an estimate of b. Values for b are tabulated in **Table 1**. For example, a reasonable estimate of how the mesh-size varies with hydroxyethyl cellulose concentration is given by $\xi = 5.99 \cdot C^{-0.75}$, where ξ is again in angstroms (Å). The dependence of mesh-size with aqueous polyethylene oxide (PEO) concentrations were not possible since data was only available for PEO immersed in acetone at room temperature. To summarize, the design of a separation medium depends on the control of mesh-size of the selected polymer. As shown above, one can select among the parameters of polymer segment length and polymer concentration to obtain a desired mesh-size.

4.4. Attributes of the Polymers Currently Used for DNA Sequencing

Table 1 compares the predictions of **Eq. 8** (*see* **Table 1**, columns 4 and 5) with measured values of the overlap concentration based on viscosity and light scattering experiments for four polymers used in DNASCE experiments ***(1–10)***. The viscosity measurements are based on departures from linearity in log-log plots of viscosity versus polymer concentration ***(2,48,49)***. Static light scattering experiments are used to determine the radius of gyration and molecular weight of 1 *M* NaCl polyacrylamide solutions ***(37)***. The values obtained from **Eq. 8** using viscosity measurements should be regarded as approximate due to the influence of different solution solvation properties ***(44,48)***. They tend to lie at points intermediate between the hydrodynamic and the static light scattering measurements for the polymer solutions examined (within a factor of 1.7). The radius dimensions obtained from diffusion measurements vs those obtained from static light scattering experiments can differ substantially by factors as high as 2.5 for dilute polyacrylamide solutions ***(37)***. For example, polyacrylamide (LPA) is more hydrophilic than poly(dimethyl acrylamide) (PDMA), yet, if one compares the overlap concentration ($^{mhs}C^*$) predicted for each polymer system (**Table 1**, column 7, 0.63% for LPA vs 0.15% for PDMA), one might conclude that LPA is more hydrophobic. This is not possible because hydrophilic polymers are more extended in solution than comparably sized hydrophobic polymers and therefore become entangled at lower concentrations.

It is intended that **Table 1** be used for estimation purposes and rough comparisons. For example, **Table 1**, column 7 shows the overlap concentration ($^{mhs}C^*$) predicted for each polymer system assuming the polymers had either a molecular weight of 1 MDa or contained 14,000 repeating units. The large size of hydroxyethylcellulose (HEC) repeating monomer unit clearly manifests itself with low $^{mhs}C^*$ values (0.05%) compared with the other polymers. It is difficult to make a substantive argument concerning the difference between chain stiffness (persistence-values) values obtained for LPA and PDMA from Kurata's Tables alone *(38)* (*see* last column of **Table 1**). The variation in viscosity measurements is due to the different temperatures at which the measurements are taken (e.g., LPA at 25°C, and PDMA at 30°C), which affects the measured persistence-length (chain stiffness) value. In the following subheadings, we will compare actual DNASCE operating conditions for different polymer systems to better appreciate the role of these effects.

5. DNA Migration in Polymer Solutions

5.1. Mobility and Separation–Theoretical Background

Basic "reptating" migration of DNA in a semi-dilute polymer solution is illustrated in **Fig. 4**. Under the influence of an electric field, a flexible DNA chain moves through a polymer network (**Fig. 4A**) modeled by fixed obstacles O_1, O_2, and so on. These obstacles may be thought of as a two-dimensional representation of polymer entanglements. Whereas the DNA chain cannot cross the obstacles, it can execute an undulatory motion that effectively moves mass forward, snake-like, in the electric field direction. Since the DNA cannot cut through the network obstacles, it may be thought to undergo this motion in a tube of order ξ (**Fig. 4B**). As time passes, the DNA is able to wriggle partially out of the tube region from α to β in **Fig. 4B** into a newly created tube region from γ to δ. De Gennes called this type of motion "reptation." This simplified picture of reptation may be described at a finer level of detail with the so-called "blob picture" (*see* **Fig. 4C**) *(42,44)*. At low electric fields, the DNA ends are weakly biased and orient the new tube sections (created during reptation) toward the field direction. The motion in the tube is characterized by a mobility $\mu_,$ inversely proportional to the length of the DNA chain. At higher electric fields, the leading terminal of the DNA fragment approaches complete alignment with the field such that no further field effects occur, i.e., the field effect becomes saturated. The relationship between mobility (μ) and size dependence (N) has been concisely stated *(50)*:

$$\mu \propto \mu_o \cdot \xi^2 \cdot [1/(3N) + e^2/27 + \ldots] \quad (11)$$

in which, μ_o is the DNA free-solution mobility, ξ is the mesh size, N is the DNA base-size and e is an electric-field reduced variable, $e = E/E_\xi$. E_ξ is the electric field intensity for which the drop in potential energy of a molecule of size R_g across mesh size ξ is equal to the thermal energy ($E_\xi = 2kT/qR_g$, q ~ effective charge). The exponent of e in **Eq. 11** is typically less than 2 *(34)*. The appropriate mesh size according to the "blob picture" is simply 2.86 times the mesh size from neutron scattering studies *(22,34,44)* (*see* **Fig. 4C**).

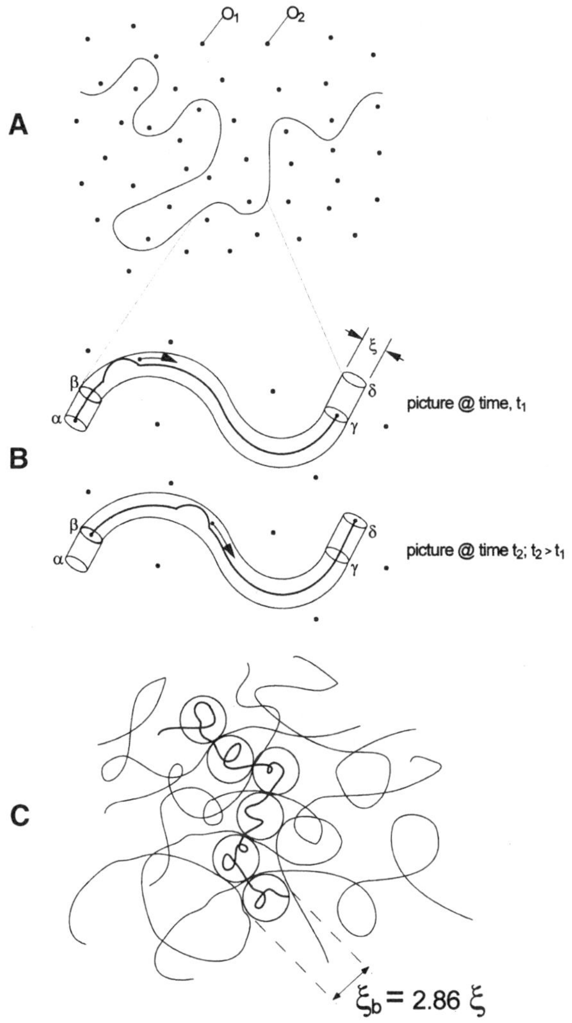

Fig. 4. Schematic descriptions of biased reptation movement of DNA in entangled polymer solutions. **(A)** Random lattice description of DNA reptation: A flexible DNA chain moves through a polymer network modeled by fixed obstacles, O_1, O_2, and so on, under the influence of an electric field. Theses obstacles may be thought of as a two-dimensional representation of polymer entanglements. Although the DNA chain cannot cross the obstacles, it can execute an undulatory motion that effectively moves mass forward, snakelike, in the electric field direction ***(36,42)***. **(B)** "Tube" description of DNA reptation: Since the DNA cannot cut through the network obstacles, it may be thought to undergo reptation in a tube of order ξ. As time passes, the DNA is able to wriggle partially out of the tube region from α to β into a newly created tube from region γ to δ. **(C)** "Blob" description of DNA reptation: At a finer level of detail in which the DNA chain is visualized as a succession of independent subunits, or "blobs" of size ξ_b undergoing reptation through the polymer network. Inside one blob, the DNA does not interact with other portions of the chain ***(36,42,44)***.

In practice, one strives to maintain the effect of the first term in **Eq. 11**, which gives the linear relationship between arrival time at a fixed detector for a given DNA base size (*see* **Fig. 1C**). Loss of separation at high electric field occurs because of the second term in **Eq. 11**, which places a limit on the speed (electric field strength) used to sequence DNA. The general optimization strategy that workers have adopted becomes clear: delay the onset of the high field, for which there is no DNA size dependence of mobility, by reducing the electric field. This effectively extends the size dependent region for longer read lengths with the trade-off of longer run times and increased peak widths due to diffusion *(51)*. The spacing between DNA peaks obtained from **Eq. 11** ($\partial\mu/\partial N \propto \xi^2/N^2$, to first order), clearly shows the rapid loss of separation for the larger DNA bases. It also shows the basis of the well-known strategy of diluting gels (polyacrylamide or agarose) to separate long DNA fragments and for the use of more concentrated gels to separate smaller fragments.

A successful strategy for increasing throughput and read-length while maintaining resolution has been developed recently *(7,8)*. It involves preparing higher molecular weight polymers (to inhibit peak broadening) at lower polymer-solution concentration (forming larger mesh size and promoting separation of larger DNA molecules), combined with electrophoresis at elevated temperature (for faster separation, reduced peak broadening and improved separation of larger DNAs). For LPA, it was found that the separation characteristics of the medium could be maintained or enhanced with very low concentration of the polymer by substantially increasing the molecular weight of the polymer molecules *(7,8)*.

5.2. Resolution and Performance

In general, one can optimize conditions for longer read-length, or for shorter analysis time. Only a finite number of parameters are typically at one's disposal: length of capillary (L), electric field used (E), temperature (T), choice of polymer, polymer concentration, and molecular weight. The overall read-length is determined by the resolution between individual DNA bases (peaks). **Figure 5A** shows that the resolution is determined by essentially two components: the peak separation (mobility) and the peak width (peak-broadening or dispersion). **Figure 5A** also illustrates that three resolution regimes can be defined:

1. High resolution regimes, in which output bands comprise single peaks only, consisting of a single DNA base-size ($R \geq 1.0$),
2. Intermediate resolution regimes where peaks begin to coalesce, but the tops of the peaks can still be identified within a multi-peak band ($R = 0.6$–1.0), and
3. Low resolution regimes in which the peaks coalesce into a single featureless band ($R \leq 0.5$).

Very accurate base identification can be made in the areas of high resolution. In areas of intermediate resolution, base identification requires use of routines that identify slope changes within a band. These routines can be robust or may be poor depending on the quality of the algorithm and the ease of recognition of peak relief within the multi-peak band or envelope. Generally, the procedures or reagents yielding data with peaks of equal intensity also aids in the identification of bases within the intermediate resolution region. Within the low resolution regimes, the base composition can be

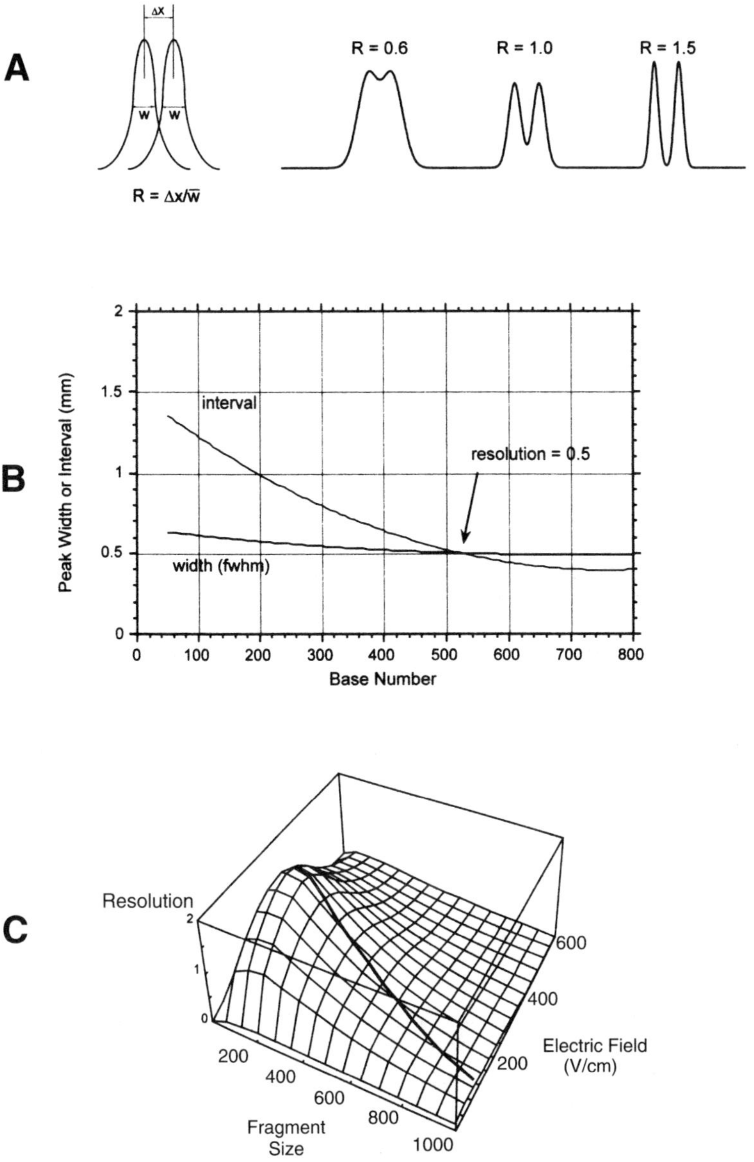

Fig. 5. **(A)** The leftmost peaks show the two attributes used to describe the fundamental dependences of resolution R: Δx, peak separation and w, peak width (R = Δx/w). To the right of this are three doublets showing the resolution morphologies of R = 0.6, 1.0, and 1.05. **(B)** A cross-over plot (*see* **Subheading 5.3.**). **(C)** Three-dimensional representation of the resolution predicted from **Eq. 15**. The resolution is plotted vs DNA fragment size and the electric field strength. The curve superimposed on the surface denotes the maximum resolution for each fragment size and shows the gradual decrease in the optimal field strength as the fragment size increases. Reproduced with permission from **ref.** *55*, Copyright (1993) American Chemical Society.

determined only with the use of algorithms that can deconvolute the individual peaks from a featureless envelope of peaks. Accurate base determination within these regimes generally requires extensive prior knowledge of peak broadening, peak separation, and mobility information. In general, the accuracy of identification of a particular base depends on the resolution regime that the base occurs in, the quality of the algorithm used for that regime, and the ability to identify and discriminate between regimes within an electrophoresis experiment.

5.3. Crossover Plot

One way to quantify the resolving power of a medium is with a "crossover plot" *(52)*, of the type shown in **Fig. 6A**,**B**. These plots convert temporal data into spatial data, and compare the peak to peak separation ("interval") to the peak-width data (width) as a function of base size. The limit of resolution is the point at which the peak separation equals the peak-width. This point is referred to as the crossover point. Crossover plots can be used to identify the polymer sieving characteristics throughout the range of the desired separation, and predict where the resolution regimes are for application of the appropriate algorithm routines. In addition, these plots can be used to identify how the components of resolution (peak-width and peak separation) are affected by electrophoresis variables.

The dominant source of peak broadening in capillary gel electrophoresis is longitudinal diffusion (σ_D) with a smaller contribution from sample electrokinetic injection σ_i *(51,53)*. These electrokinetic contributions may be neglected provided peak broadening is minimized by reducing injection time (typically $\leq$ 60 s) or even reducing the physical sample size *(54)*. An expression that illustrates the dependence of resolution, R, on the capillary length, L, the electric field, E, the separation time, t, and the DNA size, N, is:

$$R = \Delta x/(4\sigma_t) \cong \Delta\mu E{\cdot}t/(2 \cdot \sqrt{Dt}) = \Delta\mu{\cdot}L/(2\mu \cdot \sqrt{Dt})$$

with,

$$\sigma^2_t = (\sigma^2_D + \sigma^2_i + \sigma^2_T + \ldots) \qquad (12)$$

in which, "D" is the longitudinal diffusion coefficient dependent on L, E, T, N, mesh size, and arguably polymer molecular weight *(13,53)*. It should be noted that Slater et al. *(53)* have shown that the Einstein relationship between mobility and the diffusion constant (an equilibrium condition) is invalid under most relevant experimental conditions.

The general dependence of resolution on electric field strength and DNA fragment size is illustrated in the three-dimensional plot (**Fig. 5C**) for a typical set of experimental conditions. It also illustrates an optimal range of field values of ~ 150–250 V/cm *(55)*. The optimal range undoubtedly shifts under differing parameter values of L, T, N, mesh size and polymer molecular weight. While some have given theoretical estimates *(53)* and strategies for discerning the optimal field strengths *(50)*, a growing body of experimental work (*see* **Table 2**) provides a clear indication of the important range of values *(1–8,10)*. We will review this body of work later in **Subheading 6.1.**

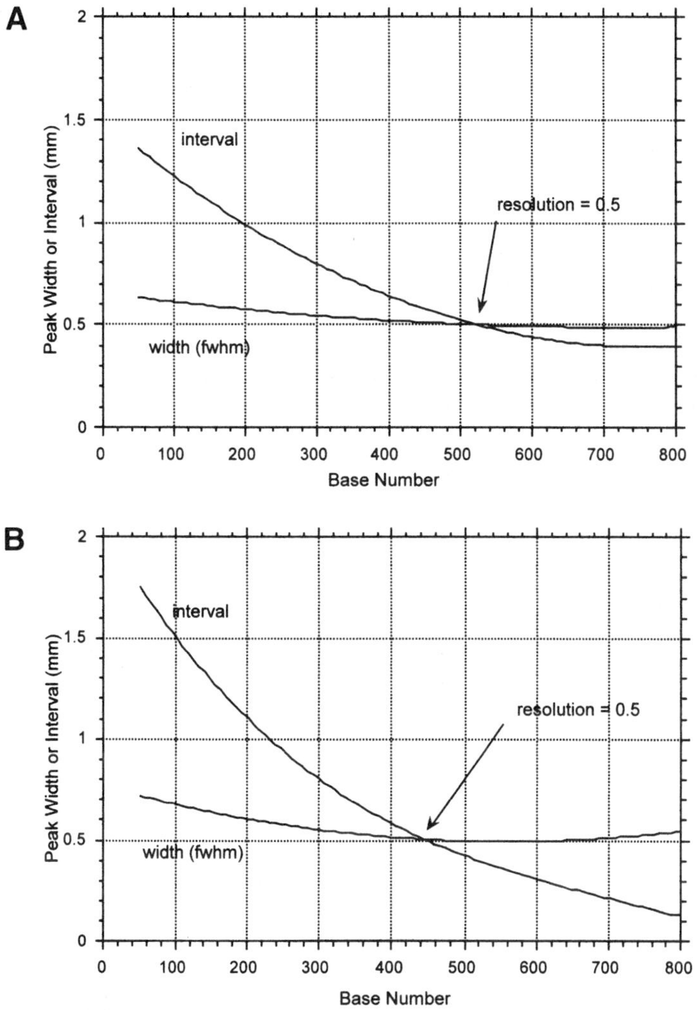

Fig. 6. **(A)** A crossover plot demonstrating the relationship of the spacing between successive DNA bands (interval) and the width of each band (at half of the maximum height) as a function of the length of the DNA fragment. The limit of resolution occurs where the interval curve intersects the width curve. This plot shows a small change in the interval values with increasing fragment size beyond the crossover point, and a small difference between the interval and width values. Under these conditions, expected band spacing and bandwidth values are predictable making deconvolution algorithms successful at determining the number of bands in an unresolved peak. **(B)** This crossover plot shows a large decrease of interval values with increasing fragment length beyond the crossover point. The interval value is highly dependent on the fragment size, and the band spacing rapidly becomes much less than the widths of the bands. Under these conditions, there is much uncertainty in the expected values for the band spacing, and consequently, deconvolution algorithms are prone to errors when determining the number of bands in an unresolved peak.

5.4. Data Analysis and Base Calling

The peak identification routines rely on resolution models that are based on plots similar to those in **Fig. 6A**,**B**. Imbedded in the algorithms are the expected DNA mobility as a function of base size, the changes in peak-width and peak separation as a function of base size, and the specific electrophoresis conditions. These parameters are obtained empirically using standardized run conditions. Much effort has been expended to determine sequence beyond the resolution limit, using deconvolution routines to extract peak composition information from unresolved peak data *(8)*. These algorithms require specific knowledge of the predicted peak widths and peak spacing beyond the resolution limit, and their utility has been enhanced by improvements in the separation media. As examples, **Fig. 6B** shows the peak separation decreasing rapidly with fragment length after the crossover point. Thus, a multiple peak envelope containing five unresolved peaks at 700 bases would not have the same width as an envelope containing three unresolved bands at 600 bases. A small deviation in the expected slope of the interval curve or an error in the expected fragment length mobility will result in a miss-assignment of the number of peaks within an unresolved envelope. In **Fig. 6A**, the interval curve has little slope past the crossover point. This constant peak spacing results in a reliable value for the expected peak spacing for hundreds of bases beyond the crossover point, and yields accurate base-calling. In general, separation media prepared from high concentrations (> 5%) of low molecular weight polymer tend to yield crossover curves similar to **Fig. 6B** *(52)*, whereas separation media prepared from low concentrations (< 4%) of high molecular weight polymer tend to yield crossover curves similar to **Fig. 6A** *(8)*.

The most desirable separation medium has a resolution limit that extends to large fragment sizes to get accurate base calls over a wide range. In addition, the medium should generate peak separation and peak spacing curves, whose slopes are very small past the crossover point, so that deconvolution routines can extract as much information as possible. Thus, long read-length in a DNA sequencing experiment is dependent on controlling the factors that affect peak width, peak separation, and peak deconvolution beyond the resolution limit. The development of DNA chain extension enzymes (DNA polymerases) that yield little peak height variation *(56,57)* has aided base identification within low-resolution multiple-base envelopes *(58–60)*. In addition, the development of high molecular weight, low concentration polymer formulations has increased the accuracy of base identification within the envelopes in the area just beyond the crossover point. These two achievements, along with the development of high sensitivity, fluorescent reporter systems such as energy transfer-dye labeled primers *(61,62)* and labeled terminators *(63)*, have led to the ability to determine > 1000 bases of sequence from a single DNASCE experiment *(8)*.

6. Comparison of Experimental Data Used with DNASCE

6.1. Useful Strategies

Numerous experiments (*see* **Table 2**) with polyacrylamide illustrate the range of parametric values commonly explored with DNASCE and indicate which values provide optimal performance. The best performance, in terms of readable bases per unit time, occurs when high molecular weight polymer at low concentration is used in conjunction with moderate electric field and elevated temperature. By reducing the polyacrylamide concentration, Salas-Solano et al. *(7,8)* attained a large mesh size that allows passage of the larger DNA fragments. The higher molecular weight polymer ensured reduction in peak broadening *(46)* and the operating concentration remained well above the overlap concentration (~ 25 times). An electric field strength of 200 V/cm was apparently optimal, i.e., low enough to inhibit the onset of oriented reptation and yet high enough for high-speed performance with excellent resolution (resolution, R = 0.6 at ~700 base read length/40 min). An elevated capillary temperature seems to play a role beyond a simple increase in the separation speed and a reduction of band compressions *(3,64,65)*. Although it is well-known that the viscosity of the medium decreases by about 2%/°C, producing a proportional decrease in separation time, an elevated temperature also increases the resolution of larger DNA fragments *(7,8,66)*. It is tempting to speculate that thermal energy may disrupt the onset of oriented reptation by reinstating the randomizing forces at work in the reptation regime. In addition, elevated temperature was shown to reduce sequence dependent migration anomalies in PDMA *(67)*.

6.2. Polydispersity and Capillary Dimensions

Comparison of the LPA experiments also sheds light on other variables thought to affect separation performance. The capillary internal diameter does not appear to play a significant role in the range of 50–100 μm, whereas the sieving capillary length of 30–50 cm yields excellent results without need of extended sieving with longer capillary lengths. At higher temperatures, a capillary with internal diameter near 50–75 μm yields slightly improved performance *(8)*. Differences in performance due to choice of buffer system appear to be negligible in the presence of sufficient denaturant (~ 7 *M* urea, pH ~ 8.0). The effect of polydispersity on the separation of DNA fragments is still not clear. Because *in situ* polymerization of polyacrylamide produces a large range of chain lengths, it has been suggested that pronounced polydispersity provides simultaneous sieving of both long and short DNA fragments *(8,68)*. Yet, experimental comparison of polymer solutions with known mixed molecular weight distributions indicates that polydispersity has little effect on resolution during DNASCE *(46)*.

6.3. Chain Stiffness

Although HEC, PEO, PDMA, and other linear polymers *(69)* may benefit from optimization strategies used with LPA, polymer hydrophobicity and chain stiffness also appear to play a role in the performance of these polymers. Remarkably, HEC, with a molecular weight of only 10^5, provides good resolution at room temperature for polymer concentrations close to the overlap concentration *(6)*! Among the four linear polymers described in **Tables 1** and **2**, the chain stiffness of HEC (persistence length ~ 8.3 nm *[48]*) far exceeds the chain stiffness of the others (e.g., PEO ~ 0.65 nm, LPA ~ 1.73 nm, PDMA ~ 1.1 nm; calculated from **ref.** *38*). Barron et al. *(70)* have suggested that if DNA separations are governed by entanglement interactions of the DNA with the polymer chains, stiffer polymers should be more useful at lower concentrations because they occupy a greater coil volume and provide stronger entanglement interactions.

The positive effect of increasing the polymer chain stiffness has been shown with modified PEO. Unmodified PEO is only useful as a separation medium at very high mol wt (8,000,000 Dalton) *(5)*. However, the addition of hydrophobic tails on each end of PEO polymer creates micellar crosslinks that stretch the backbone, in effect yielding a stiffer polymer in an aqueous medium *(71)*. Under these conditions, modified-PEO of low mol wt (35,000 Dalton) is found to have a good performance *(52)*.

6.4. Capillary Coatings

The most exciting development in recent years is the use of "self-coating" polymers for DNASCE *(1,2,5,72)*. Traditionally, different coatings are applied onto the interior walls of capillaries to reduce or eliminate electroosmotic flow (EOF) during electrophoresis *(73–76)*. In addition to the variable efficiency of producing a uniform coating *(74)*, the majority of these methods inevitably suffer degradation during repeated runs and require either the resuscitation of the homogeneous coating, or replacement of the entire capillary *(18,77)*. However, PEO *(5,78)*, PDMA *(1,2)*, and polyvinylpyrrolidone (PVP) *(79)* simply require preliminary acidification of the silanol groups before routine, between-run, replacement of the sieving polymer. This dual function of the linear polymer, inhibition of electroosmosis and DNA sieving, approaches the ideal performance that has been sought for multiple capillary systems slated for production sequencing. One hundred or more sequential separations with a single capillary have been reported using PDMA *(67)* and more than thirty separations using PEO *(5)*. Why the methylation of amide groups in LPA (to produce PDMA) inhibits electroosmosis is still not clear. Studies with methyl cellulose indicate that the dynamic adsorption of polymers on smooth fused silica surfaces probably plays a significant role *(80)*. With the use of an atomic force microscope, the surface of the inner wall of fused silica capillaries was shown to be remarkably smooth, with an average surface roughness from 2.8 Å to 6.7 Å *(81)*. It is concluded that since the surface is much smaller than the diffuse double layer, it cannot possibly influence peak shape and contribute to peak broadening in capillary electrophoresis. It is suggested that the extent to which cellulose polymers adsorb may be dependent on the polymer properties of the specific cellulose derivative *(77)*. The use of elevated temperatures during DNASCE separations highlighted a problem associated with using adsorptive coatings for suppression of electroosmosis. The loss of sequence resolution at elevated temperatures occurs when using low molecular weight PDMA, however this problem was suppressed with use of higher molecular weight PDMA *(82,83)*. The improved performance of the high molecular weight PDMA is attributed to a stronger surface adsorption capacity that typically occurs with longer chain polymers.

7. Novel Separation Polymers

Novel properties have been constructed from polymer chains composed of two or more different monomers (copolymer) *(84,85)*. Soane and coworkers *(86)* have demonstrated the use of "thermoreversible" copolymers for separation of both ssDNA and dsDNA with CE. These polymer solutions have the interesting property that, when heated, the viscosity dramatically drops, due to a phase transition, allowing easy capillary loading. Another group of interesting synthetic copolymers self-assemble into extended network structures with a well-defined pore size, which permits their use for DNASCE. These poly(ethylene glycol) polymers end-capped with micelle-forming fluorocarbon tails, form flowable aqueous networks that achieve read-lengths near to 400 bases long during DNASCE *(2,87)*.

8. Conclusions

Optimized sequence determination using DNASCE relies on a sieving medium that has good resolution over a wide DNA size range, predictable peak broadening and peak separation characteristics past the resolution limit. Combined with good deconvolution algorithms and detection design providing good signal-to-noise, spectacular performance has been achieved with DNASCE in recent years.

The development of very high molecular weight polymers used at low concentrations and elevated temperature has not only extended resolution limits and read lengths past the resolution limit but also decreased the duration of separations. Read lengths near 1000 bases have been obtained in less than 60 min by optimization of electrophoresis parameters and base-calling software *(8)*. Continued development may result in accurate base-calls approaching 1500 bases in a single run *(88)*. The recent development of "self-coating" polymer solutions that provide for both DNA sieving and inhibition of electroosmosis approaches the ideal performance that has been sought for multiple capillary systems slated for production sequencing. One hundred or more sequential separations with a single capillary have been reported for PDMA *(2)* and greater than thirty sequential separations are reported with PEO *(5)*. We anticipate that the same strategies used for LPA optimization (high molecular weight at low concentrations) will be operational for these self-coating polymers. Additional improvements of the sieving polymer may come from examination of new polymer backbone architectures or development of concentration gradient separations. It is hoped that this chapter may provide a guide to the successful methods in present use with replaceable linear polymers for DNASCE.

References

1. Johnson, B. F., Oaks, F. N., and Menchen, S. M. (1996) DNA sequencing in uncoated capillaries using a new sieving medium on the ABI prism 310 genetic analyzer, in *Eighth International Genome Sequencing and Analysis Conference*, Hilton Head, SC, USA.
2. Menchen, S., Johnson, B., Madabhushi, R., and Winnik, M. (1996) The design of separation media for DNA sequencing in capillaries, in *Progress in Biomedical Optics: Proceedings of SPIE – The International Society for Optical Engineering*, San Jose, California, **2680,** pp. 294–303.
3. Zhang, J., Fang, Y., Hou, J. Y., Ren, H. J., Jiang, R., Roos, P., and Dovichi, N. J. (1995) Use of non-cross-linked polyacrylamide for four-color DNA sequencing by capillary electrophoresis separation of fragments up to 640 bases in length in two hours. *Anal. Chem.* **67,** 4589–4593.
4. Manabe, T., Chen, N., Terabe, S., Yohda, M., and Endo, I. (1994) Effects of linear polyacrylamide concentrations and applied voltages on the separation of oligonucleotides and DNA sequencing fragments by capillary electrophoresis. *Anal. Chem.* **66,** 4243–4252.

5. Fung, E. N. and Yeung, E. S. (1995) High-speed DNA sequencing by using mixed poly(ethylene oxide) solutions in uncoated capillary columns. *Anal. Chem.* **67,** 1913–1919.
6. Bashkin, J., Marsh, M., Barker, D., and Johnston, R. (1996) DNA sequencing by capillary electrophoresis with a hydroxyethylcellulose sieving buffer. *Appl. Theor. Electrophoresis* **6,** 23–28.
7. Carrilho, E., Ruiz-Martinez, M. C., Berka, J., Smirnov, I., Goetzinger, W., Miller, A. W., Brady, D., and Karger, B. L. (1996) Rapid DNA sequencing of more than 1000 bases per run by capillary electrophoresis using replaceable linear polyacrylamide solutions. *Anal. Chem.* **68,** 3305–3313.
8. Salas-Solano, O., Carrilho, E., Kotler, L., Miller, A. W., Goetzinger, W., Sosic, Z., and Karger, B. L. (1998) Routine DNA sequencing of 1000 bases in less than one hour by capillary electrophoresis with replaceable linear polyacrylamide solutions. *Anal. Chem.* **70,** 3396–4003.
9. Chang, H.-T. and Yeung, E. S. (1995) Poly(ethyleneoxide) for high-resolution and high-speed separation of DNA by capillary electrophoresis. *J. Chromatogr. B* **669,** 113–123.
10. Grossman, P. D. (1994) Electrophoretic separation of DNA sequencing extension products using low-viscosity entangled polymer networks. *J Chromatogr. A* **663,** 219–227.
11. Fleischmann, R. D., Adams, M. D., White, O., Clayton, R. A., Kirkness, E. F., Kerlavage, A. R., et al. (1995) Whole-genome random sequencing and assembly of *Haemophilus influenzae* rd. *Science* **269,** 496–512.
12. Adams, M. A., Fields, C., and Venter, J. C. (1994) *Automated DNA Sequencing and Analysis*, 1st ed., Academic Press, London.
13. Grossman, P. D. and Co!burn, J. C. (1992) *Capillary Electrophoresis: Theory and Practice*, 1st ed., Academic Press, San Diego.
14. Li, S. F. Y. (1994) *Capillary Electrophoresis: Principles, Practice and Applications*, 1st ed., Elsevier, Singapore, Republic of Singapore.
15. Landers, J. P. (1994) *Handbook of Capillary Electrophoresis*, 1st ed., CRC Press, Boca Raton, Florida.
16. Righetti, P. G. (1996) *Capillary Electrophoresis in Analytical Biotechnology*, 1st ed., CRC Press, Boca Raton, Florida.
17. Cohen, A. S., Najarian, D. R., Paulus, A., Guttman, A., Smith, J. A., and Karger, B. L. (1988) Rapid separation and purification of oligonucleotides by high-performance capillary gel electrophoresis. *Proc. Natl. Acad. Sci. USA* **85,** 9660–9663.
18. Swerdlow, H., Dew-Jager, K. E., Brady, K., Grey, R., Dovichi, N. J., and Gesteland, R. (1992) Stability of capillary gels for automated sequencing of DNA. *Electrophoresis* **13,** 475–483.
19. Sudor, J., Foret, F., and Bocek, P. (1991) Pressure refilled polyacrylamide columns for the separation of oligonucleotides by capillary electrophoresis. *Electrophoresis* **12,** 1056–1058.
20. Cantor, C. R. and Schimmeı, P. R. (1980) *Biophysical Chemistry: Part 3*, W. H. Freeman and Company, New York.
21. Eisenberg, D. and Crothers, D. (1979) *Physical Chemistry with Applications to the Life Sciences*, Benjamin/Cummings Publishing Company Inc., Menlo Park.
22. Heller, C. (1995) Capillary electrophoresis of proteins and nucleic acids in gels and entangled polymer solutions. *J. Chromatogr. A* **698,** 19–31.
23. McDonell, M. W., Simon, M. N., and Studier, F. W. (1977) Analysis of restriction fragments of T7 DNA and determination of molecular weights by electrophoresis in neutral and alkaline gels. *J. Mol. Biol.* **110,** 119–146.

24. Slater, G. W., Rousseau, J., Noolandi, J., Turmel, C., and Lalande, M. (1988) Quantitative analysis of the three regimes of DNA electrophoresis in agarose gels. *Biopolymers* **27,** 509–524.
25. Heller, C., Duke, T., and Viovy, J.-L. (1994) Electrophoretic mobility of DNA in gels. II. Systematic experimental study in agarose gels. *Biopolymers* **34,** 249–259.
26. Lumpkin, O. J., Dejardin, P., and Zimm, B. H. (1985) Theory of gel electrophoresis of DNA. *Biopolymers* **24,** 1573–1593.
27. Lumpkin, O. J. and Zimm, B. H. (1982) Mobility of DNA in gel electrophoresis. *Biopolymers* **21,** 2315–2316.
28. Lerman, L. S. and Frisch, H. L. (1982) Why does the electrophoretic mobility of DNA in gels vary with the length of the molecule? *Biopolymers* **21,** 995–997.
29. Slater, G. W. and Noolandi, J. (1985) Prediction of chain elongation in the reptation theory of DNA gel electrophoresis. *Biopolymers* **24,** 2181–2184.
30. Slater, G. W. and Noolandi, J. (1986) On the reptation theory of gel electrophoresis. *Biopolymers* **25,** 431–454.
31. Slater, G. W. (1992) Anomalous electrophoresis, self-trapping and <<freezing>> of partially charged polyelectrolytes. *J. Phys. II* (Paris) **2,** 1149–1158.
32. Slater, G. W., Turmel, C., Lalande, M., and Noolandi, J. (1989) DNA gel electrophoresis: effect of field intensity and agarose concentration on band inversion. *Biopolymers* **28,** 1793–1799.
33. Luckey, J. A. and Smith, L. M. (1993) A model for the mobility of single-stranded DNA in capillary gel electrophoresis. *Electrophoresis* **14,** 492–501.
34. Mitnik, L., Salome, L., Viovy, J.-L., and Heller, C. (1995) Systematic study of field and concentration effects in capillary electrophoresis of DNA in polymer solutions. *J. Chromatogr. A* **710,** 309–321.
35. Kloczkowski, A. (1996). Theoretical Models for Polymer Chains, in *Physical Properties of Polymers Handbook* (Mark, J. E., ed.), AIP Press, Woodbury, New York, pp. 61–70.
36. Doi, M. (1996) *Introduction to Polymer Physics* (See, H., translator), Clarendon Press, Oxford.
37. Ying, Q., Wu, G., Chu, B., Farinato, R., and Jackson, L. (1996) Laser light scattering of poly(acrylamide) in 1 M NaCl aqueous solution. *Macromolecules* **29,** 4646–4654.
38. Kurata, M. and Tsunashima, Y. (1989) Viscosity - molecular weight relationships and unperturbed dimensions of linear chain molecules, in *Polymer Handbook* (Brandrup, J. and Immergut, E. H., eds.), Wiley-Interscience, New York.
39. Cowie, J. M. G. (1991) *Polymers: Chemistry and Physics of Modern Materials*, Blackie Academic and Professional, Glasgow.
40. Sundararajan, P. R. (1996) Theta Temperatures, in *Physical Properties of Polymers Handbook* (Mark, J. E., ed.), AIP Press, Woodbury, New York, pp. 197–226.
41. Fuchs, O. (1989) Solvents and Non-Solvents for Polymers, in *Polymer Handbook* (Brandrup, J. and Immergut, E. H., eds.), Wiley-Interscience, New York.
42. DeGennes, P.-G. (1991) *Scaling Concepts in Polymer Physics*, Cornell University Press, Ithaca.
43. Ying, Q. and Chu, B. (1987) Overlap concentration of macromolecules in solution. *Macromolecules* **20,** 362–366.
44. Viovy, J.-L. and Duke, T. (1993) DNA electrophoresis in polymer solutions: Ogston sieving, reptation and constraint release. *Electrophoresis* **14,** 322–329.
45. Quesada, M. A. (1997) Replaceable polymers in DNA sequencing by capillary electrophoresis. *Curr. Opinion Biotech.* **8,** 82–93.

46. Wu, C., Quesada, M. A., Schneider, D. K., Farinato, R., Studier, F. W., and Chu, B. (1996) Polyacrylamide solutions for DNA sequencing by capillary electrophoresis: mesh sizes, separation and dispersion. *Electrophoresis* **17,** 1103–1109.
47. Daoud, M., Cotton, J. P., Farnoux, B., Jannink, G., Sarma, G., Benoit, H., Duplessix, R., Picot, C., and Gennes, P.-G. D. (1975) Solutions of flexible polymers. Neutron experiments and interpretation. *Macromolecules* **8,** 804–818.
48. Barron, A. E., Soane, D. S., and Blanch, H. W. (1993) Capillary electrophoresis of DNA in uncross-linked polymer solutions. *J. Chromatogr. A* **652**, 3-16.
49. Grossman, P. D. and Soane, D. S. (1991) Experimental and theoretical studies of DNA separations by capillary electrophoresis in entangled polymer solutions. *Biopolymers* **31,** 1221–1228.
50. Slater, G. W., Mayer, P. and Drouin, G. (1996) Migration of DNA through gels, in *Methods in Enzymology,* (Karger, B. L. and Hancock, W. S., eds.), Academic Press, pp. 272–295.
51. Luckey, J. A., Norris, T. B., and Smith, L. M. (1993) Analysis of resolution in DNA sequencing by capillary gel electrophoresis. *J. Phys. Chem.* **97,** 3067–3075.
52. Menchen, S., Johnson, B., Winnik, M. A., and Xu, B. (1996) Flowable networks as DNA sequencing media in capillary columns. *Electrophoresis* **17,** 1451–1458.
53. Slater, G. W., Mayer, P., and Grossman, P. D. (1995) Diffusion, Joule heating, and band broadening in capillary gel electrophoresis of DNA. *Electrophoresis* **16,** 75–83.
54. Wooley, A. T. and Mathies, R. A. (1994) Ultra-high-speed DNA fragment separations using microfabricated capillary array electrophoresis chips. *Proc. Natl. Acad. Sci. USA* **91,** 11,348–11,352.
55. Luckey, J. A. and Smith, L. M. (1993) Optimization of electric field strength for DNA sequencing in capillary gel electrophoresis. *Anal. Chem.* **65,** 2841–2850.
56. Tabor, S. and Richardson, C. C. (1987) DNA sequence analysis with a modified bacteriophage T7 DNA polymerase. *Proc. Natl. Acad. Sci. USA* **84,** 4767–4771.
57. Tabor, S. and Richardson, C. C. (1995) A single residue in DNA polymerases of the Escherichia coli DNA polymerase I family is critical for distinguishing between deoxy- and dideoxyribonucleotides. *Proc. Natl. Acad. Sci. USA* **92,** 6339–6343.
58. Tabor, S. and Richardson, C. C. (1990) DNA sequence analysis with a modified bacteriophage T7 DNA polymerase: Effect of pyrophosphorolysis and metal ions. *J. Biol. Chem.* **265,** 8322–8328.
59. Ansorge, W., Zimmerman, J., Schwager, C., Stegemann, J., Erfle, H., and Voss, H. (1990) One label, one tube, Sanger DNA sequencing in one and two lanes on a gel. *Nucleic Acids Res.* **18,** 3419–3420.
60. Chen, D. Y., Harke, H. R., and Dovichi, N. J. (1992) Two-label peak-height encoded DNA sequencing by capillary gel electrophoresis: three examples. *Nucleic Acids Res.* **20,** 4873–4880.
61. Ju, J., Kheterpal, I., Scherer, J. R., Ruan, C., Fuller, C. W., Glazer, A. N., and Mathies, R. A. (1995) Design and synthesis of fluorescence energy transfer dye-labeled primers and their application for DNA sequencing and analysis. *Anal. Biochem.* **231,** 131–140.
62. Lee, L. G., Spurgeon, S. L., Heiner, C. R., Benson, S. C., Rosenblum, B. B., Menchen, S. M., Graham, R. J., Constantinescu, A., Upadhya, K. G., and Cassel, J. M. (1997) New energy transfer dyes for DNA sequencing. *Nucleic Acids Res.* **25,** 2816–2822.
63. Rosenblum, B. B., Lee, L. G., Spurgeon, S. L., Khan, S. H., Menchen, S. M., Heiner, C. R., and Chen, S. M. (1997) New dye-labeled terminators for improved DNA sequencing patterns. *Nucleic Acids Res.* **25,** 4500–4504.
64. Lu, H., Arriaga, E., Chen, D. Y., Figeys, D., and Dovichi, N. J. (1994) Activation energy of single-stranded DNA moving through cross-linked polyacrylamide gels at 300 V/cm:

Effect of temperature on sequencing rate in high-electric-field capillary gel electrophoresis. *J. Chromatogr. A* **680,** 503–510.

65. Lu, H., Arriaga, E., Chen, D. Y., and Dovichi, N. J. (1994) High-speed and high-accuracy DNA sequencing by capillary gel electrophoresis in a simple, low cost instrument: two-color peak-height encoded sequencing at 40°C. *J. Chromatogr. A* **680,** 497–501.
66. Klépárník, K., Foret, F., Berka, J., Goetzinger, W., Miller, A. W., and Karger, B. L. (1996) The use of elevated column temperature to extend DNA sequencing read lengths in capillary electrophoresis with replaceable polymer matrices. *Electrophoresis* **17,** 1860–1866.
67. Rosenblum, B. B., Oaks, F., Menchen, S., and Johnson, B. (1997) Improved single-strand DNA sizing accuracy in capillary electrophoresis. *Nucleic Acids Res.* **25,** 3925–3929.
68. Chiari, M., Nesi, M., and Righetti, P. G. (1993) Movement of DNA fragments during capillary zone electrophoresis in liquid polyacrylamide. *J. Chromatogr. A* **652,** 31–39.
69. Chiari, M. and Righetti, P. G. (1995) New types of separation matrices for electrophoresis. *Electrophoresis* **16,** 1815–1829.
70. Barron, A. E., Sunada, W. M., and Blanch, H. W. (1996) The effects of polymer properties on DNA separations by capillary electrophoresis in uncross-linked polymer solutions. *Electrophoresis* **17,** 744–757.
71. Semenov, A. N., Joanny, J. F., and Kholov, A. F. (1995) Associating polymers: equilibrium and linear viscoelasticity. *Macromolecules* **28,** 1066–1075.
72. Chiari, M. and Cretich, M. (2001) Capillary coatings: choices for capillary electrophoresis of DNA, in *Capillary Electrophoresis of Nucleic Acids,* Vol. 1 (Mitchelson, K. R. and Cheng, J., eds.), Humana Press, Totowa, NJ, pp. 125–138.
73. Cobb, K. A., Dolnik, V., and Novotny, M. (1990) Electrophoretic separations of proteins in capillaries with hydrolytically stable surface structures. *Anal. Chem.* **62,** 2478–2483.
74. Hjertén, S. and Kubo, K. (1993) A new type of pH- and detergent-stable coating for elimination of electroendosmosis and adsorption in (capillary) electrophoresis. *Electrophoresis* **14,** 390–395.
75. Schmalzing, D., Piggee, C. A., Foret, F., Carrilho, E., and Karger, B. L. (1993) Characterization and performance of a neutral hydrophilic coating for the capillary electrophoretic separation of biopolymers. *J. Chromatogr. A* **652,** 149–59.
76. Bruin, G. J. M. and Paulus, A. (1995) Biopolymer separations with capillary electrophoresis. *Anal. Meth. Instrument.* **2,** 3–26.
77. Barron, A. E., Sunada, W. M., and Blanch, H. W. (1995) The use of coated and uncoated capillaries for the electrophoretic separation of DNA in dilute polymer solutions. *Electrophoresis* **16,** 64–74.
78. Kim, Y. and Yeung, E. S. (2001) Capillary electrophoresis of DNA fragments using poly(ethylene oxide) as a sieving material, in *Capillary Electrophoresis of Nucleic Acids,* Vol. 1 (Mitchelson, K. R. and Cheng, J., eds.), Humana Press, Totowa, NJ, pp. 215–223.
79. Gao, Q. and Yeung, E. S. (1998) A matrix for DNA separation: genotyping and sequencing using poly(vinylpyrrolidone) solution in uncoated capillaries. *Anal. Chem.* **70,** 1382–1388.
80. Bello, M. S., Besi, P. D., Rezzonico, R., Righetti, P. G., and Casiraghi, E. (1994) Electroosmosis of polymer solutions in fused silica capillaries. *Electrophoresis* **12,** 623–626.
81. Barberi, R., Giocondo, M., Bartolino, R., and Righetti, P. G. (1995) Probing the inner surface of a capillary with the atomic force microscope. *Electrophoresis* **16,** 1445–1450.
82. Madabhushi, R. S. (1998) Separation of 4-color DNA sequencing extension products in noncovalently coated capillaries using low viscosity polymer solutions. *Electrophoresis* **19,** 224–230.

83. Madabhushi, R. S. (2001) DNA sequencing in noncovalently coated capillaries using low viscosity polymer solutions, in *Capillary Electrophoresis of Nucleic Acids,* Vol. 2 (Mitchelson, K. R. and Cheng, J., eds.), Humana Press, Totowa, NJ, pp. 309–315.
84. Chu, B., Liu, T., Wu, C.-H., and Liang, D. (2001) DNA capillary electrophoresis using block copolymer as a new separation medium, in *Capillary Electrophoresis of Nucleic Acids,* Vol. 1 (Mitchelson, K. R. and Cheng, J., eds.), Humana Press, Totowa, NJ, pp. 225–238.
85. Liu, Y.-J. and Rill, R. L. (2001) DNA separation by capillary electrophoroesis in lyotropic polymer liquid crystals, in *Capillary Electrophoresis of Nucleic Acids,* Vol. 1 (Mitchelson, K. R. and Cheng, J., eds.), Humana Press, Totowa, NJ, pp. 203–213.
86. Sassi, A. P., Barron, A., Alonso-Amigo, M. G., Hion, D. Y., Yu, J. S., Soane, D. S., and Hooper, H. H. (1996) Electrophoresis of DNA in novel thermoreversible matrices. *Electrophoresis* **17,** 1460–1469.
87. Menchen, S., Johnson, B., Winnik, M. A., and Xu, B. (1996) Flowable networks as equilibrium DNA sequencing media in capillary columns. *Chemistry of Materials – Am. Chem. Soc.* **8,** 2205–2208.
88. Slater, G. W. and Drouin, G. (1992) Why can we not sequence thousands of DNA bases on a polyacrylamide gel? *Electrophoresis* **13,** 574–582.
89. Personal communication, Dr. Josh Bashkin (Molecular Dynamics).
90. Personal communication, Dr. Stephen Pentoney (Beckman).

11

Sieving Matrix Selection

Tim Wehr, Mingde Zhu, and David T. Mao

1. Introduction

The charge density of nucleic acids is uniform and constant for nucleic acid molecules of varying length. Therefore, they typically migrate at the same velocity in response to an electrical potential in free solution. Capillary electrophoretic (CE) separation of nucleic acid species requires addition of a sieving matrix to the background electrolyte to retard the analytes in proportion to their molecular size *(1)*. This can be achieved by casting a crosslinked gel in the capillary, or by incorporating a linear gel material or polymer in the background electrolyte. Both of these approaches have been applied successfully to the separation of nucleic acid species ranging from short, single-stranded oligonucleotides to dsDNA of several kilobase lengths. The selection of the appropriate sieving matrix will depend upon the size of the molecules that are to be separated, the resolution required, and the capabilities of the analytical instrumentation. This chapter will review the various sieving matrices which have been employed for CE separation of nucleic acids, identify the appropriate matrix for specific applications, and provide practical suggestions for preparation and use of sieving media for CE.

1.1. Gel-Filled Capillaries

Much of the earlier DNA separations in capillary gel electrophoresis were performed on gel-filled capillaries since it is a simple extension of traditional slab-gel electrophoresis. The gel matrix, most commonly polyacrylamide, can be cast inside a small diameter capillary in the same manner as slab gels, and yet provides many advantages including fast analysis time, automation, and extraordinary resolution. A separation efficiency of 30 million theoretical plates, which is beyond reach of any other existing separation technique, has been reported for the analysis of DNA fragments *(1,2)*.

From: *Methods in Molecular Biology, Vol. 162:*
Capillary Electrophoresis of Nucleic Acids, Vol. 1: Introduction to the Capillary Electrophoresis of Nucleic Acids
Edited by: K. R. Mitchelson and J. Cheng © Humana Press Inc., Totowa, NJ

1.1.1. Selection of Gel Porosity

Gel-filled capillaries are prepared in a similar fashion to slab gels by polymerizing the monomer solution *in situ*. The pore size is determined by % T and % C, in which % T is the total monomer concentration and % C is the concentration of the crosslinker as a percentage of total monomer concentration. A typical concentration of a gel-filled capillary for DNA separation is from 3% to 10% T, and from 0% to 5% C. Although the pore size can be controlled by the monomer concentration alone without any crosslinker (linear polyacrylamide), it is suggested that a crosslinked gel would last longer than a linear gel *(3)*. The separation of DNA fragments in capillary gel electrophoresis is ultimately determined by several factors including the pore size of the gel and the length of the DNA fragments. The resolution is inversely proportional to the square root of the pore size *(4)*. For small DNA fragments such as oligonucleotides, a better separation can be achieved on a capillary gel of higher gel concentration (giving smaller pore size) such as a 5% T, 5% C polyacrylamide gel, compared to a typical 3% T, 3% C polyacrylamide gel. As shown in **Fig. 1**, the longer migration time of oligonucleotides on the 5% T, 5% C gel is evident. A typical 5% T, 5% C polyacrylamide gel-filled capillary is capable of baseline separation of oligonucleotides up to 150-bases length (**Fig. 2**). However, the resolution is drastically reduced for the separation of larger DNA fragments. A gel-filled capillary of larger pore size such as 3% T, 3% C should be the choice for larger DNA fragments such as restriction enzyme digest products and sequencing reaction products.

1.1.2. Selection of Gel Matrix Material

Whereas polyacrylamide has been the most popular matrix for slab-gel electrophoresis, it frequently exhibits a lack of chemical and mechanical strength during use for CE, in which the practice of multiple injections on the same capillary is demanded for both practical and economical reasons. Many alternative gel matrices have been studied in the past ten years including several substituted polyacrylamides. Scholsky et al. **(5)** concluded in their study that methacrylamide and its *N*-substituted derivatives could be potential replacements for acrylamide, because they are chemically more stable, less toxic, and the gel sieving properties are similar to that of acrylamide. Dolnik et al. *(6)* made high concentration gel-filled capillaries, which are difficult to prepare with acrylamide, using a HydroLink gel solution, a commercial substituted acrylamide mixture. Nevertheless, in most commercial gel-filled capillary products, polyacrylamide is still the choice of the gel material for many reasons. Highly purified acrylamide is widely available from many different sources, ensuring the reproducibility of the performance from batch to batch would not be affected by the quality of the raw material. The polymerization kinetics of acrylamide in slab gels are well understood, and this knowledge can be transferred directly for the preparation of gel-filled capillaries, thus simplifying the manufacturing process.

1.1.3. Techniques for Casting Gels in Fused Silica Capillaries

Following the early success of DNA separations using gel-filled capillaries, it was soon realized that it is very difficult to reproduce the results from laboratory to labora-

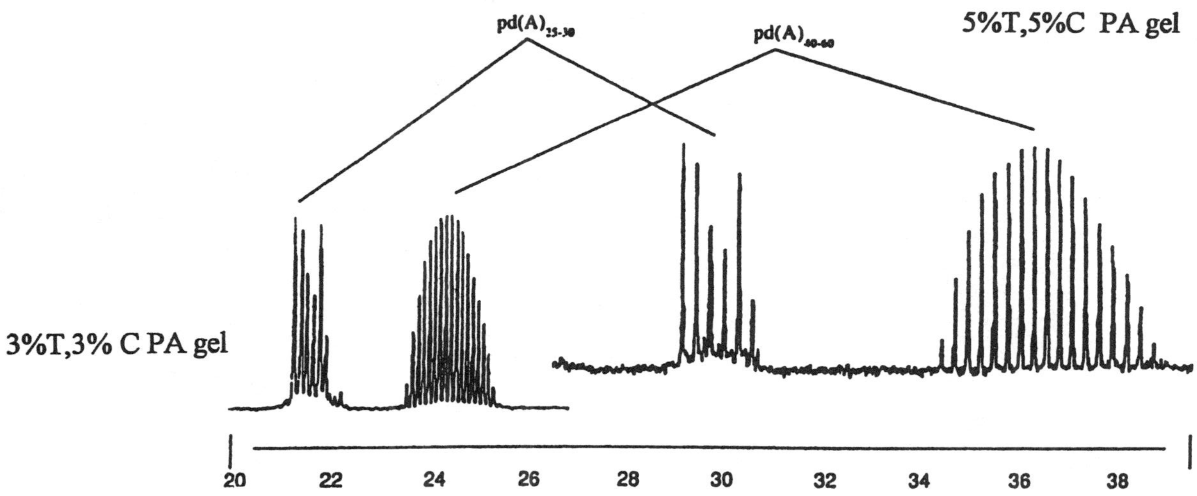

Fig. 1. Comparison of the migration speed of oligonucleotides in 5% T, 5% C and 3% T, 3 % C polyacrylamide gels. The sample is a mixture of two polydeoxyadenylic acid ladders, pd$(A)_{25-30}$ (0.01 mg/mL) and pd$(A)_{40-60}$ (0.02 mg/mL). Capillaries are 5 %T, 5 %C gel (µPAGE-5, J&W Scientific) and 3% T, 3% C gel (µPAGE-3, J&W Scientific), 75 µm id × 75 cm total length (50-cm effective length from injector to detector). Buffer: 0.1 *M* Tris-borate, pH 8.3, 7 *M* urea. Applied voltage: –18.75 kV, 4 µA. Column temperature is ambient at 21°C. Sample detection: UV at 260 nm. Sample injected electrokinetically at –5 kV, 3 s.

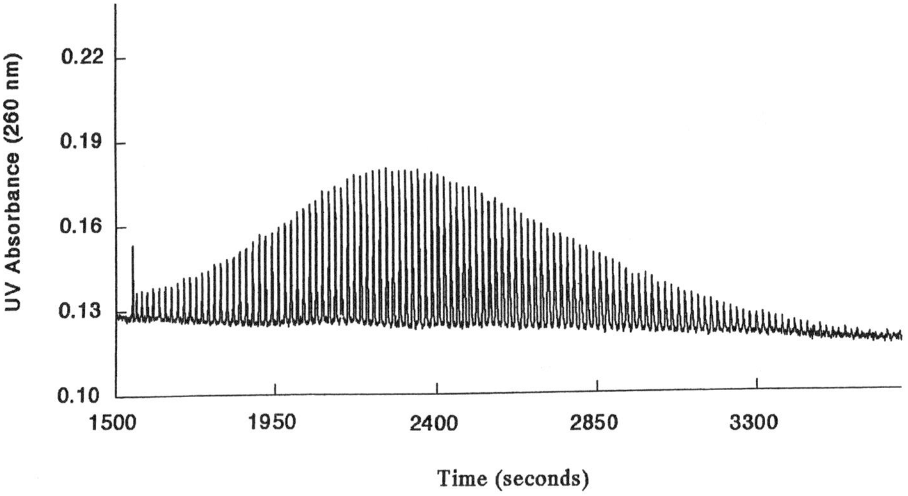

Fig. 2. Separation of 20–150-mer polydeoxyadenylic acid oligonucleotides. The sample is a polydeoxyadenylic acid ladder prepared using the terminal transferase reaction. Capillary: 5% T, 5% C (μPAGE-5, J&W Scientific), 75 μm id × 75 cm total length (50 cm from injector to detector). Buffer: 0.1 *M* Tris-borate, pH 8.3, 7 *M* urea. Applied voltage is –18 kV, 4 μA; temperature: ambient; detection: UV, 260 nm. The sample is injected electrokinetically at –5 kV, 3 s.

tory because of many practical difficulties in casting polyacrylamide gel uniformly in a small diameter capillary. The free-radical solution polymerization is highly dependent on many variables that are often hard to control, such as temperature and oxygen content. However, the most damaging factor leading to capillary failure is the shrinkage of the gel. Gel shrinkage is an inherent result of polymerization reactions and causes formation of small bubbles within the capillary. The bubbles result in current drop and in disruption of electrophoretic continuity, which render the gel-filled capillary useless for separation purposes. To overcome these difficulties, many novel methods have been developed. Karger et al. ***(7)*** used hydrophilic polymer additives to compensate for the space lost during the polymerization reaction. Drossman et al. ***(8)*** carried out the polymerization reaction under high pressure to suppress the formation of bubbles. The use of a partially polymerized solution (prepolymer), instead of a monomer solution, successfully limited the number of bonds formed during polymerization thereby reducing the degree of shrinkage ***(9)***. Schomburg ***(10)*** describes a method for the production of gel-filled capillaries, which limits bubble formation by using gamma radiation.

Bubble formation can also be avoided by using a progressive polymerization technique as described by Novotny et al. ***(11)*** and Fishel et al. ***(12)***. In this method, polymerization is initiated either by a migrating catalyst or by heat, and starts from one end of the capillary and progressively moves across the whole capillary length. The shrinkage occurring during the slow polymerization is compensated by addition from the surrounding unpolymerized monomer solution, allowing the reproducible preparation of capillaries with consistent performance. The other advantages of this technique are:

1. Multiple capillaries can be prepared simultaneously, thus reducing the manufacturing cost,
2. High concentration gels can be prepared, and
3. Various amounts of crosslinking can be included in the gel.

High concentration gel-filled capillaries with up to 25% T (total monomer concentration) have been prepared using this technique ***(6)***. The amount of crosslinking can be as high as 6% C. Although gels with crosslinking greater than 6% C can be made, the gel is opaque and larger pores than expected occur, due to the formation of dense clusters of polymerized polyacrylamide. Many gel-filled capillaries also contain urea as an additive to provide stability to the gel matrix, thereby increasing the durability of the gel during electrophoresis. Contrary to general belief, urea in the capillary gel does not denature dsDNA at the on-column ambient temperature at which runs are generally made.

1.2. Entangled Polymer Solutions

As the name implies, sieving behavior in entangled polymer solutions arises from formation of a dynamic pore system as polymer chains interact; the pore system retards the migration of polymeric analytes through the medium. The technique has been variously termed non-gel sieving, entangled polymer sieving, and dynamic sieving. Such polymer solutions have also been termed physical gels. The separation mechanism is the same in all cases. The great advantage of polymer solutions over gel-filled capillaries is the ability to replace the polymer solution between each analysis. This improves the reproducibility of separation and increases the capillary lifetime. The application of entangled polymer solutions for the separation of nucleic acids was first described by Zhu et al. ***(13)***. Today, entangled polymer solutions are routinely used for the analysis of various DNA fragments from synthetic oligonucleotide preparations, DNA amplification products and more recently, for automated DNA sequencing.

1.2.1. Mechanism of Entangled Polymer Sieving

The migration of nucleic acids through a sieving polymer matrix has been characterized as being similar to the behavior of nucleic acids in rigid crosslinked gels, i.e. dependent upon the size of the analyte relative to the effective pore diameter of the polymer network. The migration of short DNA molecules (<300 bp) is considered to behave according to the Ogston model ***(14)***. This model assumes that DNA molecules behave as unperturbed spheres and that separation is achieved by "sieving," i.e., smaller molecules migrate faster through the matrix, since they can enter a larger number of pores than longer DNA molecules. For larger molecules, the migration behavior undergoes a transition from a sieving regime to a "reptation" regime ***(15)***. In this case, the size of the DNA exceeds the effective pore size, but molecules continue to migrate through the matrix in a snake-like fashion, in which the DNA mobility is inversely proportional to molecular size. At high electric field strengths, the stretching of the

DNA molecules as they migrate through the matrix causes a transition to "biased reptation," in which migration becomes independent of chain length and resolution between molecules of different size is lost.

Grossman and Soane *(16)* first proposed a mechanism for entangled polymer sieving. In dilute solution, polymer chains are considered to behave independently with no interaction. At some concentration, termed the entanglement threshold, the polymer chains begin to interact to form a dynamic pore network. Below the entanglement threshold, the viscosity changes linearly with polymer concentration. At the entanglement threshold (where the bulk concentration of the solution reaches the same concentration of the solution within the polymer coil), the slope of a plot of the viscosity vs the log concentration curve increases, signaling the interaction between polymer chains. The effective pore diameter of the resulting polymer network above the entanglement threshold depends upon the polymer concentration, with the pore size diminishing with increasing concentration of the polymer. The optimization of an entangled polymer separation required finding a polymer concentration at which the effective pore diameter provides sufficient sieving strength to achieve the separation. Practically, this concentration is one at which the viscosity is sufficiently low to allow the rapid introduction of the polymer solution into the capillary and its replenishment between injections. Because the pore size is proportional to polymer chain length, use of shorter chains can be used to obtain smaller pores, at reasonable solution viscosity values.

Later studies by Barron et al. *(17,18)* suggest that the migration behavior of nucleic acids in polymer solutions is different from that in gels, and fits neither the Ogston model, nor the reptation model. The size-based resolution of large DNA molecules is observed in very dilute polymer solutions, far below the entanglement threshold. This indicates that transient coupling alters the frictional characteristics of the DNA molecule, and enables size-based separations under conditions where entangled polymer networks are absent.

2. Methods

2.1. Gel-Filled Capillaries

2.1.1. Preparation of Gel-Filled Capillaries

1. A typical procedure for the preparation of gel-filled capillaries is described by Novotny et al. *(11)*. In the first step, a layer of linear polyacrylamide is covalently linked to the capillary wall to eliminate the inherent elecroosmotic flow.
2. An 80-cm length of fused silica capillary tubing (75 µm id, 360 µm od) is used with a window at 25 cm from one end.
3. The capillary is rinsed with 0.1 *M* NaOH for 30 min, and then washed with deionized water for 5 min.
4. The washed capillary was filled with a solution of 4 µL of methacryloxypropyltrimethoxysilane in 1 mL of 6 m*M* acetic acid.
5. After 1 h at room temperature, the capillary is again rinsed with deionized water for 5 min and is then emptied.
6. The capillary is then filled with a deaerated solution of 2.54% acrylamide containing 0.1% ammonium persulfate (APS) and 0.1% *N,N,N',N'*-tetramethylethylenediamine (TEMED).

7. After polymerization at ambient temperature for 30 min, the capillary is emptied then is rinsed with distilled water for 5 min.
8. The coated capillary is then filled with a deaerated monomer solution containing 4.75% acrylamide, 0.25% *N,N'*-methylene bisacrylamide (C=5%), 0.1 *M* Tris-borate buffer (pH 8.3), 0.1% APS and 7 *M* urea.
9. Two ends of the capillary are immersed in two separate vials containing 5 mL of the same monomer solution as in the capillary.
10. The electrodes were connected to a low-voltage power supply, 5 μL of TEMED is added to the anode side of the vial, and a potential of 800 V is applied across the capillary. The polymerization is completed in 8–15 h as indicated by the formation of gel at the cathode end vial.
11. The cathodic and anodic vials are then replaced by vials containing the running buffer, which consists of 0.1 *M* Tris-borate and 7 *M* urea.
12. The capillary is washed free of unreacted ions by a stepwise increase in voltage: 2 kV for 3 h, 5 kV for 2 h and 10 kV for 1 h.
13. The capillary can now be stored until use.

2.1.2. Operation and Storage of Gel-Filled Capillaries

1. Although a gel-filled capillary can provide ultra-high resolution for DNA separation, it remains as a delicate column to handle.
2. Part of the reasons are that the polyacrylamide gel is not chemically stable for long term use, and the gel matrix conformation changes (shrinks or swells) depending upon many variables, such as the temperature, the buffer concentration, and the pH of the gel.
3. The shrinking and swelling of the gel leads to the formation of bubbles and results in poor electrophoretic resolution.
4. In order to avoid changing the temperature too quickly, care should be taken both during storage and conditioning of the capillary. Capillaries stored at room temperature for more than a few days will degrade rapidly.
5. When not in use for more than 10 h, the capillary should be stored in a refrigerator, but should be cooled slowly to avoid shrinkage. The storage of capillaries between two packs of thermogel reduces the thermal shock significantly, minimizing the damage to the gel structure and increasing the working-life of the capillary.
6. Similarly, a capillary taken from a refrigerator should be warmed up to room temperature slowly.

2.1.3. Conditioning of Gel-Filled Capillaries

1. Conditioning is necessary for new capillaries and for capillaries taken from storage. Conditioning removes the polyacrylamide degradation products from the gel, and ensures uniformity of buffer throughout the entire capillary.
2. The conditioning is achieved by increasing the applied voltage stepwise over a period of ~30 min: for example, 10 min at 100 V/cm, 10 min at 150 V/cm, 5 min at 200 V/cm, and finally 5 min at 250 V/cm.
3. Additional electrophoresis time may be required if the baseline is not stable.

2.1.4. Preventing the Formation of Bubbles

1. The primary failure mode of a gel-filled capillary is due to the formation of bubbles. Some bubbles form due to improper gel preparation. If the freshly cut ends of the capillary are allowed to dry, the gel matrix will retreat back into the capillary and an air gap

forms within seconds. This gap grows into a series of bubbles during electrophoresis and prevents samples from being injected into the capillary.

2. To prevent gel dehydration and to maintain aqueous continuity in the capillary, after surface scoring, break the capillary in a drop of the separation buffer.
3. After breaking, both inlet and outlet ends are immediately immersed into the buffer reservoir.
4. In systems where capillary installation is difficult, the application of glycerol to the ends after breakage may give extra min to install the capillaries.
5. High internal temperature, or local heating causes the gel to degrade quickly. A capillary gel operated at 25°C will last much longer than at 40°C.
6. Sample impurities such as high salt concentration and high molecular weight template can create a region of local heating, and reduce the capillary lifetime. This typically occurs at the injection end.
7. The applied voltage should keep below 250 V/cm to avoid the excess Joule heating.

2.1.5. Troubleshooting with Gel-Filled Capillaries

1. Drifting baseline:
   a. Improper instrument temperature control: Perform a diagnostic test.
   b. Incompatible buffer: Replace the running buffer with the manufacturer supplied-buffer and recondition the capillary.
   c. Fingerprint left on the window portion during installation: Wipe the window portion with a methanol-wetted tissue and reinstall the capillary.
2. Shifting migration time:
   a. Buffer depletion: Replace the buffer periodically as suggested by the manufacturer.
   b. Incompatible buffer: Use a manufacturer's buffer suggested.
3. Current drop:
   a. Bubble formation: Inspect the capillary for bubble.
4. No electrokinetic injection of sample:
   a. Bubble formation: Inspect the ends of the capillary for a possible void generated during capillary installation.
   b. Sample containing a high salt concentration: Remove excess salts using either membrane dialysis or other desalting devices.
5. Spikes:
   a. Voltage leakage: Clean electrodes to remove any accumulated salts and urea.
6. Distorted peaks:
   a. Overloading of sample: Reduce the injection voltage and injection time.
   b. DNA self complementarity: Denature the DNA in deionized formamide before injection onto the capillary.

2.2. Liquid Sieving Gels

2.2.1. Linear Polyacrylamide

1. Linear polyacrylamide (LPA) has been widely used as a sieving matrix for nucleic acids. Three approaches have been employed.
2. LPA may be polymerized *in situ* within the capillary and anchored to the capillary wall by covalent attachment.
3. Alternatively, it can be polymerized *in situ* within capillaries precoated with an appropriate material (typically LPA, using the procedure of Hjertén ***[19]***). It is necessary to use the *in situ* polymerized techniques when the desired LPA concentration is sufficiently high as to prevent the easy introduction and replacement of liquid polymer at reasonable

pressures. Similar to crosslinked gel-filled capillaries, the lifetime of *in situ* LPA-filled capillaries is limited.

4. For this reason noted in **step 2**, it is preferable to polymerize replaceable LPA matrices in bulk amounts and introduce it into capillaries under pressure.
5. Since LPA matrices with high % T are highly viscous, manual syringe or high-pressure pumps are typically employed for high-pressure filling, and the capillaries are used for multiple runs.
6. Alternatively, use of lower viscosity LPA solutions readily enable automatic replacement of the sieving matrix after each CE run, resulting in improved separation reproducibility and increased column lifetime.

2.2.1.1. Polymerization of LPA

1. One advantage of LPA is the ability to control both the viscosity and chain length of the matrix by manipulating the polymerization conditions. This control is not possible when using off-the-shelf commercial polymer formulations.
2. The sieving properties of LPA solutions can be controlled by the monomer concentration (%T), the concentration of catalyst (APS, TEMED), polymerization temperature and the use of chain transfer agents, such as isopropanol.
3. The disadvantages of LPA are the sensitivity of the reaction to polymerization conditions, such as the presence of oxygen. Degassing is recommended.
4. Because of the presence of residual toxic monomer and unreacted reagents in the final polymer, dialysis of bulk LPA solutions or preconditioning of *in situ* polymerized capillaries is necessary and the toxicity of the unpolymerized acrylamide.
5. Because of the hydrophilic character and the flexibility of polyacrylamide, high concentrations are required for adequate sieving strength, with 3–10% T solutions employed for separation of dsDNA fragments and 6% T solutions typically used for DNA sequencing.
6. Early studies on the use of LPA for separations of DNA used capillaries prepared by "*in situ* polymerization within the capillary" and attachment of the polymer to the wall via a bifunctional silane ***(20)***. Restriction fragments from 72–1353 bp and single-stranded oligonucleotides from 40–60 bases in length are separated at LPA concentrations of 6–12% T.
7. *In situ* polymerization within coated capillaries and the use of replaceable LPA solutions are simpler approaches and both have been used for separation of DNA restriction fragments, both for the separation of oligonucleotides and for DNA sequencing.

2.2.1.2. Optimizing LPA Polymerization

1. Ruiz-Martinez et al. ***(21)*** optimized the LPA polymerization conditions for DNA sequencing in LPA-coated capillaries using replaceable polymer solutions. It is observed that the viscosity of the replaceable matrix increases with decreasing APS and TEMED concentrations, reflecting fewer chain initiations and longer polymer chains.
2. Optimal polymerization conditions are: 3.5 *M* urea, 30% formamide, 0.02% (w/v) APS, 0.1% (v/v) TEMED. These conditions result in a 6% T LPA preparation of M_w 1×10^6 which can be introduced into the capillary within 2–3 min at a pressure of 1250 psi.
3. Replacement of the LPA matrix after each run avoids the accumulation of urea and formamide degradation products within the capillary. Using this system, sequence read lengths of 350 bases were achieved within 30 min.
4. A lower-viscosity replaceable LPA system for sequencing was developed by Grossman ***(22)***. The performance of a LPA solution of 6% T (M_w 339,000, M_n 100,000), with a viscosity of 150 cP is comparable to the performance of a crosslinked slab gel of 4% T, 5% C in a commercial automated sequencer. The lower calling selectivity/base in the CE

system is compensated by a higher separation efficiency, and read lengths of 450 bases are achieved at field strengths of 200 V/cm.

5. The longer read lengths are probably due in part to the reduced onset of biased reptation at the low field strength ***(23)***.

2.2.2. Poly(N-Substituted Acrylamides)

2.2.2.1. N-Acryloylaminoethoxy Ethanol (AAEE)

1. Chiari et al. ***(39,40)*** have developed a novel substituted acrylamide monomer, AAEE, that has been applied both as a capillary coating and as a sieving agent for nucleic acid separations. Poly(AAEE) is more hydrophilic than polyacrylamide, and is also much more resistant to hydrolysis under alkaline conditions than polyacrylamide.
2. The sieving properties of linear poly(AAEE) solutions have been compared to LPA solutions for separations of DNA restriction fragments from 57–12,216 bp in length.
3. The polymer-filled capillaries are precoated with linear poly[AAEE].
4. The gel filled capillaries are prepared by the introduction of a solution mixture of the monomer polymer, APS and TEMED into the capillary and *in situ* polymerization allowed to occur, followed by conditioning at low voltage to remove residual unreacted catalyst.
5. The optimum concentration for resolution of restriction fragments in the 80–6557 bp range is 10–12% T. This is higher than the optimum concentration of 6% T observed for LPA with the same analytes.
6. The separation performance is stable for over 35 1-h analyses. One disadvantage of this monomer is its tendency to autopolymerize ***(40)***.

2.2.2.2. N-Acryloylamino Propanol (AAP)

1. More recently, Righetti and Gelfi ***(41)*** have developed another monomer, AAP, which exhibits the advantages of AAEE: high hydrophilicity and resistance to alkaline hydrolysis, but with a reduced tendancy for autopolymerization.
2. Poly(AAP) provided good resolution of DNA restriction fragments when polymerized *in situ* at a concentration of 8% T, similar to poly(AAEE).
3. An added advantage of poly(AAP) is its stability at elevated temperatures up to 65°C, compared to both poly(AAEE), LPA, and derivatized celluloses. The high-temperature stability makes poly(AAP) a good candidate for use in procedures for the detection of point mutations at elevated temperatures.
4. Unfortunately, neither AAEE nor AAP are yet commercially available as a monomer or as polymerized products.

2.2.3. Derivatized-Cellulose Polymers

1. Water-soluble alkylated celluloses include methyl cellulose (MC), hydroxyethyl cellulose (HEC), hydroxypropyl cellulose (HPC), and hydroxypropyl methyl cellulose (HPMC). These materials exhibit several advantages over LPA when used to prepare entangled polymer solutions for nucleic acid separations.
2. First, they are commercially available and can be used directly for preparation of sieving matrices, thus avoiding the complications of polymerization and capillary conditioning necessary with the use of LPA.
3. Second, they are available in a range of polymer chain lengths so that an appropriate cellulose polymer may be selected for any particular application.
4. Third, the longer cellulose polymers provide satisfactory mesh sizes at low viscosity, enabling rapid between-run replenishment.

5. Fourth, the cellulose polymers often exhibit a strong affinity for the fused silica surface of the capillary inner wall. When absorbed to the wall of untreated capillaries, they increase the local viscosity at the surface and reduce the electroosmotic effects *(23,24)*. When the cellulose polymers are used with covalently-attached capillary coatings, the alkylated celluloses probably further reduce the residual electroosmotic flow (EOF), and thus help to protect the wall coating.
6. Fifth, these materials are both inexpensive and nontoxic.

2.2.3.1. Optimization of HEC Polymers

1. The optimization of DNA separations using HEC solutions has been extensively investigated by Barron et al. *(18)* for two HEC polymers with M_n 24,000–27,000 (M_w 138,581) and M_n 90,000–105,000 (M_w 1,315,000) respectively, using mixtures of restriction fragments ranging in size from 72–23,130 bp.
2. The optimum polymer concentrations for the shorter HEC polymer of M_w 138,581 (entanglement threshold 1.8% w/w) at which the best separation was obtained increases with decreasing DNA size.
3. For example, the optimum concentration for resolving 1078- and 1353-bp fragments was 0.5%, whereas the optimum concentration for resolving 872- and 1078-bp fragments was 0.7%.
4. This shorter HEC polymer was not effective for separating fragments larger than 1 kb, and at low concentrations of polymer DNA fragments greater than 603 bp could not be resolved.
5. For the longer HEC of M_w 1,315,000 (entanglement threshold 0.37% w/w), the resolution of larger DNA fragments (1353–23,130 bp) was optimal at polymer concentrations of 0.05–0.07%, with the optimal concentration decreasing with increasing DNA size.
6. The separation of large DNA fragments could be achieved at extraordinarily low polymer concentrations (0.00125%). The separation of small fragments was poor at polymer concentrations above the entanglement threshold, and the resolution of very small fragments (<200 bp) could not be achieved at concentrations within acceptable viscosity limits.
7. These studies suggest that the optimum HEC concentration represents that at which sufficient polymer molecules are present to provide a high probability of entanglement coupling with the DNA fragment of interest, but below the concentration at which DNA is forced into an elongated field-oriented conformation.
8. The optimal HEC concentration will be highly dependent on both the DNA and the HEC length, as illustrated in **Fig. 3**. It should be noted that all of these studies are performed using uncoated capillaries in the presence of strong electroosmotic flow, which enhanced DNA separations, particularly for very dilute polymer solutions *(26)*.
9. When using coated capillaries with no EOF, the optimal separation conditions would require higher HEC concentrations than used with uncoated capillaries.
10. Mixed HEC solutions have been used for separations of PCR-amplified human VNTR sequences for forensic analysis *(28)*. A blend of 0.3% HEC (M_n 40,000), and 0.28% HEC (M_n 140,000–160,000) can separate fragments in the 200–1000 bp range such as amplified VNTR ladders of the D1S80 alleles.
11. Low-viscosity HEC solutions have been applied to separations of short single-stranded oligonucleotides. Khan et al. *(29)* used a 4% solution of HEC (1–50 × 10^4 M_w, 9 mPa·S for a 2% w/v solution) to achieve baseline resolution of $p(dA)_{12-24}$. The use of relatively high concentration of short-chain HEC provides an effective sieving matrix for short oligonucleotides, while longer-chain HEC polymers at lower concentration did not achieve effective sieving.

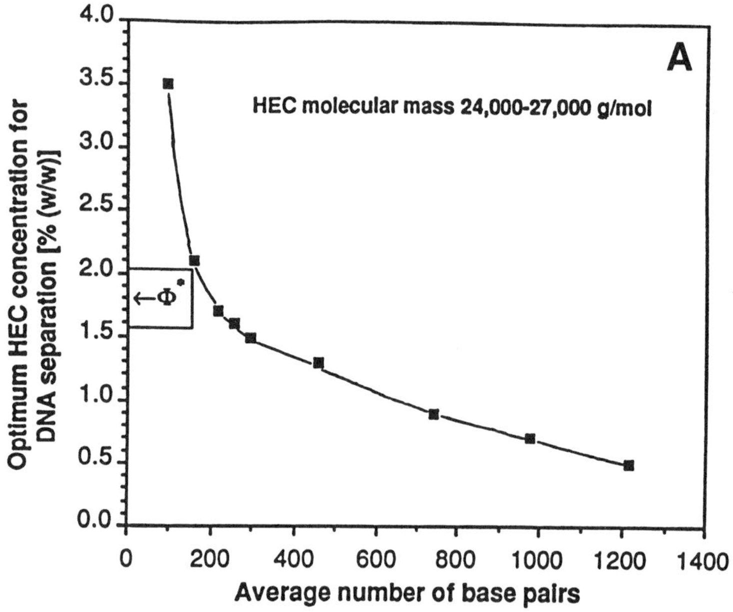

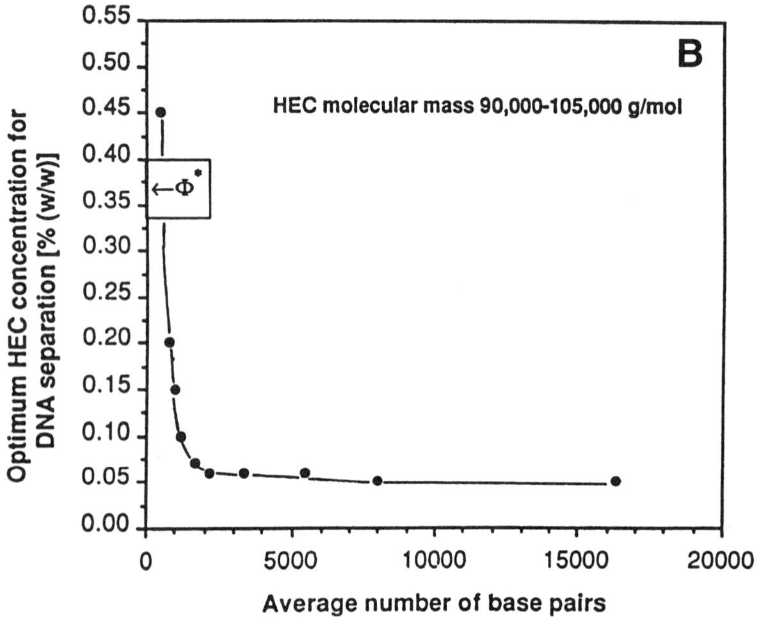

Fig. 3. The optimum HEC concentration for DNA separation, plotted as a function of the length (in base pairs) of the "average" DNA molecule being separated, in solutions of: **(A)** HEC 24,000–27,000 g/mol, and **(B)** HEC 90,000–105,000 g/mol. Each point represents the HEC concentration at which the largest difference in the electrophoretic mobilities of two similarly sized DNA restriction fragments is obtained. The overlap threshold contentrations (ϕ^*) of these two HEC samples are also shown. Reprinted from **ref. *18***.

2.2.3.1.1. Sequencing with HEC

1. HEC solutions have been used successfully for separation of Sanger sequencing products. Bashkin et al. ***(30)*** used a 2% solution of HEC (90,000–105,000 M_n) with 6 *M* urea, 10% formamide in 1X TBE buffer.
2. The formamide is added to help melt out compressions in sequencing data. This solution viscosity was low enough to be easily pumped into the capillary in less than 1 min at 400 psi.
3. The sequence resolution is equivalent to that obtained with crosslinked polyacrylamide, and allowed read lengths of up to 500 bases.
4. The HEC and urea solution is initially stirred overnight with an ion-exchange resin to remove carbamyl ions, prior to the addition of formamide and TBE. This procedure reduced the electrophoresis current and improved the separation performance.

2.2.3.2. Optimization of Hydroxypropyl Cellulosic Polymers

1. Entangled polymer separations of DNA restriction fragments using hydroxypropylcellulose (HPC) have also been investigated by Barron et al. ***(27)***. This polymer exhibits characteristics similar to those of HEC.
2. Large DNA fragments can be resolved at HPC concentrations well below the entanglement threshold, and the separation of small fragments (<310 bp) improves with increasing polymer concentration.
3. Similar to observations using HEC, the characteristics of separation with HPC did not to fit the Ogston model or the reptation model, but rather appear to occur by a mode of transient entangled coupling.
4. HPC of $M_w \geq 300,000$ provided better separation of DNA fragments greater than 1 kb, while HPC of $M_w \leq 100,000$ was unable to resolve DNA ≥ 1 kb.
5. HPC of $M_w = 300,000$ provided the best compromise for separation of DNA fragments across a wide range of sizes.
6. At comparable concentrations, high molecular weight HEC provides a better resolution than high molecular weight HPC, and equivalent separations can only be obtained at still higher concentrations of HPC. This is attributed to the more hydrophilic character of HEC, which allows it to adopt a more extended, stiffer conformation in solution with increasing likelihood of entangled coupling, than HPC.

2.2.4. Poly(Ethylene Oxide) (PEO) and Poly(Ethylene Glycol) (PEG)

2.2.4.1. Preparation of PEO Solutions

1. PEO solutions are prepared by gradually adding polymer to buffer held at 85–90°C and stirred at high speed to produce a homogeneous solution.
2. To prevent entrapment of bubbles in high-viscosity solutions within the capillary, the capillary is prefilled with a low concentration (0.5%) solution of PEO prior to the introduction of the separation matrix.
3. The advantages of PEO polymers compared to cellulose-based sieving agents includes the ease of polymer solution preparation, longer run lifetimes, and high batch-to-batch reproducibility.
4. Uncoated capillaries are used with PEO sieving matrices and the column is purged between runs with 0.1 *N* HCl to reduce the EOF. This simple treatment to protonate the surface silanols of the capillary is ineffective for LPA-filled capillaries, perhaps because the presence of residual viscous LPA prevents the regeneration of the silica surface.

2.2.4.2. DNA Sizing

1. Chang and Yeung ***(31)*** investigated the use of PEO over a range of polymer lengths for the separation of DNA restriction fragments from 18–1353 bp.
2. Using individual polymers, lower molecular weight PEO (600,000–2,000,000 M_r) provided better separation of small DNA fragments (<80 bp), whereas larger DNA fragments (80–400 bp) were better resolved with larger polymers (5,000,000–9,000,000 M_r).
3. For comparable performance, the concentrations of short-chain polymers need to be higher than separations using long-chain polymers.
4. The use of mixed polymer solutions was recommended for separation of a broad range of DNA fragments, since the mixed PEO solutions appear to provide polymer networks with random pore sizes.
5. A blend of 300,000, 600,000, 1,000,000, 2,000,000, 5,000,000, and 6,000,000 M_r PEO polymers, each at 0.7% concentration, provided satisfactory separation across a fragment size range from 18–587 bp, and exhibited a lower viscosity compared to other PEO combinations.

2.2.4.3. DNA Sequencing Using PEO Polymer Matrix

1. A binary mixture PEO polymer solution has been applied to separation of the products of the Sanger DNA sequencing reaction, and its performance compared to a commercial LPA sieving matrix ***(32)***.
2. A mixture of 1.4% (600,000 M_n) PEO and 1.5% (8,000,000 M_n) PEO is employed, instead of the 0.7% solution, 6-component PEO mixture used for restriction fragment separation.
3. The binary mixture exhibited similar performance to the more complex PEO mixture but has significantly lower viscosity.
4. The resolution of sequencing fragments from 28–420 bases was similar to that seen using a 6% T LPA (700,000–1,000,000 M_n) solution, but with markedly lower viscosity and increased separation speed than achieved using the LPA matrix.

2.2.5. Modified PEG

1. A modified PEG material has been developed for DNA sequencing in a commercial capillary-based DNA sequencer ***(33)***.
2. This material consists of a 35,000 M_w PEG derivatized to introduce C_6F_{13} or C_8F_{17} fluorocarbon groups at the polymer termini. The end-capped PEG self-assembles into micelles, which aggregate to form a continuous network.
3. A (1:1) mixture of 35 K PEG with C_6F_{13} and C_8F_{17} endgroups at a total concentration of 6% w/v provides satisfactory sieving performance for the resolution of sequence reads up to 450 bases.
4. PEG has an operational disadvantage due to its high viscosity (10,000 cP) which necessitates a pressure of 1500 psi and a filling time of 15 min for a 47-cm capillary.

2.2.6. Liquified Agarose

1. Liquified agarose solutions held above their gelling temperature can be introduced into capillaries under pressure and applied to the separation of double-stranded DNA ***(34–36)***.
2. Low-melting hydroxyethylated agaroses have been employed for the separation of DNA restriction fragments from 75–12,216 bp using thermostat-controlled, LPA-coated capillaries with a matrix of 1.7% agarose at 40°C.
3. Base-line resolution is only achieved for fragments smaller than 1 kbp, and separations are equivalent to those obtained with a 0.9% HPMC solution of 4000 cP.

4. Agarose solutions are prepared by boiling and then maintained at 50°C prior to introduction into the capillary.
5. Some commercial instruments do not have the capability of heating the inlet and outlet vessels, and these should contain only buffer solution without agarose.
6. Although a 15% hydroxyethylated agarose solution remains a stable liquid at room temperature, practically an operating temperature of 40°C is required to reduce the UV background noise arising from light scattering from the liquid gel at lower temperatures.

2.2.7. Poly(Vinyl Alcohol) (PVA)

1. PVA has been used as a capillary coating for gas chromatography and more recently has been proposed as a coating to reduce EOF and wall adsorption in CE separations of biomolecules ***(37)***. This polymer has been evaluated as a sieving agent for CE separations of DNA restriction fragments ***(38)***.
2. The 3% (w/w) PVA solution is introduced into bare fused silica capillaries and acts as both a surface-deactivating agent to reduce EOF and as the sieving agent.
3. The separation performance deteriorates after a few runs and can not be recovered by typical capillary regeneration procedures. This is attributed to the self-aggregation of PVA polymers, both in solution and at the silica wall surface.
4. The short-term performance of PVA as a sieving agent is improved by the addition of *n*-propanol to the buffer and by the use of LPA-coated capillaries, but the overall long-term performance of this agent is not satisfactory.
5. The use of PVA as a sieving agent for nucleic acids is not generally recommended.

2.3. Other Polymers

2.3.1. Trevigel

1. A commercial sieving matrix developed for slab-gel electrophoresis (Trevigel 500®, Trevigen, Inc., Gaithersburg, MD) has been evaluated by Siles et al. ***(42)*** for separation of dsDNA by CE. This material is a blend of polysaccharides containing AgaCryl®.
2. It was prepared by dissolving the dry material in buffer at 95°C, then introducing it into the capillary under pressure at 60°C.
3. The optimum separations of DNA fragments from 100–7000 bp are achieved at a TG-500 mass fraction of 0.18% in Tris-phosphate-EDTA buffer.
4. At this concentration, the sieving matrix could be replenished between runs at modest pressure (14 psi).
5. Borate buffers cannot be used because of the interaction of the borate with the polyol sieving polymers.
6. In contrast to entangled polymer solutions of the derivatized celluloses, an increase in the mass fraction of TG-500 polymer increases the resolution of both small and large DNA fragments, suggesting the polymer may have wide applications over a range of DNA sizes.

2.3.2. Block Copolymer

1. Wu et al. ***(43,44)*** have evaluated a PEO-PPO-PEO triblock copolymer as a replaceable sieving medium for CE. This polymer consists of a central block of poly(propylene oxide) (PPO), and end blocks of PEO.

Table 1
Commercially Available Entangled Polymer Sieving Systems for Nucleic Acid Separations

Supplier	Product	Application	Composition and performance	Ref.
Beckman Coulter	eCAP® ssDNA 100-R Kit	Unmodified oligonucleotide and phosphorothioate oligonucleotide analysis	Single-base resolution from 10 to 100 bases. Includes LPA solution, Tris-borate EDTA buffer, 7 *M* urea, and 2 polyacrylamide-coated capillaries.	
	eCAP® dsDNA 20,000 Kit	dsDNA fragment analysis	Separation of dsDNA fragments from 1000 to 20,000 bp. Includes LPA sieving matrix and polyacrylamide-coated capillary.	
	Separaton gel for CEQ® 2000 DNA Analysis System	DNA sequencing	Linear polyacrylamide in denaturing buffer. LOR of 300 bases in 1 h, 800 bases in 3 h.	
Bio-Rad Laboratories	CE Oligonucleotide Analysis Kit	Oligonucleotide analysis	Baseline resolution from 4–40 bases. Supplied with 2 LPA-coated capillaries.	***45***
	CE dsDNA 1000 Analysis Kit	dsDNA fragment analysis	Baseline resolution of 100-bp ladder to 1 kb. Supplied with 2 LPA-coated capillaries.	***46***
	CE dsDNA 4000 Analysis Kit	dsDNA fragment analysis	Partial resolution of 200-bp ladder to 4 kb. Supplied with 2 LPA-coated capillaries.	***46***
Hewlett Packard	Oligonucleotide Kit	Oligonucleotide analysis	Separation of oligonucleotides from 5 to 80 bases. Oligonucelotide polymer solution A recommended for unmodified oligonucleotides and polymer solution B recommended for phosphothioate antisense oligonucleotides. Supplied with PVA-coated capillary.	
	dsDNA Kit	dsDNA fragment analysis	Recommended for separations in the range of 75–10,000 bp. Supplied with 1 CEP-coated capillary.	
Molecular Dynamics	MegaBACE Gel Matrix, HEC	DNA sequencing	3% hydroxyethyl cellulose, 6 *M* urea, 10% formamide in 0.75X TBE buffer. LOR of 350–550 bases in 90 min run (120 min injection-to-injection).	
	MegaBACE Long Read Matrix	DNA sequencing	3% LPA, 7 *M* urea in Tris-TAPS buffer. LOR of up to 1000 bases in 400 min, 500–600 bases in 90 min.	

Table 1 *(continued)*

Supplier	Product	Application	Composition and performance	Ref.
PE Corp. Applied Biosystems	GeneScan polymer	SSCP analysis	Poly(dimethyl acrylamide) supplied as 7% stock solution. SSCP analysis of lengths up to 740 bp using 3% polymer in 1X TBE + 10% glycerol.	***47,48***
		Fast cycle analysis genotyping with STRs	2% polymer in 0.6X GeneScan buffer, 7 min cycle time for fragments up to 500 bp, at 5-nt resolution.	***54***
	POP-4 polymer		Single base resolution to 250 bp, sizing precision <0.15-nt standard deviation, analysis time <30 min. Polymer solution contains 4% poly(dimethyl acrylamide), 8 *M* urea, 5% 2-pyrrolidinone, 1 m*M* EDTA. Used with uncoated capillaries. Capillary lifetime of 100 injections. Compatible with temperatures up to 60°C.	***49,50***
	POP-6 polymer		6.5% poly(dimethyl acrylamide), 8 *M* urea in 100 m*M* TAPS. LOR of 600 bases in 2 h at 42°C with resolution of 0.59 in an uncoated capillary. Capillary lifetime of 100 sequencing runs.	***51–53***

2. The amphiphilic character of the block copolymer causes self-assembly into micelles at elevated temperature, which increases the hydrophobicity of the PPO block.
3. Thus, at low temperature the unimeric polymer exhibits low viscosity and can be introduced into the capillary. Once the capillary is filled it can be brought to a higher temperature at which formation of overlapping micelles yields a gel-like medium with sieving properties.
4. A PEO-PPO-PEO triblock copolymer solution in 1X TBE with1 μg/mL ethidium bromide has been used for the separation of DNA restriction fragments from 89–1560 bp.
5. The polymer solution was introduced at 4°C, then the capillary is raised to room temperature to induce the formation of micelles.

2.3.3. Commercial Entangled Polymer Matrices

1. Entangled polymer matrices are available from several vendors and are listed in **Table 1** (*see* previous pages).
2. These polymers are sold as lyophilized powders, as concentrated stock solutions with buffer diluents, or as "ready-to-use" electrophoresis buffers.
3. The polymer type, size, and concentration may be tailored for specific applications such as mutation detection, DNA genotyping, or DNA sequencing.
4. It should be noted that several of these products are designed specifically for use in dedicated CE-based analyzers, and the polymer solutions have a viscosity that is not compatible with the replenishment systems of other CE instruments.
5. For a theoretical description of the use of sieving polymers, we recommend readers to **ref.** ***55***.

References

1. Issaq, H. J. (2001) Parameters affecting capillary electrophoretic separation of DNA, in *Capillary Electrophoresis of Nucleic Acids,* Vol. 1 (Mitchelson, K. R. and Cheng, J., eds.), Humana Press, Totowa, NJ, pp. 189–199.
2. Guttman, A., Cohen, A. S., Heiger, D. N., and Karger, B. L. (1990) Analytical and micropreparative ultrahigh resolution of oligonucleotides by polyacrylamide gel high-performance capillary electrophoresis. *Anal. Chem.* **62,** 137–143.
3. Cordier, Y., Roch, O., Cordier, P., and Bischof, R. (1994) Capillary gel electrophoresis of oligonucleotides: prediction of migration times using base-specific migration coefficients. *J. Chromatogr. A* **680,** 479–489.
4. Slater, G. W., Mayer, P., and Grossman, P. D. (1995) Diffusion, Joule heating, and band broadening in capillary gel electrophoresis of DNA. *Electrophoresis* **16,** 75–80.
5. Scholsky, K. M., Mao, D. T., and Turner, K. A., (1992) Synthetic oligonucleotide analysis using substituted acrylamide capillary electrophesis. ACS Polymer Preprints 1146–1147.
6. Dolnik, V., and Novotny, M. V. (1993) Separation of amino acid homopolymers by capillary gel electrophoresis. *Anal. Chem.* **65,** 563–567.
7. Karger, B. L. and Cohen, A. S. (1989) Capillary gel electrophoresis columns, US patent 4,865,707.
8. Drossman, H., Luckey, J. A., Kostichka, A. J., D'Cunha, J., and Smith, L. M. (1990) High-speed separation of DNA sequencing reactions by capillary electrophoresis. *Anal. Chem.* **62,** 900–903.
9. Holloway, R. R. (1992) Capillary electrophoresis columns and method of preparing the same. US patent 5,167,783.

10. Schomburg, G. M., Lux,J. A., and Yin, F.-F. (1992) Production of polyacrylamide gel filled capillaries for capillary gel electrophoresis. US Patent 5,141,612.
11. Novotny, M. V., Dolnik, V., and Cobb, K. A. (1990) Capillary Gels formed by spatially progressive polymerization using migrating initiator. US patent 5,080,771.
12. Fishel, L. A., Rosenberg, B., and Juckett, D. A (1996) Method of preparing gel containing capillaries. US patent 5,522,974.
13. Zhu, M., Hansen, D. L., Burd, S., and Gannon, F. (1989) Factors affecting free zone electrophoresis and isoelectric focusing in capillary electrophoresis. *J. Chromatogr.* **480**, 311–319.
14. Ogston, A.G. (1958) The spaces in a uniform random suspension of fibers. *Trans. Faraday Soc.* **54**, 1754–1757.
15. Lerman, L. S. and Frisch, H. L. (1982) Why does the electrophoretic mobility of DNA in gels vary with the length of the molecule? *Biopolymers* **21**, 995–997.
16. Grossman, P. D. and Soane, D. S. (1991) Capillary electrophoresis of DNA in entangled polymer solutions. *J. Chromatogr.* **559**, 257–266.
17. Barron, A. E., Soane, D. S., and Blanch, H. W. (1993) Capillary electrophoresis of DNA in uncross-linked polymer solutions. *J. Chromatogr. A* **652**, 3–16.
18. Barron, A. E., Blanch, H. W., and Soane, D. S. (1994) A transient entanglement coupling mechanism for DNA separation by capillary electrophoresis in ultradilute polymer solutions. *Electrophoresis* **15**, 597–615.
19. Hjertén, S. (1985) High-performance electrophoresis: elimination of electroendosmosis and solute adsorption. *J. Chromatogr.* **347**, 191–198.
20. Heiger, D., Cohen, A. S., and Karger, B. L. (1990) Separation of DNA restriction fragments by high performance capillary electrophoresis with low and zero crosslinked polyacrylamide using continuous and pulsed electric fields. *J. Chromatogr.* **516**, 33–48.
21. Ruiz-Martinez, M., Berka, J., Belenkii, A., Foret, F., Miller, A. W., and Karger, B. L. (1993) DNA sequencing by capillary electrophoresis with replaceable linear polyacrylamide and laser-induced fluorescence. *Anal. Chem.* **65**, 2851–2858.
22. Grossman, P. D. (1994) Electrophoretic separation of DNA sequencing extension products using low-viscosity entangled polymer networks. *J. Chromatogr. A* **663,** 219–227.
23. Best, N., Arriaga, E., Chen, D. Y., and Dovichi, N. J. (1994) Separation of fragments up to 570-bases in length by use of 6% T non-cross-linked polyacrylamide for DNA sequencing in capillary electrophoresis. *Anal. Chem.* **66,** 4063–4067.
24. Hjertén, S. (1967) Free zone electrophoresis. *Chromatogr. Rev.* **9**, 122-219.
25. Bello, M. S, de Besi, P., Rezzonico, R., Righetti, P. G., and Casiraghi, E. (1994) Electroosmosis of polymer solutions in fused silica capillaries. *Electrophoresis* **15,** 623–626.
26. Barron, A. E., Sunada, W. M., and Blanch, H. W. (1995) The use of coated and uncoated capillaries for the electrophoretic separation of DNA in dilute polymer solutions. *Electrophoresis* **16,** 64–74.
27. Barron, A. E., Sunada, W. M., and Blanch, H. W. (1996) The effects of polymer properties on DNA separations by capillary electrophoresis in uncross-linked polymer solutions. *Electrophoresis* **17**, 744–757.
28. Isenberg, A. R, McCord, B. R., Koons, B. W., Budowle, B., and Allen, R. O. (1996) DNA typing of a polymerase chain reaction amplified D1S80/amelogenin multiplex using capillary electrophoresis and a mixed entangled polymer matrix. *Electrophoresis* **17,** 1505–1511.
29. Khan, K., van Schepdael, A., and Hoogmartens, J. (1996) Capillary electrophoresis of oligonucleotides using a replaceable sieving buffer with low viscosity-grade hydroxyethyl cellulose. *J. Chromatogr. A* **742,** 267–274.

30. Bashkin, J., Marsh, M., Barker, D., and Johnston, R. (1996). DNA sequencing by capillary electrophoresis with a hydroxyethylcellulose sieving buffer. *App. Theo. Elect.* **6,** 23–28.
31. Chang, H.-T. and Yeung, E. S. (1995). Poly(ethyleneoxide) for high-resolution and high-speed separation of DNA by capillary electrophoresis. *J. Chromatogr. B* **669,** 113–123.
32. Fung, E. N. and Yeung, E. S. (1995) High-speed DNA sequencing by using mixed poly(ethyene oxide) solutions in uncoated capillary columns. *Anal. Chem.* **67,** 1913–1919.
33. Menchen, S., Johnson, B., Winnik, M. A., and Xu, B. (1996) Flowable networks as DNA sequencing media in capillary columns. *Electrophoresis* **17,** 1451–1459.
34. Bocek, P. and Chrambach, A. (1991) Capillary electrophoresis of DNA in agarose solutions at 40°C. *Electrophoresis* **12,** 1059–1061.
35. Bocek, P. and Chrambach, A. (1992) Capillary electrophoresis in agarose solutions: Extension of size separations to DNA of 12-kb in length. *Electrophoresis* **13,** 31–34.
36. Palm, A. K. (2001) Capillary electrophoresis of DNA fragments with replaceable low-gelling agarose gels, in *Capillary Electrophoresis of Nucleic Acids,* Vol. 1 (Mitchelson, K. R. and Cheng, J., eds.), Humana Press, Totowa, NJ, pp. 279–290.
37. Belder, D. and Schomburg, G. (1992) Modification of silica surfaces for CZE by adsorption of non-ionic hydrophilic polymers or use of radial electric fields. *J. High Res. Chromatogr.* **15,** 686–693.
38. Kleemiss, M. H., Gilges, M., and Schomburg, G. (1993) Capillary electrophoresis of DNA restriction fragments with solutions of entangled polymers. *Electrophoresis* **14,** 515–522.
39. Chiari, M., Micheletti, C., Nesi, M., Fazio, M., and Righetti, P. G. (1994) Towards new formulations for polyacrylamide matrices: N-acryloylaminoethoxyethanol, a novel monomer combining high hydrophilicity with extreme hydrolytic stability. *Electrophoresis* **15,** 177–186.
40. Chiari, M., Nesi, M., and Righetti, P. G. (1994) Capillary zone electrophoresis of DNA fragments in a novel polymer network: Poly(N-acryloylaminoethoxyethanol). *Electrophoresis* **15,** 616–622.
41. Righetti, P. G. and Gelfi, C. (1997) Recent advances in capillary electrophoresis of DNA fragments and PCR products in poly(*N*-substituted acrylamides). *Anal. Biochem.* **244,** 195–207.
42. Siles, B. A., Collier, G. B., Reeder, D. J., and May, W. E. (1996) The use of a new gel matrix for the separation of DNA fragments: a comparison study between slab gel electrophoresis and capillary electrophoresis. *App. Theo. Elect.* **6,** 15–22.
43. Wu, C., Liu, T., and Chu, B. (1997) Characterization of the PEO-PPO-PEO triblock copolymer and its application as a separation medium in capillary electrophoresis. *Macromolecules* **30,** 4574–4583.
44. Chu, B., Liu, T., Wu, C.-H., and Liang, D. (2001) DNA capillary electrophoresis using block copolymer as a new separation medium, in *Capillary Electrophoresis of Nucleic Acids,* Vol. 1 (Mitchelson, K. R. and Cheng, J., eds.), Humana Press, Totowa, NJ, pp. 225–238.
45. Talmadge, K., Zhu, M., Olech, L., and Siebert, C. (1996) Oligonucleotide analysis by capillary polymer sieving electrophoresis using acryloylaminoethoxyethanol-coated capillaries. *J. Chromatogr.* **744,** 347–354.
46. Talmadge, K., Tan, A. K., and Zhu, M. (1997) DNA fragment analysis by capillary polymer sieving using poly(acryloylaminoethoxyethanol)-coated capillaries. J. *Chromatogr.* **781,** 335–345.
47. Inazuka, M., Wenz, H.-M., Sakabe, M., Tahira, T., and Hayashi, K. (1997) A streamlined mutation detection system: multicolor post-PCR fluorescence labeling and single-strand conformational polymorphism analysis by capillary electrophoresis. *Genome Res.* **7,** 1094–1103.

48. Atha, D., Wenz, H.-M., Morehead, H., and Tian, J. (1998) Detection of p53 point mutations by single strand conformational polymorphism: analysis by capillary electrophoresis. *Electrophoresis* **19,** 172–179.
49. Wenz, H.-M., Robertson, J. M., Menchen, S., Oaks, F., Demorest, D., Scheibler, D., Rosenblum, B. B., Wike, C., Gilbert, D. A., and Efcavitch, J. W. (1998) High-precision genotyping by denaturing capillary electrophoresis. *Genome Res.* **8,** 69–80.
50. Lazaruk, K., Walsh, P. S., Oaks, F., Gilbert, D., Rosenblum, B. B., Menchen, S., Scheibler, D., Wenz, H.-M., Holt, C., and Wallin, J. (1998) Genotyping of forensic short tandem repeat (STR) systems based on sizing precision in a capillary electrophoresis instrument. *Electrophoresis* **19,** 86–93.
51. Rosenblum, B. B., Oaks, F., Menchen, S., and Johnson, B. (1997) Improved single-strand DNA sizing accuracy in capillary electrophoresis. *Nucleic Acids Res.* **25,** 3925–3929.
52. Madabhushi, R. S. (1998) Separation of 4-color DNA sequencing extension products in noncovalently coated capillaries using low viscosity polymer solutions. *Electrophoresis* **19,** 224–230.
53. Madabhushi, R. S. (2001) DNA sequencing in noncovalently coated capillaries using low viscosity polymer solutions, in *Capillary Electrophoresis of Nucleic Acids,* Vol. 2 (Mitchelson, K. R. and Cheng, J., eds.), Humana Press, Totowa, NJ, pp. 309–315.
54. Wenz, H.-M., Dailey, D., and Johnson, M. D. (2001) Development of a high-throughput capillary electrophoresis protocol for DNA fragment analysis, in *Capillary Electrophoresis of Nucleic Acids,* Vol. 2 (Mitchelson, K. R. and Cheng, J., eds.), Humana Press, Totowa, NJ, pp. 3–17.
55. Heller, C. (2001) Influence of polymer concentration and polymer composition on capillary electrophoresis of DNA, in *Capillary Electrophoresis of Nucleic Acids,* Vol. 1 (Mitchelson, K. R. and Cheng, J., eds.), Humana Press, Totowa, NJ, pp. 111–123.

12

Parameters Affecting Capillary Electrophoretic Separation of DNA

Haleem J. Issaq

1. Introduction

Since its introduction in 1930 by Tiselius *(1)*, slab-gel electrophoresis (SGE) has been utilized as an important separation technique in molecular biology for the resolution of DNA fragments, polymerase chain reaction (PCR) products, polynucleotides, proteins, and other compounds. Capillary electrophoresis (CE) which was introduced in 1981 in its modern and practical instrumental format by Jorgenson and Lukacs *(2)* for the separation of charged molecules, has been adapted by analytical biochemists for the same applications that SGE can perform, but faster and cheaper with high resolution. SGE allows the application of multiple samples on the same slab gel. Array capillary electrophoresis and microfabricated, capillary array electrophoresis microplates (microchips) allow the simultaneous sequencing of 96 DNA samples using the same instrument. CE offers high resolution, speed, automation of sample handling and analysis, uses small sample volumes (nL), and produces minimum waste. Also, CE with laser-induced fluorescence (LIF) allows sensitive detection in the zeptomole range. Today CE has been applied for the analysis of oligonucleotides, restriction digests, point mutations, DNA-protein and protein-drug interactions, and PCR products *(3–23)*.

In this chapter, we will discuss the effect of different parameters on the migration time, the resolution, and the speed of analysis of DNA fragments and PCR products. These parameters include the column length, the applied voltage, the column temperature, the gel type and concentration, and the buffer ionic strength.

2. Materials

1. CE was performed using a Beckman P/ACE 5200 unit equipped with a LIF detector.
2. Fluorescence excitation was provided by a 488-nm argon-ion laser (ILT, Salt Lake City, UT). The laser light was guided to the LIF module via an optical fiber.
3. Excitation power was approx 5 mW, which was measured at the fiber tip.

From: *Methods in Molecular Biology, Vol. 162:*
Capillary Electrophoresis of Nucleic Acids, Vol. 1: Introduction to the Capillary Electrophoresis of Nucleic Acids
Edited by: K. R. Mitchelson and J. Cheng © Humana Press Inc., Totowa, NJ

4. The fluorescence light was detected after passing through 488-nm cutoff and 520-nm interference filters.
5. Separations were performed in a 50-μm id DB-17 capillary (J&W, Folsom, CA) mounted on the Beckman capillary cartridge
6. The electric field for CE separation was provided by a Glassman high-voltage power supply (Whitehouse Station, NJ).
7. The capillary was mounted on a *x-y* translational stage for precise movement.
8. The argon-ion laser beam was focused onto the capillary with a biconvex lens. Fluorescence was collected at 90° from the excitation beam with a 10X microscope objective.
9. After passing through a 488-nm interference filter, the fluorescence was detected by a photomultiplier tube (PMT, Oriel, Stratford, CT). The PMT current was monitored by a picoammeter (Keithley, Cleveland, OH). The output from the picoammeter was fed to a Beckman 406 data acquisition module.

3. Methods

1. A hydroxyethyl cellulose (HEC) buffer, a polyacrylamide (PAA) buffer, and a commercially available, replaceable gel buffer for CE (Sigma Chemical, St. Louis, MO) are used for DNA separations.
2. The Sigma gel buffer consists of a proprietary sieving medium prepared in 260 m*M* Tris-borate-EDTA buffer (TBE) at pH 8.5. In some experiments, the Sigma buffer is diluted to 60% with deionized water before use.
3. HEC (M_r 90,000–105,000) and 10% polyacrylamide (PAA, M_r 70,000–1,000,000) are obtained from Polysciences, Inc. (Warrington, PA). The HEC and PAA gel buffers are prepared in Tris-borate buffer containing 1 m*M* EDTA. For on-column DNA fluorescence labeling, 1 μL of the intercalating dye YO-PRO-1 in 1 m*M* DMSO (Molecular Probes, Eugene, OR) was added per milliliter of buffer, giving a final concentration of 1 μ*M*.
4. The high-pressure setting of the P/ACE unit was changed from the default of 20 psi to 40 psi to facilitate the flushing process.
5. To avoid contamination of samples by the running buffer, the capillary and the electrode surfaces are cleaned for a few seconds by dipping the ends in water, before each sample is electrokinetically injected.
6. The effective 50-μm id, DB-17 capillary column lengths are, either 20, 7, 2, or 1 cm (*see* **item 7**).
7. A home built system similar to that used previously ***(24)*** is employed with capillaries shorter than 7 cm (the 1- and 2-cm capillaries).
8. Beckman System Gold software controls the Beckman CE system and the 406 module. The rate of data sampling is 20 Hz with a rise time of 0.1 s.
9. Detailed experimental conditions are described in the legends to **Figs. 1–5**.

3.1. Column Type

1. Column type (coated or uncoated) and the length, the applied voltage, the column temperature, the buffer type and its concentration are each parameters that influence not only the migration times of DNA molecules, but also the resolution between different molecules. It is known that the use of an uncoated or a coated capillary, i.e., electroosmotic flow is suppressed in the latter coated capillary, and also the pH of the buffer will influence the migration times of DNA.
2. An increase in the applied voltage would lead to:
   a. An increase in the resolution between DNA molecules,

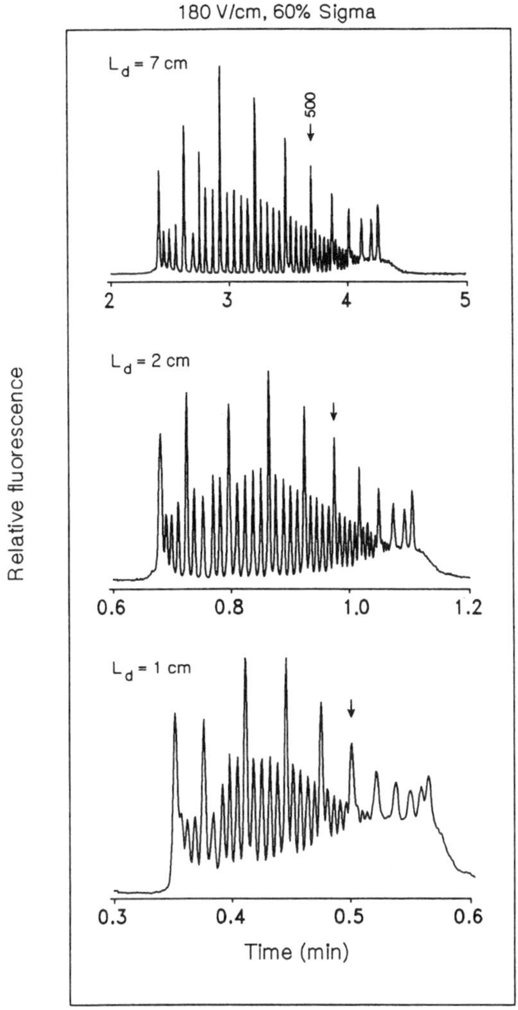

Fig. 1. Electropherogram of the separation of a 20- and 100-bp DNA standard solution in 10 m*M* Tris-borate buffer containing 0.1 m*M* EDTA, pH 8.0, and 1 μL/mL YO-PRO-1 intercalting fluorescent dye (1 m*M* in DMSO). Detection was achieved via an argon-ion laser λ_{ex} = 488 nm and a 520-nm interference filter. A 50-μm id DB-17 coated fused silica capillary of 7-cm, 2-cm, and 1-cm length filled with Sigma DNA replaceable gel buffer was used with an applied voltage of 180 V/cm. Reprinted from **ref. *17***.

 b. An increase in the Joule heating, which, if not efficiently dissipated will lead to lowering of the resolution between molecules, and
 c. A decrease in migration time of DNA. These parameters will be discussed in detail in **Subheading 3.**

3. Since DNA separation involves replaceable gels (*see* **Subheading 3.5.**), it is wise to use neutrally coated capillaries to eliminate any charges on the inner surface of the capillary,

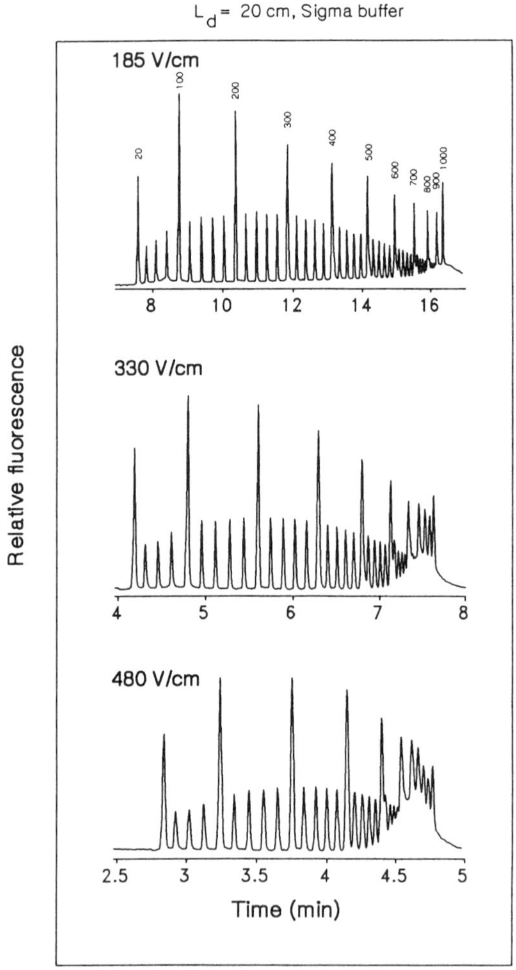

Fig. 2. Electropherogram of the separation of a 20- and 100-bp DNA standard solution in 10 m*M* Tris-borate buffer containing 0.1 m*M* EDTA, pH 8.0, and 1 µL/mL YO-PRO-1 intercalting fluorescent dye (1 m*M* in DMSO). Detection was achieved via an argon-ion laser λ_{ex} = 488 nm and a 520-nm interference filter. A 20 cm long × 50 µm id, DB-17 coated fused silica capillary filled with Sigma DNA replaceable gel buffer was used with an applied voltage of 185, 330, and 480 V/cm. Reprinted from **ref. *17***.

in order to prevent the bulk flow of the gel from the capillary. The replaceable gel should stay stationary in the capillary during the separation process.

4. We use a commercially coated, DB-17, capillary in the present work. Coated capillaries are found to give superior performance of DNA fragment separations when compared with uncoated fused silica capillaries ***(25)***. It was also observed that better resolution is obtained when narrower diameter capillaries are used, due to the more efficient Joule heat dissipation.

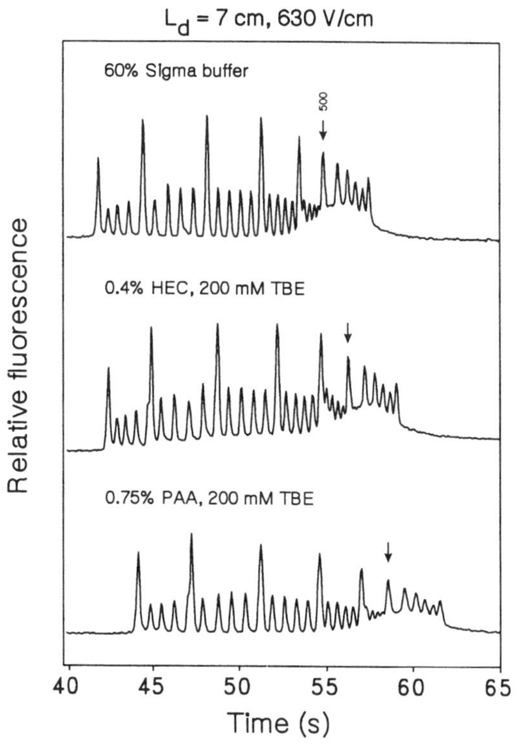

Fig. 3. Electropherogram of the separation of a 20- and 100-bp ladder using different replaceable gel buffers, a 7-cm effective length D-17 capillary and an applied voltage of 630 V/cm. Reprinted from **ref. *17***.

3.2. Separation of a 20-bp Ladder

1. The effect of capillary length on the separation of 20-bp DNA ladder is studied. **Figure 1** shows a series of electropherograms that compare the separation of the standard DNA solution using 7-, 2-, and 1-cm effective length capillaries under identical experimental conditions.
2. The results show no appreciable difference in resolution of the DNA fragments between the 7-cm long and the 2-cm long capillaries.
3. However, the resolution of the same mixture using a 1-cm capillary is not as good as desired. This is due to the fact that we have not optimized the injection and separation procedures.
4. It is notable however, that the use of a shorter capillary contributed to a considerable savings in the length of analysis time; 4.5 min for 7-cm capillary, to 1.2 min for the 2-cm capillary, respectively. Although these parameters were not at their optimum it was possible in the past to use the 1-cm capillary to resolve the two *TGF*-β_1 PCR products ***(18)***. A short effective length capillary of 1–2 cm can be used to extend the size separation range of DNA fragments at low field strength.

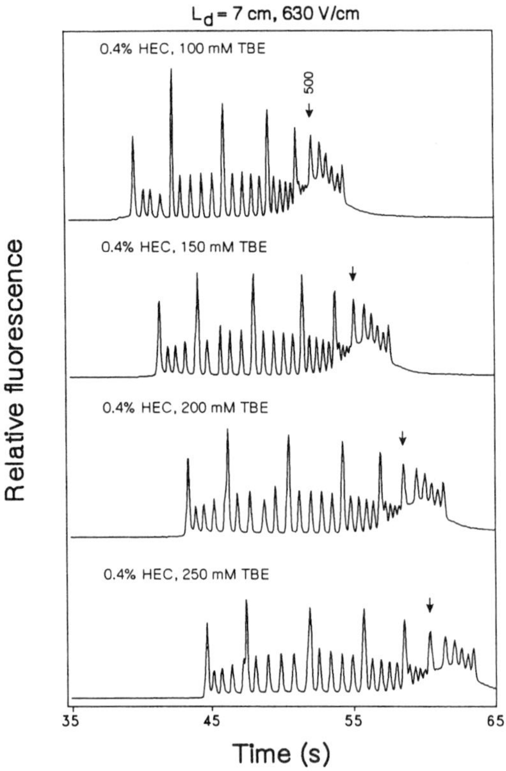

Fig. 4. Electropherogram of the separation of a 20- and 100-bp ladder using different buffer concentrations and 0.4 hydroxyethyl cellulose replaceable gel buffer, 7-cm effective length capillary and applied voltage of 630 V/cm. Reprinted from **ref. *17***.

3.3. Effect of Applied Voltage

1. Applied voltage plays an important role in CE. Increasing the applied voltage will:
   a. Increase the resolution;
   b. Increase the Joule heating;
   c. Result in faster migration times; and
   d. Shorten the overall run time of the experiment.
2. To study the effect of applied voltage on the separation of DNA fragments we used a liquid cooled cartridge, which efficiently eliminated the effects of Joule heating except at extremely high applied voltages. A plot of Ohm's Law, i.e., field strength vs current showed that above 480 V/cm, deviation from linearity was observed, which means that there is nonefficient dissipation of the heat generated inside the capillary.
3. **Figure 2** is a series of electropherograms of a mixture of the 20- and 100-bp DNA fragments standard solution obtained at three different field strengths using a 20-cm capillary and the Sigma buffer. When the applied voltage is increased from 185 V/cm to 480 V/cm two things occur:

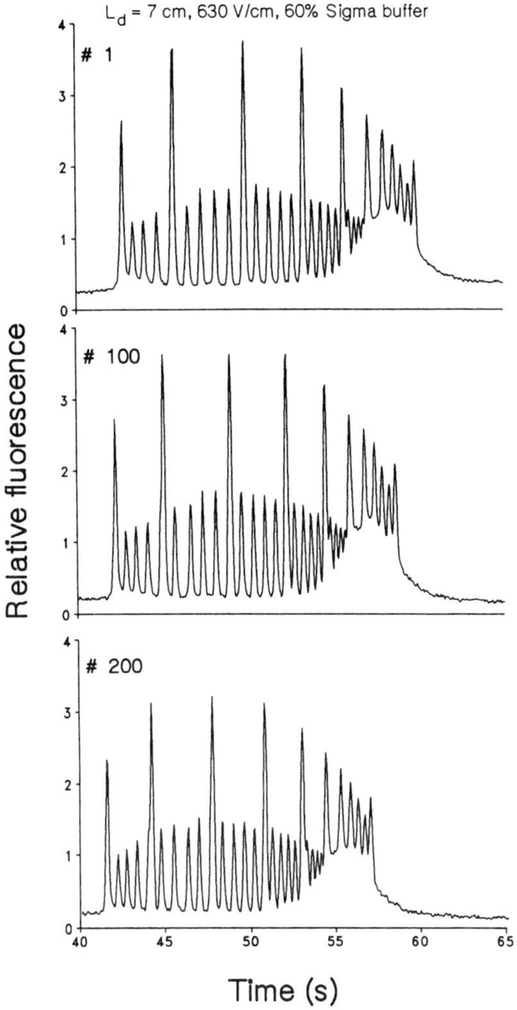

Fig. 5. Reproducibility of the 20- and 100-bp ladder after 1, 100, and 200 analyses using the Sigma replaceable DNA gel buffer. Before each injection the capillary was rinsed with the running buffer for 30 s. Other conditions as in **Fig. 4**. Reprinted from **ref. *17***, with the permission of Wiley-VCH.

a. The time of analysis reduces from 16 min to 5 min, and
b. A deterioration in the resolution of fragments greater than 500 bp occurs, with virtually no effect on the resolution of fragments below 500 bp.

4. This is due to the fact that small DNA fragments migrate in the gel as a mobile coil, explained by the "Ogston model." Whereas larger fragments migrate via a different mode explained by the "biased reptation model" in which DNA stretch and align themselves in the direction of the field ***(26)***. These modes of migration of extended polymers differ greatly in their form and in their dependence on experimental conditions ***(26,27)***.

5. These results agree with those observed by Gutman et al. *(20)*, in which the resolution of large fragments decreases at higher voltages. The peak efficiency decreased for fragments below 500 bp when high voltages are applied. For example, for the 260-bp fragment, the column efficiency dropped by 67% when field strength is increased from 185 to 480 V/cm. This decrease in efficiency at higher voltages may be attributed to the reptation mode of DNA migration and to temperature effects, where DNA molecules align themselves in the direction of the field (they are stretched) rather than move through the gel as a coil. The DNA fragments, which move through the gel as a coil, will allow for higher resolution of large DNA fragments. Most PCR products for molecular diagnostics are under 500 bp, therefore, high voltages can be used to achieve good and meaningful separations.

3.4. Effect of Temperature

1. Experimental temperature in CE affects the migration times of analytes, with higher temperatures resulting in faster migration times, due to the decrease in the viscosity of the running buffer. The effect of temperature on the separation of DNA fragments is studied. Three temperatures are used, 25, 40, and 50°C.
2. We find that although the migration times of dsDNA fragments become shorter, 10.5 min at 25°C to 6.5 min at 50°C at higher temperatures, resolution is not appreciably affected, and baseline resolution at the three temperatures is realized *(28)*.

3.5. Effect of Gel Type

1. The use of replaceable gels, polymer solutions, for the size-selective separations of DNA fragments has been reported. The advantage of the use of replaceable polymer solutions over gel-filled capillaries is that the capillary is filled and emptied after each analysis, which prevents cross contamination.
2. Many types of gels have been used for DNA separations *(23,26)*; however, the most commonly used polymer solutions are made of linear polyacrylamide (LPA) or cellulose derivatives. In our hands, HEC gave more reproducible results than LPA. Also, HEC solutions are less toxic, cheaper, and less viscous than LPA solutions.
3. The effect of gel type on the resolution of DNA fragments is evaluated. Three different gels are used: Sigma DNA separation buffer, 0.4% HEC and 0.75% PAA. **Figure 3** shows that the resolution is comparable in all the three gel matrices. Roed et al. *(23)* also evaluate the effect of polymer concentration in terms of separation efficiency, resolution and separation time, and report that the best performance is obtained using a polymer concentration of 0.5% for all polymers, independent of the length of unit polymer molecules, or the type of polymer.

3.6. Effect of Buffer Concentration

1. The effect of the buffer ionic strength in which the sieving gel is made on the resolution of analytes is also studied. For a given set of conditions (*see* **Fig. 4**), where the buffer concentration is increased from 100 to 250 m*M*, the migration times increased by 20% and the resolution improves.
2. It is important, therefore to pay attention to the ionic strength, for the sake of DNA fragment resolution, for the duration of the experiment, and for Joule heat output. It is also important to know what gel concentration is being used, because the gel concentration and viscosity will affect all of these parameters, the resolution, the duration of analysis, and the pressure necessary to fill the capillary with matrix.

3. Lower concentration gels are used to resolve larger DNA fragments. Low concentration gels have another advantage, as decreasing the gel concentration means that it is easier to fill the capillary with a diluted (less viscous) gel than a more concentrated (more viscous) one, especially if the gel dilution does not affect resolution.

3.7. Reproducibility

1. The reproducibility of a particular method is a very important factor in the acceptability of the method for routine use. As mentioned earlier, we found that HEC gel buffers gave very reproducible results.
2. **Figure 5** shows the excellent reproducibility of the method after 100 and 200 injections using a 7-cm capillary, Sigma DNA replaceable gel buffer and 630 V/cm.

Acknowledgments

This project has been funded in whole or in part with Federal funds from the National Cancer Institute, National Institutes of Health, under Contract No. NO1-CO-56000. By acceptance of this article, the publisher or recipient acknowledges the right of the U.S. Government to retain a nonexclusive, royalty-free license and to any copyright covering the article. The content of this publication does not necessarily reflect the views or policies of the Department of Health and Human Services, nor does mention of trade names, commercial products, or organizations imply endorsement by the U.S. Government.

References

1. Tiselius, A. (1930) *The Moving Boundary Method of Studying the Electrophoresis of Proteins,* Inaugural Dissertation. Almqvist and Wiksells Boktryckeri AB, Uppsala.
2. Jorgenson, J. W. and Lukacs, K. D., (1981) Zone electrophoresis in open-tubular glass capillaries. *Anal. Chem.* **53,** 1298–1302.
3. Effenhauser, C. S., Paulus, A., Manz, A., and Widmer, H. M., (1994) High-speed separation of antisense oligonucleotides on a micromachined capillary electrophoresis device. *Anal. Chem.* **66,** 2949–2953.
4. Oefner, P. J. and Bonn, G. K. (1992) Comparative study of capillary zone electrophoresis and high-performance liquid chromatography in the analysis of oligonucleotides and DNA. *J. Chromatogr.* **625,** 331–340.
5. Maschke, H. E., Frenz, J., Belenkii, A., Karger, B. L., and Hancock, W. S. (1993) Ultrasensitive plasmid mapping by high performance capillary electrophoresis. *Electrophoresis* **14,** 509–514.
6. Gelfi, C., Cremoresi, L., Ferrari, M., and Righetti, P. G. (2001) Point mutation detection by temperature-programmed capillary electrophoresis, in *Capillary Electrophoresis of Nucleic Acids,* Vol. 2 (Mitchelson, K. R. and Cheng, J., eds.), Humana Press, Totowa, NJ, pp. 73–88.
7. Chan, K. C., Whang, C-W., and Yeung, E. S. (1993) Separation of DNA restriction fragments using capillary electrophoresis. *J. Liq. Chromatogr.* **16,** 1941–1962.
8. Janini, G. M., Fisher, R. J., Henderson, L. E., and Issaq, H. J. (1995) Application of capillary zone electrophoresis for the analysis of proteins, protein-small molecules and protein-DNA interactions. *J. Liq. Chromatogr.* **18,** 3617–3628.
9. Manabe, T., Chen, N., Terabe, S., Yohda, M., and Endo, I. (1994) Effects of linear polyacrylamide concentrations and applied voltages on the separation of oligonucleotides

and DNA sequencing fragments by capillary electrophoresis. *Anal. Chem.* **66,** 4243–4252.

10. Fung, E. N. and Yeung, E. S. (1995) High-speed DNA sequencing by using mixed poly(ethylene oxide) solutions in uncoated capillary columns. *Anal. Chem.* **67,** 1913–1919.
11. Butler, J. M., McCord, B. R., Jung, J. M., Wilson, M. R., Budowle, B., and Allen, R. O. (1994) Quantitation of polymerase chain reaction products by capillary electrophoresis using laser fluorescence. *J. Chromatogr. B* **658,** 271–280.
12. Vincent, U., Patra, G., Therasse, J., and Gareil, P. (1996) Quantitation of polymerase chain reaction-amplified dna fragments by capillary electrophoresis and laser-induced fluorescence detection. *Electrophoresis* **17,** 512–517.
13. Williams, S. J., Schwer, C., Krishnarao, A. S. M., Heid, C., Karger, B. L., and Williams, P. M. (1996) Quantitative competitive polymerase chain reaction: Analysis of amplified products of the HIV-1 *gag* gene by capillary electrophoresis with laser-induced fluorescence detection. *Anal. Biochem.* **236,** 146–152.
14. Ueno, K. and Yeung, E. S. (1994) Simultaneous monitoring of DNA fragments separated by electrophoresis in a multiplexed array of 100 capillaries. *Anal. Chem.* **66,** 1424–1431.
15. Mathies, R. A. and Huang, X. C. (1992) Capillary array electrophoresis: An approach to high speed, high-throughput DNA sequencing. *Nature* **359,** 167–169.
16. Stebbins, M. A., Davis, J. B., Clark, B. K., and Sepaniak, M. J. (1996) Optimization of separation conditions in the size-selective capillary electrophoretic separation of DNA fragments. *J. Micro. Sep.* **8,** 485–494.
17. Issaq, H. J., Chan, K. C., and Muschik, G. M. (1997) The effect of column length, applied voltage, gel type and concentration on the CE separation of DNA fragments and PCR products. *Electrophoresis* **18,** 1153–1158.
18. Chan, K. C., Muschik, G. M., Issaq, H. J., Garvey, K. J., and Generlette, P. L (1996) High speed screening of polymerase chain reaction products by capillary electrophoresis. *Anal. Biochem.* **243,** 133–139.
19. Chan, K. C., Muschik, G. M., and Issaq, H. J. (1997) High-speed electrophoretic separation of DNA fragments using a short capillary. *J. Chromatogr.* **695,** 113–115.
20. Guttman, A., Wanders, B., and Cooke, N. (1992) Enhanced separation of DNA restriction fragments by capillary gel electrophoresis using field strength gradients. *Anal. Chem.* **64,** 2348–2351.
21. Gutman, A. (1998) Rehydratable polyacrylamide gels for capillary electrophoresis. *J. Liq. Chromatogr. and Rel. Technol.* **21,** 1249–1258.
22. Wooley, A. T. and Mathies, R. A. (1994) Ultra high speed DNA fragment separations using microfabricated capillary array electrophoresis chips. *Proc. Natl. Acad. Sci. USA* **91,** 11,348–11,352.
23. Roed, L., Arsky, I., Lundanes, E., and Greibrokk, T. (1998) Rapid and reproducible CE separations of double-stranded DNA fragments in a simple methyl cellulosic sieving system. *Chromatographia* **47,** 125–134.
24. Chan, K. C., Janini, G. M., Muschik, G. M., and Issaq, H. J. (1993) Pulsed UV laser-induced fluorescence detection of native peptides and proteins. *J. Liq. Chromatogr.* **16,** 1877–1890.
25. Strege, M. and Lagu, A. (1991) Separation of DNA restriction fragments by capillary electrophoresis using coated fused silica capillaries. *Anal. Chem.* **63,** 1233–1236.
26. Slater, G. W., Desruisseaux, C., and Hubert, S. J. (2001) DNA separation mechanisms during electrophoresis, in *Capillary Electrophoresis of Nucleic Acids,* Vol. 1 (Mitchelson, K. R. and Cheng, J., eds.), Humana Press, Totowa, NJ, pp. 27–41.

27. Clark, B. K., Nickles, C. L., Morton, K. C., Kovac, J., and Sepaniak, M .J. (1994) Rapid separation of DNA restriction digests using size selective capillary electrophoresis with application to DNA fingerprinting. *J. Microcol. Sep.* **6,** 503–513.
28. Issaq, H. J., Xu, H., Chan, K. C., and Dean, M. C (1999) Effect of temperature on the separation of DNA fragments by HPLC and CE: a comparative study. *J. Chromatogr. B*, in press.

III

Electrophoresis Matrix: Applications for DNA Fragment Sizing

13

DNA Separation by Capillary Electrophoresis in Lyotropic Polymer Liquid Crystals

Yingjie Liu and Randolph L. Rill

1. Introduction

Electrophoresis is a critical separation method of molecular biology and biotechnology. Conventional electrophoresis of nucleic acids is conducted in gels, consisting of crosslinked polyacrylamide or agarose, that provide mechanical support and quench convection. A polymer network is also essential for separations, since the electrophoretic mobilities of nucleic acids in simple buffers depend only weakly, or not at all, on molecular size *(1)*. In general terms, polymer networks separate by sorting molecules that differ in size and shape. The actual sorting mechanism depends on the size and flexibility of the particular nucleic acid, and the nature and concentration of the polymer support *(2–7)*.

1.1. Capillary Electrophoresis in Replaceable Polymer Solutions

Electrophoresis in thin capillaries effectively dissipates heat, which allows application of high-voltage gradients and produces much more efficient separations than conventional electrophoresis *(8–11)*. Capillary "gel" electrophoresis (CGE) of DNA is now done almost exclusively with replaceable solutions of polymers that form random coils such as linear polyacrylamide, hydroxyalkyl celluloses, or poly(ethylene oxide), reviewed in *(2,12–18)*. Chains of these random coil polymers entangle when the concentration is sufficiently high, creating dynamic networks that sort molecules in ways more or less like crosslinked polymer gels *(2–7,12–18)*. The replacement of the polymer solution between electrophoresis runs enhances run reproducibility and avoids the problems associated with the physical and chemical instabilities of crosslinked gels.

The ease of capillary refilling depends on the sizes of nucleic acids being separated, since the size range of the analyte determines the most suitable polymer concentration. Solutions of modest polymer concentration separate ssDNA or dsDNA molecules ranging in size from a few hundred to many thousand nucleotides, or base pairs (bp), respectively. These solutions are fluid enough to be loaded into capillaries by apply-

From: *Methods in Molecular Biology, Vol. 162:*
Capillary Electrophoresis of Nucleic Acids, Vol. 1: Introduction to the Capillary Electrophoresis of Nucleic Acids
Edited by: K. R. Mitchelson and J. Cheng © Humana Press Inc., Totowa, NJ

ing low to medium pressure. Optimum separations of smaller analytes such as oligonucleotides are more problematic as they require polymer solutions of high concentration and viscosity. Thus, the upper range of polymer concentration which can be used is ultimately limited by the high pressures required for loading the capillary with the gel solution ***(14)***.

1.2. Noncrosslinked, Replaceable Gels of Polymer Liquid Crystals

A totally different, easily replaceable polymer medium with a temperature-adjustable viscosity was recently described for CGE separations of small to moderate size nucleic acids ***(19–26)***. This medium is a self-forming liquid crystalline gel phase, consisting of large, spherical micelles of a commercially available polymer surfactant, Pluronic® F127. Pluronic F127 is a triblock copolymer of poly(ethylene oxide) (PEO) and poly(propylene oxide) (PPO) with an approximate molecular formula $(EO)_{100}(PO)_{70}(EO)_{100}$ and molar mass of 13,000 (*see* **Note 1**). Pluronic F127 forms a "liquid-crystal" when the solution becomes so crowded that the micelles are forced to pack together into a semiregular lattice ***(19,27–31)***. The phase is gel-like because the micelles cannot easily slip past one another, despite the absence of crosslinks. The micelles of Pluronic F127 are packed with local face-centered cubic symmetry ***(19)***. The hydrophobic PPO micelle core has a diameter of about 9 nm, whereas the surrounding brush of hydrated PEO chains has an effective diameter that decreases from about 9–7 nm as the polymer concentration is increased ***(19,27,30,31)***, as illustrated in **Fig. 1**.

Concentrated gels of Pluronic F127 can be formed within capillaries because the viscosity of Pluronic F127 solutions decrease dramatically with decreasing temperature (*see* **Note 2**). When the polymer concentration and temperature are above critical values the less polar PPO segments desolvate and aggregate, creating micelles. The temperature dependence of the critical micelle concentration is very strong ***(27,29–32)***. Chilled solutions of Pluronic F127 at concentrations from approx 14–25% (w/v) are low viscosity liquids, and can be introduced into capillaries by mild pressure applied to a conventional hand syringe. The solutions form self-supporting gels within a few minutes, upon warming to room temperature. The temperature at which gelation occurs increases with decreasing polymer concentration. For example, a 25% solution forms a "gel" at just above 0°C, whereas a 14% solution "gels" at about 21°C.

1.3. Application of Pluronic F127 Gels in the CGE of Nucleic Acids

Pluronic F127 gels complement more conventional random polymer solutions as separation media of nucleic acids by CGE. The gel excels in the separation of single-stranded nucleic acids smaller than 40 nucleotides (nt), and in the separation of double-stranded nucleic acids smaller than about 800 bp. Perhaps the most promising application of Pluronic F127 is for CGE separation of small oligonucleotides, typically of the sizes used as therapeutic agents and as primers for polymerase chain reaction (PCR) reactions, or for molecular structural studies. Conventional random polymer solutions that are effective for separations of oligonucleotides <40 nt are difficult to introduce into capillaries because of their high viscosities ***(14)***.

Figure 2 demonstrates that the resolution of oligomers from 4 nt to about 27 nt long on Pluronic F127 gels by CGE is much better than the single nucleotide level ***(20,21)***.

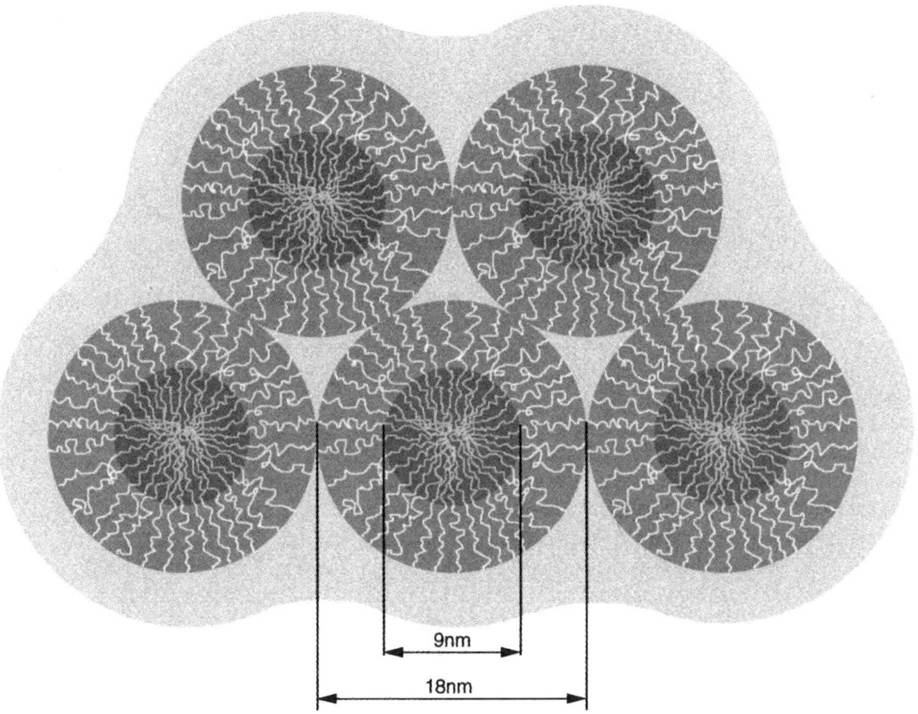

Fig. 1. Schematic of a plane of Pluronic® F127 micelles in the liquid crystalline phase of a 20% solution. The central, darker grey region represents the approximate hard-core dimensions of the hydrophobic poly(propylene oxide) or $(PO)_{70}$ micelle core. The lighter grey region represents the approximate extent of the hydrated poly(ethylene oxide), or $(EO)_{100}$ micelle "brush." Individual $(EO)_{100}$ chains (curly lines) are flexible random coils, but will be more extended than an unperturbed random coil due to crowding near the micelle core surface. Interactions between micelles are expected to be repulsive ***(36)***. Reprinted from Rill, R. L., Locke, B. R., Liu, Y., and Van Winkle, D. H. Pluronic® copolymer liquid crystals: unique, replaceable media for capillary gel electrophoresis. Copyright (1998), with permission from Elsevier publishing.

Contaminant polymers created during oligonucleotide synthesis are easily detected. Oligonucleotide conformers such as hairpin, unstructured, and duplex forms are also readily distinguished ***(26)***. The conformers migrate in the same order (hairpin is fastest) in both Pluronic and conventional polyacrylamide gels. However, the size range of oligonucleotides that can be acceptably separated in Pluronic F127 gels is short, ranging from about 4–60 nt, regardless of the polymer concentration ***(20–22)***.

A convenient feature of Pluronic F127 gels is their ability to separate a variety of nucleic acid types. The CGE procedures and typical gel formulations used to separate oligonucleotides can also be used to separate modest length dsDNA, supercoiled plas-

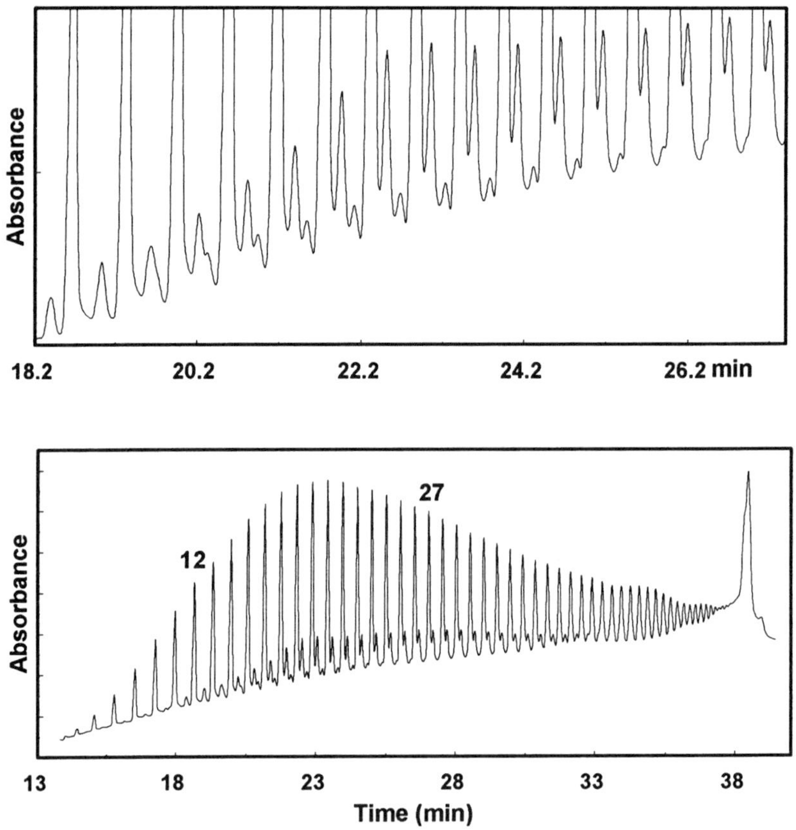

Fig. 2. CGE separation of components of "polyU" in 20% Pluronic F127. The position of U_{12} and U_{27} are indicated above their respective peaks by numerals. Two contaminants are observable between each of the oligonucleotides from about 15–27 nt long (inset). Elec–trophoresis is performed in 20% (w/w) Pluronic F127 in Tris-borate-EDTA buffer at 25°C, 500 V/cm, and an effective column length of 30 cm). Reprinted from Rill, R. L., Locke, B. R., Liu, Y., and Van Winkle, D. H. Pluronic® copolymer liquid crystals: unique, replaceable media for capillary gel electrophoresis. Copyright (1998), with permission from Elsevier publishing.

mids, and small RNA molecules such as tRNA and 5S ribosomal RNA ***(19,20,22,26*** and unpublished observations). With these larger nucleic acids, optimization of the CGE conditions is recommended as small variations in the temperature or the gel concentration may give modest improvement in separation efficiency.

2. Materials

2.1. Hardware

1. CE is normally performed using commercially available instruments such as the Beckman P/ACE system, which provide all the components necessary for sample introduction, electrophoresis, and detection.

2. Detect DNA with moderate sensitivity by monitoring absorbance at 254 or 260 nm (280 nm with less sensitivity) (*see* **Note 3**).
3. DNA may be detected with highest sensitivity using the laser-induced fluorescence (LIF) detector by including ethidium bromide in the gel mixture. Ethidium exhibits enhanced fluorescence when bound to DNA (*see* **Note 4**).
4. There are several alternative dyes to ethidium bromide that offer higher sensitivity for DNA detection. Some examples are ethidium dimer, TOTO, POPO, and the SYBER dyes from Molecular Probes.

2.2. Capillary Selection

1. Commercial, surface-coated capillaries with low electroosmotic flow (EOF) and surface affinity such as 75 μm id Celect-N columns from Supelco, Inc. are used without pretreatment *(33)*. A total capillary length of 37 cm, yielding an effective length to the detector of about 30 cm is sufficient for most applications.
2. Ordinary fused silica capillaries can be used if they are prewashed with 1 *M* HCl *(23)* (*see* **Note 5**).

2.3. Pluronic F127 Gel Medium

1. Pluronic F127, manufactured by BASF Performance Chemicals (Mount Olive, NJ) is available commercially (e.g., from Sigma/Aldrich) and can be used without further preparation (*see* **Note 2**).
2. Prepare the polymer solution using the highest quality reagents.
3. The "5X TBE" buffer is prepared as a concentrated stock, consisting of 0.45 *M* Tris(hydroxymethyl) aminomethane (Tris), pH 8.3, 0.45 *M* boric acid, 10 m*M* Na_2EDTA in HPLC-grade or similar purity water. The pH of the stock solution should be close to pH 8.3. A fivefold dilution of the stock, or "1X TBE" (90 m*M* Tris-HCl, pH 8.3, 90 m*M* boric acid, 2 m*M* Na_2EDTA) is used as the gel buffer and electrolyte during CGE.
4. If the pH is much different, sodium borate may have been mistakenly used instead of boric acid, or Tris-HCl used instead of the free base. Take care.
5. Store the 5X TBE buffer in a refrigerator to retard bacterial growth. A small amount of cacodylic acid (2 m*M* dimethylarsinic acid) can be added to further inhibit bacteria if the buffer is to be stored for a prolonged period.
6. Pluronic F127 solution (20%) in "1X TBE." Prepare a 100 mL vol by mixing 20 mL of stock "5X TBE" and 20 g of dry Pluronic F127 with 60 mL of "HPLC-grade" or comparable purity water that has been prechilled in a refrigerator, or ice bath (*see* **Note 6**).
7. The polymer is only soluble in a chilled solution and will take some time to dissolve. The simplest procedure is to cover the mixture and place it for 12–16 h on a magnetic stirrer or shaking device in a cold box (4–5°C). Alternatively, the mixture can be placed in an ice bath and swirled or stirred occasionally until the polymer is uniformly dissolved. Transfer the polymer solution to a well-sealed bottle and store in a refrigerator.
8. If fluorescence detection will be used, add 10 μL of a 10,000-fold concentrated solution of ethidium bromide to the chilled polymer solution (*see* **Note 7**). Cover the bottle with aluminum foil or transfer to a brown bottle to protect the dye from photobleaching.
9. **SAFETY NOTE**: Ethidium bromide and other DNA binding dyes are mutagens and potential carcinogens (*see* **Subheading 2.1.**). Handle and dispose of these dyes with appropriate care.

2.4. Nucleic Acid Standards

Several useful DNA standards are commonly available from commercial sources.

1. Single-stranded oligonucleotide series: dT_{12-18}, dT_{19-25}, and "polyU" (Sigma). Polymers of dT have the virtues of greater stability, compared to oligoribonucleotides and have a low tendency to form secondary structure.
2. Oligonucleotides that have been highly concentrated, dried, or exposed to organic solvents (such as oligonucleotides eluted from reverse phase materials) often assume aggregated or other nonnative conformations. Reproducible results often cannot be obtained unless the sample is heated to 90°C and allowed to air cool before injection (*see* **Note 8**).
3. dsDNA: "Nx123 bp ladder" DNA standard from Sigma. Similar DNA standards, including NX50, NX100, and NX250 ladders are also available from other suppliers such as Promega, or US Biochemicals (*see* **Note 9**).

3. Methods

3.1. Instrument Setup

1. Program or otherwise set the capillary temperature to 30°C, set the absorbance detector to 260 nm (or 254 nm if filter selected) and the "run voltage" to produce ~500 V/cm (i.e., 18,500 V for a 37-cm capillary).
2. The conductivity of a 20% Pluronic gel is substantially lower than that of other polymer solutions used for CGE, allowing application of high voltage fields. We have successfully used fields as high as 650 V/cm.
3. Nucleic acids migrate toward the positive electrode (anode), so electrophoresis is performed in the reversed polarity mode (opposite the polarity for capillary zone electrophoresis [CZE]). This is a manual adjustment of the Beckman P/ACE 5000.
4. Place "1X TBE" in vials in appropriate locations to serve as the electrolytes.

3.2. Capillary Filling

1. Chill the 20% Pluronic F127 solution to < 10°C so that it will become sufficiently fluid to flow into the capillaries. It is convenient to store solutions and fill capillaries in a walk-in cold room, when available. Solutions can also be chilled and manipulated in a refrigerator or ice bath.
2. Carefully degas the Pluronic® solution prior to filling the capillaries, or bubbles will be released as the gel warms and little or no electrophoresis will occur. Some procedures used typically for degassing semi-dilute aqueous solutions are not effective for Pluronic solutions.
3. The following procedure is very reliable: Apply a vacuum to the polymer solution using an aspirator for 10–15 min. Keep the solution cold and agitated during this process, either by stirring on a magnetic stir plate or by intermittent hand swirling.
4. The solution must be centrifuged to drive bubbles to the surface. Residual bubbles are completely removed in a microcentrifuge at full speed ~5500*g* for 1 min for < 2 mL of polymer, for larger volumes a low speed centrifugation in a table-top or clinical centrifuge operated at ~800*g* for 5–10 min is sufficient. Capillaries can be filled from a single polymer solution for several hours after degassing, often to the end of the working day, unless the solution is strongly agitated.
5. Fill the capillary with cold Pluronic F127 solution using modest hand pressure applied by a conventional syringe. The simplest arrangement is to connect the syringe needle and capillary with a short piece of narrow bore tygon or similar flexible tubing. Solutions can either be pushed or pulled through the capillary, but care most be taken to avoid development of air spaces at the capillary ends (*see* **Note 10**).

6. Immediately mount the capillary in the instrument after filling and immerse the ends in electrolyte buffer. The capillary should be exposed to room temperature for 5–10 min before use to allow time for complete formation of a gel (*see* **Note 11**).

3.3. Sample Preparation and Injection

1. Dissolve the dsDNA in 0.1X TBE buffer. Dissolve the oligonucleotides in water, or 0.1X TBE (*see* **Notes 8** and **9**). It is convenient to prepare and store samples in small conical tubes (e.g., microcentrifuge tubes) that fit upright within the standard Beckman vials. Concentrations of 1–2 mg total analytes/mL are suitable with absorbance detection at 254 nm if there are no more than 10–20 components. Lower DNA concentrations can be used with fluorescence detection.
2. Cap the sample tube and heat the oligonucleotide solution to 90–100°C for 5 min. If the sample is floated on boiling water, a small hole should be punched in the tube cap with a syringe needle to prevent pressure buildup.
3. Allow the sample to cool to room temperature before electrokinetic injection at 10 kV for 5 s.

3.4. Capillary Refilling

1. Chill the capillary cartridge in a cold room or refrigerator until the solution becomes fluid, typically about 5 min. Flush the capillary with 1 mL of 1X TBE buffer, delivered with a hand syringe.
2. Refill the capillary with polymer solution as described above.
3. Gelation occurs over a temperature range of only a few degrees when the temperature is changed slowly (*see* **Note 9**). The solution remains fluid, but becomes substantially more viscous when the temperature is 1–2°C below the critical temperature.
4. Some hysteresis may occur in the fluid-gel-fluid cycle, and the temperature required to convert gel back to fluid solution may be lower than the original gelation temperature. The extent of the hysteresis depends on the rate of heating or cooling.

3.5. Optimization of DNA Separation

1. A 20% gel operated at 30°C produces the highest resolution of dT oligomers, separating dT_{12-24} to much better than single nucleotide resolution in ≤ 20 min.
2. Fragments of dsDNA in the length range from about 50–800 bp are well separated on the same concentration of Pluronic F127 gel, but a 17% gel provides essentially equivalent quality of resolution and requires less time ***(19,20,23)***.
3. DNA fragments of 1000–3000 bp are poorly separated, and the DNA mobility becomes independent of length above 3000 bp ***(20,26)***.
4. The size range of effective separation is modestly sensitive to the gel concentration for both ssDNA and dsDNA (*see* **Note 12**).
5. Electrophoresis times can be reduced by a factor of two or more, while retaining resolution that is acceptable for many applications, by increasing the run temperature to 50°C or by reducing the gel concentration ***(21,26)***.
6. The viscosity of a solution at a concentration adjusted to be only slightly above the critical concentration, about 14% Pluronic F127 solution, is low enough at room temperature (21–22°C) so that capillaries can filled automatically on the P/ACE instrument ***(26)***.
7. Gel formation is induced in a 14% Pluronic F127 solution by subsequently raising the temperature to 30°C. Once optimized, this procedure increases the convenience of the method and can greatly reduce the turnover time of multiple analyses.
8. Capillaries can be used several times without replacing the gel if the polymer solution is well degassed, though the migration time of nucleic acids increase somewhat with

increasing use. The highest reproducibility of migration time is attained when the Pluronic gel is replaced after each electrophoretic run *(23)*.

9. Pluronic F127 solutions in aqueous TBE buffer are suitable for electrophoresis of nucleic acids in their folded states under nondenaturing conditions (*see* **Note 12**). Nucleic acids can also be electrophoresed under standard denaturing conditions, as concentrated Pluronic F127 solutions in 7 *M* urea are still capable of forming the liquid crystalline gel phase. Chu and co-workers have shown that the gel phase in urea is suitable for electrophoresis of denatured DNA *(23–25)*.

4. Notes

1. Pluronic F127 is one of highest molecular weight members of a large family of triblock copolymers with the general formula $(EO)_x(PO)_y(EO)_x$, where $(EO)_x$ is poly(ethylene oxide) or PEO, and $(PO)_y$ is poly(propylene oxide) or PPO. Pluronic is a registered trade name of BASF, but PEO-PPO-PEO polymers are available from other manufacturers under different trade names. To our knowledge Pluronic F127 is the only PEO-PPO-PEO polymer reported for use in capillary electrophoresis.
2. Pluronic copolymers are distinctly different from other triblock polymers used previously for CGE that have a hydrophilic central block flanked by hydrophobic blocks *(15,34,35)*. Aqueous solutions of the latter polymers form thermally reversible, crosslinked, infinite networks when the hydrophobic blocks associate *(36)*.
3. The absorbance of a 20% solution of Pluronic F127 is negligible at 254 nm, but it increases sharply below 240 nm, preventing DNA detection at shorter wavelengths.
4. The fluorescence intensity of DNA-bound ethidium bromide is reduced in Pluronic gels compared to polyacrylamide, perhaps because ethidium can be sequestered within the micelles.
5. The work reported from our laboratory was performed using coated capillaries, but we have confirmed reports *(19,23,24)* that high reproducibility can be obtained using capillaries briefly prewashed with 1 *M* HCl and rinsed with buffer before introducing the Pluronic solution.
6. Polymer concentrations in some earlier reports *(20–22)* were given in units of grams polymer per 100 mL buffer solution, a common method for describing the formulation of crosslinked polyacrylamide gels.
7. When using fluorescence detection it should be kept in mind that the electrophoretic mobility of nucleic acids is reduced by bound dye. This effect may be either DNA sequence dependent or conformation dependent, or be effected by the particular dye used.
8. Most oligonucleotide samples continue to yield reproducible electropherograms over periods of days or weeks after a single heating, if protected from degradation and stored as a liquid. Some samples may require reheating, especially if they contain troublesome sequences such as G_n or $(GC)_n$ tracts, or are stored frozen, which can produce high local concentrations. When dealing with nominally single-stranded oligonucleotides or polynucleotides it is wise to recall that a wide variety of conformational variants and aggregated forms are possible. Conformational rearrangements can be slow, so that non-equilibrium conformations or aggregates may remain present after samples are transferred to typical buffers unless steps are taken to actively disrupt base pairing interactions and to reestablish equilibrium.
9. It is not advisable to dissolve dsDNA in pure water because the temperature for strand separation is dramatically lowered by the absence of charge screening by coelectrolytes.

10. It is very important to avoid evaporation of water from the capillary ends since gel formation is not reversible with decreasing temperature when the Pluronic concentration becomes too high. The gel is then difficult or impossible to replace. A filled cartridge can be stored with the ends immersed in Pluronic solution.
11. Preparations of Pluronic F127 contain a distribution of different triblock copolymer chain lengths. Thus, the critical gelation temperature and concentration may vary modestly from one polymer lot (batch) to another because of differences in the distribution of polymer sizes. Preliminary optimization should be established for each new batch of the polymer.
12. The separation mechanisms operating on Pluronic gels differ in some respects from those of random polymer networks. One manifestation of this difference is the ability of a single gel to separate by length, both short, single-stranded oligonucleotides and also dsDNA fragments up to several hundred base pairs long. Another difference is the narrow size range for effective separation on Pluronic® gels in comparison to polyacrylamide gels. Resolution is lost in Pluronic® gels because the larger nucleic acids migrate faster than expected, at a rate independent of size for polynucleotides greater than 100 nt, or for double-stranded DNA greater than 3000 bp *(20,22,26)*. Care must be taken in interpreting electrophoretic patterns of double stranded DNA fragments smaller than 100 bp. Anomalous patterns are obtained for standard 10- and 25-bp ladders because the dependence of mobility on dsDNA length inverts when the length falls below the micelle diameter (E. Filenova and Rill, unpublished observations).

Acknowledgments

We thank Brian Ramey and Drs. Bruce Locke and David Van Winkle of Florida State University. This work was made possible in part by support from Beckman Instruments, NSF grant BES-951381 and the FSU Center for Materials Research and Technology (MARTECH).

References

1. Stellwagen, N. C. (1997) The free solution mobility of DNA. *Biopolymers* **42,** 687–703.
2. Grossman, P. D. and Soane, D. S. (1991) Experimental and theoretical studies of DNA separations by capillary electrophoresis in entangled polymer solutions. *Biopolymers* **31,** 1221–1228.
3. Viovy, J.-L. (1993) DNA electrophoresis in polymer solutions: Ogston sieving, reptation and constraint release. *Electrophoresis* **14,** 322–329.
4. Hubert, S. J. and Slater, G. W. (1995) Theory of capillary electrophoretic separations of DNA-polymer complexes. *Electrophoresis* **16,** 2137–2142.
5. Viovy, J.-L. (1996) Reptation theories of electrophoresis. *Mol. Biotechnol.* **6,** 31–46.
6. Slater, G. W. (1998) Recent developments in DNA electrophoretic separations. *Electrophoresis* **19,** 1525–1541.
7. Slater, G. W., Desruisseaux, C., and Hubert, S. J. (2001) DNA separation mechanisms during electrophoresis, in *Capillary Electrophoresis of Nucleic Acids,* Vol. 1 (Mitchelson, K. R. and Cheng, J., eds.), Humana Press, Totowa, NJ, pp. 27–41.
8. Cohen, A. S., Najarian, D. R., Paulus, A., Guttman, A., Smith, J. A., and Karger, B. L. (1988) Rapid separation and purification of oligonucleotides by high-performance capillary gel electrophoresis. *Proc. Natl. Acad. Sci. USA* **85,** 9660–9663.
9. Baba, Y., Matsuura, T., Wakamoto, K., and Tsuhako, M. (1991) Comparison of high-performance liquid chromatography with capillary gel electrophoresis in single-base resolution of polynucleotides. *J. Chromatogr.* **558,** 273–284.

10. Karger, B. L. (1992) Capillary electrophoresis. *Curr. Opin. Biotechnol.* **3,** 59–64.
11. Karger, B. L., Chu, Y.H., and Foret, F. (1995) Capillary electrophoresis of proteins and nucleic acids. *Annu. Rev. Biophys. Biomol. Struct.* **24,** 579–610.
12. Heiger, D. N., Cohen, A. S., and Karger, B. L. (1990) Separation of DNA restriction fragments by high performance capillary electrophoresis with low and zero crosslinked polyacrylamide using continuous and pulsed electric fields. *J. Chromatogr.* **516,** 33–48.
13. Tietz, D., Aldroubi, A., Pulyaeva, H., Guszczynski, T., Garner, M. M., and Chrambach, A. (1992) Advances in DNA electrophoresis in polymer solutions. *Electrophoresis* **13,** 614–616.
14. Chiari, M., Nesi, M., Fazio, M., and Righetti, P. G. (1992) Capillary electrophoresis of macromolecules in 'syrupy' solutions: facts and misfacts. *Electrophoresis* **13,** 690–697.
15. Barron, A. E., Soane, D. S., and Blanch, H. W. (1993) Capillary electrophoresis of DNA in uncrosslinked polymer solutions. *J. Chromatogr.A* **652,** 3–16.
16. Chrambach, A. and Aldroubi, A. (1993) Relative efficiency of molecular sieving in solutions of four polymers. *Electrophoresis* **14,** 18–22.
17. Fung, E. N. and Yeung, E. S. (1995) High-speed DNA sequencing by using mixed poly(ethylene oxide) solutions in uncoated capillary columns. *Anal. Chem.* **76,** 1913–1919.
18. Kim, Y. and Yeung, E. S. (2001) Capillary electrophoresis of DNA fragments using poly(ethylene oxide) as a sieving material, in *Capillary Electrophoresis of Nucleic Acids,* Vol. 1 (Mitchelson, K. R. and Cheng, J., eds.), Humana Press, Totowa, NJ, pp. 215–223.
19. Wu, C., Liu, T., Chu, B., Schneider, D.K. and Graziano, V. (1997) Characterization of the PEO-PPO-PEO triblock copolymer and its application as a separation medium in capillary electrophoresis. *Macromolecules* **30,** 4574–4583.
20. Rill, R. L., Locke, B. R., Liu, Y. and Van Winkle, D. (1998) Electrophoresis in lyotropic polymer liquid crystals. *Proc. Natl. Acad. Sci. USA* **95,** 1534–1539.
21. Liu, Y., Locke, B. R., Van Winkle, D. H., and Rill, R.L. (1998) Optimizing CGE of oligonucleotides in liquid crystalline Pluronic® F127. *J. Chromatogr. A* **817,** 367–375.
22. Rill, R. L., Locke, B. R., Liu, Y., and Van Winkle, D. H. (1998) Pluronic® copolymer liquid crystals: unique, replaceable media for capillary gel electrophoresis. *J. Chromatogr. A* **817,** 287–295.
23. Wu, C., Liu, T., and Chu, B. (1998) Viscosity-adjustable block copolymer for DNA separation by capillary electrophoresis. *Electrophoresis* **19,** 231–241.
24. Liang, D. and Chu, B. (1998) High speed separation of DNA fragments by capillary electrophoresis in poly(ethylene oxide)-poly(propylene oxide)-poly(ethylene oxide) triblock polymer. *Electrophoresis* **19,** 2447–2453.
25. Chu, B., Liu, T., Wu, C.-H., and Liang, D. (2001) DNA capillary electrophoresis using block copolymer as a new separation medium, in *Capillary Electrophoresis of Nucleic Acids* (Mitchelson, K. R. and Cheng, J., eds.), Humana Press, Totowa, NJ, pp. 225–238.
26. Rill, R. L., Liu, Y., Ramey, B.A., Van Winkle, D. H., and Locke, B. R. (1999) Capillary gel electrophoresis of nucleic acids in Pluronic® F127 copolymer liquid crystals. *Chromatographia* **49,** S65–71.
27. Mortensen, K., Brown, W., and Norden, B. (1992) Inverse melting transition and evidence of three dimensional cubatic structure in a block-copolymer micellar system. *Phys. Rev. Lett.* **68,** 2340–2343.
28. Linse, P. and Malmsten, M. (1992) Temperature-dependent micellization in aqueous block copolymer solutions. *Macromolecules* **25,** 5434–5439.
29. Linse, P. (1993) Phase behavior of poly(ethylene oxide)-poly(propylene oxide) block copolymers in aqueous solution. *J. Phys. Chem.* **97,** 13,896–13,902.

30. Wanka, G., Hoffmann, H., and Ulbricht, W. (1994) Phase diagrams and aggregation behavior of poly(oxyethylene)-Poly(oxypropylene)- Poly(oxyethylene) triblock copolymers in aqueous solutions. *Macromolecules* **27,** 4145–4149.
31. Mortensen, K. and Talmon, Y. (1995) Cryo-TEM and SANS microstructural study of Pluronic® polymer solutions. *Macromolecules* **28,** 8829–8834.
32. Malmsten, M. and Lindman, B. (1992) Self-assembly in aqueous block copolymer solutions. *Macromolecules* **25,** 5440–5445.
33. Huang, M., Bigelow, M., and Byers, M. (1996) Applications of a self-assembled bilayer coating on a fused-silica capillary surface for capillary electrophoresis. *American Laboratory* (October), 32–36.
34. Sassi, A. P., Barron, A., Alonso-Amigo, M. G., Hion, D. Y., Yu, J. S., Soane, D. S., and Hooper, H. H. (1996) Electrophoresis of DNA in novel thermoreversible matrices. *Electrophoresis* **17,** 1460–1469.
35. Magnusdottir, S., Viovy, J.-L., and Francois, J. (1998) High resolution capillary electrophoretic separation of oligonucleotides in low-viscosity, hydrophobically end-capped polyethylene oxide with cubic order. *Electrophoresis* **19,** 1699–1703.
36. Leibler, L. and Pincus, P. A. (1984) Ordering transition of copolymer micelles. *Macromolecules* **17,** 2922–2924.

14

Capillary Electrophoresis of DNA Fragments Using Poly(Ethylene Oxide) as a Sieving Material

Yongseong Kim and Edward S. Yeung

1. Introduction

Capillary electrophoresis (CE) has been recognized as a precision tool, which is an excellent analytical technique with high sensitivity, high resolution, and has a robust ability for quantitation of the amount of analyte. It also has proven to be a versatile tool for the separation of an enormous range of molecular sizes, from small molecules such as inorganic cations, anions, and drugs, through to large molecules such as peptides, proteins, carbohydrates, polysaccharides, DNA fragments, and even whole chromosomes.

Among many methods developed using CE, DNA analysis using capillary gel electrophoresis (CGE) is now a powerful tool in many research areas including biochemistry, molecular biology, and clinical chemistry. Studies such as polymerase chain reaction (PCR) product analysis *(1,2)*, DNA restriction fragment analysis *(3,4)*, DNA fingerprinting *(5,6)*, and DNA sequencing *(7–9)* have been successfully performed by CGE. Especially, CGE for DNA sequencing is a promising technique to meet the goals of the Human Genome Project (HGP) because of high speed and high-throughput capabilities. A small id capillary column, typically of 50–100 μm, can employ an electric field up to 1200 V/cm with the low current generated during the electrophoresis run and with efficient Joule heat dissipation. CE provides a 20–30-fold increase in separation speed over conventional slab-gel electrophoresis (SGE).

In CGE, the choice of polymer solution as sieving matrix (or gel) is very important for the appropriate application because the sieving power and the migration behavior of a DNA fragment is primarily determined by the physical and chemical properties of the gel material. Initially for CGE, conventional SGE matrices such as agarose and crosslinked polyacrylamide were employed. The separation of DNA fragments up to 700 bases in 230 min with a crosslinked polyacrylamide gel has been reported *(10)*. However, these matrices have become less attractive due to gel inhomogeneity, the

From: *Methods in Molecular Biology, Vol. 162:*
Capillary Electrophoresis of Nucleic Acids, Vol. 1: Introduction to the Capillary Electrophoresis of Nucleic Acids
Edited by: K. R. Mitchelson and J. Cheng © Humana Press Inc., Totowa, NJ

short lifetime of the capillary, and bubble formation during the electrophoresis run. A replaceable polymer such as linear polyacrylamide (LPA) has been employed more recently for DNA sequencing *(11)*. It has been shown that a replaceable polymer solution could extend the lifetime of the column and reduce the formation of bubbles under high electric field. Another advantage is that a fresh medium can be easily introduced into the separation column for each electrophoresis run by cycles of column washing and reloading with fresh matrix. In addition to LPA, methyl cellulose *(12)*, hydroxyalkyl cellulose *(13)*, polyhydroxy- and poly(ethylene glycol methacylate) *(14)*, and poly(vinyl alcohol) *(15)* also have been successfully employed for DNA separations. Entangled polymer solutions are also useful for the separation of biological samples, because the dynamic coating of the polymer solution onto the capillary–wall surface minimizes any possible solute-wall interactions *(16)*. Several new entangled polymer materials have been described: one that consisted of poly(ethylene glycol) end-capped with a micelle–forming fluorocarbon is recently reported to self-assemble in water into equilibrium structures with a well-defined mesh size *(17)*, and a temperature-controlled polymer matrix for CGE and SGE was also reported *(18)*.

Thus, one approach to further increase the speed and throughput of DNA analysis is the identification of better sieving matrices. Several physical characteristics such as high-separation efficiency, good chemical stability and tolerance to high electric field, low background signal in both absorbance and fluorescence, ease of polymer matrix preparation, good reproducibility, and resistance to high temperature are important practical factors that must be considered before new matrices are employed. Poly(ethylene oxide) (PEO) is such a material (*see* **Note 1**) that has been well characterized by studies of its physical characteristics such as, its phase behavior *(19)*, morphology *(20,21)*, solution dynamics *(22)*, micelle formation *(23)*, and rheological phenomena *(24)*. We find that PEO provides a homogeneous solution structure, is easy to prepare, has high resolution, and provides for high-speed CE analysis of DNA fragments. PEO has been successfully applied for the separation of dsDNA *(4)* and for ssDNA, and DNA sequencing *(25)*. PEO and other sieving matrices have also been employed with pulsed-field CE formats for the separation of very large DNA molecules *(26–28)*. **Figure 1** shows an example of the separation of DNA sequencing of a sample for up to 800 bases using PEO solution as a separation matrix.

2. Materials

2.1. CE System and Laser-Induced Fluorescence Detection of DNA Fragments

1. CE systems are commercially available, however, it may be beneficial to build a system in-house for better PEO gel loading (*see* **Note 2**). To build the system at the laboratory, it is necessary to purchase several electrical and optical components. Instructions are provided below for system construction.
2. Use a high-voltage power supply for electrophoresis runs (*see* **Note 3**). **CAUTION**—The electrophoresis and detection systems should be enclosed in a sheet-metal box with a safety interlock for high-voltage applications. This interlock is an essential component for user safety, because the electrophoresis run usually requires 10–20 kV.

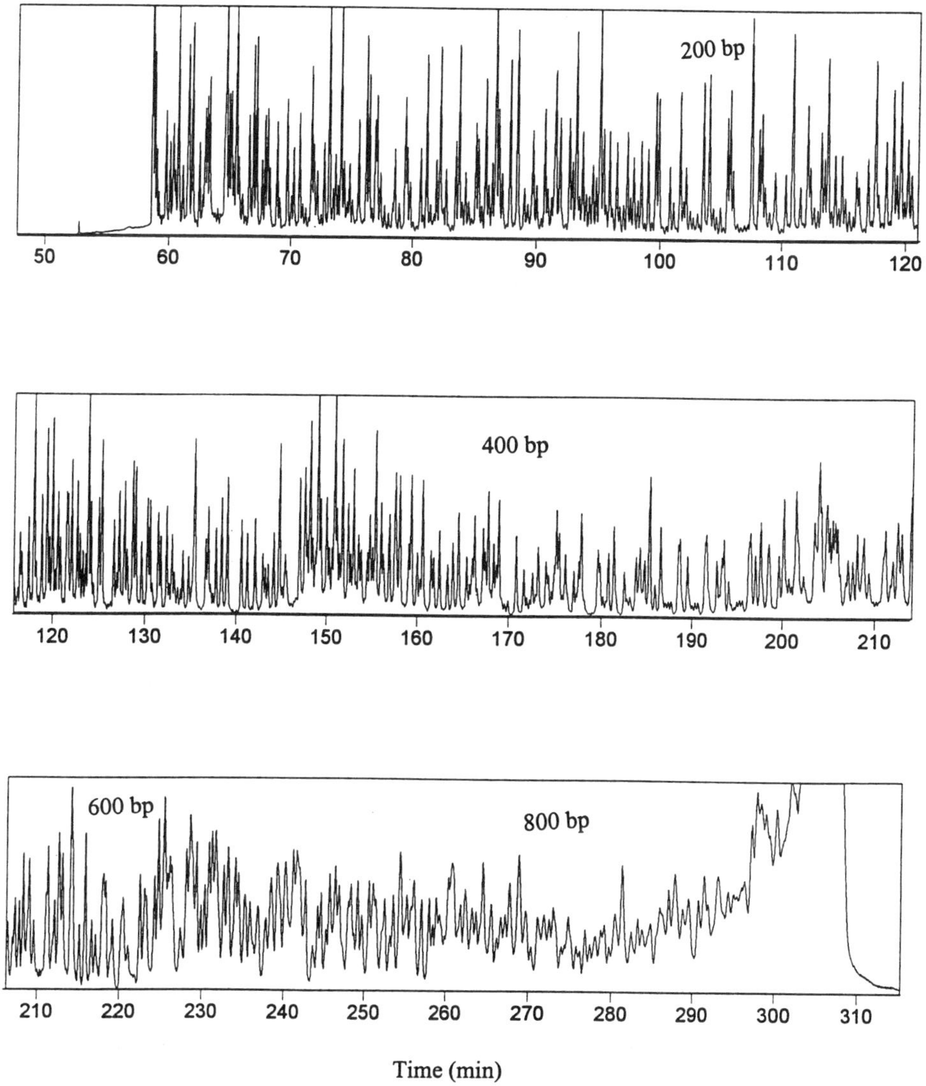

Fig. 1. Electropherograms of pGEM/U sequencing sample up to 800 bp (counting from the primer peak). Electrophoresis conditions are: 1X TBE buffer with 3.5 *M* urea, 75 V/cm field strength, gel concentration at 1.5% mol wt 8,000,000 and 1.4% 600,000 PEO, 50-cm effective capillary length, 30-s injection at running voltage.

3. Dip the two ends of a capillary into two buffer reservoirs and apply the high voltage in the sample injection side.
4. Use a 1.5-mW He-Ne laser with 543.6-nm output for excitation (*see* **Note 4**). About 1 cm of the window should be prepared at the detection side of the capillary by burning with a flame. Focus the laser line to the center of the capillary at the window region using a 20X

microscope objective. Collect the fluorescence signal through photomultiplier at 90° to the capillary. Use two RG 610 filters to block scattered laser light.

5. The signal transfer and data storage can be controlled by an analog-to-digital (A/D) interface board through a 10-kΩ resistor and an IBM compatible computer at data collection frequency of 4 Hz. Current monitoring inside the capillary using a digital volt-meter is an advisable option. In order to monitor the current, the volt-meter should be connected to a 10-kΩ resistor located at the detection side of the capillary.

2.2. Capillary and CE Reagents

1. The capillary and reagents described can be used for both ssDNA and dsDNA. For specific conditions of use, *see* **Notes 5** and **6**.
2. The capillary is fused silica with 75 µm id and 365 µm od (Polymicro Technologies, Phoenix, AZ).
3. Tris(hydroxymethyl) aminomethane (THAM): 90 m*M*.
4. Ethylenediaminetetraacetic acid (EDTA): 2 m*M*, 20 m*M*.
5. Boric acid: 90 m*M*.
6. Pure urea: 3.5 M (*see* **Note 7**).
7. Formamide.
8. HCl: 2.5 m*M*.
9. PEO, mol wt 8,000,000 and 600,000 (Aldrich, Milwaukee, WI).
10. DNA restriction fragments (*see* **Note 5**).
11. DNA sequencing samples (*see* **Note 6**).

2.3. The Buffer and PEO Gel Preparation

1. Prepare the buffer by dissolving 90 m*M* THAM, 90 m*M* boric acid, 2 m*M* EDTA (1X TBE), and 3.5 *M* urea in deionized water at the final pH of 8.2 (*see* **Note 8**).
2. Prepare the gel by gradually adding 200 mg of PEO, mol wt 8,000,000 and 150 mg of PEO, mol wt 600,000 in 10 mL of the buffer at room temperature (*see* **Note 9**). The polydispersity of the polymer is not known, but we found that polydisperse solutions are in fact advantageous (*see* **Note 10**).
3. Use a magnetic stirring bar at a high-speed setting during the addition of PEO to achieve a homogeneous solution (*see* **Note 11**).
4. After addition, stir the solution for another 12 h. Then, degas the resulting gel by centrifuging the bottle for 3 h in the bench top centrifuge (5000*g*) (*see* **Note 12**).
5. Following centrifugation, store the gel in the bottle at room temperature, as no exposure to air is allowed. PEO gel in solution degrades considerably if it is exposed to air, probably because of oxidation and light-induced cleavage of PEO chain.
6. Different PEO concentrations are preferable for the separation of different size ranges of DNA. For example, 3.5% of PEO, mol wt 2,000,000 gave a reasonable resolution in between 70–1350 bp, and a single-base resolution is obtained around 100 bp sizes with 2.5% PEO, mol wt 8,000,000 *(4)*.

3. Methods

3.1. Capillary Preconditioning and PEO Gel Loading

1. Prepare a bare capillary of total length of 55 cm, with an effective length (from injection to detection) of 50 cm (*see* **Note 13**). No capillary surface coating is required for PEO gel (*see* **Note 14**).

2. Flush the capillary with the 2.5 m*M* HCl solution for 0.5 h before PEO gel loading (*see* **Note 15**).
3. Push PEO gel though one end of the capillary with high pressure (400 psi) for 20–30 min by using a pressure injection-flush device (*see* **Note 16**).
4. Place both ends of the capillary in vials containing the running buffer: 1X TBE with 3.5*M* urea.
5. Apply a field of 5.5 kV for 20 min to electrophoretically equilibrate the column.

3.2. DNA Sample Injection and Electrophoresis Run

1. Prepare a 1:1 mixture of formamide: 20 m*M* EDTA.
2. Add 4 μL of the formamide, EDTA mixture to the vial containing DNA sequencing sample (*see* **Note 17**).
3. Heat the DNA sample at 95°C for 3 min.
4. Immediately after heating the sample, place the sample vial in an ice-box in order to prevent the renaturation of the DNA.
5. Move the injection end of the capillary from the running buffer to the sample vial.
6. Inject the sample at 150 V/cm for 10 s.
7. Return the injection end of the capillary from the sample vial to the running buffer.
8. Store the DNA sample in a refrigerator at 4°C for the next use. Store it in freezer at –20°C for long-term use (*see* **Note 18**).
9. Apply high voltages for the field strength of 100–300 V/cm.
10. Monitor the current of the electrophoresis. It should be between 1 μA and 50 μA for good separation.

3.3. DNA Sample Detection

1. Our fluorescence detection system is optimized for ABI dyes (FAM, JOE, TAMRA, and ROX) (Perkin-Elmer-ABI, Foster City, CA).
2. For DNA sequencing, four dideoxy nucleotides are labeled with one of four fluorophores samples. The dyes are FAM, JOE, TAMRA, and ROX, which are attached to ddATP, ddGTP, ddTTP, and ddCTP, respectively. TAMRA and ROX produce reasonable signal intensities with the laser line of 543.6 nm, however, FAM and JOE are not detected well. A similar power of the argon-ion laser at 488-nm line will produce fairly good signal distributions for all four dyes (*see* **Note 8**).
3. For dsDNA detection, either ethidium bromide (EB), or TOTO-1 (Molecular Probes, Boulder, CO) are used with proper cutoff filters and collection optics.
4. For DNA restriction samples, simply dilute the sample 10–100 times with 1X TBE buffer. For detection, add mono-intercalating dye EB, at a concentration of 1 mg/mL to 3 mg/mL in the running buffer. Alternatively, add the *bis*-intercalating dye, TOTO-1, at the concentration of 1 μg/mL in the sample vial and incubate for at least 30 min before injection (see **Note 17**).

3.4. Column Regeneration

This step is very important for the longevity of the capillary column.

1. Once the electrophoresis run is complete, turn the high voltage off.
2. Push the PEO gel out of the capillary column by using the pressure injection-flush device (*see* **Note 16**) for 15–30 min (400 psi). Replace PEO gel tube in the pressure injection-flush device with deionized water.

3. Flush the HCl solution through the capillary for 20–30 min at high pressure. A 500 µL of 2.5 m*M* HCl solution is loaded in the pressure injection-flush device this time.
4. Replace the HCl solution with deionized water and flush the system for 10 min.
5. Load the PEO gel solution for 20–30 min.
6. In the long-term, place the both ends of the capillary in a vial containing deionized water (*see* **Note 19**).

4. Notes

1. PEO is commercially available with various molecular weights from 100,000 to 8,000,000. It is synthesized from ethylene oxide in gas phase and inhibited with <500 ppm butylated hydroxytoluene (BHT). It is in the form of a white powder and can be stored at room temperature for more than 6 mo. Once a gel is prepared with a certain concentration, the storage period decreases significantly. A maximum storage time of 7 d is allowed for good gel performance. Make sure that there is no air contact. PEO tends to be degraded faster with air contact. Storage under inert gas is recommended.
2. Commercial CE instruments require a modification at the sample injection port for the loading of PEO matrix. Because a PEO solution has higher viscosity than a normal, capillary zone electrophoresis buffer such as TBE or phosphate, the pressure injection-flush device is essential.
3. Any high-voltage power supplies designed for CE experiments can be employed. We prefer Trek 20/20 (Trek 20/20, Medina, NY), because it provides better flexibility for voltage setting. Since it is not a high-voltage power supply but a voltage amplifier, an additional low voltage source should be connected to the Trek 20/20. The Trek 20/20 is used for pulsed-field capillary electrophoresis as well as conventional CE ***(26)***.
4. If separate four-color detection is required, more than one photomultiplier is needed.
5. Any commercial DNA restriction fragment standards are acceptable. DNA restriction fragments can also be prepared by using restriction enzyme for any dsDNA substrates. No further purification step is required for restriction fragments as long as they are dissolved in the proper buffer for the CE run.
6. DNA sequencing samples can be either prepared at the laboratory or purchased from commercial vendor. For other dyes used for DNA sequencing (Big dyes, energy transfer dyes, BODIPY dyes), it may be nesessary to change the laser wavelength and collection optics depending on excitation and emission maxima.
7. Usually, 7 *M* urea is used for the denaturation of DNA sequencing samples. With the PEO matrix, it turned out that 3.5 *M* urea is sufficient for denaturation, and this buffer is prepared for DNA sequencing samples.
8. The buffer for analysis of DNA restriction fragments can be prepared in a similar way by dissolving 90 m*M* THAM, 90 m*M* boric acid, and 2 m*M* EDTA in deionized water. Alternatively, Tris-acetic acid and Tris-TAPS buffers can be also employed at concentrations similar to the TBE buffer.
9. This PEO combination is designed for DNA sequencing samples. For DNA restriction fragments, different combinations of PEO are possible. For example, 3.5% of PEO, mol wt 2,000,000 gave a reasonable resolution in between 70–1350 bp and a single-base resolution was obtained around 100 bp sizes with 2.5% of PEO, mol wt 8,000,000 ***(4)***.
10. Generally, mono-disperse sieving matrices yield better separations in high performance liquid chromatography (HPLC) and size exclusion chromatography. However, for linear polymer matrices in CGE, the separation mechanism might not be simple sieving but friction-induced differentiation ***(18)***. This idea is supported by results obtained with a

mixed PEO system, where two molecular weights of PEO are mixed together to give good separation for broad range of DNA sizes.

11. The resolution obtained in separations of the DNA samples is strongly dependent upon the quality of the gel and its preparation methods. A stirring speed setting of 6 or 7 for corning stirrer is in good range. If the stirring speed is too slow, then a relatively inhomogeneous PEO gel is formed, which will produce a low-resolution electropherogram. If the stirring speed is too fast, then long PEO chains tend to be broken and the gel only works for small DNA fragment sizes. Also, it is important to make sure that PEO powder is completely solvated in the beginning of stirring. Once the gel is formed, slow stirring for an additional 3 h before degassing may help for the generation of a good quality PEO gel.
12. A viscous PEO gel entraps large number of bubbles formed during the stirring period. Removal of the bubbles is critical for separation performance. If there are any residual bubbles during the CGE run, separation efficiency suffers and the current inside the capillary is unstable. Bubbles could be also removed by vacuum or suction.
13. The total and effective length of the capillary can be customized for DNA size range, instrument configuration, and so on.
14. A PEO gel has an ability to coat the capillary surface by a dynamic coating process. Electroosmotic flow is markedly reduced by PEO coating. Other polymers such as LPA and cellulose derivatives require the permanent coating of the surface. If considerable electroosmotic flow exists, resolution suffers and migration speed becomes slow.
15. Preparation of a new capillary requires washing with methanol, then a flush with 0.1 *N* NaOH for 20 min. Then wash the capillary with 0.1 *N* HCl for an additional 20 min. For the next electrophoresis run and subsequent runs, a wash with 2.5 m*M* HCl is enough for column cleaning.
16. It is simple to construct the device that allows pressure control for gel loading and washing. Standard HPLC fittings can be used. A nest should be made in the device to hold PEO gel tube (500-µL tube size). Usually, less than 500 µL of PEO is loaded. Put the cover on top of the device and fix it with screws. Two holes are necessary for capillary and pressure control. Make a small hole at top of the device for capillary introduction with fitting. Make another hole for gas inlet to provide high pressure (up to 1000 psi). The N_2 gas tank produces enough pressure to push PEO gel into the capillary. Check the viscosity of the solution at the other side of the capillary to ensure that PEO loading is complete.
17. For DNA restriction samples, simply dilute the sample 10–100 times with 1X TBE buffer, and for detection, add mono-intercalating dye, EB, or *bis*-intercalating dye, TOTO to the buffer or DNA sample, respectively. No heat treatment is required for DNA restriction fragments.
18. Up to four injections are allowed. DNA sequencing sample degrades if stored at room temperature for more than 8 h. Even though it lasts longer if it is frozen, the DNA sequencing sample will slowly degrade. On the other hand, DNA restriction fragments tend to be stable for longer periods of time.
19. The separation efficiency remains the same for up to 60 runs. After 60 runs, the resolution suffers due to degradation of the column. Replacement of the matrix with new PEO gel does not solve this problem. This capillary degradation might be caused by, either the partial trapping of the DNA fragments by PEO gel, or by a surface change by PEO or urea. If the capillary performance decreases after a few runs, try to clean the capillary with water at high speed by using a high pressure device such as a liquid chromatography (LC) pump.

Acknowledgments

The Ames Laboratory is operated for the U.S. Department of Energy by Iowa State University under Contract No. W-7405-Eng-82. This work was supported by the Director of Science, Office of Biological and Environmental Research, and by the National Institutes of Health.

References

1. Vincent, U., Patra, G., Therasse, J., and Gareil, P. (1996) Quantitation of polymerase chain reaction-amplified DNA fragments by capillary electrophoresis and laser-induced fluorescence detection. *Electrophoresis* **17,** 512–517.
2. Lu, W., Han, D. S., and Andrieu, J. M. (1994) Multitarget PCR analysis by capillary electrophoresis and laser-induced fluorescence detection. *Nature* **368,** 269–271.
3. Kim, Y. and Morris, M. D. (1994) Separation of nucleic acids by capillary electrophoresis in cellulose solutions with mono-and bis-intercalating dyes. *Anal. Chem.* **66,** 1168–1174.
4. Chang, H. and Yeung, E. S. (1995) Poly(ethyelene oxide) for high-resolution and high-speed separation of DNA by capillary electrophoresis. *J. Chromatogr. B* **669,** 111–123.
5. Zhang, N. and Yeung, E. S. (1996) Genetic typing by capillary electrophoresis with the allelic ladders as an absolute standard. *Anal. Chem.* **68,** 2927–2931.
6. Michael, A. M., Devamey, J. M., Davis, P. A., Smith, J. K., and Girard, J. E. (1998) Spectral measurement of intercalated PCR-amplified short tandem repeat alleles. *Anal. Chem.* **7,** 4514–4519.
7. Kheterpal, I. and Mathies R. A. (1999) Capillary array electrophoresis DNA sequencing. *Anal. Chem.* **71,** 31A–37A.
8. Smith, L. M. (1991) High-speed DNA sequencing by capillary gel-electrophoresis. *Nature* **349,** 812–813.
9. Salas-Solano, O., Carillho, E., Kotler, L., Miller, A. W., Goetzinger, W., Sosic, Z., and Karger, B. L. (1998) Routine DNA sequencing of 1000 bases in less than hour by capillary electrophoresis with replaceable linear polyacrylamide solutions. *Anal. Chem.* **70,** 3996–4003.
10. Kamahori, M. and Kambara, H. (2001) Capillary array electrophoresis analyzer, in *Capillary Electrophoresis of Nucleic Acids,* Vol. 2 (Mitchelson, K. R. and Cheng, J., eds.), Humana Press, Totowa, NJ, pp. 271–287.
11. Ruiz-Martinez, M. C., Berka, J., Blelnkii, A., Foret, F., Miller, A. W., and Karger, B. L. (1993) DNA sequencing by capillary electrophoresis with replaceable linear polyacrylamide and laser-induced fluorescence detection. *Anal. Chem.* **65,** 2851–2858.
12. Crehan, W. A. M., Rasmussen, H. T., and Northrop, D. M. (1992) Size-selective capillary electrophoresis (SSCE) separation of DNA fragments. *J. Liq. Chromatogr.* **15,** 1063–1080.
13. Grossman, P. D. and Soane, D. S. (1991) Capillary eletrophoresis of DNA in entangled polymer solutions. *J. Chromatogr.* **559,** 257–266.
14. Zewert, T. and Harrington, M. (1992) Polyhydroxy and polyethyleneglycol (meth) acrylate polymers-physical properties and general studies for their use as electrophoresis matrices. *Electrophoresis* **13,** 817–824.
15. Kleemib, M. H., Gilges, M., and Schomburg, G. (1993) Capillary electrophoresis of DNA restriction fragments with solutions of entangled polymers. *Electrophoresis* **14,** 515–522.
16. Zhu, M., Hansen, D. L., Burd, S., and Gannon, F. (1989) Factors affecting free zone electrophoresis and isoelectric focusing in capillary electrophoresis. *J. Chromatogr.* **480,** 311–319.
17. Menchen, S., Johnson, B., Winnik, M. A., and Xu, B. (1996) Flowable network as DNA sequencing media in capillary columns. *Electrophoresis* **17,** 1451–1459.

18. Sassi, A. P., Barron, A., Alonso-Amigo M. G., Hion, D. Y., Yu, J. S., and Soane, D. S. (1996) Electrophoresis of DNA in novel thermoreversible matrices. *Electrophoresis* **17,** 1460–1469.
19. Balijepalli, S. and Schultz. J. M. (1996) Phase behavior and morphology of poly(ethylene oxide) blends. *Macromolecules* **29,** 6601–6611.
20. Chen, J. H., Cheng, S. Z. D., Wu, S. S., Lotz, B., and Wittmann, J. C. (1995) Polymer decoration study in chain folding behavior of solution-grown poly(ehtylene oxide) crystals. *J. Polym. Sci. B. Polym. Phys.* **33,** 1851–1855.
21. Cheng, S. Z. D. and Chen, J. H. (1991) Nonintegral and integral folding crystal-growth in low-molecular mass poly(ethylene oxide) fractions. 3. Linear crystal-growth rates and crystal morphology. *J. Polym. Sci. B. Polym. Phys.* **29,** 311–327.
22. Bieze, T. W. N., van der Maarel, J. R. C., Eisenbach, C. D., and Leyte, J. C. (1994) Polymer dynamics in aqueous poly(ethylene oxide) solutions-an NMR study. *Macromolecules* **27,** 1355–1366.
23. Merdas, A., Gindre, M., Ober, R., Nicot, C., Urbach, W., and Waks, M. (1996) Nonionic surfactant reverse micelles of C(12)E(4) in dodecane: temperature dependence of size and shape. *J. Phys. Chem.* **100,** 15,180–15,186.
24. Briscoe, B., Luckham, P., and Zhu, S. (1996) Rheological study of poly(ethylene oxide) in aqueous salt solutions at high temperature and pressure. *Macromolecules* **29,** 6208–6211.
25. Kim, Y. and Yeung, E. S. (1997) Separation of DNA sequencing fragments up to 1000 bases by using poly(ethylene oxide)-filled capillary electrophoresis. *J. Chromatogr. A* **781,** 315–325.
26. Kim, Y. and Morris, M. D. (1994) Pulsed field capillary electrophoresis of multi-kilobase length nucleic acids in dilute methyl cellulose solutions. *Anal. Chem.* **66,** 3081–3085.
27. Heller, C., Magnúsdóttir, S., and Viovy, J.-L. (2001) Robust field inversion capillary electrophoretic separation of long DNA fragments, in *Capillary Electrophoresis of Nucleic Acids,* Vol. 1 (Mitchelson, K. R. and Cheng, J., eds.), Humana Press, Totowa, NJ, pp. 293–305.
28. Morris, M. D., Schwinefus, J. J., and de Carmejane, O. (2001) Pulsed-field capillary electrophoresis separation of large DNA fragments, in *Capillary Electrophoresis of Nucleic Acids,* Vol. 1 (Mitchelson, K. R. and Cheng, J., eds.), Humana Press, Totowa, NJ, pp. 307–321.

15

DNA Capillary Electrophoresis Using Block Copolymer as a New Separation Medium

Benjamin Chu, Tianbo Liu, Chunhung Wu, and Dehai Liang

1. Introduction

The development of automated DNA sequencing has drawn the attention of many research groups in recent years, especially for the high-throughput demand of the Human Genome Project *(1,2)*. Automation of DNA sequencing based on standard electrophoresis is the most commonly accepted and used approach. Capillary electrophoresis (CE) using replaceable polymer solution as a rapid separation medium with high-resolution separations can be automated relatively easily when compared with slab-gel electrophoresis. The development of capillary array electrophoresis (CAE) has further enhanced the throughput of DNA sequencing. The application of replaceable polymer solutions over permanent crosslinked gels as a separation medium made CE the best conventional analytical method at this time *(3)*, because gels tend to be more difficult to handle and have limited column lifetime.

Several polymer solutions have been used successfully in CE, e.g., linear polyacrylamide *(4–6)*, poly(*N,N*-dimethyl acrylamide) *(7)*, hydroxyalkyl cellulose *(8,9)*, and polyoxyethylene *(10–13)*. In order to obtain high resolution, high viscosity solutions with substantial chain entanglements are usually used. However, it is often difficult to inject and to replenish such a separation medium because of their high viscosity. Moreover, in order to suppress electroosmotic flow (EOF) and to prevent analyte adsorption, inner wall coatings with polymers *(14,15)* are usually inevitable. The stability and the reproducibility of coatings that involve multistep reactions are also difficult to control. Before complete automation of CE, especially for CAE, many improvements are needed. For example, a better separation medium that can provide the high-separation resolution but is also easy to handle can be very useful.

Within the last several years, our research group developed another kind of separation medium for dsDNA fragments—amphiphilic block copolymer micelles, represented by a commercial Pluronic sample F127 (PEO_{99}-PPO_{69}-PEO_{99}, $E_{99}P_{69}E_{99}$ with

From: *Methods in Molecular Biology, Vol. 162:*
Capillary Electrophoresis of Nucleic Acids, Vol. 1: Introduction to the Capillary Electrophoresis of Nucleic Acids
Edited by: K. R. Mitchelson and J. Cheng © Humana Press Inc., Totowa, NJ

E and P being polyoxyethylene and polyoxypropylene, respectively) ***(16–18)***. It has unique viscosity-adjustable property that makes it easy to be injected into capillary tubes and it also has a coating effect that makes the experiments easier to be handled. Similar systems were also studied by Rill et al. ***(19–21)*** and by Liu et al. ***(22)***.

In this chapter, the experimental procedures are described for the use of a triblock copolymer F127 ($E_{99}P_{69}E_{99}$, with E and P being polyoxyethylene and polyoxypropylene, respectively) in 1X TBE buffer as a separation medium for DNA CE. In aqueous solution, F127 copolymer chains tend to self-assemble into micelles that can pack into a gel-like structure with a sieving ability at high polymer concentrations. In comparison with commercial separation media such as polyacrylamide, a triblock copolymer F127 is easily injected into capillaries at low temperatures due to its viscosity-adjustable properties. Furthermore, no coating of the inner-wall of the capillary is necessary. The use of a solution of 21.2% (w/v) of F127 in 1X TBE to successfully separate dsDNA fragments and oligonucleotide sizing markers is illustrated.

2. Materials

1. F127 polyol (obtained as a gift from the BASF Corp., NJ).
2. Tris(hydroxymethyl)aminomethane (Tris) (Sigma, >99%), boric acid (Aldrich, >99.9%), EDTA (Sigma, >99%).
3. Ethidium bromide (Sigma).
4. ΦX174 DNA-*Hae*III digest and pBR322 DNA-*Bst*NI digest (New England Biolabs, Inc., Beverly, MA).
5. ΦX174 DNA-*Hinc*II digest and oligonucleotide sizing marker (Pharmacia Biotech, Piscataway, NJ).
6. Silica capillary (id/od = 98 μm/364 μm, Polymicro Technologies, AZ).

3. Methods

3.1. Micelles of Copolymer F127

1. F127 powder is dissolved in 1X TBE buffer (89 m*M* Tris-base, pH 8.3, 89 m*M* boric acid, 2 m*M* EDTA in deionized water at around 0°C) (*see* **Note 1**).

3.1.1. Physical Methods to Characterize the Gel-Like System

1. The viscosity of F127 in 1X TBE buffer has a very strong temperature dependence ***(16)***. A 21.2% (w/v) solution of F127 in 1X TBE buffer has a viscosity of only about 30–40 centipoise at temperatures around 4°C, at which the fluid can be transferred easily by pressure injection into capillaries. However, at temperatures above 15°C, the viscosity suddenly increases to hundreds of centipoise suggesting that the system has become gel-like and immobile. As shown in **Fig. 1**, a simple phase diagram can be constructed to clarify the "solution" region and "gelation" region of the F127/1X TBE system.
2. Laser light scattering (LLS) is used to characterize F127 micelles in dilute solutions. The critical micelle concentration (cmc) and the association number of micelles (N_w, number of copolymer chains per micelle) can be calculated from static light scattering (SLS) measurements ***(16)***. The micellization process of F127 copolymer has a very strong temperature dependence ***(18)*** in which more F127 copolymer chains are joined into one micelle at higher temperatures.

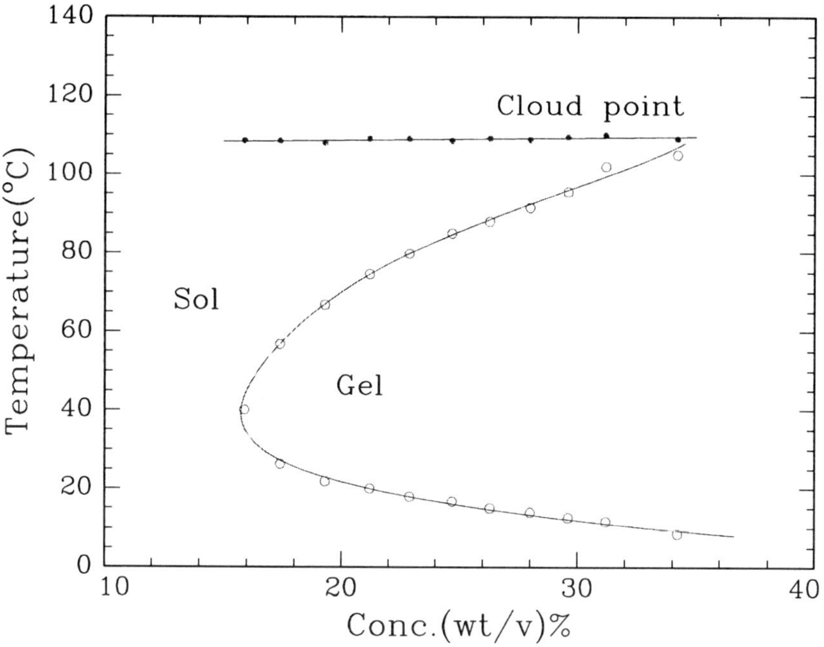

Fig. 1. Sol-gel phase diagram of F127 solution in 1X TBE buffer. Reprinted from **ref. *17***.

3. Dynamic light scattering (DLS) with CONTIN analysis ***(23–25)*** is used to characterize the micellar size in solution. F127 micelles have an average R_h (hydrodynamic radius) of 11.5 nm and a narrow size distribution ***(15)***.
4. Small-angle X-ray scattering (SAXS) measurements are used to clarify the packing patterns of gel-like structures in F127 solutions at high polymer concentrations. The F127 micelles could be packed into an ordered face-centered cubic (fcc) structure ***(16)***.
5. Atomic force microscopy (AFM) measurements are also performed to characterize the micellar morphology on the silicon surface ***(26)***. The size of the F127 micelles determined by AFM is comparable to that obtained from LLS measurements in dilute solution and by SAXS measurements at high polymer concentrations.

3.2. Capillary Electrophoresis

1. The CE instrument is shown schematically in **Fig. 2** ***(27)***. The principle is similar to that of capillary electrophoresis instruments by means of laser-induced fluorescence (LIF) detection. The dye was added into 1X TBE buffer solution that would be used to fill the reservoirs. These dye molecules could combine with DNA fragments and move along with them in an applied electric field.

3.2.1. Capillary Preparation

1. A 13-cm long fused silica capillary is flushed with 1 mL of 1 *N* HCl by use of a 1-mL syringe over a period of 10 min in order to clean the inner surface.
2. The HCl solution remaining inside the capillary is expelled by a low-pressure flush with air from an empty cylinder.

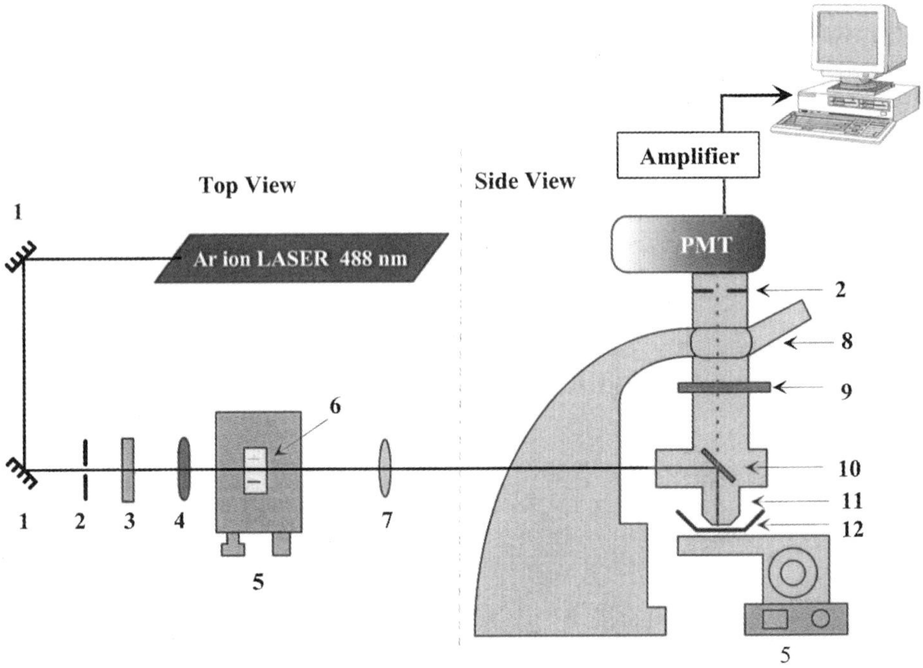

Fig. 2. Schematic diagram of the laser-induced fluorescence capillary electrophoresis setup: (1), mirrors; (2), pinholes; (3), shutter; (4), attenuator; (5), translation stage; (6), mask; (7), lens; (8), eyepiece; (9), 605DF50 bandpass emission filter (Omega, Vermont); (10), 505DRLPO2 dichroic mirror (Omega, Vermont); (11), 40X objective lens (Nikon, Canada); (12), electrophoresis tank for capillary tubing. Reprinted from **ref.** ***27***.

3. A detection window of 2-mm width is opened approx 10 cm from the cathodic end by carefully stripping off the polyimide-coating using a razor blade ***(16,17)***.

3.2.2. Preparation of Separation Medium

1. Both the cathode and anode reservoirs (2.6-mL plastic vials with caps) are filled with 1X TBE running buffer containing 1 µg/mL ethidium bromide.
2. Two small holes are drilled on the cap of each vial, one for the electrode and the other for the capillary to dip into the reservoir.
3. The F127 block copolymer solution is slowly pumped at 4°C into the silica capillary using a 50-µL syringe (*see* **Notes 2** and **3**).
4. The capillary tubing is then assembled on the CE apparatus (illustrated in **Fig. 3**) and the capillary temperature is raised to room temperature. The F127 solution then becomes "gel-like" over about 5 min (*see* **Note 4**).

3.2.3. Prerun and Injection of DNA Sample

1. The capillary gel is prerun at a constant electric field strength of 200 V/cm for 20–30 min, the time depending on the electrophoretic mobility of the dye in this separation medium. The prerun is used to introduce the fluorescent dye into the gel-like F127 material, and also to allow the electrophoresis current to stabilize (*see* **Note 5**).

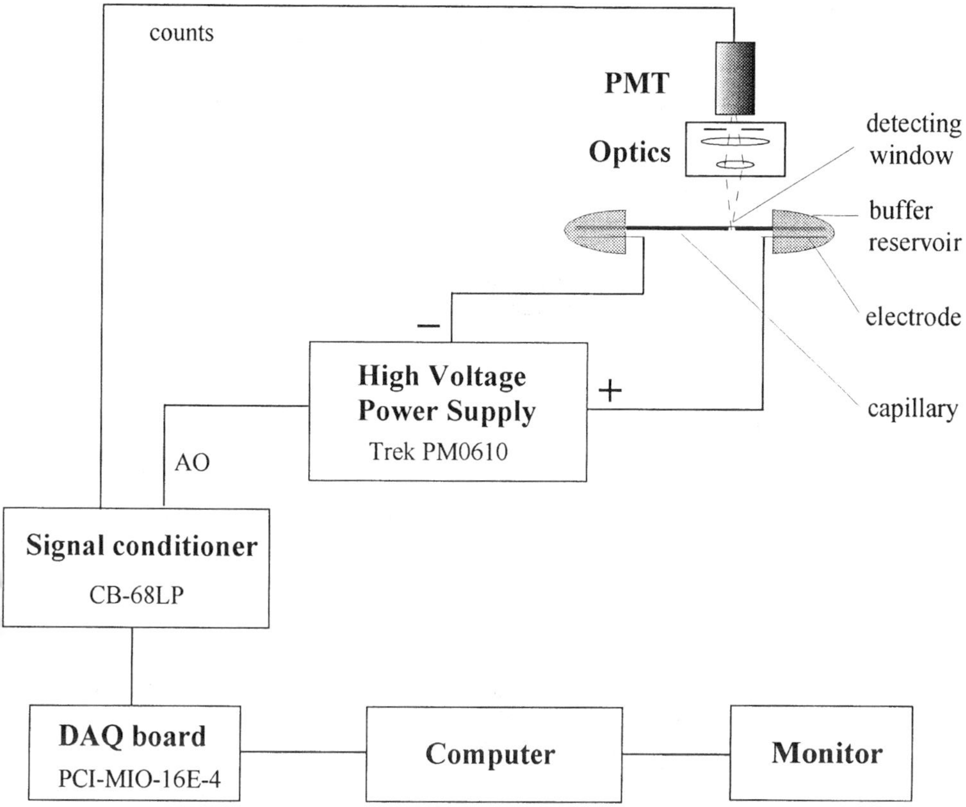

Fig. 3. A diagram of the CE apparatus, with associated LIF detection apparatus.

2. After the prerun, the reservoir at the cathodic end is removed and replaced by another smaller vial containing a 2-µL sample of 10-µg/mL DNA (e.g., ΦX174DNA-*Hae*III). The DNA stock solution (200 µg/mL) is diluted and warmed to room temperature before use. The DNA stock is stored at –5°C.
3. The DNA sample is electrokinetically injected into the capillary at an electric field strength of 300 V/cm for 5 s.
4. Then the sample vial is removed and replaced by a buffer reservoir.
5. The high voltage is generated by two computer-controlled Quatech interface boards, a WSB-100 digital pattern generator adapter, a WSB-A12M waveform synthesizer module (Akron, OH) and a Trek PM0610 amplifier (*see* **Fig. 3**).

3.2.4. DNA Electrophoresis and Signal Detection

1. The electric field strength is 200 V/cm during the run.
2. Our laboratory-built LIF detection CE instrument uses an argon-ion laser operating at a wavelength of 488 nm with a nominal 5-mW output power. The fluorescence from the DNA-ethidium bromide complex is collected at the window on the capillary by a Zeiss

epi-illumination fluorescence microscope coupled to a Hamanatsu R298 photomultiplier tube (PMT). A 605DF50 band-pass emission filter (Omega Optical, Bratleboro, Vermont) is used to collect the fluorescence signals from the DNA-ethidium bromide complex.

3. Visual alignment through the microscope is necessary in order to fix the capillary to a proper position, where the strongest fluorescence signal could be detected (*see* **Note 5**). This step must be accomplished within the first few minutes (before the fastest DNA fragment passes through the detecting window), so that the DNA signals are not affected. When a DNA fragment passes through the detecting window a sudden increase in the fluorescence intensity is recorded by the PMT. The fluorescence signals are transferred to a personal computer (PC) and shown on the screen. In order to prevent any disturbance from outside light sources the entire apparatus is surrounded by a black curtain.
4. To replace the separation medium for a subsequent run, the F127 block copolymer separation medium is cooled down to about 4°C so that the gel-like material can be returned to the solution state. It can then be extruded using 1 mL of distilled water. The capillary tubing is then rinsed with 1 mL of 1 *N* HCl for 10 min to reprotonate the silica surface, before introducing the fresh copolymer solution.

3.3. F127 Copolymer Gel as a Separation Medium for dsDNA

3.3.1. Effect of F127 Block Copolymer Concentration

1. **Figure 4** shows a typical electropherogram for the separation of ΦX174-*Hae*III DNA by F127 in 1X TBE buffer. At an appropriate F127 concentration, all the DNA fragments could be clearly separated within 30–40 min.
2. The polymer concentration had a very strong effect on the separation resolution. At very low F127 concentrations <16.0% (w/v), the system did not have any separation ability, because of the large micellar distance and the low N_w that provided a mesh size too large for sieving DNA fragments.
3. In contrast, at very high F127 concentrations (28.0% [w/v]), the resolution again became worse and the elution time became prolonged. The optimal polymer concentration for DNA separation should be around 21% (w/v).

3.3.2. Separation Mechanism of F127 Gel-Like Material

1. The effect of F127 polymer concentration on the separation resolution can reveal some information on the separation mechanism of this system. The DNA electrophoretic mobility μ has a relation with copolymer concentration C ***(28,29)***:

$$\log \mu = \log \mu_0 - KC(r + R)^2 = \log \mu_0 - K_r C \quad (1)$$

in which, μ_0 is the DNA-free solution electrophoretic mobility, K is a constant of proportionality, r is the thickness of the polymer strands, R is the radius of the DNA sphere, and K_r is a retardation coefficient.

2. Several different models are available to try to explain the behavior of DNA fragments in the separation medium. The Ogston sieving mechanism ***(30)*** assumes that DNA molecules are undeformed spherical particles passing through a random network with an average characteristic mesh size. Obviously, it can be used only for small DNA particles at low applied electric field.
3. The reptation model ***(31)*** describes the migration of a flexible DNA coil in a polymer network as a snake-like, head first movement and concludes that the DNA electrophoretic mobility is inversely proportional to the DNA fragment size ($\mu \propto 1/N$).

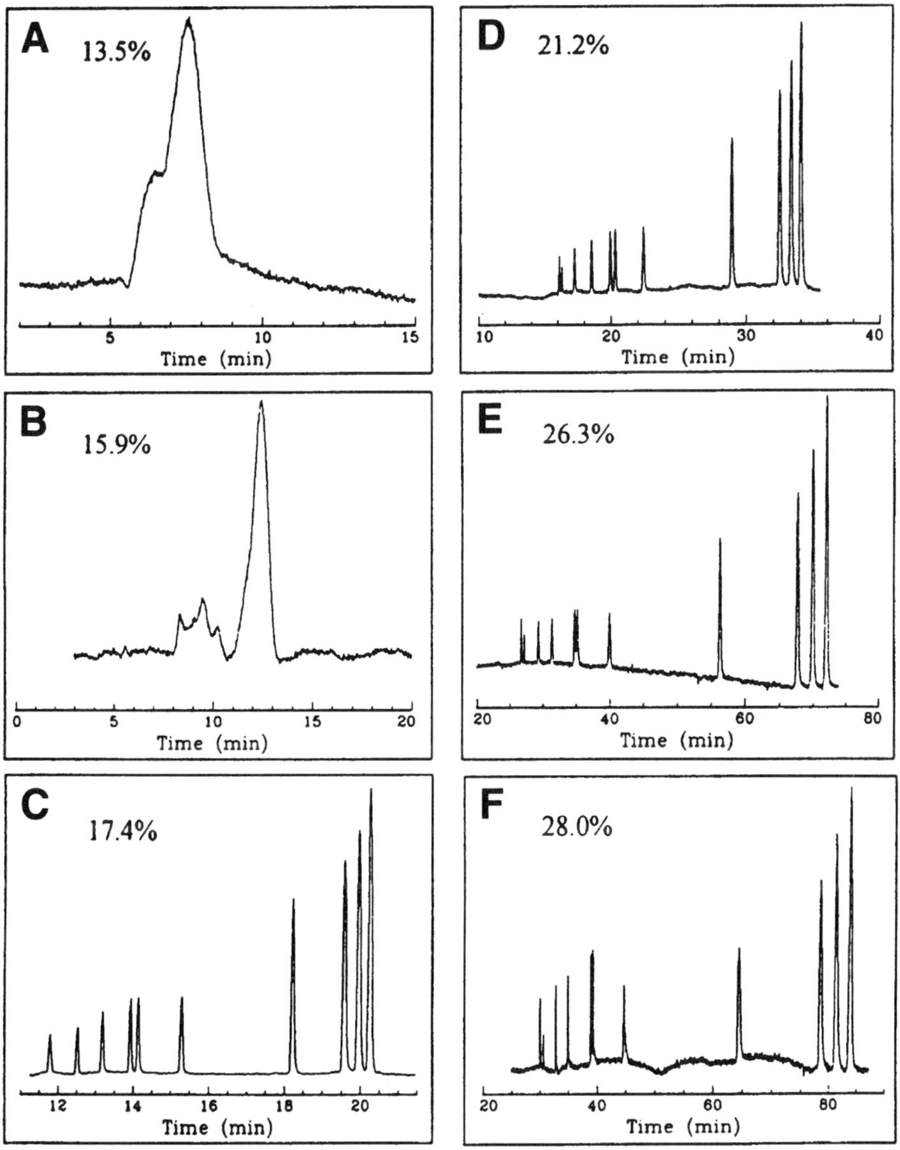

Fig. 4. Electropherograms of ΦX174 DNA-*Hae*III digest obtained by using 21.2% (w/v) F127 as the separation medium. Curves demonstrate the effects of F127 concentration on DNA fragment separation. The peaks indicated from the right side are: 1353, 1078, 872, 603, 310, 281, 271, 194, 72, 118 bp. Reprinted from **ref. *17***.

4. The bias reptation model ***(32)*** explains the mobility to be independent of DNA size for stretched DNA coils induced by a large electric field and by using a small mesh size formed by the polymer network.

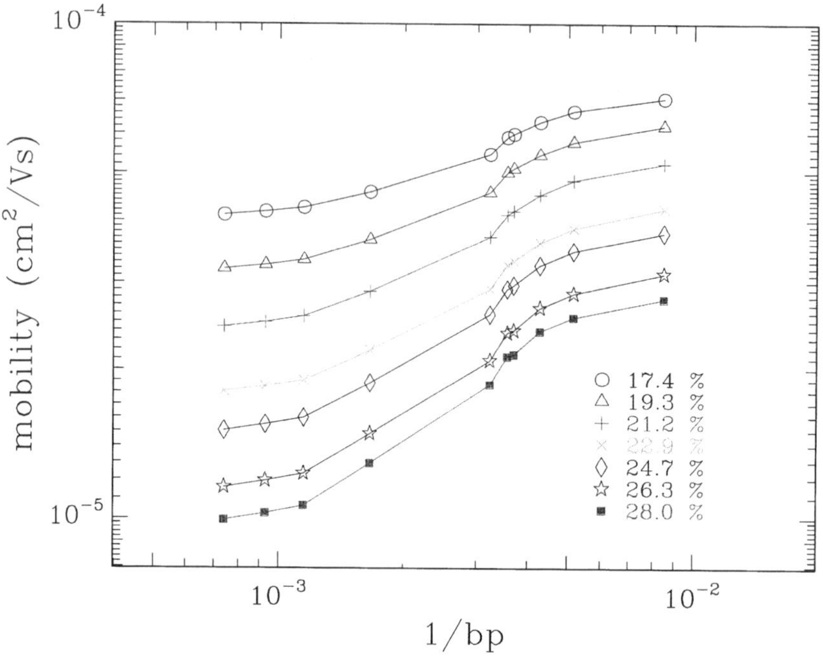

Fig. 5. Log-log plot of DNA electrophoretic mobility vs the reciprocal of DNA size obtained at different F127 concentrations and 25°C. Reprinted from **ref. *17***.

5. To test whether the reptation model is applicable to our results, we plot the data of DNA mobility versus the reciprocal of DNA size obtained at different F127 concentrations in a log-log scale as shown in **Fig. 5**. The solid lines connecting the data points are used to guide the eyes. The above models could not successfully explain the nature of the curves. We can conclude that the behavior of DNA electrophoretic migration in the F127 block copolymer gel does not obey the existing model mechanisms.
6. This observation is not surprising if we consider that there were at least four different domains each with distinct equivalent mesh sizes in our gel-like system. The domains include: the condensed PPO micellar cores, the hydrated PEO micellar shells, the entangled PEO chains in the overlapped micellar shells due to micellar close packing, and the water-rich gaps among the micellar shells.
7. Among the DNA fragment studied, the DNA chains have different conformations. This is because the chain length varied over a wide range, from less than the persistence length, to much larger than the persistence length. Furthermore, as these DNA fragments move through the different domains of the separation medium they probably adopt different types of migration mechanism, and therefore the migration is expected to exhibit a complex behavior.

3.3.3. Resolution of dsDNA in F127 Gels

1. **Figure 6** shows that single base pair resolution of dsDNA fragments is achieved with 21.2% (w/v) F127 medium. In **Fig. 6A,B**, respectively, two different restricted double-

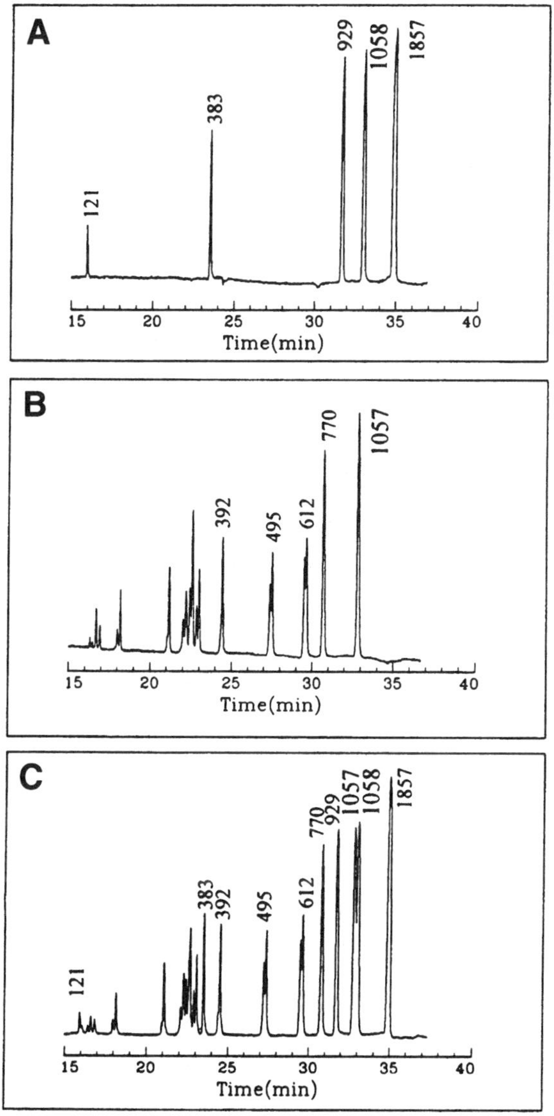

Fig. 6. Electropherograms of plasmid digests using 21.2% (w/v) F127 gel, at room temperature. Separation traces of: **(A)** pBR322 DNA-B*st*NI digest; **(B)** ΦX174 DNA-H*inc*II digest; and **(C)** the 1:1 mixture of pBR322 DNA-B*st*NI and ΦX174 DNA-H*inc*II digests. All DNA samples are at 20 μg/mL. Reprinted from **ref. *17***.

stranded DNAs, pBR322 DNA-*Bst*NI and ΦX174 DNA-*Hinc*II are separated individually. **Figure 6C** shows that the mixture of these DNAs is also resolved using 21.2% (w/v) F127 in 1X TBE buffer.

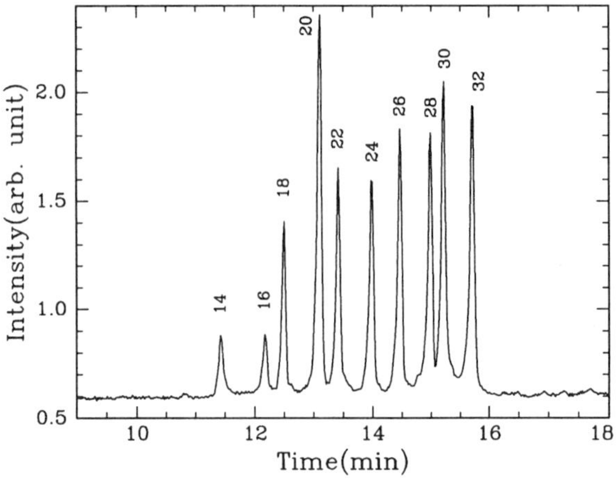

Fig. 7. Electropherogram of oligonucleotide sizing markers obtained in 21.2% (w/v) F127 gel at room temperature by using CE. The 20-base peak was used as internal standard to assign the peak position. Reprinted from **ref. *17***.

2. The DNA fragments of 1058 and 1057 bp from the two different plasmid sources can be successfully separated. However, it should be noted that since the two DNA fragments come from two separate plasmids and two restriction digests, the fragments represent entirely different DNA sequences and consequently the conformations assumed by the two fragments in solution may be different.
3. The ability of F127 to achieve separation of ssDNAs is shown in **Fig. 7**, in which 21.2% (w/v) F127 in 1X TBE was used to separate several oligonucleotide size markers ***(17)***. Although the CE is run without denaturing reagents in 1X TBE buffer under native conditions, the observed efficiency of separation suggests that single-base resolution may be achievable using this polymer system.

3.3.4. Reproducibility of the F127 System for CE

1. In order to test the reproducibility of the F127 gel-like system ***(17)***, two sets of experiments are performed.
2. First, the same gel-like material, buffer, and capillary system are used for several different runs. The migration time of the same DNA fragment becomes longer for the later runs. This degradation can usually be attributed to the recovery of EOF, due to the ionization of the silanol group and the adsorption of charged fluorescence dye onto the capillary wall. After six runs, the running current drops to about half of the original value.
3. However in another set of experiments, the F127 copolymer is removed from the capillary after each run. The inner wall of the capillary is washed by 1 mL 1 *N* HCl each time and new F127 copolymer and buffer are also used for each subsequent run. This results in the observation of stable migration times of different DNA fragments over each run, suggesting that good run-to-run reproducibility is achieved by supression of the EOF.

3.4. High-Speed DNA Separation

1. The high-speed separation of dsDNA fragments using the F127 block copolymer with a marked decrease in the run time can be achieved in two different ways ***(33)***.

2. One way is to shorten the effective capillary length. In a short capillary, the sieving ability of the separation medium vs the initial bandwidth, and the bandwidth spreading as a function of distance traveled, dominate the resolution. In order to compensate for the negative effect of the short effective capillary length on the separation resolution, a shorter DNA injection time is used to narrow the initial DNA bandwidth.
3. The second way is to increase the electric field strength. At high electric field strength, Joule heating could diminish the separation.
4. Therefore, both field strength and capillary length have to be taken into account to achieve the desired increase in electrophoresis speed, without loss of fragment resolution. All of the fragments of a ΦX174DNA *Hae*III digest can be successfully separated within ~100 s using a capillary with an 8-mm effective length and 50 μm id, and operating at 300 V/cm. All other experimental conditions are the same as those mentioned above.

3.5. Applications for Triblock Copolymers

1. This new sieving system has many unique advantages in comparison with commercial separation media, such as polyacrylamide. First, the sol-gel transition properties and ability to form viscosity-adjustable solutions make it easier for it to be injected into small bore capillaries. Second, the polyoxyethylene parts (E blocks) of F127 have the self-coating ability such that no additional coating on the inner wall of the capillary is needed.
2. The mechanism of F127 gel-like system as a separation medium cannot be explained satisfactorily by existing models, because it contains a broad distribution of mesh sizes.
3. F127 copolymer chains tend to form micellar structures in aqueous solution. At higher polymer concentrations, F127 copolymer micelles will pack together in an orderly form to create a gel-like material with certain supra-molecular structures, as the fcc in the concentration range occurs at around 21.2% (w/v). This gel-like system has a sieving ability suitable for the separation of dsDNAs up to about 1.3 kbp.
4. The stable performance of the F127 gel-like polymer material makes it useful for high-speed dsDNA separations, even in very short capillaries of several millimeters effective length.
5. Considering that the block length, the block length ratio, and the chemical nature of the blocks can be modified during synthesis of the monomer block, it is anticipated that a variety of block copolymers of differing molecular formulation would also be useful as alternative separation media in CE. Such copolymers may also be useful for high-speed separation of ssDNA for DNA sequencing.

4. Notes

1. The low temperature of 4°C is necessary because both block copolymers can be readily dissolved in aqueous media at that temperature. This results in a homogenous solution, mainly consisting of unimeric molecules. At room temperature however, the middle P block of F127 copolymer becomes hydrophobic.
2. It is important to make the capillary cold before filling it with the F127 solution. A simple way is to leave the capillary at 0°C for about 15 min. During the filling process, the capillary can be placed on the top of a cold metal block to keep the temperature stable.
3. After filling the capillary with F127 copolymer solution, it is important to check whether there are any air bubbles that could ruin the electrophoretic separation. One simple visual test is to place the capillary in a light source, any bubbles should appear as dark spots.
4. The heating history of the F127 polymer solution in the capillary tube has been found to effect the sensitive gel-like structure of copolymer. A fast rate of heating will result in a

very poor "order" to the supra-molecular structure, and consequently the efficiency of the polymer to separate analytes could be strongly impaired. It is important to allow the polymer-filled capillary to warm to room temperature slowly so that a high degree of order occurs during gelation. Keep the capillary aligned on top of the cold metal block and allow the assembly to warm slowly to the room temperature.

5. It is important that the visual alignment of the capillary window and the detector are correctly made during the prerun electrophoresis. Open the detector shutter and the sudden increase in fluorescence intensity can be detected as the dye molecules pass through the detection window.

Acknowledgment

B. Chu gratefully acknowledges the support of this research by the National Human Research Institute (2R01HG0138604).

References

1. Li, S. F. Y. (1992) *Capillary Electrophoresis: Principles, Practice and Applications.* Elsevier, New York.
2. Adams, M. A., Fields, C., and Venter, J. C. (1994) *Automated DNA Sequencing and Analysis.* Academic Press, London.
3. Ruiz-Martinez, M. C., Berka, J., Belenkii, A., Foret, F., Miller, A. W., and Karger, B. L. (1993) DNA-Sequencing by capillary electrophoresis with replaceable linear polyacrylamide and laser-induced fluorescence detection. *Anal. Chem.* **65,** 2851–2858.
4. Heiger, D. N., Cohen, A. S., and Karger, B. L. (1990) Separation of DNA restriction fragments by high-performance capillary electrophoresis with low and zero cross-linked polyacrylamide using continuous and pulsed electric-fields. *J. Chromatogr.* **516,** 33–48.
5. Guttman, A. and Cooke, N. (1991) Capillary gel affinity electrophoresis of DNA fragments. *Anal. Chem.* **63,** 2038–2042.
6. Kleemiss, M. H., Gilges, M., and Schomburg, G. (1993) Capillary electrophoresis of DNA restriction fragments with solutions of entangled polymers. *Electrophoresis* **14,** 515–522.
7. Menchen, S., Johnson, B., Madabhushi, R., and Winnik, M. (1996) *Progress in Biomedical Optics: Proceedings of Ultrasensitive Biochemical Diagnostics.* SPIE-The International Society for Optical Engineering, San Jose, CA.
8. Grossman, P. D. and Soane, D. S. (1991) Experimental and theoretical-studies of DNA separations by capillary electrophoresis in entangled polymer-solutions. *Biopolymers* **31,** 1221–1228.
9. Baba, Y., Ishimaru, N., Samata, K., and Tsuhako, M. (1993) High-resolution separation of DNA restriction fragments by capillary electrophoresis in cellulose derivative solution. *J. Chromatogr. A* **653,** 329–335.
10. Chang, H.-T. and Yeung, E. S. (1995) Poly(ethylene oxide) for high-resolution and high-speed separation of DNA by capillary electrophoresis. *J. Chromatogr. B* **669,** 113–123.
11. Fung, E. N. and Yeung, E. S. (1995) High-speed DNA-sequencing by using mixed poly(ethylene oxide) solutions in uncoated capillary columns. *Anal. Chem.* **67,** 1913–1919.
12. Kim, Y. and Yeung, E. S. (2001) Capillary electrophoresis of DNA fragments using poly(ethylene oxide) as a sieving material, in *Capillary Electrophoresis of Nucleic Acids,* Vol. 1 (Mitchelson, K. R. and Cheng, J., eds.), Humana Press, Totowa, NJ, pp. 215–223.
13. Iki, N. and Yeung, E. S. (1996) Non-bonded poly(ethylene oxide) polymer-coated column for protein separation by capillary electrophoresis *J. Chromatogr. A* **731,** 273–282.

14. Schmalzing, D., Piggee, C. A., Foret, F., Carrilho, E., and Karger, B. L. (1993) Characterization and performance of a neutral hydrophilic coating for the capillary electrophoretic separation of biopolymers. *J. Chromatogr. A* **652,** 149–159.
15. Bello, M. S., DeBesi, P., Rezzonico, R., Righetti, P. G., and Casiraghi, E. (1994) Electroosmosis of polymer-solutions in fused-silica capillaries *Electrophoresis* **12,** 623–626.
16. Wu, C., Liu, T., Chu, B., Schneider, D. K., and Grazano, V. (1997) Characterization of the PEO-PPO-PEO triblock copolymer and its application as a separation medium in capillary electrophoresis *Macromolecules* **30,** 4574–4583.
17. Wu, C., Liu, T., and Chu, B. (1998) Viscosity-adjustable block copolymer for DNA separation by capillary electrophoresis *Electrophoresis* **19,** 231–241.
18. Wu, C., Liu, T., and Chu, B. (1998) A new separation medium for DNA capillary electrophoresis: self-assembly behavior of Pluronic polyol $E_{99}P_{69}E_{99}$ in 1X TBE buffer. *J. Non-Cryst. Solids* **235–237,** 605–611.
19. Rill, R. L., Locke, B. R., Liu, Y., and Van Winkle, D. H. (1998) Electrophoresis in lyotropic polymer liquid crystals. *Proc. Natl. Acad. Sci. USA* **95,** 1534–1539.
20. Rill R. L., Liu Y., Van Winkle D. H., and Locke B. R. (1998) Pluronic copolymer liquid crystals: unique, replaceable media for capillary gel electrophoresis. *J. Chromatogr. A* **817,** 287–295.
21. Rill, R. L. and Liu, Y.-J. (2001) DNA separation by capillary electrophoroesis in lyotropic polymer liquid crystals, in *Capillary Electrophoresis of Nucleic Acids,* Vol. 1 (Mitchelson, K. R. and Cheng, J., eds.), Humana Press, Totowa, NJ, pp. 203–213.
22. Liu, Y., Locke, B. R., Van Winkle, D. H., and Rill, R. L. (1998) Optimizing capillary gel electrophoretic separations of oligonucleotides in liquid crystalline Pluronic F127. *J. Chromatogr. A* **817,** 367–375.
23. Provencher, S. W. (1979) Inverse problems in polymer characterization - direct analysis of polydispersity with photon correlation spectroscopy. *Makromol. Chem.* **180,** 201–209.
24. Provencher, S. W. (1982) A constrained regularization method for inverting data represented by linear algebraic or integral-equations. *Comput. Phys. Commun.* **27,** 213–227.
25. Provencher, S. W. (1982) CONTIN - a general-purpose constrained regularization program for inverting noisy linear algebraic and integral-equations. *Comput. Phys. Commun.* **27**, 229–242.
26. Wu, C., Liu, T., White, H., and Chu, B. (2000) Atomic force microscopy study of E99P69E99 triblock copolymer chains on silicon surface. *Langmuir* **16,** 656–661.
27. Wu, C., Quesada, M. A., Schneider, D. K., Farinato, R., Studier, F. W., and Chu, B. (1996) Polyacrylamide solutions for DNA sequencing by capillary electrophoresis: mesh sizes, separation and dispersion. *Electrophoresis* **17,** 1103–1109.
28. Slater, G. W., Rousseau, J., Noolandi, J., Turmel, C., and Lalande, M. (1988) Quantitative-analysis of the 3 regimes of DNA electrophoresis in agarose gels. *Biopolymers* **27,** 509–524.
29. Slater, G. W., Desruisseaux, C., and Hubert, S. J. (2001) DNA separation mechanisms during electrophoresis, in *Capillary Electrophoresis of Nucleic Acids,* Vol. 1 (Mitchelson, K. R. and Cheng, J., eds.), Humana Press, Totowa, NJ, pp. 27–41.
30. Slater, G. W. and Noolandi, J. (1989) The biased reptation model of DNA gel-electrophoresis - mobility vs. molecular-size and gel concentration. *Biopolymers* **28,** 1781–1791.

31. Lerman, L. S. and Frisch, H. L. (1982) Why does the electrophoretic mobility of DNA in gels vary with the length of the molecule. *Biopolymers* **21,** 995–997.
32. Lumpkin, O. J., Dejardin, P., and Zimm, B. H. (1985) Theory of gel-electrophoresis of DNA. *Biopolymers* **24,** 1573–1593.
33. Liang, D. and Chu, B. (1998) High speed separation of DNA fragments by capillary electrophoresis in poly(ethylene oxide)-poly(propylene oxide)-poly(ethylene oxide) triblock polymer *Electrophoresis* **19,** 2447–2453.

16

DNA Analysis Under Highly Denaturing Conditions in Bare Fused Silica Capillaries

Karel Klepárník, Zdeňka Malá, and Petr Boček

1. Introduction

Denaturing electrophoresis has widely been used for DNA sequencing and mutation detection. Specific electromigration behavior of completely or partially dissociated DNA molecules is the prerequisite of high-selective separations, based on their size and/or conformation differences. Various denaturing techniques have been developed for both slab gel (SGE) and capillary electrophoresis (CE) formats. The highest possible denaturing ability of the background electrolytes (BGE) of both the sample solution and electrophoresis buffer is used for sequencing, and for analyses of restriction fragment length polymorphism (RFLP), or for length polymorphism of fragments amplified by polymerase chain reaction (PCR-FLP) *(1–5)*. In the technique of single-strand conformation polymorphism (SSCP) analysis, completely denatured samples are loaded into a native sieving medium where ssDNA fragments adopt a conformation that is determined by their nucleotide sequence *(6,7)*. Thus, not only complementary strands, but also strands carrying mutations are separated under optimum conditions. The SSCP technique is very sensitive, and even the point mutations, i.e., the substitutions of a single nucleotide in a sequence, can be detected. In the constant denaturant (CDCE) *(8,9)* or denaturing gradient capillary electrophoresis (DGCE) techniques *(10,11)*, a native sample is loaded into a sieving medium with a moderate denaturing ability. Consequently, dsDNA molecules dissociate according to their melting temperature, which sensitively reflects the mutations in the DNA sequence. The optimum denaturing power of a BGE is controlled by the temperature (constant or programmed) of the run, and by the concentration of denaturing agent(s). Based on the extent of dissociation, mutant DNA sequences are separated reproducibly from wild-type sequences, with high resolution.

Most effort, however, has been put into the development of DNA sequencing systems *(12–19)*. As the separation selectivity of ssDNA fragments is higher than the

From: *Methods in Molecular Biology, Vol. 162: Capillary Electrophoresis of Nucleic Acids, Vol. 1: Introduction to the Capillary Electrophoresis of Nucleic Acids*
Edited by: K. R. Mitchelson and J. Cheng © Humana Press Inc., Totowa, NJ

separation selectivity of their ds analogs, electrophoresis in denaturing buffers provides a better resolution between fragments which differ in size *(20)*. Thus, both RFLP and PCR-FLP analyses are faster and more reliable under denaturing conditions than native conditions. The DNA sequencing makes the most severe demands on systems for size-based separations. Therefore, the separations of sequencing fragments are exclusively performed under denaturing conditions. The number or the maximum length of completely resolved sequencing fragments, which differ in size by a single nucleotide, determines the sequence read length. The minimum difference in ssDNA sequencing fragments can be resolved under denaturing conditions for lengths up to 1000 nt *(15)*. In contrast, single base-pair resolution of dsDNA fragments can hardly be attained under native conditions for fragment lengths over 300 bp *(21)*.

In some instances, only highly denaturing conditions can be applied successfully for a reliable analysis. Some DNA fragments form a considerable amount of heteroduplexes after PCR amplification, or show a sequence-specific anomalous electrophoretic migration. Heteroduplexes comigrate with the specific products and, thus, deteriorate the peak resolution. The anomalous migration affects the accuracy and precision of the fragment size determination when the standard addition technique is used. In contrast, fully denatured ssDNA fragments, i.e., without any intra- and intermolecular interactions, may be expected to migrate independently, strictly according to their lengths. Various organic compounds have been tested as denaturing agents toward DNA *(22)*. The denaturing effectiveness of organic denaturants is not always adequate if the guanine and cytosine rich regions are present in a nucleotide sequence *(23,24)*. Urea is the denaturing agent most frequently used in CE, typically in a concentration range from 3.5 to 8.3 *M*. Notably, the addition of up to 8 *M* urea to the BGE may be insufficient to fully denature some anomalously migrating ssDNA fragments. Usually, elevated temperature is used to solve this problem and fully denature the DNA *(1–3,15–17,19)*. Nevertheless, there is still the demand for a highly efficient, denaturing electrolyte for use in CE.

1.1. Denaturation at pH Extremes

A more reliable denaturing process can be achieved using either acidic or alkaline conditions. At pH ≤2.0, adenine-thymine and even guanine-cytosine pairs are completely disrupted at room temperature *(25,26)*. However, the use of acidic buffers for DNA electrophoresis is impractical, since the dissociation of phosphate groups is reduced and, therefore, electrophoretic mobilities decrease greatly. Alkaline electrolyte would potentially be convenient, since the denaturing ability is also very high and the dissociation of phosphate groups is at the maximum. DNA fragments can be separated in solutions at pH 12.0–13.0 *(27,28)* without a risk of their destruction, as verified by the "comet assay" technique widely used in the genetic research *(29–31)*. Denaturing agarose electrophoresis in alkaline conditions is used frequently in a SGE format to analyze ssDNA. The sizes of both strands of DNA can be determined. Under the denaturing conditions, the presence of DNA single-strand breaks, alkali-labile sites and double-strand breaks can be visualized under microscope, as the DNA migrates away from the nucleus to form a characteristic "comet" appearance. Alkaline denaturing electrolytes have not been used in capillaries, since the stability of covalently bonded coatings decreases rapidly at pH values over 11.0 and the dissociation of silanol groups at the silica wall, the cause of electroosmotic flow (EOF), increases.

1.2. CE in Alkaline Buffers

In several of our publications, we described the excellent properties of alkaline denaturing electrolytes *(32,33)*. Narrow capillaries provide very convenient parameters that are advantageous for the application of the highly electric conductive alkaline electrolytes, such as high electric resistance and the ability to dissipate Joule heat efficiently. Thus, electric field strength up to 300 V/cm can still be applied in capillaries of 50 µm id. Alkaline electrophoresis on slab gels could only be carried out at <0.25 V/cm, which makes the technique rather time consuming. We demonstrated using model DNA standards and clinical samples that the enormous denaturing power of the alkaline electrolyte improved the resolution between separated fragments. Moreover, the migration of ssDNA molecules is significantly faster under alkaline conditions due to their increased effective charge. The electrophoretic mobilities of ssDNA fragments under these conditions are even slightly higher when compared to their dsDNA analogs at native conditions. Another interesting phenomenon is a surprising decrease in the EOF of free solutions of NaOH at pH values of 11.0–13.0 and a temperature of 40°C in bare fused silica capillaries. Consequently, the EOF mobilities of 2% agarose solutions in 0.01 *M* NaOH are even lower, and varied within experimental error around a mobility of $6 \times 10^{-10}\ m^2V^{-1}s^{-1}$.

In this chapter, the excellent separation properties of a new highly denaturing electrolyte system based on the 0.01 *M* solution of NaOH are presented, and the subsequent laboratory methodology is described, together with some theoretical considerations. The practical utility of the method is demonstrated through the detection of short tandem repeat polymorphism in the *endothelin 1* gene. Thus, the optimized analyses proves to be applicable for DNA diagnostics with PCR amplified fragments, in a case in which the DNA is resistant to denaturation and readily creates amplification heteroduplexes.

2. Materials

1. Fragments of pBR322 DNA-*Alu*I digest (MBI Fermentas, Vilnius, Lithuania) are used as the model mixtures.
2. All the CE experiments are performed using the BioFocus 3000 system (Bio-Rad, Hercules, CA) with the absorbance detection at 260 nm.
3. Capillaries are installed in a user-assembled cartridge and kept at a constant temperature of 40°C.
4. Fused silica capillaries of 50 µm and 100 µm id (367 µm od) are obtained from Polymicro Technologies (Phoenix, AZ).
5. The inner walls of the 100-µm id capillaries were coated by linear polyacrylamide (LPA) according to the Hjertén's procedure *(34)*. The total length of the capillary is 34.6 cm, and the effective length is 4.6 cm.
6. The inner walls of 50-µm id capillaries are uncoated. The total length of the capillary is 34.6 cm, and the effective length is 30 cm.

7. A 2% solution of low-melting-point agarose (FMC Bioproducts, Rockland, ME) in BRE is used as the sieving medium.

3. Methods

Even if the use of alkaline electrolyte in the denaturing CE makes the laboratory procedure easier, there are some important practical principles, which must be followed so as to get reliable and reproducible results in a reasonable period of time. Since any CE instrumentation is amenable to denaturing electrophoresis, only the specially optimized procedures for the capillary pretreatment, preparation of a separation medium, and sample denaturing are described.

3.1. Separation Capillary

1. Many procedures for the coating of capillary inner walls have been designed to suppress both EOF and an adsorption of analytes. All coatings based on covalently bonded polymers, via Si-O- ***(34–36)*** or Si-C- ***(37)*** bonds, will definitely suffer from the alkali hydrolysis, at pH values higher than 10.0 and at elevated temperatures.
2. Recently, Liu et al. ***(38)*** published the results of DNA denaturing electrophoresis in capillaries coated with an epoxy resin. The epoxy coating is stable over pH range of 2.0–12.0, with practically undetectable EOF of running buffer above pH 7.0 ***(39)***. However, at pH 13.0, the resolution of DNA fragments is decreased, probably due to the destruction of epoxy coating.
3. In this respect, dynamic coatings of the capillary wall would be more convenient. Such additives to the BGE are based on solutions of polymers or surfactants. The polymers cover the capillary wall noncovalently, and increase the solution viscosity at the wall, or compensate for the negative electrostatic charge ***(40–43)***.
4. A further way to avoid difficulties with coatings is to use bare (uncoated) capillaries, where the EOF dominates the electrophoresis. Here, DNA fragments migrate selectively against the faster migration of the bulk EOF, which transports them to the detector in the order of their descending size. This counter-flow technique has also been used for the separations of DNA fragments ***(44,45)***.
5. In these cases, the EOF mobility is evaluated with the help of a neutral marker, which possesses an adequate extinction coefficient and/or fluorescence quantum yield. The most common markers are mesityloxide, phenol, acetophenone, and pyridine for absorbance detection, and coumarin, 4-nitroaniline, riboflavine, and umbelliferone for fluorescence detection.
6. Since the alkaline electrolytes are highly conductive, capillaries of 50 µm id or narrower are needed if electric field strengths higher than 100 V/cm are to be used.
7. As noted in **Subheading 1.2.**, the EOF of free alkaline electrolytes in bare fused-silica capillaries is substantially reduced, and the negligible EOF of 2% agarose solutions in 0.01 *M* NaOH is acceptable for high-efficiency separations.

3.2. Treatment of Bare Fused Silica Capillaries

1. When working with highly aggressive alkaline electrolytes, a partial destruction of outer polyimide coating at the capillary ends inserted in electrode chambers might be expected. Therefore, the coatings at the injection and outlet points must be removed. Usually 5-mm long bare ends of the capillary are enough to avoid any problems with loose pieces of coating.

2. Before any treatment, capillary should be fixed in a holder of CE system. The procedure of preparation is very simple and fast:
   a. Rinse a new capillary with 1 *M* NaOH for 2 h and then with 0.01 *M* NaOH for 2 min.
   b. Before each analysis run, rinse the capillary with 0.01 *M* NaOH for 2 min and then for 2 min with the separation medium, i.e., a sieving polymer solution in 0.01 *M* NaOH.
3. For storage overnight, the capillaries are filled with distilled water, or a free solution of 0.01 *M* NaOH.
4. It is not recommended to leave solutions of polysaccharides inside capillaries. They slowly form an inhomogeneous gel layer at capillary wall, even at a temperature above the gelling point. This causes a dramatic decrease in the separation efficiency ***(46)***. However, the gel layer can easily be removed by flowing distilled water through the capillary while held at a temperature above the melting point of the gel.
5. The best separations are obtained with capillaries, run at temperatures of 40°C. The use of elevated temperature in DNA analysis brings several advantages. Except for an increased denaturing ability of the separation medium, faster and more selective separations of a larger range of ssDNA fragments can be performed ***(19,46)***. Therefore, it is very convenient to have good temperature control of outer wall of the capillary during denaturing CE.
6. Use of high concentrations of NaOH at elevated temperature does not affect the transparency of the detection window, even over the long term.

3.3. Denaturing Separation Medium

1. Only solutions of polysaccharides, such as low melting point agarose BRE (FMC Bioproducts, Rockland, ME) and hydroxyethyl cellulose, relative molecular weight 105,000 (Polysciences, Warrington, PA), have been reported as replaceable sieving media for denaturing CE of DNA ***(32,33,38)***.
2. Agarose must be fully dissolved before addition of the alkaline electrolytes.
3. The dissolution of polysaccharides is not a trivial procedure. Slowly add the agarose powder to the Erlenmeyer flask from a weigh boat while swirling as much as possible so that it gets completely dispersed. With material like this, clumping is virtually irreversible. It is helpful to use the buffer or distilled water, which is chilled. After dispersion, let the mixture stand for 5–10 min to give the agarose time to fully hydrate. Then you can heat the mixture to speed up the dissolution process. Low melting point agaroses melt rapidly at a temperature over 50°C and are relatively stable in the state of sol at a laboratory temperature and concentrations below 3% (w/v).
4. The sieving buffer must be prepared as follows: Since the heating of an alkaline agarose solution causes hydrolysis of the polysaccharide molecules, the agarose must be completely dissolved in distilled water and allowed to cool, prior to the addition of NaOH. To avoid a heating of agarose solution with hydroxide, a 3% (w/v) stock solution of agarose BRE is mixed with a 0.1 *M* solution of NaOH. The concentrations are then adjusted to give solutions of 2% (w/v) agarose in 0.01 *M* hydroxides.

3.4. DNA Sample Preparation

3.4.1. Denaturation in Organic Chaotrope

1. The total denaturing of a DNA sample prior to the separation process is a substantial procedure. An injection of denatured samples provides more efficient and reproducible results. Only the organic compounds, formamide, dimethylformamide and dimethylsulphoxide have been reported as denaturing agents for sample treatment before CE separations.

2. The procedure is as follows: Take 10 µL of a sample and evaporate at a laboratory temperature under vacuum. Vacuum centrifugation is recommended for this purpose, since a massive bubble formation might spread the sample over vial walls if no centrifugal forces are applied.
3. Dissolve the dried sample in 10 µL of organic denaturing solvent, e.g., formamide, and shake well to be sure that all sample is dissolved. Spin for a while to remove all droplets from the vial wall.
4. Heat the sample solution in a thermocycler at 98°C for 2 min. Then chill it immediately on ice. Such a treated sample can be used for many repeated injections. After 4 h, the heating step should be repeated, since the ssDNA fragments tend to reassociate even at a low temperature.

3.4.2. Denaturation in Alkaline Conditions

1. Sample denaturing under alkaline conditions is much simpler: Add 1 µL of 0.1 *M* NaOH solution to 9 µL of a sample, and then shake the mixture at a laboratory temperature for 5 min. The sample solution can immediately be injected into the capillary and stored at a laboratory temperature between analyses.
2. Various concentrations of NaOH solution are tested in a DNA sample. A concentration of 0.01 *M* NaOH proves to be satisfactory for the complete denaturing of all model or clinical samples used in our previous study ***(33)***.
3. Use of a higher concentration of NaOH is also advantageous if the "sample stacking" technique is used for the preconcentration of diluted solutions of DNA fragments. In this technique the leading zone of OH^- anions protects the migration of less mobile DNA molecules, which form a very sharp and concentrated zone at the beginning of the separation ***(47)***.

3.5. Separation Properties of Alkaline Electrolyte

1. Several physico-chemical properties of the alkaline electrolyte have been investigated.
2. The EOF mobility in a bare fused silica capillary is evaluated as a function of pH.
3. The electrophoretic mobilities and resolution of model ssDNA fragments denatured by dimethylformamide (DMF), 0.01 *M* HCl, 0.01 and 0.1 *M* NaOH are compared.
4. An improvement in the resolution is the result of the optimization of an alkaline BGE composition, with respect to the separation efficiency.

3.5.1. EOF in a Bare Fused Silica Capillary

1. The mobility of EOF is determined by the ζ potential, and by viscosity of an electrolyte ***(47)***.
2. Hence, all known methods for EOF suppression are based on a reduction of inner capillary wall surface charge and/or an increase in the electrolyte viscosity in close proximity to the wall. It is commonly accepted that electroosmosis in bare fused silica capillaries increases with increasing pH, due to the more extensive dissociation of the silanol groups.
3. Thus, the high EOF of the alkaline electrolyte is expected, and the counter-flow electrophoresis in bare capillaries is assumed to be the useable mode. Nevertheless, mobilities of EOF at pHs > 12.0 and at a temperature of 40°C have not been reported in the literature.
4. To our surprise, we found that the EOF of this system is small ***(32,33)***. Consequently, we are able to work in a direct mode of electrophoresis.
5. To explain the extraordinary low EOF mobility in the silica capillary with completely dissociated silanol groups (pK $\cong$ 2) ***(48)***, we studied the dependence of EOF on pH in the alkaline region.

6. The EOF is evaluated with the help of mesityloxide, a neutral marker, and with three stable organic acids, sulfosalicylic, phthalic, and benzoic, as reference anions.
7. During a single injection procedure, the three acids are injected at the cathodic end of the capillary, and mesityloxide at the anodic end, respectively.
8. Since the stability of mesityloxide at pH 12.0–13.0 is questionable, the differences in its migration times at the individual pH values are correlated with the migration times of the acids. Thus, the changes occurring in EOF can be determined reliably.
9. The results of the measurements are summarized in **Fig. 1**. The electrolytes of decreasing pH are prepared by the titration of 0.1 *M* solution of NaOH with 0.5 *M* phosphoric acid. Here, the ionic strength, *I*, is kept approximately constant in the pH range from 12.0 to 8.0. The calculated values of *I* for the particular pH are introduced in the caption to **Fig. 1**.
10. A surprisingly low EOF has been observed at pH 11.0–13.0, at a temperature of 40°C. This fact might be explained by the existence of a highly viscous thin layer at the silica capillary wall, which is formed by dissolved $Si(OH)_4$. The orthosilicic acid has a strong tendency to polymerize when it is concentrated *(48)*. This layer probably does not allow for a molecular transport to occur in the vicinity of the wall, and causes a shift of the slipping plane further away from the wall to the lower value of the ζ potential.
11. The EOF mobilities of agarose solutions are even lower. The solutions of 2% agarose in 0.01 *M* NaOH, KOH, and LiOH are tested. The lowest value of EOF is attained for NaOH, which varies within experimental error, around a mobility of 6 ξ 10^{-10} $m^2V^{-1}s^{-1}$ (*see* **Note 1**).

3.5.2. DNA Separation

1. Highly alkaline solutions of fully dissolved, linear polymers, based on solutions of polysaccharides, have remarkable separation properties which have recently been described *(32,33,38)*. These replaceable sieving media are stable even in 0.1 *M* NaOH and provide for excellent separation selectivity for ssDNA fragments. The advantages of alkaline denaturing are evident by comparing the separations achieved under various denaturing conditions.
2. The effects of various denaturing conditions on the separation of a model sample (pBR322 DNA-*Alu*I digest) are presented in **Fig. 2**. For the sake of comparison, record "2A" shows a separation under native conditions. The electrophoresis is performed in a 2% solution of agarose with 0.1 *M* Tris-0.1 *M* TAPS, pH 8.3. Denaturing electrophoreses in the same electrolyte, but with the addition of 7 *M* urea are shown in records "2B" and "2C". Prior to the analysis, the samples are denatured in solutions of 0.01 *M* NaOH (2B) and 0.1 *M* NaOH (2C), to prove the reliability of the denaturing process. Electropherograms "2B" and "2C" show nearly identical results. Therefore, the concentration of 0.01 *M* NaOH was adopted for use as it was sufficient for complete denaturation, which is in accordance with data presented in the literature *(27,28)*.
3. The complete separation of all 14 fragments of the model sample in a 2% solution of agarose with 0.01 *M* NaOH and 0.0015 *M* $Na_2B_4O_5(OH)_4$ is seen on record **Fig. 2D**. The same denaturing procedure of the sample in 0.01 *M* NaOH was used here.
4. Records "2B–2D" also represent another important result, i.e., that ssDNA fragments when denatured in a solution of NaOH migrate as rapidly as dsDNA. Usually, the migration of individual strands is slower, since their charge-to-size ratios are half of those of dsDNA fragments. This is evident from records in **Fig. 2E, F** where the samples were denatured using DMF and 0.01 *M* solution of HCl, respectively. A nearly 2X increase in the migration time of ssDNA fragments is seen in comparison with the migration times of

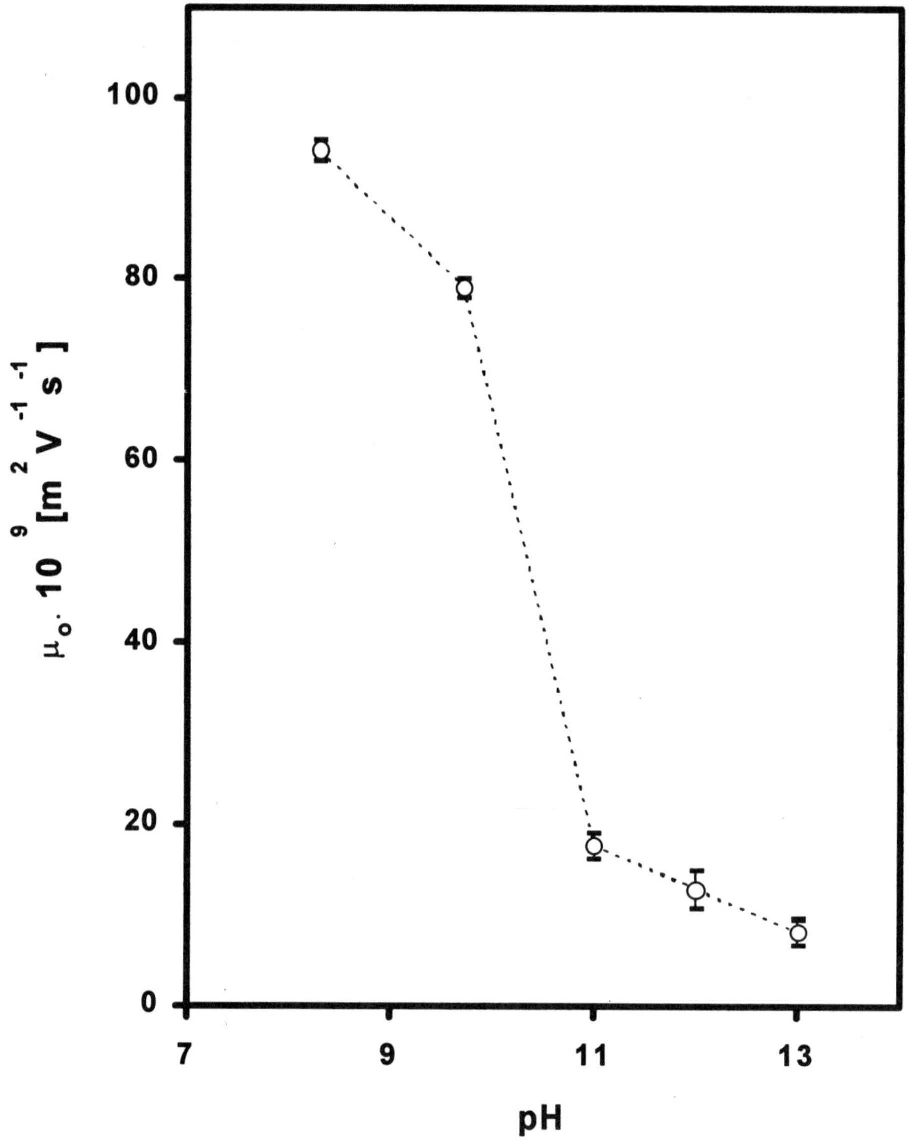

Fig. 1. Dependence of EOF in bare silica fused capillary on pH. Electrophoresis was performed at an electric field strength of 58 V/cm and temperature of 40°C in a capillary of the effective length of 4.6 (34.6) cm and id 50 mm. Error bars represent standard deviation of five consecutive measurements in a single capillary. Ionic strengths at the individual pH values are calculated to be: pH 13–0.1, pH 12–0.18, pH 11–0.16, pH 9.7–0.15, pH 8.3–0.15 mol/L. The pH values were adjusted by the titration of 0.1 *M* NaOH with 0.5 *M* phosphoric acid to the desired pH. Mesityloxide, sulfosalicylic, phthalic, and benzoic acids were used as independent EOF mobility markers. Before each run, the capillaries were rinsed with 0.1 *M* NaOH for 10 min, with H_2O for 10 min and then with a BGE for 10 min.

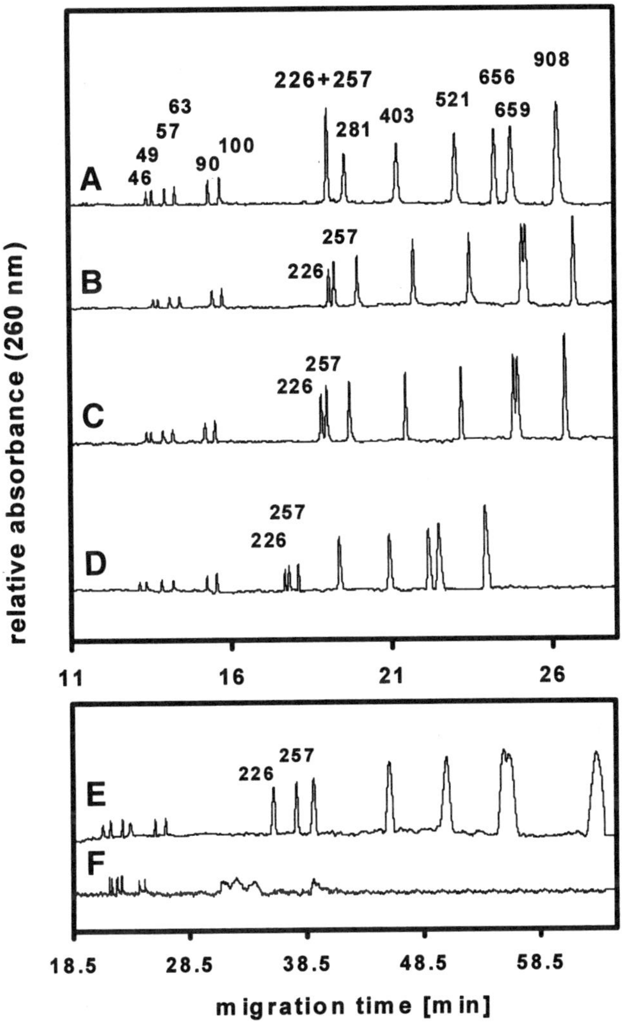

Fig. 2. Separation of standard pBR322 DNA-*Alu*I restriction fragments at various denaturing conditions. **(A)** Native dsDNA sample analyzed in 2% agarose BRE solution in 0.1 *M* Tris-0.1 *M* TAPS, pH 8.3, injected for 2 s at 58 V/cm; **(B,C)** samples denatured in 0.01 *M* (B), 0.1 *M* (C) NaOH at room temperature, analyzed in 2% agarose BRE in 0.1 *M* Tris-0.1 *M* TAPS, pH 8.3, with 7 *M* urea, injected for 5 s at 58 V/cm; **(D)** sample denatured in 0.01 *M* NaOH at room temperature, analyzed in 2% agarose BRE in 0.01 *M* NaOH with 0.0015 *M* $Na_2B_4O_5(OH)_4$, pH 12.0, injected for 2 s at 58 V/cm; **(E)** sample evaporated, dissolved in DMF, heated for 2 min and chilled on ice, analyzed as in case B, injected for 7 s at 145 V/cm; **(F)** 5 mL of sample mixed with 5 mL of 0.02 *M* HCl at room temperature, analyzed as in case B, injected for 15 s at 145 V/cm. Electrophoresis at 145 V/cm, temperature 40°C, capillary: 30 (34.6) cm. (A–F) Coated by linear polyacrylamide, 100 μm id, (D) bare fused silica , 50 μm id. Fragment sizes in number of nucleotides are over the peaks. Before each analysis run, the capillaries are rinsed: (1) for 3 min with water for the separation of DNA fragments in coated capillaries at pH 8.3, and then for 3 min with the BGE. (2) Capillaries are rinsed for 5 min with 0.01 *M* NaOH and then for 3 min with the BGE, for the separation of DNA fragments at pH 12.0.

Table 1
Comparison of the Experimentally Observed Mobilities of Some Standard DNA Fragments Under Various Denaturing Conditions[a]

	$.10^9$ [$m^2V^{-1}s^{-1}$]					
nt	A native	B NaOH - Urea	C NaOH - Urea	D NaOH - NaOH	E DMF - Urea	F HCl - Urea
46	25.72	25.30	25.82	26.30	16.44	16.05
100	21.99	21.93	22.14	22.24	12.99	14.05
281	17.65	17.32	17.55	19.14	8.80	10.18
521	15.00	14.72	14.89	16.57	6.81	
908	13.19	12.93	13.05	14.46	5.42	

[a]Data is taken from **Fig. 2A–F**.

peaks in electropherograms "2A–2D". Both separations "2E" and "2F" are performed in the same electrolyte with 7 *M* urea at 40°C, as with separation "2B."

5. Analysis shown in **Fig. 2E** represents a typical procedure optimized for DNA sequencing by CE. Denaturation at highly acidic pH is not suitable for DNA. Record "2F" shows very broad zones of fragments 226, 257, 281, and 403 nt and peaks of longer fragments are absent. We can speculate that the long DNA fragments were destroyed in such a low pH.
6. The electrophoretic mobilities of chosen DNA fragments under conditions given for "2A–2F" are compared in **Table 1**. The extraordinarily fast migration of fragments denatured in solutions of NaOH is another positive property of a highly alkaline BGE, in addition to a high-denaturing effectiveness and a low electroosmotic flow in the bare capillary.
7. There is no doubt that DNA fragments analyzed in the environment with 7 *M* urea at 40°C, and also in 0.01 *M* NaOH at the same temperature, are denatured. Therefore, the fast migration of those fragments denatured by NaOH can be interpreted in terms of an increased effective charge. Moreover, the fast migration of the DNA fragments denatured in the NaOH solution prior to analysis, and then separated at pH 8.3 seems to be a result of permanent molecular changes affecting dissociation of nucleotides (**Fig. 2B,C**) (*see* **Note 2**).

3.5.3. Free Base Dissociation

1. The dissociation extent of free DNA bases in alkaline electrolytes is demonstrated in **Fig. 3**. It can be deduced from dissociation constants, pK'_a, of guanine [9.2; 12.3], adenine [9.8], thymine [9.9; >13] and cytosine [12.2] that they become negatively charged at pH values greater than 9.0 *(49)*. At pH 8.3, a common pH for DNA separations, the bases are neutral and insoluble in water. **Figure 3A** shows the free-solution migration of the four bases in 0.1 *M* NaOH. In accordance with their respective pK'_a values, guanine which dissociates in two steps (the first proton is released completely, the second one partially) has the highest mobility. Adenine and thymine, which each carry one charge, comigrate, while the partially dissociated cytosine has the slowest mobility.
2. Similar results are shown in records of **Fig. 3B–F**. Here, the separation of the bases, performed in 2% agarose solution with 0.1 *M* NaOH, is faster than in free solution (record "3A"). Records "B–E" are introduced for identification of the peaks. It can be seen that the migration times are shorter than in free NaOH, due to the minimal EOF of the agarose

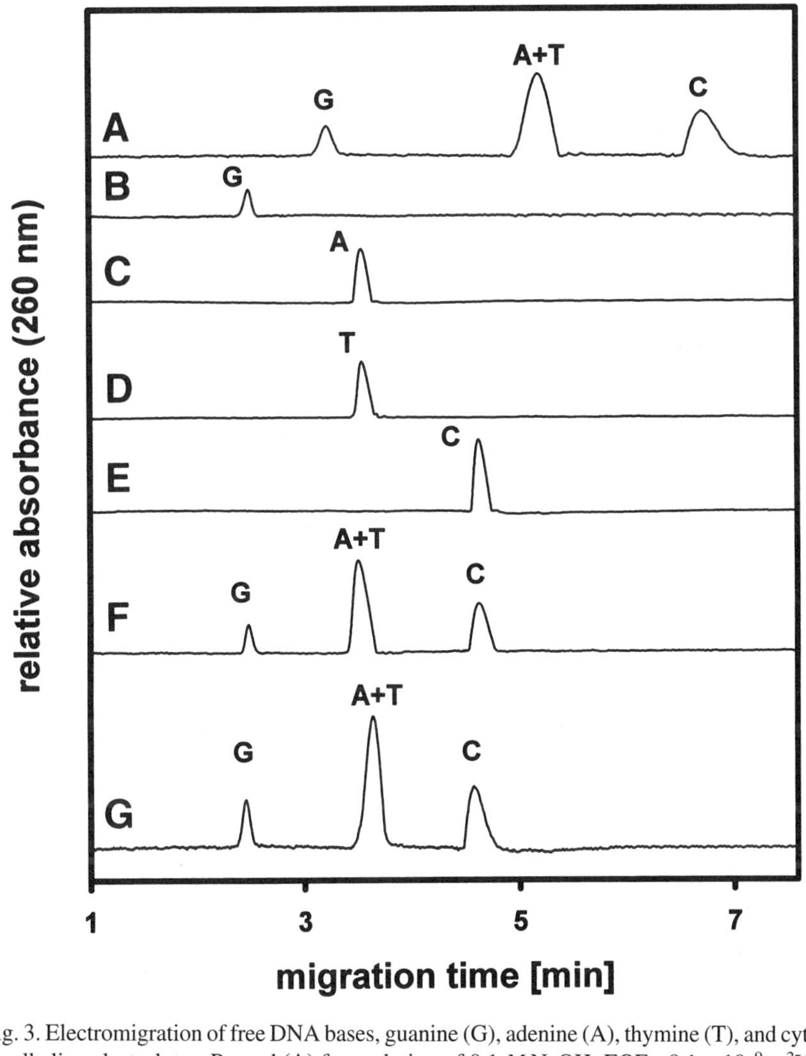

Fig. 3. Electromigration of free DNA bases, guanine (G), adenine (A), thymine (T), and cytosine (C) in alkaline electrolytes. Record **(A)** free solution of 0.1 *M* NaOH; EOF - 9.1×10^{-9} $m^2V^{-1}s^{-1}$. **(B–F)** 2% agarose solution with 0.1 *M* NaOH; EOF < 0.5×10^{-9} $m^2V^{-1}s^{-1}$. **(G)** 2% agarose solution with 0.01 *M* NaOH. Electrophoresis at 58 V/cm, temperature 40°C, bare fused silica capillary: 4.6 (34.6) cm, 50 μm id. Samples injected for 2 s at 16 V/cm.

solution. The EOF of the free solution was estimated to be 9.1×10^{-9} $m^2V^{-1}s^{-1}$. Therefore, if the net migration times of peaks in record "A" are calculated, values of 2.62, 3.81, and 4.60 min are obtained, which are within an experimental error identical with migration times of the peaks in records B–F, i.e., 2.49, 3.69, and 4.63 min. It can also be concluded that the migration of the four DNA bases in agarose solution is not affected by an eventual dissociation of the polysaccharide molecule.

3. Record “3G” shows the separation in 2% agarose solution with 0.01 *M* NaOH. Migration times are identical to those in “3B–F” even if a slower migration of guanine, cytosine could be expected. Some permanent molecular changes affecting the pK'_a values after the dissolution in highly alkaline electrolyte are observed only in the case of guanine. Thus, only guanine injected from the solution of 0.1 *M* NaOH migrates through a 100 m*M* Tris-100 m*M* TAPS, pH 8.3 (*see* **Note 2**).

3.5.4. Separation Efficiency

1. It has been found that the amount of tetraborates added to the alkaline BGE affects the separation efficiency markedly *(32,33)*. Sodium tetraborate ($Na_2B_4O_5(OH)_4$, which is frequently but incorrectly named as $Na_2B_4O_7$), is a common constituent of buffers with the maximum capacity at pH 10.0–12.0. The dependence of the separation efficiency of ssDNA fragments on the concentration of sodium tetraborate is presented in **Fig. 4**. The results are based on the separation of the model sample in 2% agarose solution with 0.01 *M* NaOH and $Na_2B_4O_5(OH)_4$. Here, the separation efficiency expressed in the number of height equivalents to a theoretical plate (HETP) *N*, has been calculated according to the relationship: $N = 5.545.(t/w_{1/2})^2$, where *t* is migration time and $w_{1/2}$ is the peak width at half of the peak height. The maximum separation efficiency of the shortest fragments was 2,300,000 HETP/m at a concentration of 0.0015 *M* $Na_2B_4O_5(OH)_4$.
2. The separation efficiencies decrease with the molecular mass. This fact can be explained by a larger electromigration dispersion of longer molecules. Their mobilities differ from those of coions more substantially and, therefore, the zones can be expected to be more dispersed.
3. The effect of tetraborates on the separation of the model ssDNA fragments is seen in **Fig. 5**. Here, separations in two electrolytes are compared: 0.01 *M* NaOH (record “5A”) and 0.01 *M* NaOH with $Na_2B_4O_5(OH)_4$ at the optimum concentration of 0.0015 *M* (record “5B”). The better resolution of the zones in record “5B” is due to the improved separation efficiency, whereas the selectivity remains unaffected (*see* **Note 3**).

3.6. Application in DNA Diagnostics

1. New cost-effective and reliable methods that are amenable to automated handling are especially attractive in clinical applications. The technologies to characterize disease-causing mutations require robust, reproducible, and accurate protocols for measuring DNA fragment lengths.
2. The alkaline electrolyte system may be used for the detection of short tandem repeat polymorphism in *endothelin 1* gene. The polymorphism consists of three adjacent regions CT, CA, GC, and seems to play a role in the regulation of transcription and is assumed to be associated with the origins of hypertension.
3. The fragments are amplified from the *endothelin 1* gene of heterozygous individuals, and two fragments which differ in size are the specific amplification products. **Figure 6** summarizes the records of electrophoreses under native (**Fig. 6A**) and under various denaturing conditions (**Fig. 6B, C**). Whereas electrophoresis under native conditions (**Fig. 6A**) provides a partial separation of the ssDNA fragments which differ by 10 bp, the sample denatured in DMF and analyzed in the denaturing environment of 7 *M* urea at 40°C (**Fig. 6B**) is lost into many nonspecific peaks, or heteroduplexes. The separation is dramatically improved if the sample is denatured in 0.01 *M* NaOH (**Fig. 6C, D**). These electropherogram traces indicate an inadequate denaturation of GC rich DNA fragments by

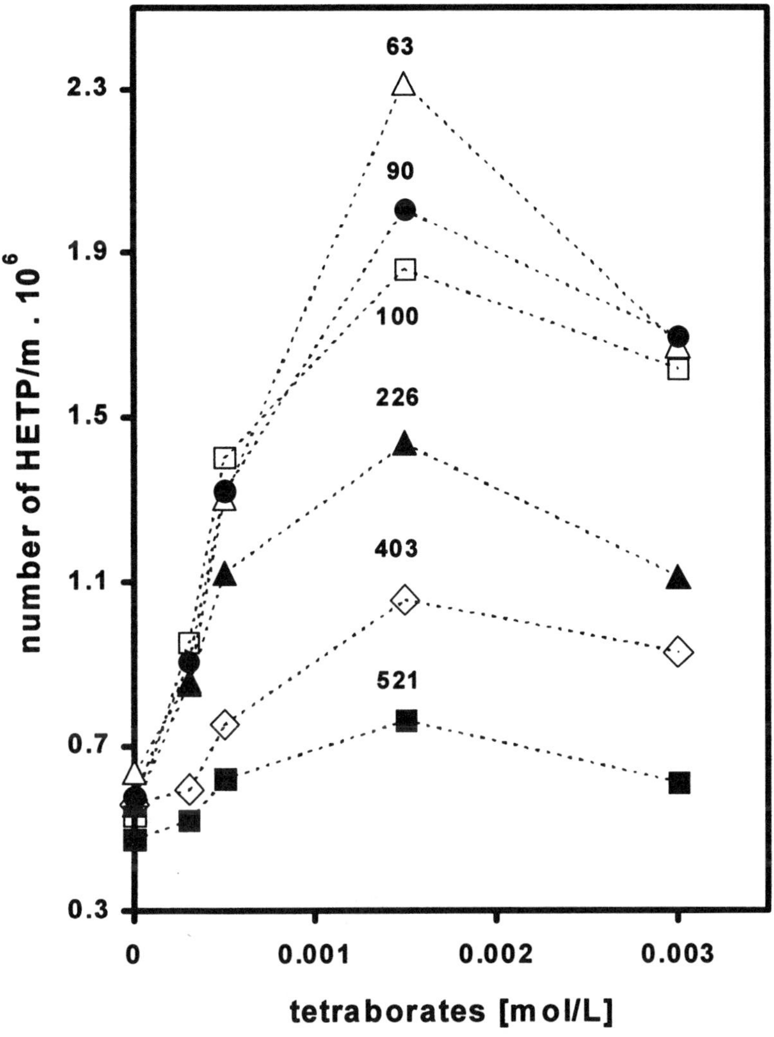

Fig. 4. The dependence of the separation efficiencies (number of HETP/m) of chosen pBR322 DNA-*Alu*I restriction fragments on the molar concentration of $Na_2B_4O_5(OH)_4$ in BGE. Electrophoresis at 289 V/cm; temperature 40°C; bare fused silica capillary 30 (34.6) cm, 50 µm id; BGE: 2% agarose BRE solution in 0.01 *M* NaOH; injection: 3 s at 58 V/cm. Sample denatured in 0.01 *M* NaOH at room temperature. Fragment sizes in number of nucleotides at the respective plots.

DMF. The broad zone of heteroduplexes is still obvious if the separation is performed at 30°C (**Fig. 6C**), but completely disapears at 40°C (**Fig. 6D**). The last electropherogram (**Fig. 6E**) shows the analysis of the sample under the optimum composition of the alkaline electrolyte. The GC region of repeats gives the fragments a resistance to the complete

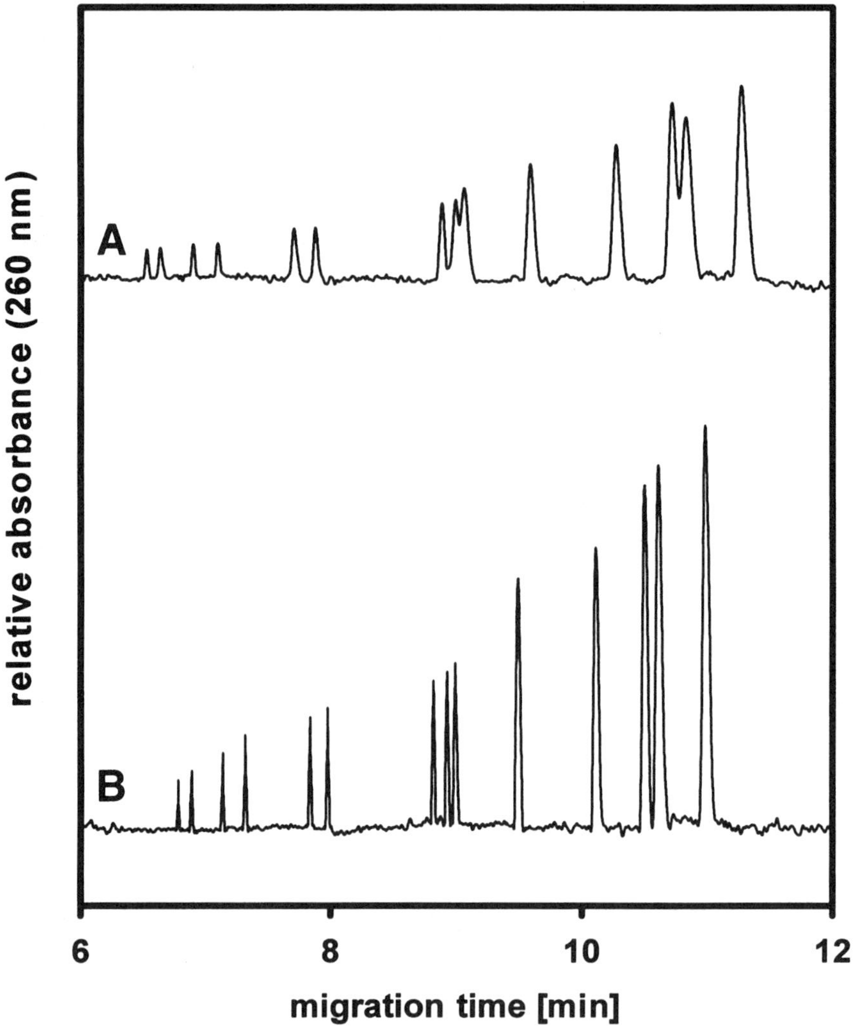

Fig. 5. The effect of the optimum addition of tetraborates to the BGE on the separation of the standard fragments. **(A)** 2% agarose BRE in 0.01 *M* NaOH, **(B)** 2% agarose BRE in 0.01 *M* NaOH with 0.0015 *M* $Na_2B_4O_5(OH)_4$. Conditions of electrophoresis are the same as shown in **Fig. 4**.

denaturation, and also the tendency to create heteroduplexes. The molecular sizes of the PCR products are determined by the calibration method of standard addition. Fragments with known sequences amplified from the same region of the gene served as standards (*see* **Note 4**).

4. **Figure 7** demonstrates the feasibility of a fast DNA diagnostics, based on the detection of short tandem repeat polymorphism using the optimized denaturing electrophoresis conditions, in short, bare fused silica capillaries (*see* **Note 5**).

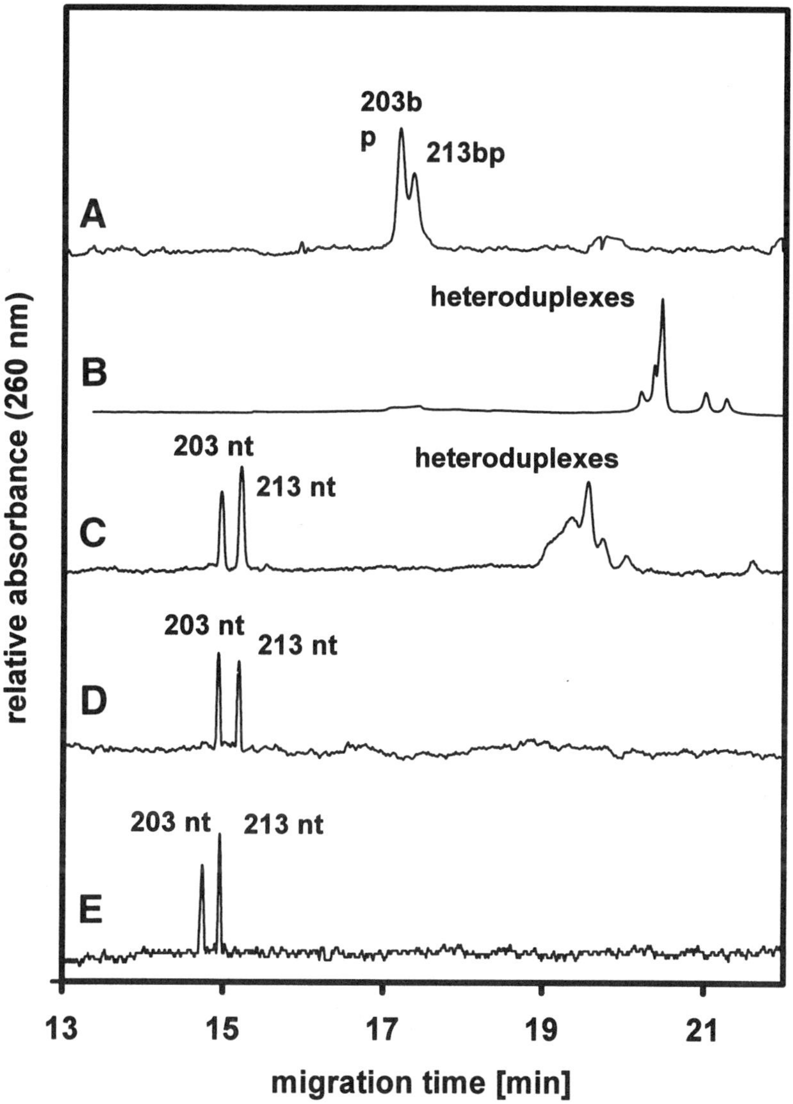

Fig. 6. Comparison of the separations of PCR products amplified from the *endothelin 1* gene under native and various denaturing conditions. **(A)** Native CE: injection 5 s at 145 V/cm, analyzed in 2% agarose BRE in 0.1 *M* Tris-0.1 *M* TAPS, pH 8.3. **(B)** Sample is evaporated, then dissolved in DMF, heated for 2 min and chilled on ice. It is analyzed at 40°C in 2% agarose BRE in 0.1 *M* Tris-0.1 *M* TAPS, pH 8.3, 7 *M* urea, injected for 8 s at 145 V/cm. **(C)** Sample denatured in 0.01 *M* NaOH and analyzed at 30°C, other conditions as at B. **(D)** Sample analyzed at 40°C, other conditions as at B. **(E)** Sample denatured in 0.01 *M* NaOH, injected for 5 s at 145 V/cm and analyzed at 40°C in 2% agarose BRE with 0.01 *M* NaOH and 0.0015 *M* $Na_2B_4O_5(OH)_4$. Electrophoresis at 145 V/cm; capillaries: (A–D) coated by linear polyacrylamide, 100 μm id; (E) bare fused silica, 50 μm id.

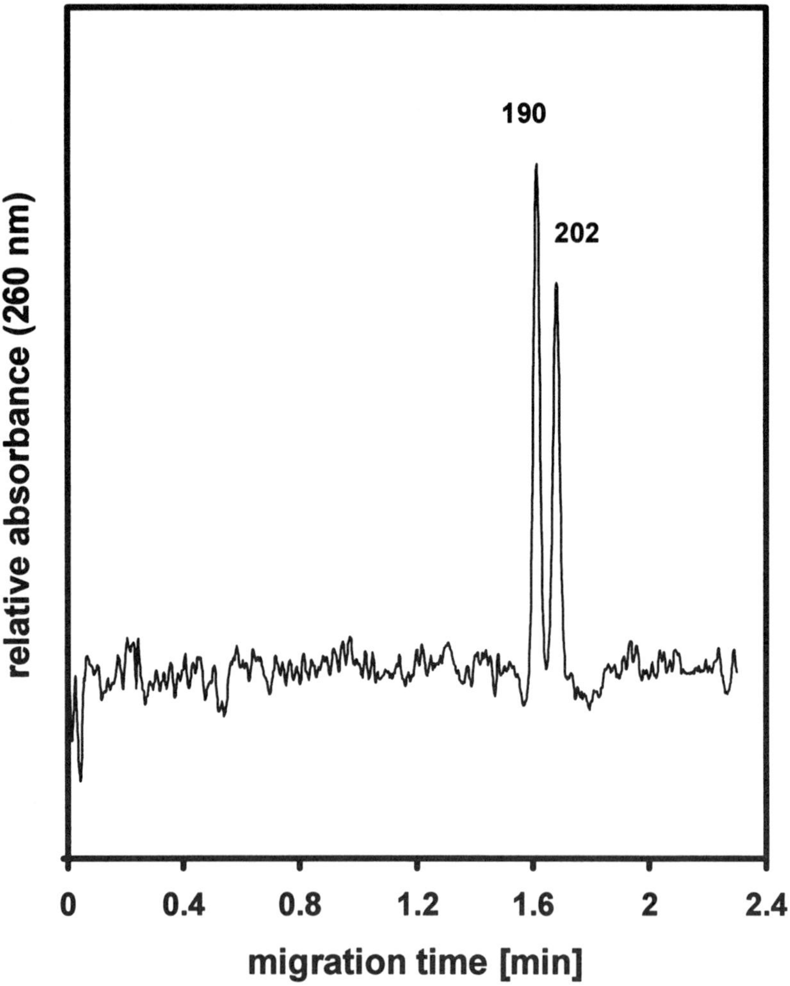

Fig. 7. Denaturing electrophoresis of PCR products from the *endothelin 1* gene in a short capillary. Sample injected 5 s at 145 V/cm; capillary 4.6 (34.6) cm; all other conditions are as described in **Fig. 6E**.

4. Notes

1. It is commonly accepted that the EOF mobility of electrolytes in silica capillaries increases with increasing pH. The results show that the opposite is true at pH values higher than 10.0 (**Fig. 1**). The extraordinary weak EOF of highly alkaline solutions in bare fused silica capillaries could be explained by the existence of a thin layer of orthosilicic acid, $Si(OH)_4$, at the capillary wall. A highly viscous thin layer of dissolved silica probably causes a shift of the slipping plane further away from the wall to a lower value of the ζ potential. As the EOF mobility of a 2% solution of agarose at pH 12.0 and a temperature

of 40°C is low (6×10^{-10} $m^2V^{-1}s^{-1}$), bare silica fused capillaries can be used for the separation of ssDNA fragments.

2. The ssDNA fragments denaturated in NaOH solutions (**Fig. 2B–D**) migrate at the same velocity or faster than dsDNA (**Fig. 2A**). Their fast migration can be interpreted in terms of an increased effective charge of the DNA molecule denatured by NaOH. Even the fragments denatured by NaOH prior to analysis but analyzed in denaturing environment at pH 8.3 migrate as fast as in alkaline electrolyte. Compare record "2B" to records "2C" and "2D" in **Fig. 2**. This fact seems to be a result of permanent changes in molecules which affect the dissociation equilibria, or pK'_a values, of nucleotides.
3. The addition of tetraborates to the alkaline background electrolyte improves the separation efficiency. At the optimum concentration of 0.0015 *M* of sodium tetraborate, $Na_2B_4O_5(OH)_4$, in the BGE, separation efficiency as high as 2,300,000 HETP/m were reached for short DNA fragments (**Fig. 4**). The positive effect of tetraborates on the separation efficiency is difficult to explain due to their very complex equilibria in solutions.
4. A 2% w/v solution of low-melting-point agarose in 0.01 *M* NaOH and 0.0015 *M* $Na_2B_4O_5(OH)_4$ proves to be an excellent denaturing electrolyte for the separation of DNA fragments, such as the products of PCR amplification which create heteroduplexes easily and are resistant to denaturation. The optimized composition of the electrolyte, with respect to the denaturing ability and separation efficiency, is convenient for implementation in clinical DNA diagnostics.
5. NaOH is a very strong denaturing agent and even the DNA fragments with GC rich regions, which are resistant to the denaturing in urea, dissociate completely in 0.01 *M* NaOH. Very well resolved and sharp zones of model fragments (**Fig. 2D**) suggest the absence of any inter- or intramolecular aggregations in the alkaline electrolyte. Even the peaks of fragments amplified from GC rich regions of *endothelin 1* gene are sharp (**Fig. 6E**). This is a prerequisite of high-throughput genetic analyses in short capillaries, where the analysis time can be reduced by a factor of 100 when compared to classical SGE (**Fig. 7**).

Acknowledgments

Preparation of this chapter has been supported by Grants No. 203/00/0772 and 303/00/0928 of the Grant Agency of the Czech Republic.

References

1. Wenz, H., Robertson, J. M., Menchen, S., Oaks, F., Demorest, D. M., Scheibler, D., Rosenblum, B. B, Wike, C., Gilbert, D. A., and Efcavitch, J. W. (1998) High-precision genotyping by denaturing capillary electrophoresis. *Genome Res.* **8,** 69–80.
2. Rosenblum, B. B., Oaks, F., Menchen, S., and Johnson, B. (1997) Improved single-strand DNA sizing accuracy in capillary electrophoresis. *Nucleic Acids Res.* **25,** 3925–3929.
3. Schmalzing, D., Koutny, L., Adourian, A., Chisholm, D., Matsudaira, P., and Ehrlich, D. (2001) Genotyping by microdevice electrophoresis, in *Capillary Electrophoresis of Nucleic Acids,* Vol. 2 (Mitchelson, K. R. and Cheng, J., eds.), Humana Press, Totowa, NJ, pp. 163–173.
4. Wang, Y., Wallin, J. M., Ju, J., Sensabaugh, G. F., and Mathies, R. A. (1996) High-resolution capillary array electrophoretic sizing of multiplexed short tandem repeat loci using energy-transfer fluorescent primers. *Electrophoresis* **17,** 1485–1490.
5. Mansfield, E. S., Vainer, M., Enad, S., Barker, D. L., Harris, D., Rappaport, E., and Fortina, P. (1996) Sensitivity, reproducibility, and accuracy in short tandem repeat genotyping using capillary array electrophoresis. *Genome Res.* **6,** 893–903.

6. Kuypers, A. W., Willems, P. M., Vanderschans, M. J., Linssen, P. C. M., Wessels, H. M. C., Debruijn, C. H., Everaerts, F. M., and Mensink, E. J. (1993) Detection of point mutations in DNA using capillary electrophoresis in a polymer network. *J. Chromatogr. B* **621,** 149–156.
7. Hebenbrock, K., Williams, P. M., and Karger, B. L. (1995) Single strand conformational polymorphism using capillary electrophoresis with two-dye laser-induced fluorescence detection. *Electrophoresis* **16,** 1429–1436.
8. Muniappan, B. P. and Thilly, W. G. (1999) Application of constant denaturant capillary electrophoresis (CDCE) to mutation detection in humans. *Genet. Anal.* **14,** 221–227.
9. Coller, H. A., Khrapko, K., Torres, A., Frampton, M. W., Utell, M. J., and Thilly, W. G. (1998) Mutational spectra of a 100-base pair mitochondrial DNA target sequence in bronchial epithelial cells: a comparison of smoking and nonsmoking twins. *Cancer. Res.* **58,** 1268–1277.
10. Righetti, P. G. and Gelfi, C. (1998) Analysis of clinically relevant, diagnostic DNA by capillary zone and double-gradient gel slab electrophoresis. *J. Chromatogr. A* **806,** 97–112.
11. Gelfi, C., Cremoresi, L., Ferrari, M., and Righetti, P. G. (2001) Point mutation detection by temperature-programmed capillary electrophoresis, in *Capillary Electrophoresis of Nucleic Acids,* Vol. 2 (Mitchelson, K. R. and Cheng, J., eds.), Humana Press, Totowa, NJ, pp. 73–88.
12. Liu, S., Shi, Y., Ja, W. W., and Mathies, R. A. (1999) Optimization of high-speed DNA sequencing on microfabricated capillary electrophoresis channels. *Anal. Chem.* **71,** 566–573.
13. Kheterpal, I. and Mathies, R. A. (1999) Capillary array electrophoresis DNA sequencing. *Anal. Chem.* **71,** 31A–37A.
14. Tan, H. and Yeung, E. S. (1998) Automation and integration of multiplexed on-line sample preparation with capillary electrophoresis for high-throughput DNA sequencing. *Anal. Chem.* **70,** 4044–4053.
15. Salas-Solano, O., Carrilho, E., Kotler, L., Miller, A. W., Goetzinger, W., Sosic, Z., and Karger, B. L. (1998) Routine DNA sequencing of 1000 bases in less than one hour by capillary electrophoresis with replaceable linear polyacrylamide solutions. *Anal. Chem.* **70,** 3996–4003.
16. Klepárník, K., Berka, J., Foret, F., Doskar, J., Kailerova, J., Rosypal, S., and Bocek, P. (1998) DNA cycle sequencing of a common restriction fragment of *Staphylococcus aureus* bacteriophages by capillary electrophoresis using replaceable linear polyacrylamide. *Electrophoresis* **19,** 695–700.
17. Schmalzing, D., Adourian, A., Koutny, L., Ziaugra, L., Matsudaira, P., and Ehrlich, D. (1998) DNA sequencing on microfabricated electrophoretic devices. *Anal. Chem.* **70,** 2303–2310.
18. Marsh, M., Tu, O., Dolnik, V., Roach, D., Solomon, N., Bechtol, K., Smietana, P., Wang, L., Li, X., Cartwright, P., Marks, A., Barker, D., Harris, D., and Bashkin, J. (1997) High-throughput DNA sequencing on a capillary array electrophoresis system. *J. Capillary Electrophor.* **4,** 83–89.
19. Klepárník, K., Foret, F., Berka, J., Goetzinger, W., Miller, A. W., and Karger, B. L. (1996) The use of elevated column temperature to extend DNA sequencing read lengths in capillary electrophoresis with replaceable polymer matrices. *Electrophoresis* **17,** 1860–1866.
20. van der Schans, M. J., Kuypers, A. W. H. M., Kloosterman, A. D., Janssen, H. J. T., and Everaerts, F. M. (1997) Comparison of resolution of double-stranded and single-stranded DNA in capillary electrophoresis. *J. Chromatogr.* **772,** 255–264.

21. Pariat, Y. F., Berka, J., Heiger, D. N., Schmitt, T., Vilenchik, M., Cohen, A. S., Foret, F., and Karger, B. L. (1993) Separation of DNA fragments by capillary electrophoresis using replaceable linear polyacrylamide matrices. *J. Chromatogr. A* **652,** 57–66.
22. Leving, L., Gordon, J. A., and Jencks, W. P. (1963) The relationship of structure to the effectiveness of denaturing agents for deoxyribonucleic acid. *Biochemistry* **2,** 168–175.
23. Cortadas, J. and Subirana, J. A. (1977) The incomplete denaturation of DNA in N,N-dimethyl formamide. *Biochim. Biophys. Acta* **476,** 203–206.
24. Blake, R. D. and Delcourt S. G. (1996) Thermodynamic effects of formamide on DNA stability. *Nucleic Acids Res.* **24,** 2095–2103.
25. Lando, D. Y., Haroutiunian, S. G., Kulba, A. M., Dalian, E. B., Orioli, P., Mangani, S., and Akhrem, A. A. (1994) Theoretical and experimental study of DNA helix-coil transition in acidic and alkaline medium. *J. Biomol. Struct. Dyn.* **12,** 355–366.
26. Tajmir-Riahi, H. A., Ahmad, R., Naoui, M., and Diamantoglou, S. (1995) The effect of HCl on the solution structure of calf thymus DNA: a comparative study of DNA denaturation by proton and metal cations using Fourier transform IR difference spectroscopy. *Biopolymers* **35,** 493–501.
27. Ranhand, J. M. (1985) The enrichment of plasmid DNAs, in bacterial cell lysates, using an alkline-pH procedure that does not permanently denature them. *Prep. Biochem.* **15,** 121–123.
28. Camien, M. N. and Warner, R. C. (1986) Denaturation of covalently closed circular DNA. Kinetics, comparison of several DNAs, mechanism and ionic effects. *J. Biol. Chem.* **261,** 6026–6033.
29. Vaghef, H., Wisen, A. C., and Hellman, B. (1996) Demonstration of benyo(a)pyrene-induced DNA damage in mice by alkaline single cell gel electrophoresis: evidence for strand breaks in liver but not in lymphocytes and bone marrow. *Pharmacol. Toxicol.* **78,** 37–43.
30. Miyamae, Y., Iwasaki, K., Kinae, N., Tsuda, S., Murakami, M., and Sasaki Y. F. (1997) Detection of DNA lesions induced by chemical mutagens using the single-cell gel electrophoresis (comet) assay. 2. Relationship between DNA migration and alkaline condition. *Mutat. Res.* **393,** 107–113.
31. Fairbairn, D. W., Olive, P., and O'Neil, K. L. (1995) The comet assay: a comprehensive review. *Mutat. Res.* **339,** 37–59.
32. Malá, Z., Klepárník, K., Havač Z., and Bocek, P. (1998) New electrolyte system for single-stranded DNA separations by electrophoresis in uncoated capillaries, in *11th International Symposium on Capillary Electroseparation Techniques*, Abstracts volume, Venice, Italy, pp. 70.
33. Malá, Z., Klepárník, K., and Bocek, P. (1999) Highly alkaline electrolyte for single-stranded DNA separations by electrophoresis in bare silica capillaries. *J. Chromatogr. A.* **853,** 371–379.
34. Hjertén, S. (1985) High-performance electrophoresis. Elimination of electroendoosmosis and solute adsorption. *J. Chromatogr.* **347,** 191–198.
35. Hjertén, S. and Kubo K. (1993) A new type of pH- and detergent-stable coating for elimination of electroendoosmosis and adsorption in (capillary) electrophoresis. *Electrophoresis* **14,** 390–395.
36. Goetzinger, W. and Karger, B. L. (1996) *International Patent Application WO* 96/23220.
37. Coob, K. A., Dolník, V., and Novotny, M. (1990) Electrophoretic separations of proteins in capillaries with hydrolytically stable surface structures. *Anal. Chem.* **62,** 2478–2483.
38. Liu, Y. and Kuhr, W. (1999) Separation of double- and single-stranded DNA restriction fragments: Capillary electrophoresis with polymer solution under alkaline conditions. *Anal. Chem.* **71,** 1668–1673.

39. Liu, Y., Fu, R., and Gu, J. (1996) Epoxy resin coatings for capillary zone electrophoretic separation of basic proteins. *J. Chromatogr. A* **723,** 157–167.
40. Fung, E. N. and Yeung, E. S. (1995) High-speed DNA sequencing by using mixed poly(ethylene oxide) solutions in uncoated capillary columns. *Anal. Chem.* **67**, 1913–1919.
41. Preisler, J. and Yeung, E. S. (1996) Characterization of nonbonded poly(ethylene oxide) coating for capillary electrophoresis via continuous monitoring of electroosmotic flow. *Anal. Chem.* **68,** 2885–2889.
42. Towns, J. K. and Regnier, F. E. (1992) Impact of polycation adsorption on efficiency and electroosmotically driven transport in capillary electrophoresis. *Anal. Chem.* **64,** 473–478.
43. Yeung, K. K. C. and Luci, Ch. A. (1997) Suppression of electroosmotic flow and prevention of wall adsorption in capillary zone electrophoresis using zwitterionic surfactants. *Anal. Chem.* **69,** 3435–3441.
44. Barron, A. E., Soane, D. S., and Blanch, H. W. (1993) Capillary electrophoresis of DNA in uncross-linked polymer solution. *J. Chromatogr. A* **652,** 3–16.
45. Barron, A. E., Blanch, H. W., and Soane, D. S. (1994) A transient entanglement coupling mechanism for DNA separation by capillary electrophoresis in ultradilute polymer solution. *Electrophoresis* **15,** 597–615.
46. Klepárník, K., Fanali, S., and Bocek, P. (1993) Selectivity of separation of DNA fragments by capillary zone electrophoresis in low-melting-point agarose sol. *J. Chromatogr.* **638,** 283–292.
47. Klepárník, K. and Bocek, P. (1991) Theoretical background for clinical and biomedical applications of electromigration techniques. *J. Chromatogr.* **569,** 3–42.
48. Iler, R. K. (1979) *The Chemistry of Silica*, Wiley, New York.
49. Gerald, D. and Fasman, P. D. (1976) *Handbook of Biochemistry and Molecular Biology,* CRC Press, Boca Raton, FL.

17

Capillary Electrophoresis with Glycerol as an Additive

Keith R. Mitchelson and Jing Cheng

1. Introduction

1.1. Entangled Solution Sieving Matrices

Two kinds of semi-liquid separation media are frequently used for the separation of DNA by high performance capillary electrophoresis (CE), namely dynamic or entangled free solution polymer sieving matrix, and dilute liquid-gel solutions. Polymer solutions may be defined in three different functional concentrations: dilute, semidilute, and entangled. In dilute solutions, the polymer molecules are physically separate; as the concentration of polymer increases the molecules begin to interact, and at a critical concentration the polymers begin to physically interact to form an entangled physical network *(1–3)*. These media have the ability to act as size selective sieving agents that partition DNA molecules, but since the media are liquid the injection of fresh media and the removal of expended media are facilitated. The media are typically neutral high molecular weight polymers and polymer sugars, which provide a high uniformity of electrophoretic medium if capillary electroosmotic flow (EOF) effects are suppressed. For the purpose of this review, particular attention will be drawn to entangled free solution capillary electrophoresis (ESCE) using Tris-borate-EDTA (TBE) buffers, with glycerol as an additive. For DNA separations, many different ESCE sieving media and dilute liquid-gel solutions have been used (**Table 1**): polysaccharide polymers such as, hydroxypropyl cellulose (HPC) *(2–4)*, hydroxypropyl methyl cellulose (HPMC) *(5–7)*, methyl cellulose *(8,9)*, hydroxyethyl cellulose (HEC) *(3,7,10–15)*, dextran *(16,17)*, TreviSol *(18)*, semi-liquid agarose *(19,20)*, and galactomannans *(1)*. Nonsaccharide polymers such as linear polyacrylamides and substituted polyacrylamides *(3,7,21–26)*, linear and branched poly(ethylene oxides), and polyvinyl polymers *(27)* are also employed for a wide variety of separations purposes.

1.2. DNA Migration in Sieving Matrices

Several authors *(2,4,28)* have shown that the migration of DNA in entangled polymer solutions occurs by a mechanism of biased reptation with fluctuations. At high

From: *Methods in Molecular Biology, Vol. 162:*
Capillary Electrophoresis of Nucleic Acids, Vol. 1: Introduction to the Capillary Electrophoresis of Nucleic Acids
Edited by: K. R. Mitchelson and J. Cheng © Humana Press Inc., Totowa, NJ

polymer concentration the highly entangled solutions achieve good DNA separations over a small range of DNA fragment sizes, however dilute or unentangled solutions can also separate DNA molecules, but less effectively and over a wider range of DNA sizes. In video-microscopic studies of CE fields during migration ***(4,28)***, the DNA molecules can be seen to become extended and also to form typical "U" shapes as the DNA "pulleys" itself through the polymer media. Unlike conventional gels, in entangled solutions the DNA "pulley" at the "U" shape which is itself moving downfield due to migration of the sieving media under EOF. Hence, factors which influence the degree of entanglement of the ESCE media, or affect the migration of the sieving media due to EOF may influence the mobility and separation of DNA molecules ***(1)***.

1.3. Tetraborate Buffer Systems

1.3.1 Chemical Interactions Between Borate and Diol Polymers

There are chemical effects on the performance of some ESCE media due to use of borate buffer and glycerol that would be expected to affect both DNA resolution and retention times. It is well known that an increase in solution conductivity results from complex formation between borate and polyhydroxy (polyol) compounds ***(29–31)***. In these complexation reactions, tetrahydroxyborate ions $B(OH)_4$- react with polyols (where $n = 0$ or 1) with a net release of a proton as shown in equilibria [1–3] in **Fig. 1**:

Tetraborate can form either monomeric or dimeric complexes with 1,2-diols or 1,3-diols (equilibria [2] and [3]) when the hydroxyl groups are oriented in the appropriate geometry ***(30,32)***. In the latter case, boric acid forms a dimeric 1:2 borate:diol spirane complex with borate as the central linkage between diols ***(29,32)***. Mono- and dicomplex borate esters can form simultaneously in aqueous solution, and their molar ratio is affected by the relative concentration of polyol molecules and borate ions, and by the presence of substituents on the polyol. Hydroxyalkyl methyl celluloses such as HPMC are not completely substituted at all hydroxyl groups, with hydroxypropyl (7–12%) and methoxy (28–30%) substitution of hydroxyl groups ***(33)***. Thus, the formation of the HPMC polymer network may result in part from complexes between boric acid and diol groups on HPMC saccharide units ***(29,30,32)***.

1.3.2. Diol Polymer Matrices

A number of widely used polymeric macromolecules used as ESCE media posses reactive hydroxyl groups. Dextran is an unsubstituted linear glucose polymer, whereas methyl cellulose, hydroxypropyl cellulose (HPC), 2-hydroxyethyl cellulose (HEC) are also not fully substituted and some hydroxyl groups remain free. The substitution levels for methyl cellulose, HPC, and HEC are ~30, ~70, and ~50%, respectively ***(33)***. Both HPC and HEC substituants also possess free terminal hydroxyl groups, which could also participate in complexes with tetraborate. Thus, a wide variety of the sugar polymers used as ESCE and liquid-gel media would be expected to form either 1,2-diols or 1,3-diols in monomeric or dimeric complexes with tetraborate. The entangled polymer network formed in polysaccharide ESCE-borate buffer solutions would be the sum of a looped network ***(1,2,11)*** and a crosslinked polymer network of borate-didiol complexes (*see* **Fig. 2A**) ***(5,29)***. The effects of crosslinking of charged borate

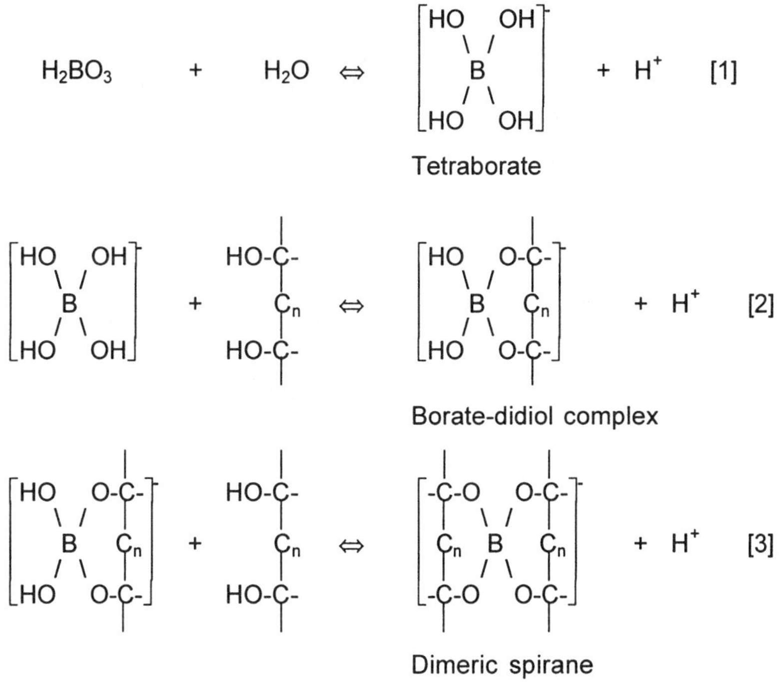

Fig. 1. Tetraborate (equilibrium [1]) can form either monomeric or dimeric complexes with 1,2-diols or 1,3-diols (equilibrium [2] and [3]) when the hydroxyl groups are oriented in the appropriate geometry. In equilibrium [3] boric acid forms a dimeric 1:2 borate:diol spirane complex with borate as the central linkage between diols. Mono- and dicomplex borate esters can form simultaneously in aqueous solution, where their molar ratio is affected by the relative concentration of polyol molecules and borate ions and the presence of substituents on the polyol. Reprinted with permission from **ref. *5***. Copyright (1994), American Chemical Society.

to neutral polymers are complex, in which the intrinsic fluidity and porosity of the media would be decreased, whereas the electromobility of the media toward the anode would be increased against the net EOF flow toward the cathode. The increase in the intrinsic viscosity of the ESCE solution would act to decrease both electromigration and diffusion of the DNA fragments, and would improve the efficiency of separation of DNA.

Such a borate containing buffer system is partially chemically entangled, and should be considered a "dynamic complexation system" ***(34)***, in which one or more additives to an ESCE system interact to influence the viscosity, rigidity, and fluidity of the entangled polymer. Both charged borate molecules complexed onto the sieving matrix and charged species released during the dynamic complexation interactions also influence the EOF and analyte migration time. Soane and colleagues ***(2,11)*** have shown that such an entangled polymer solution forms a temporarily intermingled mesh, transiently achieving a network with all possible pore sizes, with a polymer relaxation time several orders of magnitude greater than the residence time of the migrating

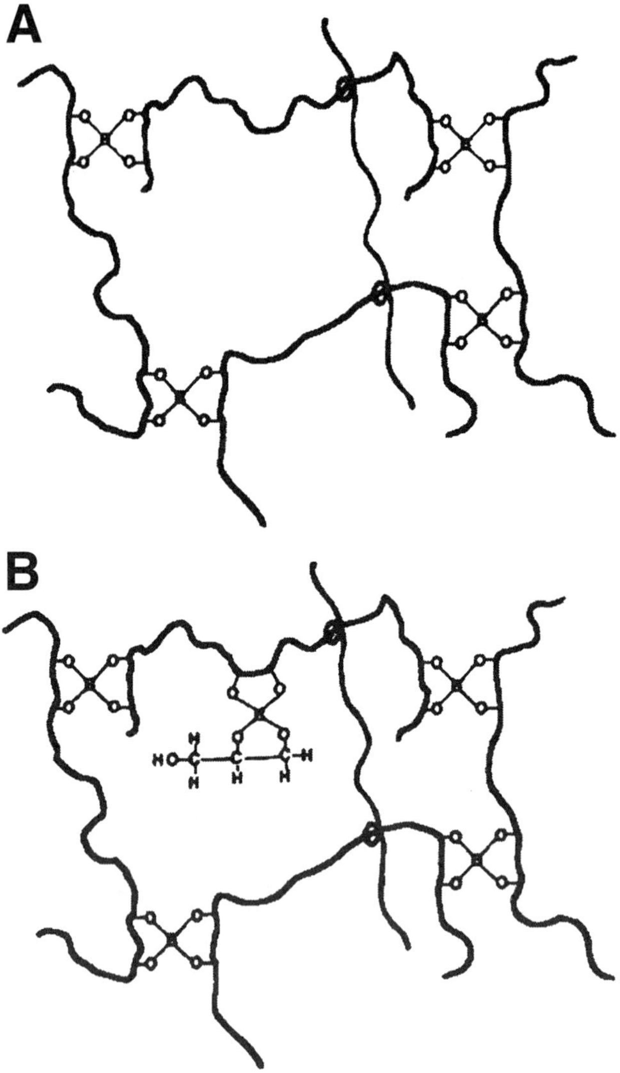

Fig. 2. (**A**) shows 1:2 borate:didiol crosslinks between HPMC polymers in an entangled HPMC-borate buffer solution. Entangled (looped) HPMC molecules are also shown. (**B**) 1:2 borate:didiol complexes between glycerol:HPMC:borate which may form in an entangled glycerol-HPMC-borate buffer system. Reprinted with permission from **ref.** *5*. Copyright (1994), American Chemical Society.

analyte molecules. Unlike agarose-borate *(35)* and PVA-borate *(26)* systems, which form solid gels at low temperatures, systems of polymers such as HPMC *(5,6)*, HPC *(3,4)*, and HEC *(10,12)* with borate buffer remain as entangled solutions, which do not gel. This suggests that borate-stabilization of entangled polymers is not fixed and that crosslinking is not as extensive as the interactions which stabilize solid gels. It is likely that the complexation between borate molecules and the polymer are transient and that exchange between complexed borate and free borate occurs as the polmer molecules relax and new entanglements reform. The viscosity of such polysaccharides in TBE is greater than in TAE buffer ***(35)***. The stability of the complexation system is dependent upon several factors, such as pH, the concentration of borate, ionic strength, and temperature ***(15,29,32,35)***.

1.3.3. Enhanced Effective Sieving

An entangled system, stabilized chemically by borate, may have an increased relaxation time because of a greater stability of polymer entanglement, and as a consequence an increased sieving power. In fact, the ability of borate to chemically interact in solution with polysaccharides such as hydroxyethyl cellulose ***(10)*** agarose and hydroxyethyl agarose ***(35)*** is shown by the reduced effective sieving in electrophoresis buffers lacking borate, such as Tris-phosphate buffer or Tris-acetate, respectively. Consideration of the interactions between polyhydroxylated sugar-polymer molecules and borate such as in ESCE and agarose-gel filled CE is important, as examples of reduced DNA separation efficiency in TreviSol media with TBE buffer compared to Tris-phosphate have been reported ***(18)***.

1.4. Glycerol-Borate Interactions

Glycerol may also form monomeric and dimeric complexes with boric acid. The current generated during electrophoresis runs increases along with the increase in the amount of glycerol added ***(5,29)***. When the percentage of glycerol is less than 5%, the increase in current is nearly proportional to the increase of glycerol, and current gradually flattens at glycerol levels above 5%. This increased current is probably due to protons released during reaction between glycerol and borate. The improved separation of DNA fragments in an ESCE buffer-glycerol system is due, not only to the intrinsic viscosity effects of glycerol, but also to an interaction between the glycerol and other components in the buffer.

1.4.1. Glycerol Viscosity

Jorgenson and Lukacs ***(36)*** demonstrated that improvement in the separation of DNA fragments could be explained by the changes in two parameters: electrophoretic mobility μ and molecular diffusion coefficient D, which both directly affect the resolution (R_s). Several authors ***(34,37)*** have both shown that the electrophoretic mobility μ is inversely proportional to the viscosity of the separation medium. Since the molecular diffusion coefficient D is also inversely proportional to the viscosity, the decrease in values of both μ and D would contribute to an increase in the resolution R_s. The addition of glycerol to an ESCE buffer would add to the intrinsic viscosity of the ESCE solution, decreasing both electromigration and diffusion of the DNA fragments and improving the separation efficiency, independent of borate-mediated chemical interactions of glycerol with polyhydroxylated matrices.

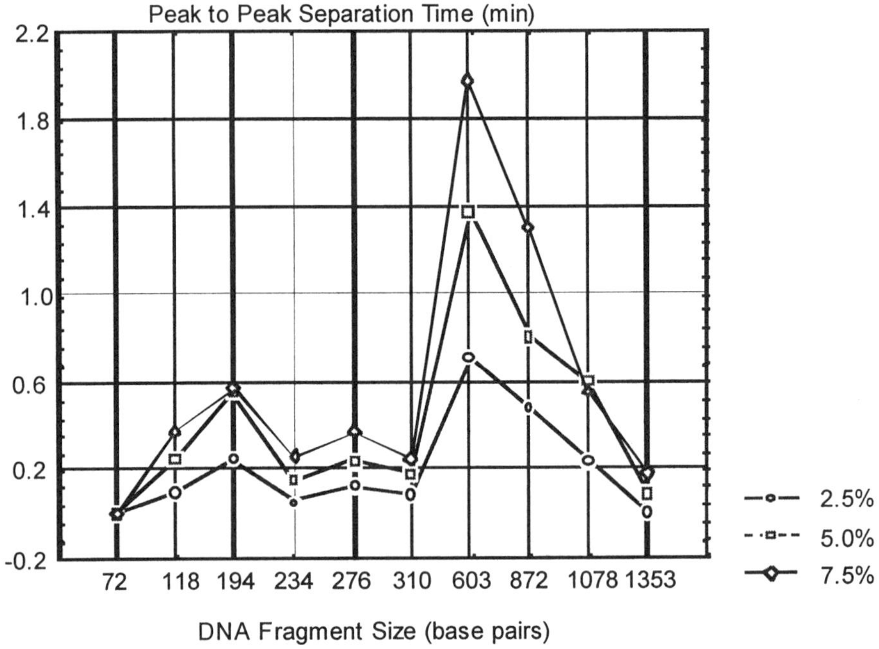

Fig. 3. Peak to peak increase in migration time against *Hae*III digested ϕX174 DNA fragment sizes (bp). Migration time for each DNA fragment peak in buffer lacking glycerol was subtracted from migration times obtained in buffers containing 2.5, 5.0, and 7.5% (v/v) glycerol, respectively. Reprinted with permission from **ref. *5***. Copyright (1994), American Chemical Society.

1.4.2. Glucose Partially Models Glycerol Effects

Graphic representation of the effect of increasing the percentage concentration of glycerol from 2.5 to 7.5% on the separation of dsDNA *Hae*III restriction fragments of ϕX174 is shown in **Fig. 3** and **Fig. 4A**. When the percentage concentration of glycerol is lower than 5% the improvement on the separation is effective for DNA fragments ranging from 100 bp up to 1.3 kb. When the glycerol concentration is 5% or more, separation improves mainly for fragments from 230 bp to 1.3 kb, resulting in the separation of the 872-, 1078-, and 1353-bp fragments to base line level. In contrast, glucose is a nonviscous polyol, which is also able to interact with borate and form borate-didiol complexes ***(38)***. Glucose also retards migration of the ϕX174 fragments (**Fig. 4B**), confirming that the effect of glycerol is due in part to the chemical interactions between the borate buffer, the glycerol, and the HMPC media. Although glycerol improves the resolution of the 603-bp fragment from adjacent fragments markedly, increasing glucose concentration imparts a uniform increase to the retention time of each restriction fragment, with the increase becoming more pronounced with increasing fragment length.

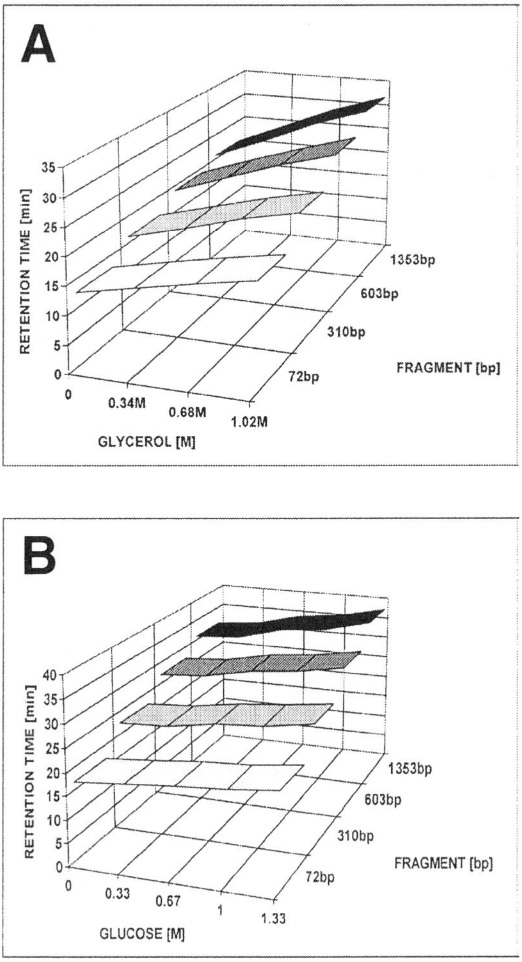

Fig. 4. Capillary retention time of different *Hae*III digested fX174 DNA fragments (bp) vs the concentration of: **(A)** glycerol in TBE buffer; **(B)** glucose in TBE buffer. Both glucose and glycerol can form didiol crosslinks with borate and HPMC. In TBE-glycerol, the retention time for each DNA fragment increased sharply with increasing glycerol, with increasing fragment size. For DNA fragments separated in TBE-glucose, retention time increased less markedly than with an equivalent concentration of glycerol. Slowed DNA migration in TBE-glucose is likely due to didiol crosslinking between borate, HPMC and glucose, whereas with glycerol-TBE the elevated solution viscosity due to glycerol may further impede DNA mobility (*see* **Notes 1–4**).

1.4.3. Borate-Mediated Dynamic Complexation with Glycerol

It is likely that the known interaction between borate and hydroxyethylated agarose models the interactions that occur with other entangled polyol systems: borate causes an increase in the rigidity of the agarose gel, an increase in the agarose bundle radius and a decrease in the inter-fiber spacing *(35,39)*. When the glycerol is introduced into an ESCE buffer such as HPMC, dimeric spirane complexation reactions may also occur between the glycerol, borate, and the HPMC (**Fig. 2B**). We previously proposed two models for this interaction *(5)*. The first model proposes the attachment of glycerol molecules to the HPMC polymers at borate-monodiol intermediates by reaction of glycerol with the free hydroxyl groups of the tetraborate. The second model proposes the formation of glycerol-borate multimers which may further crosslink HPMC complexes to reduce the effective pore size of the polymer network. The complexation between borate molecules and the polymer may be transient, or exchange may occur with free borate in the buffer; similarly, exchange may occur between complexed glycerol and free glycerol. The stability of the complexation system is dependent upon the concentration of an additional factor: glycerol.

1.5. Factors Effecting Dynamic Complexation

1.5.1. Borate Ion Concentration

The molar ratio of complex borate esters is affected by the relative concentration of polyol molecules and borate ions, and by substituents on the polyol. The effect of borate ion concentration has been exploited to manipulate the crosslinking of thermoreversible PVA-borate gels, as borate-diol interactions are favored at alkaline pH *(26,30)*. Increasing the concentration of borate also increases the ionic strength of the buffer system and as a consequence the mobility of DNA decreases because of "charge screening" which reduces the electric force exerted upon the analyte *(40)*. The improved resolution of DNA fragments with increasing concentration of borate buffer in a HEC polymer system *(15)* may also be due to improved effective sieving because of increased borate-polymer crosslinking.

1.5.2. Electroosmotic Flow

Stellwagen *(39)* notes that the markedly increased EOF observed in cast agarose gels run in TBE buffer is due to the formation of borate-agarose complexes increasing the net negative charge of the agarose gel fibers. Many of the sugar polymers used as ESCE and liquid-gel media would be expected to form either monomeric or dimeric complexes with tetraborate. This would decrease the intrinsic electromigration of the polymer media against the EOF migration of the charged double layer at the capillary surface, and accelerate the electromigration of negatively charged analytes such as DNA toward the positive (anode) detector end of the capillary.

1.6. Borate Interactions with Other Sieving Media

1.6.1. Polyacrylamide, Borate, and Glycerol

Gelfi et al. *(7,23–25)* reported the development of new hydroxylated polyacrylamine polymers, poly (*N*-acryloylamino propanol) (AAP), and poly(*N*-acroyloylaminoethoxy ethanol) (AAEE), with high resistance to alkaline hydrolysis. It was observed that the separation of DNA fragments in a novel polymer network, AAP was superior to conventional polyacrylamide, in which an 8% solution of poly(AAP) can achieve peak-resolution of the 123/124 bp adjacent pair of DNA fragments in marker

V *(23)*. The authors proposed that the distal-OH group in the AAP molecule can form transient H-bonds with the DNA double helix. Dynamic crosslinking between borate and the diols of poly(AAP) could also be expected to increase polymer crosslinking, which could also contribute to the improved sieving properties and higher plate number of this polymer in TBE buffer compared to TAE buffer *(24)*. Notably, Simò-Alfonso et al. *(24)* observed that intrinsic viscosity of poly(AAP) is higher in TAE buffer than in TBE buffer, and this difference is reversed by 6 *M* urea and the viscosity of both polymers was markedly increased. Although these phenomena are not fully understood, borate-polymer complexes would increase both the hydration and electrorepulsion between polymer chains, which could suppress polymer interchain hydrogen-bonding and hydrophobic-bonding interactions *(41)*. Urea, on the other hand, is a hydrogen bond denaturant which would strongly reduce polymer hydration and hydrogen-bonding to water, and could increase the hydrogen-bonding and hydrophobic interactions between polymer chains in both TBE and TAE which would also contribute to the markedly increased viscosity *(42)*. The proximity of polymer chains in TBE/urea could promote borate-hydroxyl interactions, leading to further viscosity increases. Glycerol has also been added to linear polyacrylamide/TBE buffers to suppress polyacrylamide polymerization *(22,26)*, which limits buffer viscosity and facilitates buffer handling. SSCP of DNA performed in this media *(22)* show sharp ssDNA peaks compared to the diffuse peaks in media lacking glycerol, also suggesting a stabilization of ssDNA structure and uniformity of migration of the DNA isomers.

1.6.2. Capillary Coating

It would also be prudent to consider possible interactions between borate and capillary coatings used to suppress EOF. Typically neutral polymers such as polyacrylamide and polysiloxane are used to suppress formation of double ionic layer between negatively charged silanol residues in the capillary wall and a layer of positive counter ions from the electrophoresis buffer. New *N*-substituted acrylamides *(7,23–25)*, such as AAP and AAEE, although neutral, possess hydroxyl groups which might interact chemically with borate under certain conditions, recreating a transiently charged capillary surface, increasing the EOF and decreasing DNA mobility. However, EOF of such coated capillaries is extremely low, inconsistent with significant levels of borate complexation and insufficient to effect DNA mobilities *(43)*.

1.7. Borate Interactions with Nucleic Acids

1.7.1. DNA

The formation of borate crosslinks to *cis*-diol also has a direct ramification for solutes possessing such groups. DNA and RNA are polyols with 2-deoxyribose and ribose polymer backbones, respectively. Although DNA would not be expected to interact strongly with borate, Stellwegen et al. *(43)* observed approx 20% higher free solution mobility of DNA in TBE buffer than in TAE buffer. These authors ascribe this effect to transient interactions between DNA and borate which result in an increase in the net negative charge of the DNA and hence an increased mobility. The symmetry of DNA fragment peaks was often compromised in free TBE buffer, possibly because of rapidly exchanging complexation between DNA and borate, resulting in a spectrum of charge densities and molecular weights for each DNA fragment. In contrast, Cohen et al. ***(44)*** found better DNA fragment resolution in TBE buffers containing denaturants compared to Tris-phosphate buffer, which the authors also ascribed to borate-DNA complexation affecting DNA charge and conformation. Strutz and Stellwagen ***(20)*** found the apparent [extrapolated] free solution mobilities of DNA to be ca. 7% greater in agarose/TBE buffer than in agarose/TAE buffer (3.2 and 3.0) $\times 10^{-4}$ cm_2/V·s, respectively). The very similar DNA mobilities observed in the two buffer media suggest that the borate ions in TBE buffer primarily form complexes with the galactose residues in the agarose gel fibers, rather than with the migrating DNA molecules.

1.7.2. RNA

Katsivela ***(45)*** reported effects of borate buffer and glycerol on the separation of low molecular weight RNAs. RNA separation in 0.5% HMPC required borate ion concentrations between 50–500 m*M* with optimal resolution and peak sharpness at 350 m*M*, while the addition of 5% glycerol reduced peak resolution. Borate would interact chemically with the ribose hydroxyl groups ***(45)*** and could potentially increase RNA mobility, however it may also facilitate transient "dynamic complexation" of RNA to the both glycerol and the entangled HPMC sieving media ***(34)***. The effects could also be due, in part, to the dynamic complexation between glycerol, borate, and HPMC, which reduces the effective pore size of the polymer network ***(5)***. Comparisons of HPCE separations of RNA in TBE and non-TBE buffers have not been reported.

1.8. Applications of Dynamically Complexed Buffers

1.8.1. DNA Size Discrimination

Buffer systems employing borate are used in many ESCE applications, often in conjunction with glycerol. Glycerol has been added to ESCE buffers to improve the efficiency of resolution of DNA molecules distinguished either by variation in length (dsDNA sizing). Glycerol can improve the performance of an ESCE system, improving the resolution between closely migrating restriction fragments up to ~1500 bp ***(5)***. This size range is suitable for many PCR-based HPCE studies, and polysaccharide/borate buffer systems have been used for such as PCR-restriction fragment length polymorphism analysis (PCR-RFLP) ***(10,12,15,46,47)***, short, multiplexed PCR products ***(6)***, and simple repeat (STR-PCR) alleles ***(8,13,14)*** (*see* **Table 1**). ESCE with a HPMC/TBE buffer and glycerol can also resolve the short (~25 bp) double-strand primers from the 50-bp dsDNA product of the ligase chain reaction ***(48)***. It also allowed the base-line separation of nested sets of multiplex PCR-amplified gene fragments in the size range ~113–535 bp on silicon microchip separation devices in less than 5 min ***(49)***.

1.8.2. Conformational Differences in DNA Molecules

Glycerol added to ESCE buffers also improves the efficiency of resolution of DNA molecules distinguished by conformational polymorphism due to mismatches between

Table 1
Application of Glycerol Containing Buffers for DNA Analysis by CE[a]

Assay	Matrix	Buffer	Detector	Comment	Ref.
dsDNA PCR-RFLP	HPMC	TBE/glycerol±ethidium	UV, LIF	Glycerol improves resolution	***5,6,46,47,49***
dsDNA PCR-RFLP	0.2–1% HEC	TBE	UV	Good resolution	***2,4,10,11***
STR-PCR PCR-RFLP	0.5% MC	0.5–2.5X TBE TAE	YoPro, LIF, no dye, UV	Better resolution in TBE	***8,9,15***
STR-PCR PCR-RFLP	1.25% HEC	TBE 7 *M* urea	no dye, UV	Good resolution of several STR	***14***
Supercoil plasmid	0.1–0.3% HEC	0.5–2.5X TBE	UV	Good resolution	***54***
dsDNA LCR	HPMC	TBE/glycerol	LIF	Good resolution 25 and 50 bp	***48***
dsDNA PCR-RFLP	0.5% PEO	TBE/dyes	LIF	Good resolution	***27,55***
dsDNA PCR-RFLP	Liquid	TBE		Better resolution in TBE	***20,39***
	Aragose	TAE			
dsDNA PCR-RFLP	Trevisol	TAPS	UV	Good resolution	***18***
HPA	HPMC	TBE/glycerol±ethidium	UV, LIF	Glycerol improves resolution	***50***
SSCP	HPMC	TBE/glycerol±ethidium	UV. LIF	Glycerol stablizes confomers	***51***
SSCP	LPA	TBE±glycerol	UV	Good resolution	***17,22,52,57***
dsDNA PCR-RFLP	LPA	TBE±ethidium	UV	Good resolution	***21,24,44***
RNA	HPMC	TBE 7 *M* urea	UV	Glycerol diminishes resolution	***45***

[a]Some comments on the effects of glycerol and borate on DNA resolution are noted.

small DNA variant molecules: as seen in conformational difference assays such as heteroduplex polymorphism assay (HPA) ***(50)*** and conformational polymorphism of single strands of DNA (SSCP) ***(17,51,52)***. Interestingly, Kukita et al. ***(53)*** reported that the beneficial effects of glycerol on SSCP sensitivity are due to the reduction of the buffer pH by protons released during the reaction of glycerol and borate ions. The sensitivity of PAGE-SSCP can be greatly improved by electrophoresis in low pH buffer systems (without glycerol). Dilute and ultra-dilute HEC/TBE solutions, at concentrations well below the point of entanglement, are also used to distinguish between linear and supercoiled plasmids and to separate between plasmids with different extents of supercoiling ***(54)***, and to affect separation of larger linear DNA molecules ***(1,28)***.

1.8.3. Ethidium Bromide and Intercalating Dyes

Ethidium bromide preferentially intercalates between CG base pairs, which results in an increase in the atomic distance between successive base pairs that distorts the sugar backbone and decreases the pitch of the double helix. This changes the charge per unit length and the structural rigidity of the molecule, whereas increasing the contour length and decreasing the persistence length of the DNA as a consequence ***(10,21,46,47)***. Mobility may also be influenced by the CG content of individual DNA fragments as the cation-charged nitrogen of ethidium bromide also interact with the phosphate groups of DNA, reducing the net charge proportionate with the extent of intercalation. Limited amounts of ethidium bromide are preferentially intercalated into GC-rich fragments and result in selectively altered mobility of dsDNA fragments. The baseline separation of the 271 bp and 281 bp *Hae*III digested ϕX174 DNA fragments is achieved ***(12,21,47,50)***, and fragments that differ by only 1–2 bp (i.e., 123 and 124 bp, or 220 and 222 bp) may be separated ***(10)***. It should be noted that some buffer systems have achieved separation of the ϕX174 fragments of 271 and 281 bp ***(9)*** and 2 bp resolution between fragments of less than 300 bp ***(14)*** without glycerol or intercalating dyes. Other intercalating dyes such as TOTO, YO-PRO, TO-PRO, Vistra Green, and propridium also increase the contour length and "stiffness" of DNA and have been reported to improve the separation between DNA fragments ***(1,6,50,55)***. Madabhushi et al. ***(55)*** could achieve sensitive detection of DNA with levels of dimeric dyes 100–500 times lower than are conventionally used by addition of 9-aminoacridine to stabilize DNA structure and assist dye binding.

2. Materials

2.1. DNAs

1. Genomic DNA is isolated using a Nucleon PhytoPure DNA isolation kit (Amersham Life Sciences).
2. ϕX174 DNA *Hae*III digest (Sigma) is diluted to a concentration of 20 ng/mL.
3. Ultrafree-MC filters (cat. no. UFC30HV100) (Millipore)
4. *Taq* DNA polymerase (Boehringer)

2.2. PCR Amplifications

1. Separate PCR reactions are performed in a total mixture of 25 mL with 50 m*M* KCl, 10 m*M* Tris-HCl, pH 8.3, 1.5 m*M* $MgCl_2$, 200 m*M* dNTPs, 25 mg gelatin, 225 n*M* primer

oligonucleotides, and 0.5 U of *Taq* DNA polymerase (Boehringer) with 5 ng of template DNA.

2.3. Capillary Electrophoresis

1. ESCE sieving media: 1X TBE, 0.5% (w/v) HPMC (Sigma H7509), 3 μ*M* ethidium bromide and 4.8% glycerol *(6)*.
2. A surface-modified fused silica capillary, DB-17 (57 cm × 100 μm, effective length 50 cm) (J & W Scientific, Folsom, CA).
3. HPCE is performed on a P/ACE 5050 HPCE system (Beckman Instruments Inc., Fullerton, CA).
4. Postrun data analysis is performed using the Gold Chromatography Data System, version 8.13 (Beckman Instruments).

2.4. HPMC Borate Buffer System

1. The buffer system consists of 90 m*M* Tris-base, 90 m*M* boric acid, 2 m*M* EDTA, pH 8.5, to which 0.5% (w/v) HPMC is added using the method recommended by Ulfelder et al. *(47)*. The viscosity of a 2% aqueous solution for this cellulose derivative (H-7509, Sigma) is 4000 cps at 25°C. Glycerol is added with different ratios, e.g., 2.5, 5.0, and 7.5% (v/v) before use and the optimal amount determined for particular DNA fragments *(5)*. The buffers are filtered using a 0.2-μm filter and are then degassed for 15 min by sonication before use.
2. Dynamically entangled buffers employing alternative sieving media to HPMC might include:
   a. 0.1–0.5% (w/w) HEC for large DNA fragments *(10,12)* or small fragments to ~1 kb *(15)*.
   b. 2–6% HEC for fine resolution of small DNA fragments at elevated temperature *(7)*.
   c. 0.2–1% HPC for fine resolution of small DNA fragments *(4)*.
   d. 0.5% (w/v) Methyl cellulose *(8,9)*.
   e. 5–10% (w/w) dextran for DNA fragments to 1.6 kb, or 15% dextran for oligonucleotides *(16)*.

3. Methods

3.1. Sizing of dsDNA by CE

1. ESCE is performed on the P/ACE system 5500 (Beckman Instruments) in the reversed polarity mode (negative potential at the injection end of the capillary). The temperature is set at 22°C and the UV absorbance is monitored at 260 nm.
2. A surface modified fused silica capillary DB-17 (57 cm × 100 μm) with effective length of 50 cm) is conditioned with 5 vol of the separation buffer mentioned above and then subjected to voltage equilibration for 15 min until a stable baseline is achieved.
3. Glycerol at 4.8% (v/v) is optimal for separation of DNA in the size range from 70 to ~1350 bp, whereas also allowing capillary flushing and loading to be achieved in reasonable times *(5,6)* (*see* **Notes 5–8**).
4. The DNA marker is electrokinetically injected into the capillary at a negative polarity of 175 V/cm for 1.5 s.
5. Separations are performed under constant voltage at 228 V/cm for 35 min.

3.1.1. Results with dsDNA Size Marker

1. **Figure 3** shows an illustrative presentation of data obtained with *Hae*III digested ϕX174 DNA where peak to peak migration change is plotted against DNA fragment size. The

separation of particular fragment sizes is improved preferentially by the addition of glycerol (*see* **Notes 1–4**).

3.2. Heteroduplex DNA

3.2.1. Preparation of DNA Heteroduplexes

1. Heteroduplex DNAs are prepared as follows: 10 mL of each of the two PCR products are mixed in a 0.5-mL Eppendorf tube, heated to 94°C for 5 min and then allowed to cool slowly to room temperature. The DNAs are then maintained at 20°C temperature until electrokinetic loading onto the capillary.

3.2.2. Heteroduplex DNA Analysis by CE

1. ESCE was performed on the Beckman P/ACE system 5500 as described above in **Subheading 3.1.**
2. Heteroduplex DNA isomers of <200 bp can be resolved by ESCE using an entangled HPMC/TBE buffer with glycerol and ethidium bromide as described in **Subheading 2.3.** ***(50)***.
3. Separations are performed under constant voltage at 228 V/cm for 30 min.

3.2.3. Analysis of Heteroduplex Retention Profiles

1. Four DNA fragments are present in a typical heteroduplex mixture: two homoduplex DNA fragments AB and A'B', and two corresponding heteroduplex DNA fragments AB' and A'B.
2. The electrophoretic migration of mismatched heteroduplex DNAs is retarded relative to homoduplex DNAs because of conformational distortion created by the mismatched region(s).
3. Homoduplex and heteroduplex PCR mixtures of DNA fragments of 124/125 bp can be separated by ESCE in glycerol/HMPC/TBE media, and two PCR-amplified DNA fragments differing by one base deletion and one base substitution (1D+1S/1B) are well resolved in less than 20 min (*see* **Notes 9–12**).
4. The separation between mismatched heteroduplex and homoduplex DNA species increased with each additional substitution mismatch or deletion mismatch.
5. The migration of homoduplex and heteroduplex molecules during ESCE was not greatly sensitive to temperature in the range of 18–22°C, although resolution between the molecular species is improved at the lower temperature (*see* **Notes 13–15**).

3.3. ssDNA

3.3.1. Methods for Preparation of ssDNA

1. The PCR products (50 µL each) are separated on a 1.0% agarose gel and are then purified using Ultrafree-MC filters (cat. no. UFC30HV100) (Millipore) as instructed by the manufacturer.
2. The purified DNAs are precipitated with three volumes of –20°C ethanol and kept at –70°C for 2 h before centrifugation. After centrifugation at 20,000*g* and 4°C for 20 min, the supernatant is removed and the pellet is dried. Each DNA pellet is then redissolved in 10 µL of 0.1X TE buffer ***(51)***.

3.3.2. CE of ssDNA

1. For CE: For 2 µL of each sample solution, 3 µL of formamide, and 5 µL of water are mixed in a Eppendorf tube and capped. The mixtures are boiled for 3 min and then snap frozen in a dry ice/acetone bath for 3 min and stored at –20°C until use.

2. Immediately before loading on to the capillary column, frozen mixtures are allowed to thaw on the P/ACE loading tray.
3. The ESCE sieving media: 1X TBE, 0.5% (w/v) HPMC and 4.8% glycerol as described in **Subheading 2.3.** ***(6,51)***. DNA may be detected by LIF using 3 µ*M* ethidium bromide, or 0.2 µ*M* TOTO-1.
3. ESCE is performed on the Beckman P/ACE system 5500, as described in **Subheading 3.1.** Separations are also performed under constant voltage at 228 V/cm for 30 min (*see* **Notes 16–19**).

4. Notes

1. The increase of glycerol in the separation buffer by steps of 2.5% results in the retardation of the DNA migration by about 4 min. The migration time increases linearly with the increasing percentage of glycerol. The amount of glycerol added may be optimised depending on the separation of DNA fragments required.
2. Glycerol and intercalating dyes such as ethidium bromide, YO-PRO or TOTO dyes are separate additives in the polymer/TBE buffer which usually result in an overall improvement in the separation of dsDNA ***(1,6,10)***.
3. Increasing the concentration of cellulose-based ESCE matrix, although reducing mobility, also reduces the size of DNA fragments which are selectively separated ***(1)***.
4. Temperature may alter the degree of dynamic crosslinking between borate and the sugar polymer matrix. Elevated temperature may decrease the resolution between DNA fragments in conventional matrices, although high concentration polymer solutions can restore DNA resolution while markedly decreasing column retention time ***(7)***.
5. When glycerol is employed as an additive in the ESCE buffer system, the resolution for all DNA fragments is improved. The most evident improvement in the separation efficiency is for DNA fragments ranging from 100 to 1.3 kb.
6. When glycerol is used as an additive to HPMC media the separation of ϕX174 fragments of 271 bp and 281 bp still cannot be achieved, although separation is improved by addition of ethidium bromide ***(50)***. Systems using cellulose-based MC and HEC entangled sieving media can achieve separation of the ϕX174 fragments of 271 bp and 281 bp ***(9)*** and the 2-bp resolution between fragments of less than 300 bp ***(14)*** respectively, without glycerol and intercalating dye additives.
7. Borate with MC and HEC media improves the resolution of simple tandem repeat PCR (STR-PCR) fragments differing by 2 bp over a fragment range to 300 bp ***(14)*** and 2 or 4 repeat units over the range 600–1000 bp ***(8)***.
8. Both **Figs. 3** and **4A** highlight the size range of DNA where the most evident separation improvement is achieved. When the concentration of glycerol is below 5%, improved separation is achieved for all DNA fragments. Further improvement can be achieved for fragments between 310 bp and about 1 kb by increasing the concentration of glycerol to 7.5%. Further addition of glycerol above 7.5% is not recommended, as the increase in viscosity prevents ready filtration of the ESCE medium.
9. Pretreatment of PCR products such as desalting and deproteinising are not necessary.
10. Desalting did increase both the detection and mobility of all heteroduplex species, but the resolution between peaks is not improved. The gain in run speed and sensitivity does not justify the additional processing and handling required to desalt the DNAs.
11. Although a single base substitution mismatch within molecules of 124 bp are not resolved by our ESCE system, the combination of a single base deletion mismatch *cis*-linked with a single base substitution mismatch can be resolved from a single base deletion mismatch

by nondenaturing ESCE. Provision of known substitution mismatch partners with an unknown DNA fragment might be used to "amplify" the distortion in reannealed heteroduplex molecules and allow rapid detection using ESCE.

12. Optimization of ESCE conditions may be required for the separation of particular heteroduplex and homoduplex DNA species. Factors such as DNA fragment length, DNA sequence, or the nature of the DNA mismatch can influence the migration of heteroduplex species.
13. The migration of homoduplex and heteroduplex molecules during ESCE was not highly sensitive to temperature in the range of 18–22°C, although resolution between the molecular species is improved at the lower temperature.
14. It may be possible to use ESCE to be able to predict the nature of mismatches in defined PCR fragments. The position of common substitution mutations can be identified by the extent to which the mutation altered migration in ESCE. Rare, novel substitution mutations could also be identifiable, by the unusual migration of their heteroduplexes formed with known deletion series standards.
15. The ESCE of heteroduplex molecules is rapid, reproducible, and simple as well as permitting superior resolution to the fractionation achieved by conventional polyacrylamide gel electrophoresis ***(50)***.
16. Conventional PAGE-SSCP is frequently run with a glycerol containing TBE buffer, which confers stability to the single-strand isomers ***(56)***. ESCE buffer containing HPMC/ TBE and glycerol can also be used to resolve the multiple SSCP isomers of purified ssDNAs ***(51)***.
17. The formation of metastable partially ssDNA oligomer complexes is consistent with the mild conditions experienced in HPCE. The ESCE-SSCP profiles obtained examining the major stable peaks of ~124 primer pair products are equivalent to the conventional PAGE-SSCP profiles. The resolution achieved by ESCE is similar, if not superior, to the fractionation achieved by PAGE.
18. The above protocol for SSCP analysis using ESCE involves minimum handling and manipulation of the DNA fragments following PCR amplification.
19. Kuyper et al. ***(57)*** and Arakawa et al. ***(58)*** reported the fractionation of SSCP by HPCE using a polyacrylamide gel medium. The assay is rapid, however the use of physical sieving media requires frequent washing and extended purging of the capillary.

Acknowledgment

The authors would like to thank Beryl Morris for encouragement and ForBio Research Pty Ltd. for support.

References

1. Sunada, W. M. and Blanch, H. W. (1997) Polymeric separation media for capillary electrophoresis of nucleic acids. *Electrophoresis* **18,** 2243–2254.
2. Barron, A. E., Soane, D. S., and Blanch, H. W. (1993) Capillary electrophoresis of DNA in uncross–linked polymer solutions. *J. Chromatogr. A* **652**, 3–16.
3. Barron, A.E., Sunada, W.M., and Blanch, H.W. (1996) The effects of polymer properties on DNA separations by capillary electrophoresis in uncross-linked polymer solutions. *Electrophoresis* **17,** 744–757.
4. Mitnik, L., Salome, L., Viovy, J.-L., and Heller, C. (1995) Systematic study of field and concentration effects in capillary electrophoresis of DNA in polymer solutions. *J. Chromatogr. A* **710,** 309–321.

5. Cheng, J. and Mitchelson, K. R. (1994) Glycerol-enhanced separation of DNA fragments in entangled solution capillary electrophoresis. *Anal. Chem.* **66**, 4210–4214.
6. Fortina, P., Cheng, J., Shoffner, M. A., Surrey, S., Hitchcock, W. M., Kricka, L. J., and Wilding, P. (1997) Diagnosis of Duchenne/Becker muscular dystrophy and quantitative identification of carrier status by use of entangled solution capillary electrophoresis. *Clin. Chem.* **43,** 745–751.
7. Gelfi, C., Perego, M., Libbra, F., and Righetti, P. G. (1996) Comparison of behavior of N-substituted acrylamides and celluloses on double-stranded DNA separations by capillary electrophoresis at 25 degrees and 60 degrees C. *Electrophoresis* **17,** 1342–1347.
8. Nishimura, A., Tsuhako, M., Miki, T., Ogihara, T., and Baba, Y. (1998) Analysis of a variable number of tandem repeats in a heart disease gene by capillary electrophoresis with laser-induced fluorescence detector. *Chem. Pharm. Bull. (Tokyo)* **46,** 294–297.
9. McGregor, D. A. and Yeung, E. S. (1995) Optimization of capillary electrophoretic separation of DNA fragments based on polymer filled capillaries. *J. Chromatogr. A.* **652,** 67–73.
10. Nathakarnkitkool, S., Oefner, P. J., Bartsch, G., Chin, M. A., and Bonn, G. K. (1992) High-resolution capillary electrophoretic analysis of DNA in free solution. *Electrophoresis* **13,** 18–31.
11. Grossman, P. D. and Soane, D. S. (1991) Capillary electrophoresis of DNA in entangled polymer solutions. *J. Chromatogr. A* **559,** 257–266.
12. Atha, D. H. (1998) Characterization of DNA standards by capillary electrophoresis. *Electrophoresis* **19,** 1428–1435.
13. Isenberg, A. R., Allen, R. O., Keys, K. M., Smerick, J. B., Budowle, B., and McCord, B. R. (1998) Analysis of two multiplexed short tandem repeat systems using capillary electrophoresis with multiwavelength fluorescence detection. *Electrophoresis* **19,** 94–100.
14. Mansfield, E. S., Wilson, R. B., and Fortina, P. (2001) Analysis of short tandem repeat markers by capillary array electrophoresis, in *Capillary Electrophoresis of Nucleic Acids,* Vol. 2 (Mitchelson, K. R. and Cheng, J., eds.), Humana Press, Totowa, NJ, pp. 151–161.
15. Issaq, H. J., Chan, K. C., and Muschik, G. M. (1997) The effect of column length, applied voltage, gel type, and concentration on the capillary electrophoresis separation of DNA fragments and polymerase chain reaction products. *Electrophoresis* **18,** 1153–1158.
16. Heller, C. (1998) Finding a universal low-viscosity polymer for DNA separation. *Electrophoresis* **19,** 1691–1698.
17. Katsuragi, K., Kitagishi, K., Mizuguchi, T., Nagashima, T., Kinoshita, M., and Kumada, H. (1997) Method for detection of epsilon-secondary structure in the precore region of human hepatitis B virus DNA using a fluorescence-based polymerase chain reaction-single-strand-conformation polymorphism technique with capillary electrophoresis. *J. Chromatogr. A* **781,** 307–314.
18. Siles, B. A. and Collier, G. B. (1996) The characterization of a new size-sieving polymeric matrix for the separation of DNA fragments using capillary electrophoresis. *J. Capillary Electrophor.* **3,** 313–321.
19. Palm, A. K. (2001) Capillary electrophoresis of DNA fragments with replaceable low-gelling agarose gels, in *Capillary Electrophoresis of Nucleic Acids,* Vol. 1 (Mitchelson, K. R. and Cheng, J., eds.), Humana Press, Totowa, NJ, pp. 279–290.
20. Strutz, K. and Stellwagen, N. C. (1998) Do DNA gel electrophoretic mobilities extrapolate to the free-solution mobility of DNA at zero gel concentration ? *Electrophoresis* **19,** 635–642.
21. Pariat, Y. F., Berka, J., Heiger, D. N., Schmitt, T., Vilenchik, M., Cohen, A. S., Foret, F., and Karger, B. L. (1993) Separation of DNA fragments by capillary electrophoresis using replaceable linear polyacrylamide matrices. *J. Chromatogr. A* **652,** 57–66.

22. Hebenbrock, K., Williams, P. M., and Karger, B. L. (1995) Single strand conformational polymorphism using capillary electrophoresis with two–dye laser–induced fluorescence detection. *Electrophoresis* **16,** 1429–1436.
23. Gelfi, C., Simò-Alfonso, E., Sebastiano, R., Citterio, A., and Righetti, P. G. (1996) Novel acrylamido monomers with higher hydrophilicity and improved hydrolytic stability: III. DNA separations by capillary electrophoresis in poly(N-acryloylaminopropanol). *Electrophoresis* **17,** 738–743.
24. Simò-Alfonso, E., Gelfi, C., Lucisano, M., and Righetti, P. G. (1996) Performance of a series of novel N-substituted acrylamides in capillary electrophoresis of DNA fragments. *J. Chromatogr. A* **756,** 255–261.
25. Righetti, P. G. (1995) Macroporous gels: facts and misfacts. *J. Chromatogr. A* **698,** 3–17.
26. Righetti, P. G. and Snyder, R. S. (1988) Thermally reversible gels in electrophoresis. I: Matrix characterization. *Appl. Theor. Electrophor.* **1,** 53–58.
27. Braun, B., Blanch, H. W., and Prausnitz, J. M. (1997) Capillary electrophoresis of DNA restriction fragments: effect of polymer properties. *Electrophoresis* **18,** 1994–1997.
28. Sunada, W. M. and Blanch, H. W. (1998) Microscopy of DNA in dilute polymer solutions. *Biotechnol. Prog.* **14,** 766–772.
29. Zittle, C. A. (1951) Reaction of borate with substances of biological interest, in *Advances in Enzymology*, Vol. 12 (Nord, F. F., ed.), Interscience Publishers, New York, NY, pp. 493–527.
30. Böeseken, J. (1949) The use of boric acid for the determination of the configuration of carbohydrates, in *Advances in Carbohydrate Chemistry,* Vol. 4 (Pigman, W. W. and Wolfrom, M. L., eds.), Academic Press, San Diego, CA, pp. 189–210.
31. Smith, J. T., Nashabeh, W., and El Rassi, Z. (1994) Micellar electrokinetic capillary chromatography with *in situ* charged micelles. *Anal. Chem.* **66,** 1119–1133.
32. Bergold, A., and Scouten, W. H. (1983) Borate chromatography, in *Solid Phase Biochemistry: Analytical and Synthetic Aspects*, Vol. 66 (Scouten, W. H., ed.), Wiley, New York, NY, pp. 149–187.
33. Sigma-Aldrich Chemical Corporation: Technical Data, St. Louis, MO.
34. Peng, X., Bowser, M. T., Britz-McKibbin, P., Bebault, G. M., Morris, J. R., and Chen, D. D. (1997) Quantitative description of analyte migration behavior based on dynamic complexation in capillary electrophoresis with one or more additives. *Electrophoresis* **18,** 706–716.
35. Peats, S., Nochumson, S., and Kirkpatrick, F. H. (1986) Effects of borate on agarose gel structure. *Biophys. J.* **49,** 91a.
36. Jorgenson, J. W. and Lukacs, K. D. (1981) Free-zone electrophoresis in glass capillaries. *Clin. Chem.* **27,** 1551–1553.
37. Shaw, D. J. (1969) *Electrophoresis.* Academic Press, London.
38. Plocek, J. and Chmelík, J. (1997) Separation of disaccharides as their borate complexes by capillary electrophoresis with indirect detection in visible range. *Electrophoresis* **18,** 1148–1152.
39. Stellwagen, N. C. (1992) Agarose gel pore radii are not dependent on the casting buffer. *Electrophoresis* **13,** 601–603.
40. Stigter, D. (1991) Shielding effects of small ions in gel electrophoresis of DNA. *Biopolymers* **31,** 169–176.
41. Schellman, J. A. (1997) Temperature, stability, and the hydrophobic interaction. *Biophys. J.* **73,** 2960–2964.
42. Schellman, J. A. (1990) A simple model for solvation in mixed solvents. Applications to the stabilization and destabilization of macromolecular structures. *Biophys. Chem.* **37,** 121–140.

43. Stellwagen, N. C., Gelfi, C., and Righetti, P. G. (1997) The free solution mobility of DNA. *Biopolymers* **42,** 687–703.
44. Cohen, A. S., Najarian, D., Smith, J. A., and Karger, B. L. (1988) Rapid separation of DNA restriction fragments using capillary electrophoresis. *J. Chromatogr.* **458,** 323–333.
45. Katsivela, E. (1997) Analysis of transfer RNA and 5S ribosomal RNA, in *Analysis of Nucleic Acids by Capillary Electrophoresis,* (Heller, C., ed), Chromatography CE Book series, (Altria, K., series ed.), Vieweg Press, Germany.
46. Schwartz, H. E., Ulfelder, K., Sunzeri, F. J., Busch, M. P., and Brownlee, R. G. (1991) Analysis of DNA restriction fragments and polymerase chain reaction products towards detection of the AIDS (HIV-1) virus in blood. *J. Chromatogr.* **559,** 267–283.
47. Ulfelder, K. J., Schwartz, H. E., Hall, J. M., and Sunzeri, F. J. (1992) Restriction fragment length polymorphism analysis of ERBB2 oncogene by capillary electrophoresis. *Anal. Biochem.* **200,** 260–267.
48. Cheng, J., Shoffner, M. A., Mitchelson, K. R., Kricka, L. J., and Wilding, P. (1996) Analysis of ligase chain reaction (LCR) products amplified in a silicon chip using entangled solution capillary electrophoresis (ESCE). *J. Chromatogr.* **732,** 151–158.
49. Fortina, P., Cheng, J., Kricka, L. J., Waters, L. J., Jacobson, S. C., Wilding, P., and Ramsey, J. M. (2000) DOP-PCR amplification of whole genomic DNA and microchip-based capillary electrophoresis, in *Capillary Electrophoresis of Nucleic Acids,* Vol. 2 (Mitchelson, K. R. and Cheng, J., eds.), Humana Press, Totowa, NJ, pp. 211–219.
50. Cheng, J., Kasuga, T., Mitchelson, K. R., Lightly, E. R., Watson, N. D., Martin, W. J., and Atkinson, D. (1994) PCR heteroduplex polymorphism analysis by entangled solution capillary electrophoresis (ESCE). *J.Chromatogr.A* **677,** 169–177.
51. Kasuga, T., Cheng, J., and Mitchelson, K. R. (2001) Magnetic bead-isolated single-strand DNA for SSCP analysis, in *Capillary Electrophoresis of Nucleic Acids,* Vol. 2 (Mitchelson, K. R. and Cheng, J., eds.), Humana Press, Totowa, NJ, pp. 135–147.
52. Inazuka, M., Wenz, H.-M., Sakabe, M., Tahira, T., and Hayashi, K. (1997) A streamlined mutation detection system: multicolor post-PCR fluorescence labeling and single-strand conformational polymorphism analysis by capillary electrophoresis. *Genome Res.* **7,** 1094–1103.
53. Kukita, Y., Tahira, T., Sommer, S. S., and Hayashi, K. (1997) SSCP analysis of long DNA fragments in low pH gel. *Hum. Mutat.* **10,** 400–407.
54. Oana, H., Hammond, R. W., Schwinefus, J. J., Wang, S. C., Doi, M., and Morris, M. D. (1998) High-speed separation of linear and supercoiled DNA by capillary electrophoresis. Buffer, entangling polymer, and electric field effects. *Anal. Chem.* **70,** 574–579.
55. Madabhushi, R. S., Vainer, M., Dolnik, V., Enad, S., Barker, D. L., Harris, D. W., and Mansfield, E. S. (1997) Versatile low-viscosity sieving matrices for nondenaturing DNA separations using capillary array electrophoresis. *Electrophoresis* **18,** 104–111.
56. Kasuga, T., Cheng, J., and Mitchelson, K. R. (1995) Metastable single strand DNA conformational polymorphism (mSSCP) analysis results in enhanced polymorphism detection. *PCR Meth. Applic.* **4,** 227–233.
57. Kuypers, A. W., Willems, P. M., van der Schans, M. J., Linssen, P. C, Wessels, H. M., de Bruijn, C. H., Everaerts, F. M., and Mensink, E. J. (1993). Detection of point mutations in DNA using capillary electrophoresis in a polymer network. *J. Chromatogr. A* **621,** 149–156.
58. Arakawa, H., Uetanaka, K., Maeda, M., and Tsuji, A. (1994) Analysis of polymerase chain reaction product by capillary electrophoresis and its application to the detection of single base substitution in genes. *J. Chromatogr* **66,** 89–98.

18

Capillary Electrophoresis of DNA Fragments with Replaceable Low-Gelling Agarose Gels

Anders K. Palm

1. Introduction

1.1. Capillary Gel Electrophoresis

Capillary electrophoresis (CE) in all its various modes is routinely used in the area of life science for the analysis of wide spectra of species ranging from small inorganic/organic molecules to multi-subunit proteins and mega-size DNA *(1)*. Capillary gel electrophoresis (CGE) is the counterpart of gel electrophoresis in the slab-gel format. Similar molecules are analyzed with CGE as with slab gels, like SDS-protein complexes and single-stranded/duplex DNA. Since these analytes have similar mass-over-charge ratio and therefore similar mobilities (for ≥10 bases in length, depending on pH and base composition) they can not be separated in a carrier-free system like capillary free-zone electrophoresis *(2)*, although some exceptions are known *(3,4)*. A size-sieving medium is therefore necessary to impart hydrodynamic friction on the migrating molecules for their separation. The constraint of the capillary format also acts to prevent convection, whereas in the slab-gel format the sieving matrix plays this role.

CGE has benefits similar to the other modes of CE and other miniaturized separation techniques. These include the need for small sample volumes, as a few µL is enough for repetitive injections. CGE has short analysis times (ranging from a few seconds to minutes), high efficiency (>100,000 plates/m is typical), on-line detection (UV/visible, fluorescence, chemiluminescence, electrochemical, nuclear magnetic resonance [NMR], mass spectroscopy [MS]), real-time data acquisition, and the apparatus is frequently automated *(5)*.

1.2. Polyacrylamide Gel CE

In its inception, crosslinked polyacrylamide gels were traditionally employed for slab-gel electrophoresis because of good separation performance *(6,7)*. Crosslinked polyacrylamide capillary gels also shows excellent separation performance and long

From: *Methods in Molecular Biology, Vol. 162: Capillary Electrophoresis of Nucleic Acids, Vol. 1: Introduction to the Capillary Electrophoresis of Nucleic Acids*
Edited by: K. R. Mitchelson and J. Cheng © Humana Press Inc., Totowa, NJ

lifetimes if properly handled. However, chemically polymerized polyacrylamide capillary gels are more difficult to make reproducibly than conventional slab gels *(8–11)*. Free-radical polymerization is highly dependent on variable factors such as the concentration of monomers, temperature, pH, oxygen content, and polymerization time *(12–14)*, which if not strictly controlled give nonreproducible polymerization and poor gels. Care must also be taken in running polyacrylamide capillary gels. This is due to several factors, such as:

a. Volume-shrinkage during polymerization giving rise to voids in the capillary which omit any current-flow,
b. Sensitivity of polyacrylamide to high voltage which induces stress on the matrix and renders it susceptible to gel-breakdown, and
c. Generation of excessive Joule heat during a run induces bubble formation, which may cause erratic or disrupted current flow.

Moreover, samples have to be highly purified to avoid particles clogging the injection end of the capillary, ruining the gel and preventing subsequent separation runs. If care is not taken these factors may shorten the lifetime of the matrix, and since the gel is either bound covalently to the capillary wall or has such a high viscosity that it can't be replaced, the capillary must be discarded. High-quality commercially manufactured crosslinked polyacrylamide capillary gels are available from suppliers like J&W Scientific.

1.3. Entangled Polymer Gel Solution CE

Capillary electrophoresis in polymer solutions was introduced in the late 1980s to overcome the problems with non-replaceable gels *(15)*. A wide range of different polymer solutions have been used successfully:

a. Derivatives of cellulose *(15–19)*;
b. Carbohydrates such as galactomannan, glucomannan, dextran, and agarose *(20–22)*;
c. Synthetic polymers such as poly(ethylene oxide), polyvinylpyrrolidone, and pluronic acids *(23–26)*; and
d. Acrylate derivatives *(27,28)*.

The polymers are employed either at moderate concentration in an entangled state, or in dilute solution (unentangled) *(29)*. The benefit of the use of such polymer solutions for CE is the low solution viscosity (preferably <100–500 cP) that allows the polymer solution to be pumped through the capillary at low pressure (1–12 bar) and therefore allows rapid replacement of capillary matrix with fresh polymer solution after each run. Other benefits of polymer solutions are, low sensitivity to high electric field strengths and excessive currents, nontoxic (except acrylate derivatives), and the high UV-transparency (except acrylate derivatives), that allow analyte detection at <260 nm. In addition, most of the useful polymers are commercially available.

The preparation and handling of most polymer solutions is also simple, requiring only complete dissolution of the polymer in an appropriate buffer. Normally, most of these polymers are used in conjunction with surface-coated capillaries *(30)*, which reduce (or eliminate) the electroosmotic induced flow of matrix during a run *(31)*. In some instances the polymer itself can act as a dynamic coat onto the fused silica capillary wall ***(18,21,23,24,28)*** which prevents electroendosmosis. Today, most reports of size-sieving of nucleic acids and SDS-proteins by CE employed polymer solutions due to the ease of preparation, good separation characteristics, and convenience of these media. However, several authors report that chemically crosslinked gels have higher efficiency and better size discrimination for nucleic acids separations than polymer solutions ***(10,32–34)***.

1.4. Low-Melting Agarose Gel CE

In order to overcome some of the practical limitations of polymer solutions for CE separations, our group applied replaceable low-melting agarose gels (LM agarose) as the sieving matrix. Agarose, which is defined as a physical gel rather than a chemically polymerized gel, unites the desirable features of both crosslinked gels and polymer solutions, that is the highest separation performance although also being readily replaceable between runs. Low-melting agaroses are simple, derivatized agaroses (e.g., methoxylated), with various degrees of substitution. The substituants impart different physical characteristics to the agarose, including reduced gel setting and remelting temperatures. Low-melting agaroses are nontoxic, easy to handle, UV-transparent, stable at high voltages and currents, and are readily available commercially.

LM agaroses have shown promise for protein separations as well as for separations of both small- and large-size dsDNA fragments ***(35–37)***. Low-melting agarose is also compatible with pulsed-field CGE because of its high chemical stability, as shown by the separation of larger DNA fragments (0.1–23 kbp).

2. Materials

1. Methoxylated low-melting agarose with a gelling point of 25.6°C, unknown EEO, (Hispanagar, Spain), or a similar agarose.
2. Fused silica capillaries of 50 µm id and 360 µm od (MicroQuartz, Germany) (*see* **Note 1**).
3. Low-size range DNA standard (42, 88, 222, 249, 279, 303, 634, 800, 1434, and 1746 bp (Bio-Rad).
4. BioFocus 3000 CE apparatus (Bio-Rad), or a similar CE apparatus capable of independent themostatic control of both the capillary and the sample/buffer compartment.
5. All other chemicals should be of highest analytical grade and are available from most commercial suppliers of fine chemicals and electrophoresis equipment.

3. Methods

3.1. Capillary Coating Procedure

1. Capillaries of final length 20 + 4.5 cm (20-cm effective separation length) are coated according to a modified Hjertén procedure ***(30)***. Briefly, a detection window (2–3 mm wide) is first made at an appropriate distance from the anodic (outlet end) of the capillary by burning off the external polyimide layer with an electrically heated wire ***(38)***. The final distance between the detection window to the capillary outlet end is 4.5 cm (using the BioFocus 3000 CE apparatus).
2. The capillary is then filled with 1 *M* NaOH and surface etched for 15 min, then etching is repeated once with fresh alkali. Flush 0.1 *M* HCl through the capillary continuously for

30 min, followed by water for 30 min using vacuum-suction (corresponds to many column-volumes flush) (*see* **Note 2**).

3. A 50% solution of 3-methacryloxypropyltrimethoxysilane (Bind-Silane) in acetone is introduced and kept static for 20 min to coat the capillary surface. Repeat the Bind-Silane coating once more, then rinse the capillary briefly with several column vol 100% acetone and then with water, and then purge the capillary dry with air.
4. Finally, a 4% (w/v) solution of deaerated acrylamide in water or buffer containing 0.1% (v/v) tetramethylethylenediamene (TEMED) and 0.1% (w/v) ammonium persulfate (APS) is sucked into the capillary. The ends of the capillary are covered with parafilm to reduce re-aeration and it is left overnight at room temperature to polymerize (*see* **Note 3**).
5. Finally, the acrylamide solution is removed from the capillary by a syringe or HPLC pump after completing the polymerization reaction (*see* **Note 4**).

3.2. Buffer Preparation

1. Titrate a solution of 0.160 *M* Tris-base and 2 m*M* EDTA with acetic acid to pH 8.0 (TAE). The same procedure is used to prepare other TAE buffers at different concentrations and pH values.

3.3. Agarose Preparation

1. A 1% (w/v) low-melting agarose solution is prepared by adding the agarose to the buffer (in a capped glass vial) at room temperature. The solution is then transferred to a heated water bath (80–90°C) and swirled until the agarose is completely dissolved (*see* **Note 5**).

3.4. Sample Preparation and Injection Procedure

1. For electrophoresis in general, both a sample with reduced salt content relative to the ionic strength of the separation buffer, as well as the ionic species present in the buffers are important for high sensitivity and high-performance separations and for reproducibility between runs ***(39)***. It is recommended that any compound extracted from a biological source be desalted before being injected (*see* **Note 6**).
2. The commercial DNA sample used here had a total concentration of 0.2 mg/mL (about 4 OD) in 10 m*M* Tris-HCl, pH 8.0. One volume of DNA solution is diluted with 3 vol of water and with 1 vol of a 1:2 dilution of electrophoresis buffer with water. There is sufficient DNA for several quantitative injections in 5 µL of this diluted standard. The lower limit of sample detectability is not known, about 5–10 µg/mL of DNA standard was sufficient (except for the 88-bp fragment) whereas preserving a high resolution and good detection (*see* **Note 7**).

3.5. Automated CGE Analysis

1. An aliquot of the warm agarose solution is transferred to a plastic vial (Eppendorf style) and placed in the sample/buffer carousel of the BioFocus 3000 CE apparatus, which should have a temperature of about 35°C to keep the agarose in a liquid state (*see* **Note 8**).
2. The agarose is then pressure injected into the capillary for 20 s with the capillary coolant liquid set at 35°C (*see* **Note 9**).
3. The temperature of the coolant is then lowered to 18°C for 5–10 min to allow the agarose solution to set, after which sample injection and electrophoretic separation can proceed (*see* **Note 10**).
4. Agarose gel replacement is also simple—the expended gel is driven out of the capillary with water at 35°C, pressure injected for 10–20 s, whereby the capillary is ready for a repeat procedure.

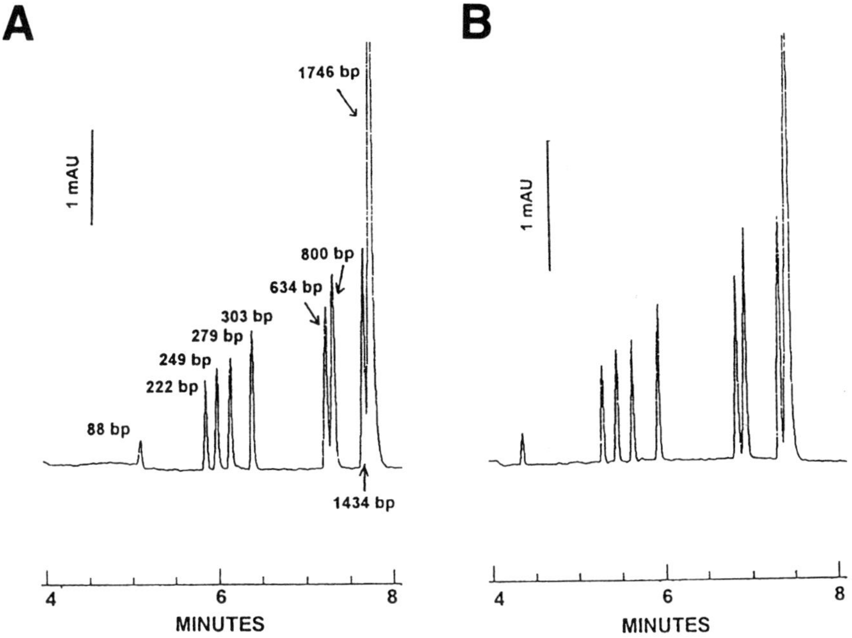

Fig. 1. CGE of DNA restriction fragments in a methoxylated agarose gel run in several TAE buffer concentrations. The resolution of DNA fragments is improved in the higher concentration buffer. Buffers: **A**, 0.040 *M* TAE and **B**, 0.160 *M* TAE, pH 8.0 containing 2 m*M* EDTA. Field strengths: A, 283 V/cm and B, 407 V/cm, respectively giving about the same analysis times. Gel: 1% agarose (gelling point 25.6°C). Capillary: id 50 μm × 30 cm (25.5 cm to detector). Sample: DNA fragments—size in bp indicated on the electropherogram (the 42-bp fragment could not be detected). Injection: 2.5 kV, 10 s; run temperature, 18°C; sample/buffer compartment, 35°C.

3.6. Analysis of Small DNA Fragments

1. The following section shows the effect of parameters such as buffer type and buffer concentration on the resolution for low-range dsDNA fragments in replaceable low-melting agarose gels at constant migration times.
2. **Figure 1A, B** shows electropherograms of DNA restriction fragments separated in a low-melting agarose matrix at two buffer concentrations, 0.040 and 0.160 *M* Tris-acetate-EDTA (TAE), respectively. It is clearly visible that the higher concentration of electrolyte gives a better resolution of all fragments. The increase in resolution was a result of both a decrease in peak width, and an increase in the time difference between adjacent fragments.
3. **Figure 2** shows that in quantitative terms, the resolution is a function of the TAE buffer concentration at constant migration time. A maximum resolution of all fragments occurs at around 0.160–0.20 *M* TAE (except for the 634/800-bp fragment, which is maximally resolved at 0.260 *M* TAE). Since the electrophoretic mobility of molecules are inversely related to the ionic strength of the electrolyte, it is necessary to apply a higher field

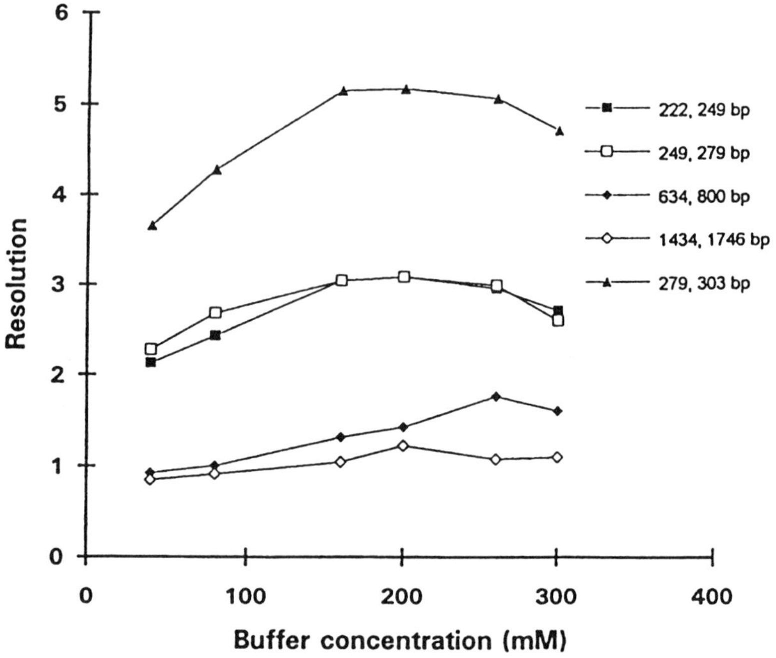

Fig. 2. The resolution of DNA in LM agarose CGE, shown as a function of the concentration of the TAE buffer at constant analysis time. Capillary: id 50 µm × 30 cm (25.5 cm to detector) with 1% agarose LM gel (gelling point 25.6°C). Injection: 2.5 kV, 10 s; run temperature, 18°C; sample/buffer compartment, 35°C. Sample DNA fragments are indicated in base pair on the electropherogram. The difference in resolution from run-to-run was < 0.05 for each fragment pair. The 1746-bp fragment had a migration time of 7.8 ± 0.2 min. Reprinted from **ref.** ***37***.

strength with an increase in buffer concentration in order to preserve constant migration times, also consequently with large increases in Joule heating. A field strength of 283 V/cm (current: 19 µA) and 407 V/cm (current: 90 µA) was employed for the 0.040 and 0.160 *M* TAE buffers, respectively.

4. High currents such as 90 µA generate excessive Joule heat which might lessen the efficiency and thereby the resolution of the separation ***(39)***. The use of narrow-bore capillaries (e.g., 50 µm id) in combination with efficient liquid cooling is sufficient to dissipate the excessive heat generated. Despite the temperature gradient (<0.5°C) created between the capillary axis and its inner wall during electrophoresis with 0.160 *M* TAE buffer, the separation efficiency increases compared to runs with more dilute buffers ***(37)*** (*see* **Note 11**).

3.7. Optimization of Electric Field Strength

1. Small DNA fragments of similar length are often difficult to separate. Good resolution can be achieved in LM-agarose CE by careful optimization of buffer strength and applied field without compromising run times.
2. **Figure 3A–C** shows the resolution for three different DNA fragment pairs (1746/1434, 800/634, and 303/279 bp) to be a function of applied field strength for 0.040 and 0.160 *M*

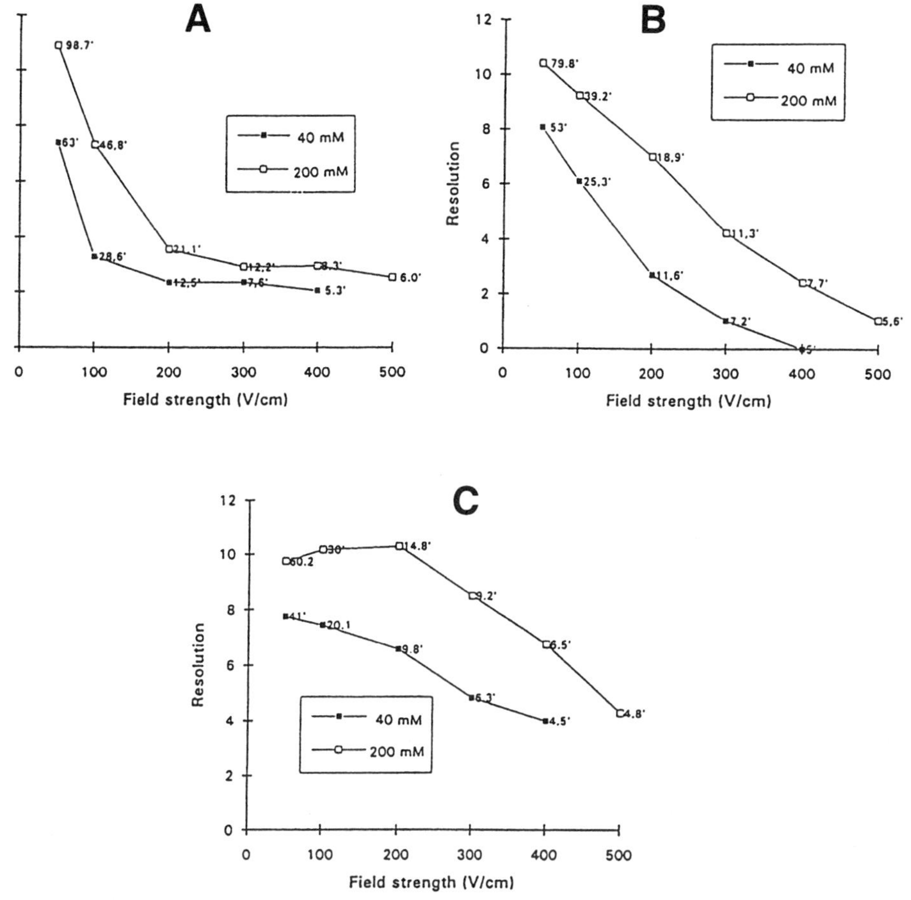

Fig. 3. The resolution of DNA in LM agarose CGE as a function of the field strength, at two different buffer concentrations. Capillary: id 50 µm × 30 cm (25.5 cm to detector) with 1% agarose LM gel (gelling point 25.6°C). Injection: 2.5 kV, 10 s; run temperature, 18°C; sample/buffer compartment, 35°C. Buffers: 0.04 *M* and 0.20 *M* TAE, pH 8.0. DNA fragment pairs are: **(A)** 1434/1746 bp; **(B)** 634/800 bp; **(C)** 279/303 bp are indicated on the electropherogram. Migration times are in min are indicated on the curves. The decrease in resolution at higher field strengths probably reflects extension of the DNA molecules.

TAE buffers, respectively. The average migration times for each respective base pair is indicated on the lines in the graphs. It can be seen that at the higher buffer concentration a better resolution for all fragment pairs occurs, although preserving the similar migration times.

3. At higher field strengths the migration times for all fragment pairs decrease, as does the resolution between the fragment pairs. The decrease in resolution with higher field strengths probably reflects the alignment or extension of the DNA fragments with the electric field in accordance with reptation theory ***(40)***.

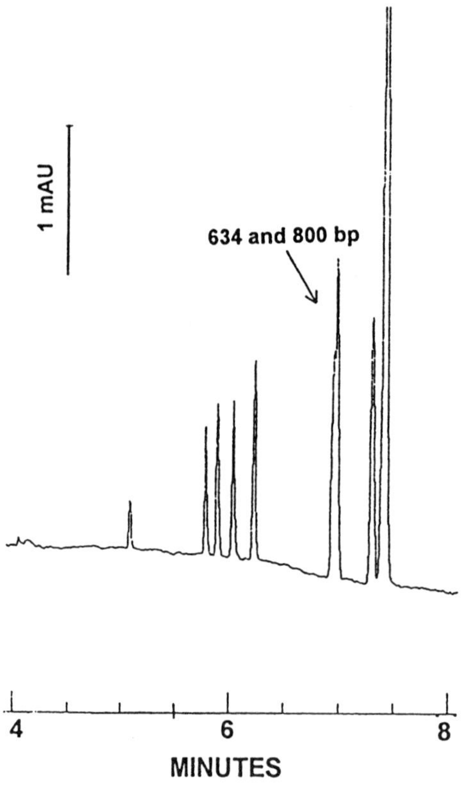

Fig. 4. CGE of DNA restriction fragments in a methoxylated agarose gel, prepared in 0.160 *M* Tris-borate buffer, 2m*M* EDTA, pH 8.5. Field strength: 350 V/cm (18 μA). Capillary: id 50 μm × 30 cm (25.5 cm to detector) with 1% agarose LM gel (gelling point 25.6°C). Injection: 2.5 kV, 10 s; run temperature, 18°C; sample/buffer compartment, 35°C. Sample DNA fragments are indicated in base pair on the electropherogram. Comparison of these data with **Fig. 1** shows that the resolution of DNA in the TBE buffer is lower than that achieved in the TAE buffer, for all DNA fragment pairs except for the 1434/1746 bp pair.

3.8. Tris-Borate-EDTA Buffer

Tris-borate-EDTA (TBE) buffer is used more commonly than TAE for nucleic acid separations both in polyacrylamide and agarose gels, and in polymer solutions.

1. **Figure 4** shows LM agarose CGE separation of DNA in 0.160 *M* TBE. The lower conductivity of 0.160 *M* TBE (350 V/cm, 18 μA) necessitates a reduction of the electric field compared with 0.160 *M* TAE (407 V/cm, 90 μA) to maintain similar migration times.
2. However, the high-resolution separations of DNA fragment pairs seen with TAE buffer is not achieved with TBE buffer, despite using optimal separation conditions and employing different buffer concentrations and pH, and keeping similar migration times.
3. In experiments using TBE buffer, only the 1434/1746-bp fragment pairs were well resolved. The 634/800-bp fragment pairs could not be separated at run times equivalent to

TAE gels, although they could be baseline resolved at a lower field strength and longer migration times.

4. Borate is known to form complexes with agarose, as well as with nucleic acid. Agarose is more crosslinked in borate buffer resulting in a general improvement in its sieving properties, which is normally beneficial in terms of an enhanced resolution for smaller DNA ***(41–44)***.
5. The reduced resolution of short DNA fragment pairs in LM agarose gels with TBE buffer compared to TAE buffer may be the result of the complexation of borate with both the sieving matrix and the analyte species, and the gain of charge by both ***(41–44)*** (*see* **Note 12**).

4. Notes

1. Commercial coated-capillaries have not been evaluated, but if compatible with agarose gels they could also be successfully used. Such coated capillaries are available from scientific suppliers such as Bio-Rad, Beckman, Hewlett-Packard, Supelco, and J&W Scientific.
2. The time schedule for NaOH, HCl, water, and Bind-Silane can be prolonged without any noticeable difference in separation performance and lifetime of the capillaries.
3. It is convenient to work with 0.5 mL of monomer solution in a 1.5–2-mL Eppendorf tube. Deaerate using vacuum until no bubbles remain by punching a hole in the cap and enclose the tube in a glass suction-bottle coupled to a vacuum unit, or He-purge for a few minutes. After the addition of TEMED and APS swirl the tube gently for about 10 s. Polymerization is inhibited in acid solutions (pH <6.5), preferably use a buffer at pH 7.0 ***(13)***.
4. If the capillaries are not intended for immediate use they can be stored in dry form for at least a few months in a fridge at +8°C. If intended for immediate use condition them in separation buffer for approx 5 min (the pressure for rinsing is 100 psi with the BioFocus 3000 apparatus). A preelectrophoresis run will also speed conditioning of the capillary surface.
5. The warm agarose solution can be stored for at least a week at +50°C without deterioration.
6. Only electrokinetic injection is possible here. See the figure legends for details of the parameters used.
7. After an electrophoresis run and before a sample injection, the capillary can be dipped briefly in water to avoid carry-over of buffer salts to the next sample.
8. The BioFocus 3000 CE apparatus has independent thermostatic temperature control of the capillary (liquid flow) and the sample/buffer compartment (air flow).
9. The purge/rinsing pressure is 100 psi (6.9 bar).
10. The agarose solution in the sample carousel may be used for several repetitive runs. Although slow evaporation of water could alter the matrix properties over time, this can be reduced by placing a few milliliters of water into the holder in which the agarose solution is located thereby raising the partial pressure of water in the compartment.
11. In order to increase the sample load and detection path length and thus enhance the sensitivity by electrophoresis in a lower strength field and/or with a lower buffer concentration, capillaries with an id of 75 or 100 μm may be successfully used.
12. It is also possible to use alternative buffers such as Tris-pentanoic acid-EDTA buffer (TPE), at a similar concentration and pH to TAE buffers. Tris-pentanoic-EDTA buffer has a 10% lower conductivity than TAE, and produces less Joule heat, and thus has a smaller thermal zone broadening at any given field strength. TPE buffer shows a similar resolution profile with DNA as TAE ***(37)***.

References

1. Handbook of Capillary Electrophoresis (1997), (Landers, J. P., ed.), 2nd ed., CRC Press, London and New York.
2. Olivera, B. M, Baine, P., and Davidson, N. (1964) Electrophoresis of the nucleic acids. *Biopolymers* **2,** 245–257.
3. Hamdan, I. I., Skellern, G. G., and Waigh, R. D. (1998) Separation of pd(GC)$_{12}$ from pd(AT)$_{12}$ by free solution capillary electrophoresis. *J. Chromatogr. A* **806,** 165–168.
4. Iki, N., Kim, Y., and Yeung, E. S. (1996) Electrostatic and hydrodynamic separation of DNA fragments in capillary tubes. *Anal. Chem.* **68,** 4321–4325.
5. Wehr, T., Zhu, M., and Mao, D. T. (2001) Sieving matrix selection, in *Capillary Electrophoresis of Nucleic Acids,* Vol. 1 (Mitchelson, K. R. and Cheng, J., eds.), Humana Press, Totowa, NJ, pp. 167–187.
6. Hjertén, S. (1983) High-performance electrophoresis: The electrophoretic counterpart of high-performance liquid chromatography. *J. Chromatogr.* **270,** 1–6.
7. Cohen, A. S., Najarian, D. R., Paulus, A., Guttman, A., Smith, J. A., and Karger, B. L. (1988) Rapid separation and purification of oligonucleotides by high-performance capillary gel electrophoresis. *Proc. Natl. Acad. Sci. USA* **85,** 9660–9663.
8. Yin, H.-F., Lux, J. A., and Schomburg, G. (1990) Production of polyacrylamide gel filled capillaries for capillary gel electrophoresis (CGE): influence of capillary surface pretreatment on performance and stability. *J. High Res. Chromatogr.* **13,** 624–627.
9. Guttman, A. and Cooke, N. (1991) Effect of temperature on the separation of DNA restriction fragments in capillary gel electrophoresis. *J. Chromatogr.* ***559,*** 285–294.
10. Nakatani, M., Skibukawa, A., and Nakagawa, T. (1994) Preparation and characterization of a stable polyacrylamide sieving matrix-filled capillary for high-performance capillary electrophoresis. *J. Chromatogr. A* ***661,*** 315–321.
11. Lindberg, P., Stjernström, M., and Roeraade, J. (1997) Gel electrophoresis of DNA fragments in narrow-bore capillaries. *Electrophoresis* **18,** 1873–1979.
12. Gelfi, C. and Righetti, P. G. (1981) Polymerization kinetics of polyacrylamide gels, II. Effect of temperature. *Electrophoresis* **2,** 220–228.
13. Caglio, S. and Righetti, P. G. (1993) On the pH dependence of polymerization efficiency. *Electrophoresis* **14,** 554–558.
14. Bio-Rad Laboratories; acrylamide polymerization - a practical approach (1987) *Bulletin 1156.*
15. Zhu, M.-D., Hansen, D. L., Burd, S., and Gannon, F. (1989) Factors affecting free zone electrophoresis and isoelectric focusing in capillary electrophoresis. *J. Chromatogr.* **480,** 311–319.
16. Baba, Y., Ishimaru, N., Samata, K., and Tsuhako, M. (1993) High-resolution separation of DNA fragments by capillary electrophoresis in cellulose derivative solutions. *J. Chromatogr. A* **653,** 329–335.
17. Kleemib, M. H., Gilges, M., and Schomburg, G. (1993) Capillary electrophoresis of DNA restriction fragments with solutions of entangled polymers. *Electrophoresis* **14,** 515–522.
18. Barron, A. E., Sunada, W. M., and Blanch, H. W. (1995) The use of coated and uncoated capillaries for the electrophoretic separation of DNA in dilute polymer solutions. *Electrophoresis* **16,** 64–74.
19. Dolnik, V. and Novotny, M. (1992) Capillary Electrophoresis of DNA fragments in entangled polymer solutions: A study of separation variables. *J. Microcol. Sep.* **4,** 515–519.

20. Izumi, T., Yamaguchi, M., Yoneda, K., Isobe, T., Okuyama, T., and Shinoda, T. (1993) Use of glucomannan for the separation of DNA fragments by capillary electrophoresis. *J. Chromatogr. A* **652,** 41–46.
21. Heller, C. (1998) Finding a universal low-viscosity polymer for DNA separation. *Electrophoresis* **19,** 1691–1698.
22. Kleparnik, K., Mala, Z., Doskar, J., Rosypal, S., and Bocek, P. (1995) An improvment of restriction analysis of bacteriophage DNA using capillary electrophoresis in agarose solution. *Electrophoresis* **16,** 366–376.
23. Fung, E. N., Pang, H.-M., and Yeung, E. S. (1998) Fast DNA separations using poly(ethylene oxide) in non-denaturing medium with temperature programming. *J. Chromatogr.* **806,** 157–164.
24. Gao, Q. and Yeung, E. S. (1998) A matrix for DNA separation: Genotyping and sequencing using poly(vinylpyrrolidone) solution in uncoated capillaries. *Anal. Chem.* **70,** 1382–1388.
25. Rill, R. L., Liu, Y., Van Winkle, D. H., and Locke, B. R. (1998) Pluronic copolymer liquid crystals: unique, replaceable media for capillary gel electrophoresis. *J. Chromatogr. A* **817,** 287–295.
26. Rill, R. L. and Liu, Y. (2001) DNA separation by capillary electrophoroesis in lyotropic polymer liquid crystals, in *Capillary Electrophoresis of Nucleic Acids,* Vol. 1 (Mitchelson, K. R. and Cheng, J., eds.), Humana Press, Totowa, NJ, pp. 203–213.
27. Chiari, M. and Righetti, P. G. (1995) New types of separation matrices for electrophoresis. *Electrophoresis* **16,** 1815–1829.
28. Chiari, M. and Melis, A. (1998) Low viscosity DNA sieving matrices for capillary electrophoresis. *Trends Anal. Chem.* **17**, 623–632.
29. Grossman, P. D. and Soane, D. S. (1991) Capillary electrophoresis in entangled polymer solutions *J. Chromatogr.* **559,** 257–266.
30. Hjertén, S. (1985) High-performance electrophoresis: Elimination of electroendosmosis and solute adsorption. *J. Chromatogr.* **347,** 191–198.
31. Chiari, M. and Cretich, M. (2001) Capillary coatings: choices for capillary electrophoresis of DNA, in *Capillary Electrophoresis of Nucleic Acids,* Vol. 1 (Mitchelson, K. R. and Cheng, J., eds.), Humana Press, Totowa, NJ, pp. 125–138.
32. Sumita, C., Baba, Y., Hide, K., Ishimaru, N., Samata, K., Tanaka, A., and Tsuhako, M. (1994) Comparative study of non-porous anion-exchange chromatography, capillary gel electrophoresis and capillary electrophoresis in polymer solutions in the separation of DNA restriction fragments. *J. Chromatogr. A* **661,** 297–303.
33. Bae, Y. C. and Soane, D. J. (1993) Polymeric separation media for electrophoresis: cross-linked systems or entangled solutions. *J. Chromatogr. A* **652,** 17–22.
34. Kamahori, M. and Kambara, H. (1996) Characteristics of single-stranded DNA separation by capillary gel electrophoresis. *Electrophoresis* **17,** 1476–1484.
35. Hjertén, S., Srichaiyo, T., and Palm, A. (1994) UV-transparent, replaceable agarose gels for molecular-sieve (capillary) electrophoresis of proteins and nucleic acids. *Biomed. Chromatogr.* **8,** 73–76.
36. Chen, N., Wu, L., Palm, A., Srichaiyo, T., and Hjerten, S. (1996) High-performance field inversion capillary electrophoresis of 0.1-23 kbp DNA fragments with low-gelling, replaceable agarose gels. *Electrophoresis* **17,** 1443–1450.
37. Palm, A. and Hjertén, S. (1996) The resolution of DNA fragments in capillary electrophoresis in replaceable agarose gels. *J. Capillary Electrophor.* **3,** 173–179.

38. Lux, J. A., Häusig, U., and Schomburg, G. (1990) Production of windows in fused silica capillaries for in-column detection of UV-absorption or fluorescence in capillary electrophoresis or HPLC. *J. High Res. Chromatogr.* **13,** 373–374.
39. Hjertén, S. (1990) Zone broadening in electrophoresis with special reference to high-performance electrophoresis in capillaries: An interplay between theory and practice. *Electrophoresis* **11,** 665–690.
40. Slater, G. W., Mayer, P., and Drouin, G. (1996) High resolution separation and analysis of biological macromolecules, in *Methods in Enzymology,* Vol. 270 (Karger, B. L. and Hancock, W. S., eds.), Academic Press, London and New York, pp. 272–295.
41. Hamelin, C. and Yelle, J. (1990) Gel and buffer effects on the migration of DNA molecules in agarose. *Appl. Theor. Electrophor.* **1,** 225–231.
42. Peats, S., Nochumson, S., and Kirkpatrick, F. H. (1986) Effects of borate on agarose structure. *Biophys. J.* **49,** 91a.
43. Zernik, J. and Lichtler, A. (1987) Borate-buffer-related effects on the electrophoretic mobility of linear DNA fragments in agarose. *BioTechniques* **5,** 411–413.
44. Mitchelson, K. R. and Cheng, J. (2001) Capillary electrophoresis with glycerol as an additive, in *Capillary Electrophoresis of Nucleic Acids*, Vol. 1 (Mitchelson, K. R. and Cheng, J., eds.), Humana Press, Totowa, NJ, pp. 259–277.

IV

Practical Applications for Large DNA Fragment Sizing and Fragment Collection

19

Robust Field Inversion Capillary Electrophoretic Separation of Long DNA Fragments

Christoph Heller, Soffia Magnúsdóttir, and Jean-Louis Viovy

1. Introduction

Electrophoretic analysis of DNA samples is a routine method in biochemistry and molecular biology. Conventional DNA separations in agarose gels are restricted up to a size of about 10 kbp. The reason for this restriction is the molecular orientation of large DNA fragments under the action of an electric field as proven by theoretical considerations *(1–3)* and experimental studies *(4,5)*. At a large molecular weight and/or strong electric fields the velocity of an oriented DNA chain becomes size-independent and the oriented DNA molecules of different size cannot be separated in continuous field gel electrophoresis.

A significant breakthrough in the separation of large DNA molecules occurred with the introduction of pulsed-field gel electrophoresis (PFGE) *(6)*. Since then, PFGE has become the most common technique used in the separation of large DNAs (up to several Mbp). The success of this technique lies in the forced reorientation of DNA chains in a new direction of the external field. As the reorientation time is size-dependent, a separation of different fragments becomes possible.

Pulsed-field electrophoresis can be carried out with parallel fields or nonparallel fields. Parallel fields are used in the so-called field inversion gel electrophoresis (FIGE) which was introduced in 1986, when it was found out that simply inverting the electric field was sufficient for the retardation of large DNA molecules *(7)*. Obviously, to achieve a net migration, the pulse time or the electric field applied in the "forward" direction has to be of a greater value than in the "backward" direction. This technique has been proven to be useful for the separation of DNA molecules with low *(8,9)*, or medium *(7,10)*, and high molecular weights *(11)*, which cover a range from a few thousand base pairs up to several Mbp. FIGE gives the opportunity to separate molecules over a wide range of sizes at a rather low resolution by using pulse time ramps, but can also achieve a very high resolution in a small window of separation by using constant pulse times *(12)*. The field inversion method has the advantage that it

From: *Methods in Molecular Biology, Vol. 162:*
Capillary Electrophoresis of Nucleic Acids, Vol. 1: Introduction to the Capillary Electrophoresis of Nucleic Acids
Edited by: K. R. Mitchelson and J. Cheng © Humana Press Inc., Totowa, NJ

can be set up with a minimum of instrumentation and that it can be easily combined with other techniques such as direct blotting electrophoresis or sequencing gels *(13,14)*.

Capillary electrophoresis (CE) is now routinely applied for the separation of double-stranded DNA fragments using polymer solutions as the separating matrix (*see* **ref. *15*** for review). However, as the upper separation limit (due to orientation) decreases with increasing electric field, this means that the rather high electric fields used in CE (up to 400 V/cm or more) decrease the range of molecular sizes that can be separated. Indeed, CE separation of DNA becomes poor above sizes of about 2 kbp. Therefore, for capillary electrophoretic separation of DNA the combination with pulsed field electrophoresis is even more interesting than for conventional slab gels. Unfortunately, up to date, no commercial instrument, able to deliver pulsed electric fields is available.

Obviously, for CE, only pulsed field regimes with parallel electric fields can be employed and field inversion CE is the method of choice (*see* **ref. *16*** for review). Several groups have reported separations of DNA standards containing fragments up to a few Mbp using this technique *(17–22)*. However, the peaks corresponding to DNA fragments of 30 kbp and larger were often found to be split in several sub-peaks *(19,23)*. Several sharp peaks are also observed when running samples containing a single species of DNA (e.g., lambda phage or T4 phage DNA). At first, these numerous and rather regularly spaced peaks were attributed to high-resolution separation of different species, but some features cast doubts on this interpretation. For instance, the occurrence of multiple peaks and the number of peaks in the electropherograms increased with increasing DNA size and/or increasing field strength. Other kinds of problems such as capillary blockage, unstable baseline and low reproducibility in the mobility where frequently observed. Finally, spurious peaks and low reproducibility were also reported in attempts to separate large DNA using constant field *(24)*, indicating that this problem is not restricted to pulsed-field CE.

Some insight into the physical mechanisms at the origin of these problems was provided by fluorescence microscopy studies of DNA solutions submitted to AC fields *(25)* and by fluorescence resonance energy transfer (FRET) microscopy *(26)*. In these studies, the application of an alternating field of typically 100 V/cm was shown to induce the separation of an initially homogeneous and monodisperse DNA solution into DNA-poor regions and vortex-like DNA-rich domains in permanent rotation. Aggregation in such conditions is favored by increasing the field strength, the DNA size and concentration, and it is further favored by a reduction in the frequency of the field alternation.

Researchers have begun to understand field-induced DNA aggregation only recently, and this has led to some confusion in the evaluation and interpretation of CE separations of large DNA. However, the aggregation can be suppressed to some extent by decreasing the electric field strength (at the expense of separation time) and by decreasing the DNA concentration. With some caution, the range of DNA that can be separated in CE can then be extended significantly by the use of pulsed fields. However, a much faster separation of the chromosomes of *Saccharomyces cerevisiae* (ranging from 250–2000 kbp) in about 10 min, was also reported by Kim et al. *(21)*, who used nonentangled HEC polymer solutions as the separation medium. In spite of a rather low resolution (some chromosomes comigrated in an unresolved band) and poor reproducibility, the high fields used in this study and the rapidity of the separation are spectacular.

Based on theoretical studies on the electrohydrodynamic instabilities, Magnúsdóttir et al. *(27)* could show that aggregation can be strongly reduced by using isoelectric buffer systems. Here, we present a protocol for the fast and robust separation of long DNA fragments (up to at least 50 kbp) in CE.

2. Materials

2.1. Buffers and DNA Samples

1. Tris-boric acid-EDTA (TBE) buffer, 1X: 89 m*M* Tris-base, 89 m*M* boric acid, 2 m*M* EDTA, pH 8.3. Make up as a 5X concentrated stock solution and dilute accordingly.
2. His-buffer: Histidine (Fluka Chemie AG, Buchs, Switzerland) 178 m*M* or 200 m*M* in water pH 7.5 (*see* **Note 1**).
3. Hydroxypropyl cellulose, Mw 1000 kDa (Sigma-Aldrich, Dorset UK), 0.7% or 0.4% dissolved in buffer.
4. DNA standards are: ΦX174/*HaeIII* (72–1656 bp); 1-kb ladder (72–12,216 bp); High Molecular Weight Marker (8271–38,416 bp), each from Life Technologies Ltd. (Paisley, UK). Lambda phage (48.5 kbp) is from New England Biolabs (Beverly, MA), and 10 kbp ladder (10–250 kbp) is from Bayoulab (Paris, France).
5. DNA labeling: Ethidium bromide (stock solution 25 m*M*) or SYBR Green I (stock solution 10,000X concentrated) from Molecular Probes Inc. (Eugene, OR) (*see* **Note 2**).
6. All solutions should be filtered through single use syringe filters with 0.45- or 0.8-µm pore size. We recommend cellulose acetate/cellulose nitrate membrane.

2.2. Electrophoresis

2.2.1. Capillaries

1. Coated fused silica capillaries, DB-17 (100 µm id) and DB-210 (75 µm id) are from J&W Scientific (Folsom, CA). (*see* **Notes 3** and **4**).

2.2.2. Electrophoresis Apparatus

1. *See* **Note 5** for important safety considerations: As devices for field inversion CE are not commercially available, existing material for constant CE must be modified (always comply with safety regulations!). In our case, we have used a P/ACE System 2100 apparatus (Beckman Instruments, Fullerton, CA) as a platform (*see* **Notes 6** and **7**). The pulses are generated by a function generator (Hameg HM 8130, Villejuif, France) and amplified 2000× by a high-voltage amplifier (Trek 20 × 20, Trec Inc., Medina, NY).
2. The output of the amplifier is connected to the inlet electrode and the outlet electrode was connected to the ground. The pulsed field is generated by applying a square formed AC pulse having equal duration and amplitude combined with a DC offset, thus creating a forward pulse three times stronger than the backward one. Alternatively, sine wave pulses (again with an offset) can be chosen.

2.2.2.1. Automation of the Pulsed-Field Power Supply

1. A low-cost PC (any "old generation" PC working in MS-DOS can do the job easily) equipped with an AD-converter card to collect the synchronizing signal from the P/ACE

is used to automate the pulsed-field power supply. A "BASIC" program memorizes the sequence of field conditions (for automated operation of a series of runs), and drives the function generator accordingly via RS232 interface. The program is specific for the communication language of the chosen function generator: the BASIC code for driving a HAMEG function generator can be obtained upon request.

2.2.2.2. Signal Detection

1. To minimize the analysis time, we use the "outlet" port of the P/ACE apparatus for injection. This results in separation distance of only 7 cm to the detector, since a capillary length of 20 cm to the detector is the minimal length that can be used in the P/ACE device when using the inlet port for injection.
2. The DNA fragments are labeled using either ethidium bromide or SYBR Green I and detected by laser-induced fluorescence (excitation argon ion line 488 nm) (*see* **Note 8**). When ethidium bromide was used, a bandpass filter with a wavelength interval of 555–655 nm was placed in front of the detector. In the case of SYBR Green I, a bandpass filter with an interval of 500–560 nm was used. In all experiments, the temperature of the cartridge containing the capillary was controlled at 25°C.

3. Methods

3.1. Experimental Protocols

1. Capillaries are cut according to the instructions for the P/ACE instrument, using a total length of 27 cm with 7 cm from the detector to the end and placed into the cartridge. The external polyimide coating is removed from the capillary to create a detection window. For this purpose a small section of the capillary is incubated in hot (ca 80°C) sulfuric acid for 15 min (*see* **Note 4**). Permanently coated capillaries DB-17 or DB-210 from J&W Scientific could be used (*see* **Note 3**).
2. The hydroxypropylcellulose powder is dissolved at room temperature in the buffer by carefully mixing the solution overnight. The solution is then filtered through a 0.8-μm filter. When using ethidium bromide (EtBr) as a fluorescent dye, 8 μL of the EtBr stock solution are added to a 20 mL polymer/buffer solution resulting in a final concentration of 10 μ*M* EtBr. However, when using SYBR Green I the affinity of the dye for DNA is strong enough to make the labeling irreversible during the time of the separation. Therefore, we could prelabel DNA by adding a diluted solution to the sample, and avoid the use of dye in the separation medium. This is beneficial in terms of cost, safety, and environment preservation.
3. The DNA samples are diluted in pure water to a concentration of 0.05–16 mg/mL, depending on the DNA standard and on the fluorescent dye used. When using SYBR Green I as a probe, the dye is added to the DNA solution prior to injection resulting in a final concentration of 0.3X *(29)*.
4. The capillary is filled with the buffer/polymer solution for 15–20 min and a baseline is recorded for 15 min using the same solution in the inlet and outlet vials to equilibrate the capillary.
5. The function generator is then programmed to the desired separation conditions. The following parameters are involved: pulse form (square, sinusoidal); frequency (Hz); amplitude (V); offset (V).

   As an example: To create an average electric field of 100 V/cm (over a 27-cm long capillary) one should give as an offset 100 V/cm which means a total of 2700 V over the capillary. Divide this number by 2000 since the signal is amplified 2000 times, which results in 1.35 V as an offset. The sign is also important: A positive offset is given if the DNA should migrate from the outlet to the inlet of the P/ACE (as in our experiments), otherwise negative offset should be used. To create a field which is three times stronger in the forward direction than in the backward direction, the amplitude should then be 4X the offset, or 5.4 V.

6. The DNA samples are injected by electrokinetic injection in a way that resembles as much to a constant field as possible. The following protocol for injection was used:

Form:	sin
Frequency:	500 kHz
Amplitude:	0.21 V
Offset:	0.25 V (18.5 V/cm)
Injection time:	8–20 s

7. Between successive runs, the capillary is simply washed with the running buffer without polymer, and then refilled with the buffer/polymer solution. At the end of each day, the capillary is washed extensively with water and the ends of the capillary closed by paraffin film (Parafilm®, 3M Corp.).

3.2. Important Considerations to Avoid DNA Aggregation

1. As noted in the introduction, during CE aggregations of DNA can occur which are detrimental to the separation. To avoid this effect, we need to know under which circumstances aggregations can occur. So far, parameters such as buffer concentration, electric field strength, pulse frequency, and polymer concentration have been identified as the main parameters influencing DNA aggregation (*see* **Note 8**).
2. As a criterion for aggregation we can use the "aggregation time" which can be measured in a special experimental setup *(27)*. The capillary is entirely filled with a solution of DNA and buffer at known concentrations and the response of the UV detector is then monitored as a function of time directly after the application of the electric field. Since the field is now applied over a homogeneous solution, the only source of variations in the detector output other than the included noise in the apparatus should originate from eventual formation of aggregates.
3. We then define the time that the aggregates take to form, T_a, as the point where the root-mean-square (rms) of the signal fluctuation has doubled as compared with the value at the beginning of signal recording. **Figure 1** shows the dependence of the aggregation time on the TBE concentration, the pulse frequency, and the polymer concentration. We can also observe that aggregation takes place only above a certain electric field threshold (**Fig. 2**). This threshold increases with buffer concentration, but is lower for TBE buffer. Therefore, we can conclude that it is advantageous to use histidine buffer instead.

3.3. Important Parameters to Achieve Separation

1. The time that a molecule takes to reorient after switching the electric field (reptation time), depends on its molecular weight and the electric field strength applied *(19,29)*. If the pulse time (forward plus backward pulse time) applied is much larger than the reptation time (i.e., for small molecules), a molecule will have sufficient time to orient itself in both directions and will reach the constant field velocity at each pulse ("plateau velocity").
2. On the other hand, if the pulse time is much smaller than the reptation time (long molecules), the DNA chain will not "feel" the alternating field, but rather an average electric field and will migrate once again with a size-independent velocity. Between these two

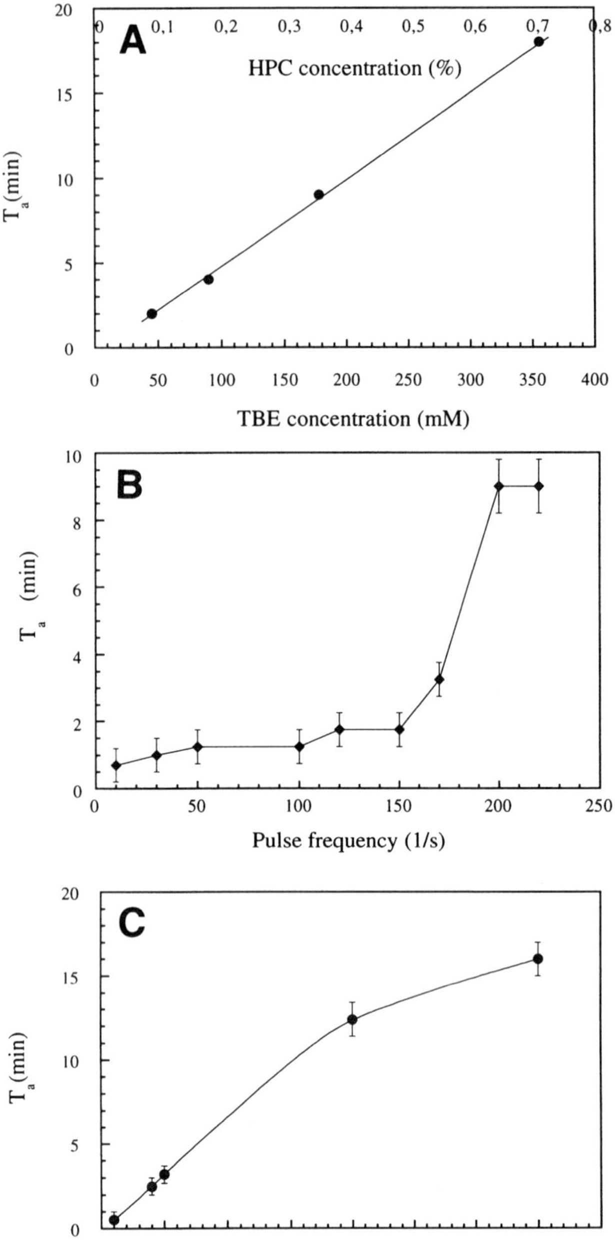

Fig. 1. Aggregation time T_a as a function of **(A)** TBE concentration, **(B)** Pulse frequency and **(C)** HPC concentration. A capillary of type DB-17, id=100 µm (total length 37 cm, length to detector 30 cm) was entirely filled with 10-µg/mL Lambda-phage DNA in 1X TBE buffer before the electric field was switched on. Electric field is (A) 200 V/cm DC, and (B, C) 432 V/cm AC square pulses with a 108 V/cm DC offset. Reproduced from **ref. *27***.

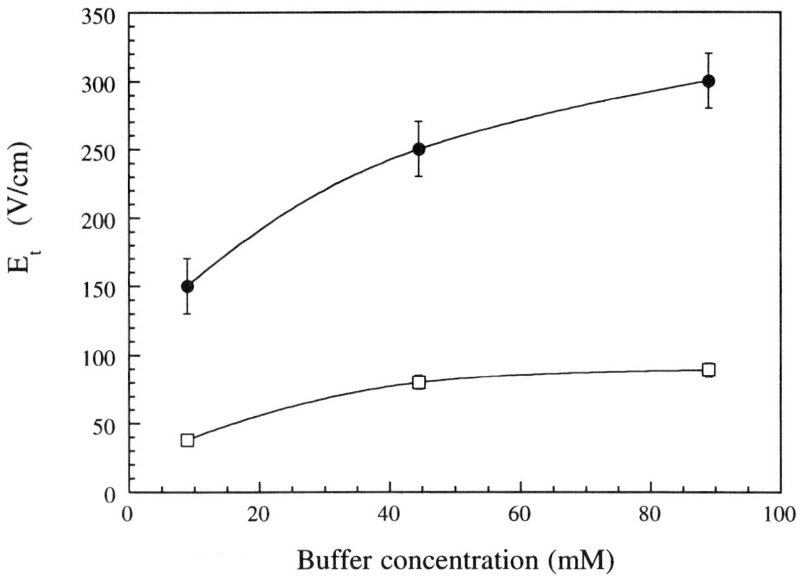

Fig. 2. Electric field threshold E_t above which aggregation occurred as a function of buffer concentration for two different solutions, TBE/DNA (□) and histidine/DNA (●). Field frequency was 0.01 Hz. The points indicate the lowest electric field at which the aggregation was observed. Other conditions as in **Fig. 1**. Reproduced from **ref.** ***27***.

extremes, a "window" of good separation is opened for molecules having a reptation time close to the pulse time.

3. We can define the optimal size N_{opt} as being the DNA molecular size at which the mobility is most sensitive to the pulse time. **Figure 3** shows a double logarithmic plot of mobility vs DNA size and we can evaluate N_{opt} by using the inflection point in the mobility curve. We can observe that the window of optimal separation shifts to larger sizes with decreasing frequency (increasing pulse time). Indeed, from a number of experiments ***(19,27)***, we obtain the following scaling law: $N_{opt} \sim t^{0.98\pm0.1} E^{1.2\pm0.2}$, with t being the pulse time and E the average electric field strength. These results are in good agreement with theoretical predictions.

3.4. Choosing Optimal DNA Separation Conditions

1. Before starting the electrophoresis experiment one should be aware of the size range to be separated. The best pulse frequency and electric fields can be determined with a few pilot experiments or by using published data ***(16,19,27)***. From experiments such as in **Fig. 3**, we see that in order to separate larger DNA, we have to lower the frequency. However, we then face the dilemma, that lower frequency increases the probability of aggregation.
2. **Figure 4** illustrates this behavior: By lowering the frequency from 10 to 2 Hz, we open the "separation window" for larger DNA fragments, but at the cost of a noisy baseline, indicating the onset of aggregation. Decreasing the electric field reduces aggregation but it also imposes an even lower optimal frequency for separating DNA in a given size

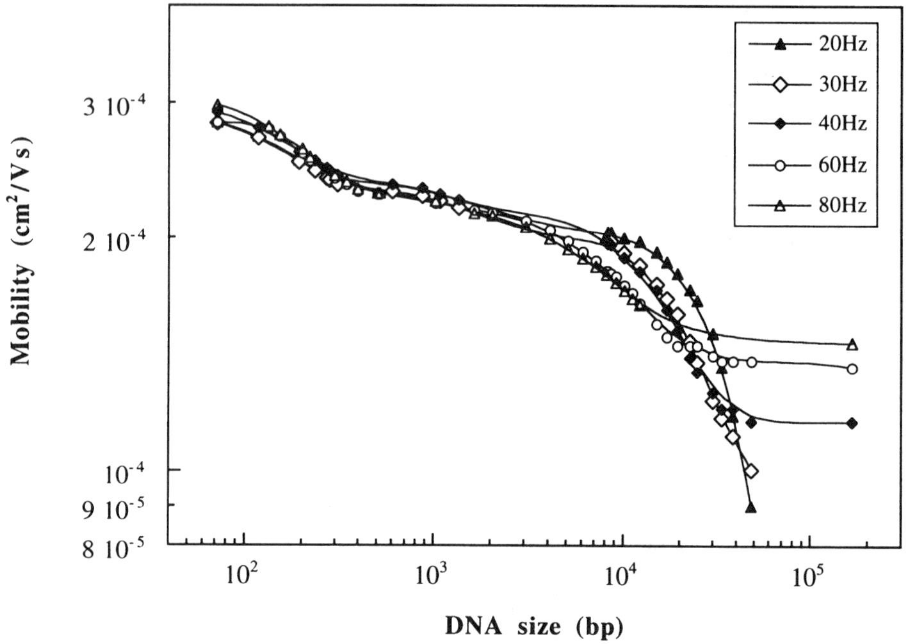

Fig. 3. Effect of pulse frequency on FICE separation plotted as mobility vs size for 20, 30, 40, 60, and 80 Hz. Running buffer: 0.7% HPC diluted in 178 m*M* histidine. Injection: 18.5 V/cm (DC-offset) for 70 s. FICE conditions: AC square pulse of 519 V/cm with a DC offset of 133 V/cm. Capillary: DB-210, 7 cm to the detector, total length 27 cm. Reproduced from **ref.** ***27***.

range. Because lowering the frequency favors the buildup of aggregation, it cannot be granted *a priori* that improvements can be obtained by lowering the electric field.

3. From **Fig. 1B**, however, one can see that the dependence of the aggregation time on frequency is weak, apart from the steep transition at approx 150 Hz. On the other hand, from experiments as shown in **Fig. 3**, we can expect that resolution/separation of DNA above 50 kbp will generally occur in the "low frequency" regime, in which aggregation is fast. This probably explains, at least in part, why there seems to be a qualitative jump in difficulty when attempting to separate DNA beyond a typical size of a few tens of kbp. Fortunately, the T_a vs frequency curve seems to remain rather flat in the low frequency regime, so that by using a low electric field strength and a frequency within this regime it should be possible to keep aggregation to a minimum. However, this will in turn lead to longer separation times. Up to a size of about 50 kbp fast and reproducible separations can be

Fig. 4. *(opposite page)* Separation of a 10-kbp ladder (0.05 ng/µL in water containing 0.3X SYBR Green I) in 0.4% HPC (M_w 1 MDa) and 200 m*M* histidine buffer (pH 7.3). Pulses are delivered at a frequency of: (**A**) 10 Hz, (**B**) 5 Hz, and (**C**) 2 Hz. Common conditions are: AC square pulse of 178 V/cm with a DC offset of 44.4 V/cm. Capillary: DB-17, id 75 mm, 7 cm to the detector, total length of 27 cm.

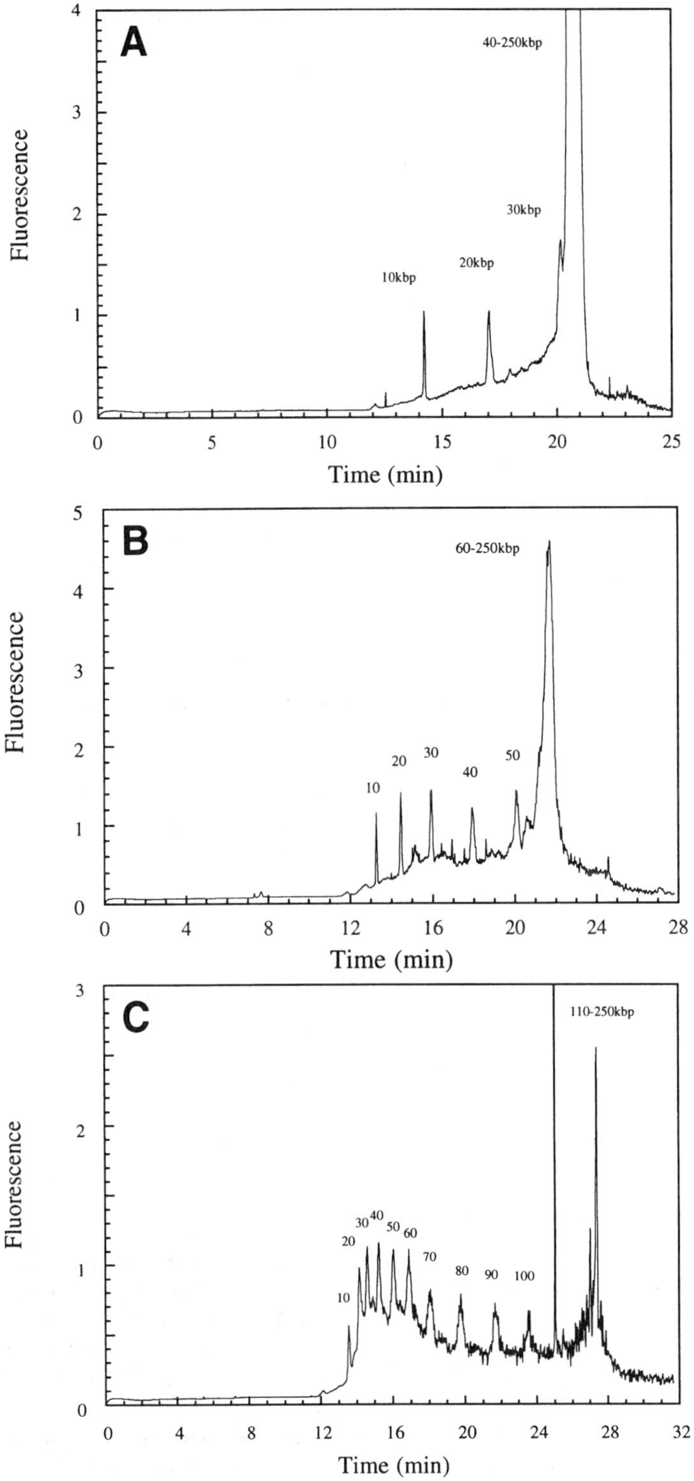
A
40-250kbp
30kbp
20kbp
10kbp
Fluorescence
Time (min)
B
60-250kbp
10
20
30
40
50
Fluorescence
Time (min)
C
110-250kbp
10
20
30
40
50
60
70
80
90
100
Fluorescence
Time (min)

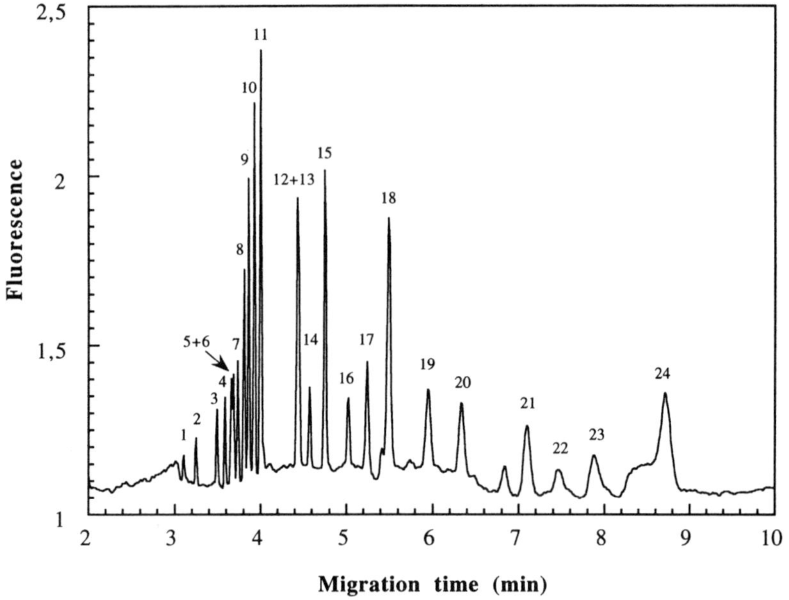

Fig. 5. Separation of a mixture of φX174/*HaeIII* fragments (5 mg/mL), high molecular weight marker (16 mg/mL) and Lambda phage DNA (2 mg/mL). Peak assignments are: (1) 72 bp, (2) 118 bp, (3) 194 bp, (4) 234 bp, (5) 271 bp, (6) 281 bp, (7) 310 bp, (8) 603 bp, (9) 872 bp, (10) 1078 bp, (11) 1656 bp, (12) 8271 bp, (13) 8612 bp, (14) 10,086 bp, (15) 12,220 bp, (16) 15,004 bp, (17) 17,057 bp, (18) 19,399 bp, (19) 22,621 bp, (20) 24,776 bp, (21) 29,942 bp, (22) 33,498 bp, (23) 38,416 bp. and (24) 48,502 bp. Frequency = 30 Hz, and all other experimental conditions as in **Fig. 3**. Reproduced from **ref.** ***27***.

achieved without any problems. **Figure 5** shows an electropherogram of DNA samples in the size range up to 50 kbp separated under optimal conditions in very short time (10 min).

4. Notes

1. Histidine stock solutions degrade rapidly. The separation performance was completely lost due to the degradation. We could use separation medium up to 1 wk, but not more, by storage at 4°C. Recent studies on degradation of histidine (Magnúsdóttir et al., submitted) recommended a small amount of fresh histidine/polymer solution is prepared daily. Because the dissolution of polymer takes a rather long time, it is recommended to prepare a polymer stock solution in pure water sufficient for a few months work (store at 4°C). Prepare the separation buffer daily by adding the relevant histidine quantity to an aliquot of this stock solution and finally just before use, add SYBR Green dye (*see* **Note 2**).
2. The SYBR Green I dye is provided as a 10,000X solution, which should be aliquoted and stored frozen. Aliquoting is in order to prevent repeated freezing and thawing and then working solutions should be diluted in TE buffer (10 m*M* Tris-HCl, 1 m*M* EDTA, pH 8.0) or another buffer such as TAE or TBE, since this dye is not stable in water alone.
3. Coated capillaries must be used for these separations. We could use fluoroalkyl-coated capillaries (DB-210, J&W Scientific) for a week of intensive use. The mobilities are

reproducible and the baseline is stable. Fresh DB-17 capillaries (phenyl-methyl coating) were also suitable, but tended to degrade faster. Probably, various types of home-made or commercial hydrophilic coatings (acrylamide, poly[vinyl alcohol], etc., *see* **ref. *30***) can be used with success, but we did not investigate this possibility, since DB-210 capillaries can be purchased in long sections and tend to be less expensive than capillaries sold specifically for CE.

4. In the case of purchasing larger quantities of capillary, the "window" must be made. The usual procedure of "burning" the polyimide coating is not suitable since it degrades the internal coating. The combination of a strong acid (sulfuric acid) and moderate temperature (60–80°C) is therefore recommended. This operation can be done for instance by threading the capillary into a small diameter glass tubing (e.g., Corning micropipet) with the desired length, filling the section by capillarity with sulfuric acid (only a few µL are necessary), and heating the ensemble with an air blower. Rinse thoroughly with cold water after the coating has dissolved, then remove the glass tubing. Cheap teflon devices allowing simultaneous treatment of several capillaries are also available commercially (Polymicro). In all cases, working with hot acid is a risk: wearing safety goggles and gloves is mandatory, and the operation should be conducted in a fume-hood adapted to acid fumes.
5. **IMPORTANT: Safety considerations to avoid the risk of electric shock.** To allow a high slew-rate (i.e., fast switching of voltage), HV amplifiers are able to deliver currents much stronger than the constant-field HV supplies conventionally used in CE machines, and thus can be lethal to humans. Therefore, extreme caution must be exerted to protect operators from electric shock. In particular, we strongly recommend that the CE apparatus, HV amplifier, and all HV connections be enclosed in a cabinet with a safety interlock on the door, connected to the "emergency switch-off" of the amplifier. Provision for air cooling, preventing the introduction of limbs or metallic elements inside the cabinet, should be included. Finally, a suitable insulation of connectors, following industrial regulations, should be adopted, and enough distance between HV elements and the walls of the cabinet should be reserved to avoid arcing.
6. Pulse-field generator: A large range of field strengths and field frequencies in the pulsed-field mode can be obtained using a TREK 20/20 driven by a programmable function generator, as described in **Subheading 3.1.** For separating large DNA in short capillaries, though, only moderate fields and low frequencies are necessary, and satisfactory results could be obtained using a cheaper system, such as the TREK 610C amplifier.
7. Pulsed-field operation in a commercial CE apparatus. This approach presents several advantages against the development of a dedicated machine, especially when:
   a. Such a machine is already available in the lab and is not used full-time.
   b. Routine operation is desirable.
   c. Reproducibility of injection is requested.

   It requires a modification of the apparatus, which is of course system-dependent. In the case of the Beckman P/ACE, the modification is very minor (it involves "dumming" of the internal high-voltage power supply, and connection of the external supply to the electrodes. The full automation features of the P/ACE can be conserved, by picking up the "enable" TTL signal driving the internal HV supply, as a synchronization signal for the external one). The modification is fully reversible. It takes only a few hours, and switching back and forth between constant field and pulsed-field modes takes less than 30 min (details applicable for the P/ACE can be obtained from the authors upon demand).

8. In the protocols presented here, we deliberately chose a combination of DNA concentration in the sample and injection conditions suitable for easy detection with conventional CE machines with "relatively low" detection sensitivity (e.g., UV absorption). With more sensitive devices (e.g., laser induced fluorescence), the quantity of DNA injected could be reduced by a factor of 10–100. From earlier studies on field-induced aggregation, reducing the concentration would be interesting to try as a way to significantly increase the range of separable DNA fragment sizes.

References

1. Lerman, L. S. and Frisch, H. L. (1982) Why does the electrophoretic mobility of DNA in gels vary with the length of the molecule? *Biopolymers* **21,** 995–997.
2. Lumpkin, O. J. and Zimm, B. H. (1982) Mobility of DNA in gel electrophoresis. *Biopolymers* **21,** 2315–2316.
3. Slater, G. W. and Noolandi, J. (1985) New biased-reptation model for charged polymers. *Phys.Rev.Lett.* **55,** 1579–1582.
4. Hurley, I. (1986) DNA orientation during gel electrophoresis and its relation to electrophoretic mobility. *Biopolymers* **25,** 539–554.
5. Jonsson, M., Åkerman, B., and Nordén, B. (1988) Orientation of DNA during gel electrophoresis studied with linear dichroism spectroscopy. *Biopolymers* **27,** 381–414.
6. Schwartz, D. C. and Cantor, C. R. (1984) Separation of yeast chromosome-sized DNAs by pulsed field gradient gel electrophoresis. *Cell* **37,** 67–75.
7. Carle, G. F., Frank, M., and Olson, M. V. (1986) Electrophoretic separation of large DNA molecules by periodic inversion of the electric field. *Science* **232,** 65–68.
8. Bostock, C. J. (1988) Parameters of field inversion gel electrophoresis for the analysis of pox virus genomes. *Nucleic Acids Res.* **16,** 4239–4252.
9. Birren, B. W., Lai, E., Hood, L., and Simon, M. I. (1989) Pulsed field gel electrophoresis techniques for separating 1- to 50-kilobase DNA fragments. *Anal. Biochem.* **177,** 282–286.
10. Heller, C. and Pohl, F. M. (1989) A systematic study of field inversion gel electrophoresis. *Nucleic Acids Res.* **17,** 5989–6003.
11. Turmel, C., Brassard, E., Slater, G. W., and Noolandi, J. (1990) Molecular detrapping and band narrowing with high frequency modulation of pulsed field electrophoresis. *Nucleic Acids Res.* **18,** 569–575.
12. Heller, C. (1994) Field Inversion Gel Electrophoresis in *Methods in Molecular Biology, Vol. 31: Protocols for Gene Analysis* (Harwood, J., ed.) Humana Press, Totowa, NJ, pp. 135–146.
13. Heller, C. and Pohl, F. M. (1989) Separation of double stranded linear DNA molecules with direct blotting electrophoresis and field inversion, in *Electrophoresis Forum '89. Proceedings of the International Meeting on Electrophoresis* (Radola, B. J., ed.), München, pp. 194–198.
14. Heller, C. and Beck, S. (1992) Field inversion gel electrophoresis in denaturing polyacrylamide gels. *Nucleic Acids Res.* **20,** 2447–2452.
15. Heller, C. (1997) The separation matrix, in *Analysis of Nucleic Acids by Capillary Electrophoresis* (Heller, C., ed.) Vieweg Verlagsgesellschaft, Wiesbaden, pp. 3–23.
16. Sudor, J. and Novotny, M. V. (1997) Pulsed-field capillary electrophoresis of large DNA, in *Analysis of Nucleic Acids by Capillary Electrophoresis* (Heller, C., ed.) Vieweg Verlagsgesellschaft, Wiesbaden, pp. 218–235.
17. Sudor, J. and Novotny, M. V. (1994) Separation of large DNA fragments by capillary electrophoresis under pulsed-field conditions. *Anal. Chem.* **66,** 2446–2450.

18. Sudor, J. and Novotny, M. (1995) The mobility minima in pulsed-field capillary electrophoresis of large DNA. *Nucleic Acids Res.* **23,** 2538–2543.
19. Heller, C., Pakleza, C., and Viovy, J.-L. (1995) DNA separation with field inversion capillary electrophoresis. *Electrophoresis* **16,** 1423–1428.
20. Kim, Y. and Morris, M. D. (1994) Pulsed field capillary electrophoresis of multikilobase length nucleic acids in dilute methyl cellulose solutions. *Anal. Chem.* **66,** 3081–3085.
21. Kim, Y. and Morris, M. D. (1995) Rapid pulsed field capillary electrophoretic separation of megabase nucleic acids. *Anal. Chem.* **67,** 784–786.
22. Kim, Y. and Morris, M. D. (1996) Ultrafast high resolution separation of large DNA fragments by pulsed-field capillary electrophoresis. *Electrophoresis* **17,** 152–160.
23. Magnúsdóttir, S. (1998) Etude par spectroscopie et electrophorese capillaire de la migration d'ADN en gel et en solution de polymeres. *Doctoral Thesis from L' Institut Curie, Section Recherche, Laboratoire Physico-Chimie-Curie (PCC)*, Paris, France.
24. Guszczynski, T., Pulyaeva, H., Tietz, D., Garner, M. M., and Chrambach, A. (1993) Capillary zone electrophoresis of large DNA. *Electrophoresis* **14,** 523–530.
25. Mitnik, L., Heller, C., Prost, J., and Viovy, J.-L. (1995) Segregation in DNA solutions induced by electric fields. *Science* **267,** 219–222.
26. Schwinefus, J. J., Wang, S. C., Hammond, R., and Morris, M. D. (1996) Fluorescence resonance energy transfer visualization of DNA aggregates formed during electrophoretic separations in ultradilute hydroxyethylcellulose. *Electrophoresis* **17,** 1110–1114.
27. Magnúsdóttir, S., Isambert, H., Heller, C., and Viovy, J.-L. (1999) Electrohydrodynamically induced aggregation during constant and pulsed field capillary electrophoresis of DNA. *Biopolymers* **49,** 385–401.
28. Skeidsvoll, J. and Ueland, P. M. (1995) Analysis of double stranded DNA by capillary electrophoresis with laser-induced fluorescence detection using the monomeric dye SYBR Green I. *Anal. Biochem.* **231,** 359–365.
29. Duke, T. A. J., Semenov, A. N., and Viovy, J.-L. (1992) Mobility of a reptating polymer. *Phys. Rev. Lett.* **69,** 3260–3263.
30. Chiari, M. and Gelain, A. (1997) Developments in capillary coating and DNA separation matrices, in *Analysis of Nucleic Acids by Capillary Electrophoresis* (Heller, C., ed.) Vieweg Verlagsgesellschaft, Wiesbaden, pp. 135–173.

20

Pulsed-Field Capillary Electrophoresis Separation of Large DNA Fragments

Michael D. Morris, Jeffrey J. Schwinefus, and Olivia de Carmejane

1. Introduction

Field-inversion capillary electrophoresis (FICE) provides many of the same benefits as the more familiar, field-inversion gel electrophoresis (FIGE). In the capillary format, field inversion allows size separation of linear dsDNA into the megabase region. Whereas FIGE separation of megabase pair (Mb) DNA takes 24 h or more, the most rapid FICE separations are complete within 4 min *(1,2)*. As with any capillary electroseparation, the technique is rapid and provides high resolution, but is analytical scale only. Preparative scale use of capillary electrophoresis (CE) is not very practical, although apparatus capable of the collection of CE fractions have been developed *(3)*.

There are other differences between separations in gels and capillaries. It is more common to use solutions of water-soluble linear polymers as sieving media in capillaries than to use crosslinked gels. The linear polymer solutions are replaceable media that are easily introduced into a capillary *(4)*. In addition, the operating electric fields are much higher than encountered in gels, because Joule heat can be efficiently dissipated *(5)*.

1.1. Theory of DNA FICE

The theory of DNA DC electrophoresis in polymer solutions is reasonably well understood. Gel electrophoresis models can be extended to the semi-dilute polymer solution regime by assuming that semi-dilute polymers behave as a dynamic network of pores *(6)*. Electrophoretic models in the dilute polymer solution regime are developed by assuming transient entanglements between a DNA molecule and one or more entangling polymer molecules *(7–9)*. The transient aggregation effectively delivers a constant increment of friction that introduces size selectivity into electrophoresis in dilute polymer solution *(10)*.

From: *Methods in Molecular Biology, Vol. 162:*
Capillary Electrophoresis of Nucleic Acids, Vol. 1: Introduction to the Capillary Electrophoresis of Nucleic Acids
Edited by: K. R. Mitchelson and J. Cheng © Humana Press Inc., Totowa, NJ

In semi-dilute polymer solutions, dsDNA longer than about 500–600 bp migrates according to the predictions of the reptation model ***(6,11–14)***. Linear DNA is envisioned moving along its backbone, constrained in a tube defined by neighboring gel or polymer molecules. The DNA chain assumes the configuration of a random walk, biased by the external electric field. At high electric fields, biased reptation of DNA molecules occurs ***(6,12–14)***. The DNA chain stretches, increasing its orientation parallel to the electric field, losing size-dependent mobility in a gel or linear polymer network. **Equation 1** describes the biased reptation mobility (μ) of a dsDNA molecule where E is the electric field and b is a function of the density of the polymer network and charge and stiffness of the migrating DNA molecule.

$$\mu \approx (1/N + bE^2) \qquad (1)$$

Reptation is an oversimplification, but a useful one. Video imaging of fluorescence-labeled dsDNA molecules ***(15–21)*** and numerical simulations ***(14,22,23)*** has shown that linear dsDNA, moves cyclically in DC electric fields. Linear dsDNA migrates as a random coil until collision with polymers or gel fibers. With DNA-polymer entanglement, the dsDNA molecule extends both arms in a U or J shape. One arm lengthens as the other shortens until the molecule slips off the obstacle and collapses back to the random coil. This extension-contraction motion is continually repeated through the gel or polymer medium. These electrophoretic dynamics are unlike the theoretical snake-like motion of reptation.

Chain extension, which can be approximated as biased reptation, limits DC electrophoresis of linear dsDNA, in semi-dilute or dilute polymer solutions, to a size of approx 30–50 kb. Pulsed electric fields in gel electrophoresis were introduced to overcome the chain extension by reorienting DNA molecules as the electric field bias changes in direction ***(24)***. Numerical simulations ***(14,22,23)*** and direct observation ***(19–21)*** have shown that the reorientation of linear dsDNA in an oscillating electric field is a complex motion.

Measurements by Sudor and Novotny ***(25,26)*** show that the extension limitation mechanism in field-inversion electrophoresis using semi-dilute polymer solutions cannot be the same as in gels. Anti-resonance, a mobility minimum at some field-inversion frequency is observed in semi-dilute linear polyacrylamide solution. Band inversion (longer DNA migrating more rapidly than shorter DNA) also did not occur. They suggested that the observed differences arise because the dynamic polymer pore structure can relieve strong entanglements on long DNA molecules. This does not happen during entanglement with a rigid network of gel fibers. However, both anti-resonance and band inversion have been successfully explained by numerical simulations and reptation theory in gels ***(22,27,28)***.

The mechanism of field inversion in dilute polymer solution is significantly different from that in semi-dilute polymer solutions or gels ***(19,20)***. Because of the lack of a rigid network, the tube constraint of reptation fails in dilute polymer solutions. The migrating DNA molecule briefly interacts with discrete polymer obstacles. Polymers become entangled transiently along the DNA chain and are dragged through solution as the DNA continues to migrate. The lifetime of these polymer entanglements along the DNA chain is a function of both polymer and DNA disengagement time ***(10)***.

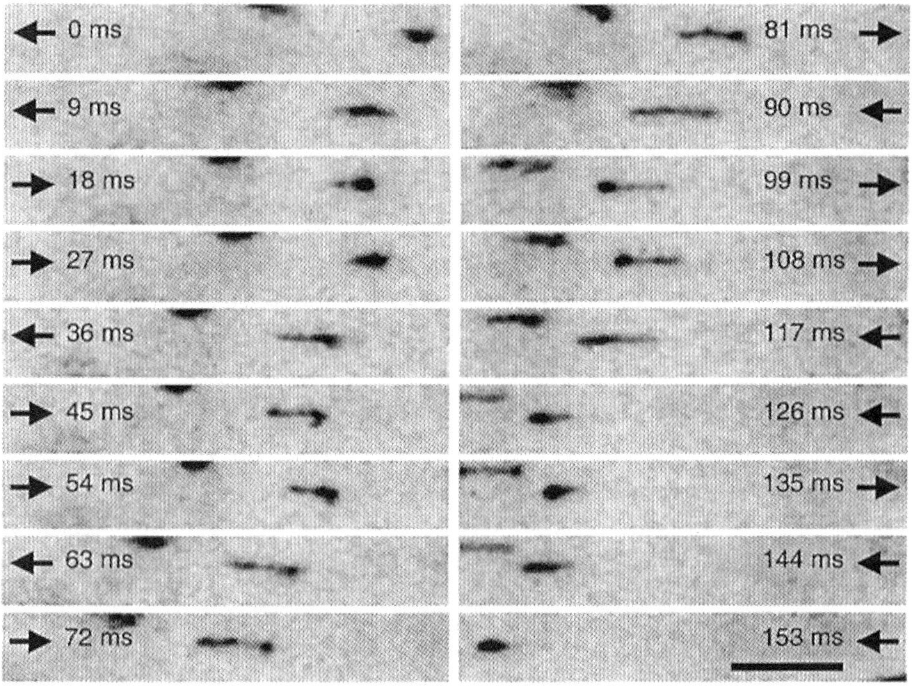

Fig. 1. Migration of λ-DNA in 0.008% HEC (M_n = 438,800), 1X TBE and 100 V/cm under field inversion conditions: 180% modulation (280 V/cm forward, –80 V/cm backward), and 35-Hz square wave. Recorded at 222 frames/s, every other frame displayed. Arrows indicate direction of migration. Scale bar = 10 μm. Reprinted with permission from **ref.** ***29***, Copyright (1998) American Chemical Society.

Although the earliest fluorescence microscopy work showed reorientation of DNA in dilute hydroxyethyl cellulose (HEC) during field inversion, these measurements were not made under realistic separation conditions ***(19)***. Nevertheless, U and J linear DNA shapes were shown to persist during field inversion. Later work showed that 167 kb T2 phage DNA did not completely reorient during field inversion under conditions closer to actual analytical separation fields and frequencies ***(20)***. Recent high frame rate (220 frames/s) video microscopy shows that during field-inversion electrophoresis of 48.5 kb phage λ DNA in dilute HEC, there is little change in DNA extension and no reorientation ***(29)***. The field-inversion extension-collapse of phage λ DNA in hydroxyethyl cellulose (HEC) is shown in **Fig. 1**. The arms of the U or J shape retract only slightly as the field is reversed. The magnitude of the retraction increases with increasing pulse period, but the time average extension of the DNA molecule is the same as with DC electric fields.

Schwinefus and Morris ***(29)*** have proposed that in dilute polymer solutions, field inversion serves to selectively remove weak polymer entanglements along the DNA chain. The effect is to decrease band broadening caused by dispersion in the average mobility of individual DNA molecules. As the entire DNA molecule begins to move backward in the electric field, it disengages from weakly bound neutral polymers that remain stationary.

This explanation is consistent with the recent theory of the widths of fragment bands (bandwidths) of biopolymers during electrophoresis ***(30,31)***. The Weiss theory explains the widths of bands and the asymmetric shapes of bands as arising from mobility dispersion caused by variations in average entanglement time of a protein (or other migrating polymer) and the entangling matrix. The model is an extension of highly successful theories of chromatographic band broadening. Numerical simulations have predicted weak or short-lived polymer interactions with DNA, during dilute polymer solution electrophoresis ***(32)***. With field reversal and retraction of the DNA arms, weak polymer interactions along the DNA chain can be removed.

High frame rate imaging of phage λ dsDNA in dilute HEC does show that DNA extension and collapse is more periodic than in DC electrophoresis, as would be expected if weak polymer entanglements were removed ***(29)***. Field inversion increases mobility, but most of the increase is simply due to Joule heating ***(33)***. The improved resolution is a consequence of reduced bandwidths ***(1,2,33–35)***. Bandwidth reduction also explains why field inversion improves resolution in sequencing and in separation of small restriction fragments ***(36–38)***.

Thus, field inversion may operate by two different mechanisms: restriction of DNA extension and constraint of DNA/polymer entanglements. Both mechanisms operate in semi-dilute and dilute polymer solutions. The first mechanism contributes more to successful field inversion separations of linear dsDNA in semi-dilute polymer solutions than constraint of DNA/polymer entanglements.

2. Materials

2.1. Apparatus for FICE

1. The electric field is provided by a high-voltage amplifier (Model 20/20; TREK, Inc., Medina, NY) controlled by a personal computer running LabView (National Instruments, Austin, TX) and containing a data acquistion board (Model PCI 6024E; National Instruments). The amplifier is controlled by the DAC output of the data acquisition board (1 V DAC output = 2-kV amplifier output) using pulse sequences written in LabView.
2. **SAFETY**: It is important to keep the high voltage end of the apparatus inside a safety chamber and to implement the amplifier interlock system, which turns off the high voltage when the chamber is opened. **The high voltage and current output of the TREK 20/20 and most other similar commercial amplifiers can be lethal.**
3. Laser-induced fluorescence (LIF) is induced with a green helium-neon laser (543 nm) output 1 mW HeNe laser (Model LHGR-0200M, Boulder, CO). Laser light is focused onto the capillary using a 10X microscope objective. Argon-ion or Helium-cadmium (He/Cd) lasers may also be used.
4. Fluorescence is collected through the microscope objective, filtered through a 600-nm long pass dielectric filter and focused onto a photomultiplier tube (Model 928; Hamamatsu, Bridgewater, NJ) and amplified by a current amplifier (Model 70710; Oriel, Stratford, CT).
5. The amplifier output is sampled and digitized 6 times/s by the data acquisition board and stored in the personal computer using a control program written in LabView. Electric

field is provided by high-voltage amplifier, controlled by personal computer, which is linked via a computer digital-analog converter (DAC) board. A typical commercial amplifier is compatible with the output of a laboratory DAC. FICE requires currents of only 10–50 μA ***(44)*** and frequency response below 1 kHz.

2.2. Capillary

1. Fused silica capillary (75 μm id, 350 μm od, polyimide coating, PolyMicro Technologies, Phoenix, AZ) is cut to 25–30-cm length. About 3 cm from one end a fluorescence observation window is formed by burning away about 1 cm of coating with a propane torch. An oxygen/hydrogen torch can be used, if available.
2. **Caution:** The removal of the polyimide coating leaves the capillary brittle and easily broken.
3. The inner wall is coated with polyacrylamide to reduce surface charge and stabilize electroosmotic flow.
4. A vacuum pump is used to pull solutions through the capillary in the following sequence of solutions:
   a. 15 min with 1 *M* sodium hydroxyde (NaOH) then,
   b. 15 min with deionized (DI) water,
   c. 15 min with methanol (MeOH),
   d. 2–3 h with a solution of 80 μL methacryloxypropyltrimethoxysilane (MPTS) in 20 mL DI water with 1 mL MeOH and one drop glacial acetic acid, then,
   e. 15 min with MeOH,
   f. 15 min with DI water.
5. The actual polymerization of PAA is accomplished by filling the capillary with degassed, acrylamide solution. Ammonium persulfate (APS) and tetramethylenediamine (TEMED) are needed to trigger the initiation (20 μL TEMED, 20 mg of APS, 700 mg of PAA, 20 mL of DI water).

3. Methods

3.1. Sieving Polymers, Buffers, and FICE Conditions

1. Different sieving polymers, buffers, and FICE conditions are used for different size ranges of DNA. The most common formulations employed by the authors and others are described. It may be necessary to modify these protocols for different DNA size ranges.
2. The threshold for aggregate formation is DNA concentration dependent. Under typical dilute polymer solution FICE conditions, the threshold is about 3–4 ng/μL.

3.1.1. DNA Sequencing (37) and DNA Fragments Under 1 kb

1. Sieving matrix: 1.5% (M_r = 8,000,000) and 1.4% (M_r = 600,000) linear PEO gel in 1X TBE with 3.5 *M* urea.
2. Field strength: 75-V/cm average DC.
3. Field inversion conditions: 125% modulation (V_{ac}/V_{dc}) (243 V/cm forward, –93 V/cm backward), and 1.2-kHz square wave.

3.1.2. DNA from 1 to 50 kb

3.1.2.1. Protocol I *(1)*

1. Sieving matrix: 0.0033% (w/w) HEC (M_n = 438 800) in 0.5X TBE (Tris-borate-EDTA buffer), 3 μg/mL ethidium bromide (EtBr).

2. Field strength: 200-V/cm average DC electric field.
3. Field-inversion conditions: 240% modulation (680 V/cm forward, –280 V/cm backward), and 35-Hz square wave.

3.1.2.2. Protocol II *(42)*

1. Sieving matrix: 1% Hydroxypropyl cellulose (HPC) (mol wt = 1 MDa) in 1X TBE, 10 µ*M* EtBr.
2. Field strength: 43.5-V/cm average DC electric field.
3. Field-inversion conditions: 200% modulation (130 V/cm forward, –43 V/cm backward), and 20-Hz square wave.

3.1.2.3. Protocol III *(25)*

1. Sieving matrix: 0.6% linear polyacrylamide (mol wt = 5–6 MDa) in 0.5X TBE.
2. Field strength: 100-V/cm average DC electric field.
3. Field-inversion conditions: 50% modulation (150 V/cm forward, 50 V/cm backward), and 15-Hz square wave.

3.1.3. DNA from 50 kb to 1.6 Mb

3.1.3.1. Protocol IV *(1)*

1. Sieving matrix: mixed polymers, 0.0035% HEC (M_n = 438,800), 0.0040% poly(ethylene oxide) (PEO) (M_n = 8 MDa), 0.5X TBE, 3 µg/mL EtBr. Since PEO is significantly heavier and longer than HEC, it is a better choice for the separation of megabase size DNA fragments. Separation of megabase DNA was not effective in HEC alone.
2. Field strength: 200-V/cm average DC electric field.
3. Field-inversion conditions: 300% modulation (800 V/cm forward, –400 V/cm backward), and 35-Hz square wave.

3.1.3.2. Protocol V *(25)*

1. Sieving matrix: 0.4% linear PAA (mol wt = 5–6 MDa), 0.5X TBE, 0.075 µg/mL EtBr.
2. Field strength: 75-V/cm average DC electric field.
3. Field-inversion conditions: 53% modulation (100 V/cm forward, 50 V/cm backward), and 30-Hz square wave.

3.1.4. Fast Separation of Megabase Size DNA, in less than 4 min **(1)**

1. Sieving matrix: 0.0033% HEC/0.00405% PEO, 0.5X TBE, 3 µg/mL EtBr.
2. Field strength: 200-V/cm average DC electric field.
3. Field-inversion conditions: The optimum pulse frequency is 35 Hz, with a modulation depth of 300% (800 V/cm forward, –400 V/cm backward). The capillary length is decreased down to 10 cm.

3.1.5. Separation of Linear dsDNA from 0.21 to 1.6 Mb

1. Sieving buffer: Dilute mixed polymer system of 0.002% (w/w) PEO ($M_n = 8 \times 10^6$) and 0.00375% (w/w) HEC (M_n = 438,800).
2. Field strength: 100-V/cm average DC.
3. Field-inversion conditions: a symmetric square wave input field, 250% modulation (V_{ac}/V_{dc}) with a pulse frequency of 14 Hz.

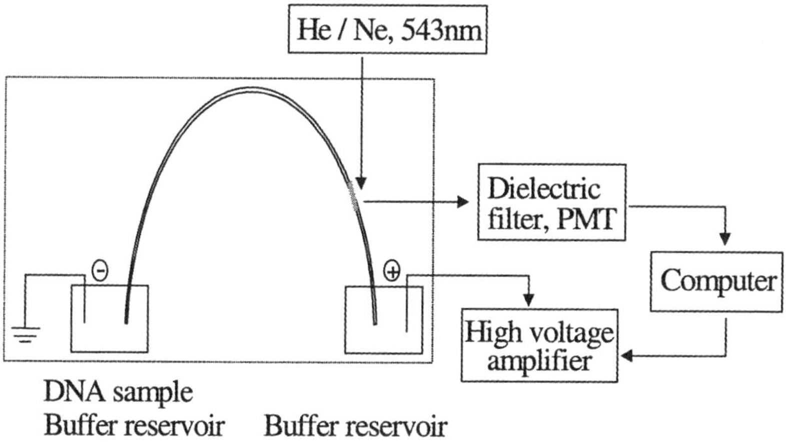

Fig. 2. Apparatus for FICE. Electric field is provided by high-voltage amplifier, controlled by personal computer. LIF (He/Ne 543 nm 1 mW), photomultiplier (PMT) detection.

3.1.6. Separation of DNA Fragments <10 kb

1. Sieving buffer: 6% Total linear polyacrylamide in 1X TBE.
2. Field strength: 100-V/cm average DC.
3. Field-inversion conditions: single frequency sine waves superimposed on a DC field. The optimum frequency is about 80 Hz.
4. Field modulation is most effective for DNA fragments above 600 bp.

3.2. FICE Conditions

1. All FICE has been performed with purpose-built instruments, such as shown in **Fig. 2** (*see* **Notes 1–4**). Except for the voltage source, instrumentation is conventional for DC CE.
2. For FICE, a high-voltage amplifier replaces the high-voltage DC power supply ***(1,2,25, 26,33–43)***. The amplifier is usually driven by a computer DAC board. Computer control allows easy implementation of almost any pulse protocol. Pulse protocols that have been used in the laboratory of the authors are shown in **Fig. 3** ***(1)***. Other laboratories have employed similar protocols ***(25,26,33–43)***.
3. As in any CE experiment, it is important to keep the high-voltage end of the capillary inside a safety enclosure with interlock that shuts off the high voltage whenever the enclosure is open.
4. Although commercial high-voltage amplifiers are quite satisfactory for FICE (*see* **Notes 5–7**), most of them have capabilities beyond what is needed for this experiment. The low-voltage input (typical 0 to ± 10 V) of a typical commercial amplifier is compatible with the output of a laboratory DAC. However, most amplifiers can output currents of 1–20 mA, depending on the model and operate at frequencies as high as 10–20 kHz. FICE requires currents of only 10–50 μA ***(44)*** and frequency response below 1 kHz. Commercial high-voltage amplifiers are larger and more expensive than necessary for the requirements of FICE.

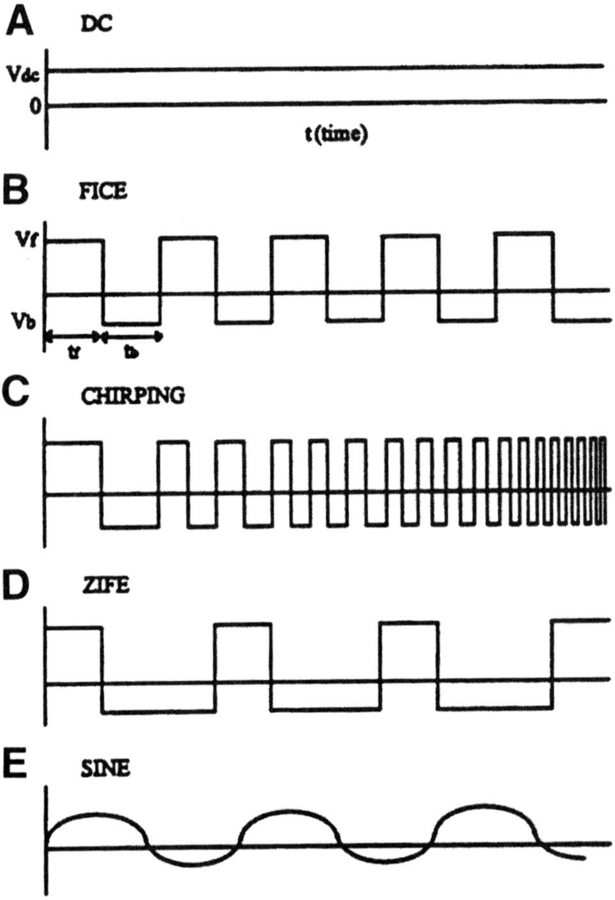

Fig. 3. Waveforms used for pulsed-field CE. **(A)**, DC field; **(B)**, FICE, with higher forward voltage than reverse, but equal pulse durations; **(C)**, chirping mode with increasing frequency; **(D)**, zero integrated field electrophoresis (ZIFE), with the product of voltage and time duration equal for both directions; and **(E)**, sine wave. Reprinted with permission from **ref.** ***1***.

5. Detection is usually by LIF after intercalation with dyes such as ethidium bromide ***(1,2,4,25,26,33–38,40)***. Low power green (543.5 nm) HeNe (helium-neon) lasers are the most popular, but a few laboratories use argon ion or HeCd (helium-cadmium) lasers. The HeNe laser is class IIIa. Eye protection is advisable, but accidental brief exposure to the beam is unlikely to cause permanent damage.
6. UV absorbance detection is less popular ***(25,26,39,41,42)***. FICE has often been performed at concentrations near or below absorbance detection limits. Please note that most absorbance detectors are not adequately electrically shielded for use near high-voltage AC circuitry, and consequently may not operate properly in a FICE experiment.

3.3. Application of FICE

1. Pulsed-field techniques have proven very powerful for the separation of large linear dsDNA where DC electric field capillary electrophoresis has failed. A wide range of different parameters in FICE promotes fast, reliable and reproducible separations of large DNA fragment mixtures. Although most reports have mentioned nucleic acid separations, FICE has been shown to improve resolution of polysaccharides as well.
2. The first report of DNA separations using FICE is by Heiger and coworkers ***(39)***. Plasmids pBR322 (4363 bp) and M13mp18 (7253 bp) DNA fragments are separated in 6% T linear polyacrylamide. The optimum resolution is obtained with a symmetric square wave of amplitude 0–300 V/cm at a frequency of 100 Hz.
3. Variations in FICE separation matrix (gel or dilute polymer solution), DNA size (DNA sequencing vs DNA mapping), and pulse protocol (waveform, modulation depth and field frequency) have been applied. Demana et al. ***(40)*** separated and resolved 72–1353 bp fragments in crosslinked polyacrylamide using single frequency sine waves superimposed on a DC field. The field modulation is most effective for DNA fragments above 600 bp and the optimum frequency is about 80 Hz.
4. Improved separations of intermediate sized DNA have been achieved with FICE. Kim and Morris ***(33)*** and Chen et al. ***(41)*** separated DNA restriction fragments up to 23 kb with FICE. Kim and Morris use dilute polymer solutions as the separation matrix, while Chen et al. use agarose gel as a sieving matrix. Both media provided baseline-resolved separations of DNA fragments.
5. Kim and Morris observe a sharpening of electrophoretic bands and an enhancement in separation resolution using FICE compared to DC CE ***(33)***. Similar band sharpening is observed with the ϕX/174 H*ae*III digest in 9% linear polyacrylamide ***(38)***. The increasing mobilities during FICE are due to the increased Joule heating of the solution. Separations are generally faster in dilute polymer solutions than in either semi-dilute or in concentrated polymer solutions because of the low viscosity of the separation medium.
6. Sudor and Novotny ***(25)***, Heller et al. ***(2,42)***, and Kim and Morris ***(1)*** obtain separations of linear dsDNA up to 50 kb with FICE. Both Sudor and Novotny ***(25)*** and Heller et al. ***(42)*** use semi-dilute polymer solutions of polyacrylamide and hydroxypropyl cellulose (HPC), respectively, and Kim and Morris ***(1)*** use dilute HEC solutions as the separation matrix.
7. The separation of megabase size DNA fragments is achieved by both Sudor and Novotny ***(25)*** and Kim and Morris ***(1,34)*** using FICE. Sudor and Novotny ***(25)*** separate megabase phage λ DNA concatamers in the size range 48.5 kbp to 1 Mb in semi-dilute polyacrylamide solutions. Optimum separation and resolution are obtained with an applied voltage of 100 V/cm in the forward direction and –50 V/cm in the reverse direction. By use of a symmetric square wave input signal at a pulse frequency of 30 Hz, the resolution of fragments improves dramatically. Besides the main fragments, there are additional smaller DNA peaks that probably correspond to DNA breakage. The authors suggest that at the field intensities used in their work, longer DNA chains might break.
8. The separation time for 1-Mb fragments is close to 3 h. In contrast, similar separations in slab gels take 24 h or more to achieve. **Figure 4** shows the separation of phage λ DNA concatamers under these optimized conditions.
9. In 1995, Kim and Morris ***(34)*** separated linear dsDNA from 0.21–1.6 Mb in dilute polymer solutions using FICE (*see* **Notes 8–10**). Their separation matrix is a mixed system of poly(ethylene oxide) (PEO, $M_n = 8 \times 10^6$) and HEC ($M_n = 438{,}800$). Separation of megabase DNA is not effective in HEC alone. Since PEO is significantly heavier and

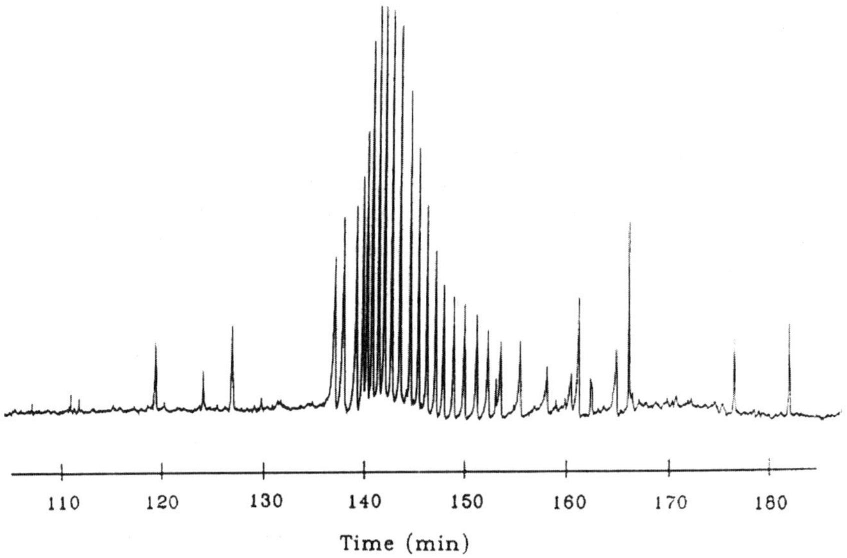

Fig. 4. Separation of λ DNA concatamers (48.5 kb to 1 Mb) by FICE in 0.4% linear polyacrylamide (mol wt = 5–6 × 10^6), 0.5X TBE, 0.075 µg/mL ethidium bromide and 100 V/cm, 400% modulation (100 V/cm forward, –50 V/cm backward), and 30-Hz square wave. Reprinted with permission from **ref. *29***, Copyright (1998) American Chemical Society.

longer than HEC, it is a better choice for the separation of megabase size DNA fragments. The smaller HEC polymer is more suited for the separation of DNA fragments up to 100–200 kbp. Optimum separation conditions are achieved with a matrix of 0.00375% (w/w) HEC + 0.002% (w/w) PEO, and a symmetric square wave input at 100-V/cm average DC field, 250% modulation (V_{ac}/V_{dc}), and with a pulse frequency of 14 Hz. Separation times are under 13 min.

10. In 1996, Kim and Morris *(1)* achieved an even faster separation time of megabase size DNAs. By increasing the average DC electric field up to 200 V/cm with a modulation depth of 300% and decreasing the length of the capillary down to 10 cm, the separation time is reduced to less than 4 min. The optimum pulse frequency for these conditions is 35 Hz. An electropherogram of this separation is shown in **Fig. 5**. Attempts to reduce the separation time yet further to 2–2.5 min were not completely successful *(45)*.
11. These separation times are the fastest yet reported in the megabase size DNA range. High-speed separations result from high field strengths, short capillaries and dilute, low-viscosity polymer solutions. According to the authors, further resolution improvement is possible through modification of the field inversion protocols.
12. FICE has recently also been applied to improvement of DNA sequencing. Kim and Young *(36,37)* achieve separation of Sanger DNA fragments up to 1000 bp long by using pulsed-field capillary electrophoresis in semi-dilute linear PEO solutions. Achieving good single base-pair resolution of DNA sequencing products over a 1-kb length is a strong function of the frequency and waveform of the pulsed fields. The capillary temperature also has a strong effect on the sequencing efficiency and speed.

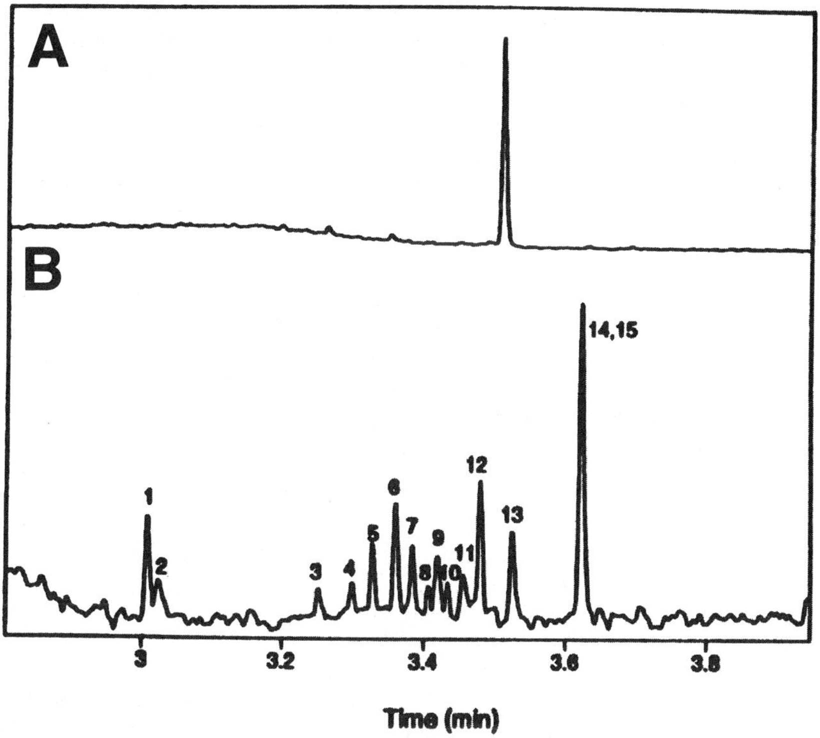

Fig. 5. Electrophoresis separation of megabase I DNA standard in 0.0033% HEC/0.00405% PEO, 0.5X TBE: **(A)**, 200 V/cm DC only; **(B)**, pulsed field, 200 V/cm DC +35 Hz square wave, 300% modulation. Reprinted with permission from **ref.** ***1***.

13. FICE has recently been shown to improve the resolution between polysaccharide molecules. Sudor and Novotny ***(43)*** reported the highly efficient separation of polydextrans in the size range 8000–2,000,000 Dalton. The application of a potential gradient along the separation capillary, periodically reversed, favors the separation of polysaccharides according to their molecular size. The potential gradient induces shape transitions of the negatively charged polysaccharides. In DC electric fields, these polymers would be stretched and separation would be impossible, similar to the situation with long DNA fragments.

4. Notes

1. One side effect of FICE is the additional Joule heating, caused by the AC component of the driving electric field. This problem was first pointed out by Demana and coworkers ***(40)***. The effect is described in **Eq. 2** in terms of the DC driving voltage, V_{dc}, the RMS component of the applied voltage, V_{rms}, and the internal resistance of the capillary, R.

$$P = (V_{dc} + V_{rms})/R \tag{2}$$

2. The expression for RMS voltage depends upon the form of the AC voltage (square wave, sine wave, or asymmetric pulses). The Joule heating caused by the AC component of the electrophoresis voltage can exceed the heating caused by the DC component under typical field inversion conditions.
3. Joule heating is difficult to model accurately *(46)*. In addition to the auto-thermal effect (thermal runaway), forced air-cooling complicates any theory or simulation. The internal temperatures of operating capillaries have been measured directly *(47)* and it has been shown that there are both longitudinal and transverse temperature gradients *(48)*.
4. Thermal effects in forced air-cooled capillaries were studied by Navin and coworkers *(38)*. Comparing isorheic (same heat dissipation) DC and FICE, they found that the mobility changes during FICE were almost entirely caused by Joule heating. With efficient forced air-cooling or with the use of heat sinks *(49,50)* capillary temperature, fluctuations of well under ±1°C are feasible. For DNA mapping, close temperature control is probably not important. However, if FICE is used to increase the number of bases called in a sequencing separation, then it is important to keep temperature fluctuations small.
5. It is difficult to maintain waveform fidelity in a capillary because of the long RC (resistance-capacitance) time constant of the capillary. The effect has been studied by Heiger and coworkers *(51)*, who modeled the capillary as a distributed RC network. The capillary/ground capacitance is about 0.05 pF/cm, or about 1.3 pF for the typical 25-cm capillary. However, the internal resistance of the filled capillary is 10^8–10^9 ohms, depending on the buffer ionic strength and capillary inner diameter. Consequently, the system time constant is 0.1–1 ms. The parasitic capillary/ground capacitance distorts waveforms above about 100 Hz.
6. The problem is exacerbated by use of an absorbance detector, which can roughly double the system capacitance. LIF detection should have no adverse effects. In conventional fluorescence detection schemes, there is either no detector component in direct contact with the capillary, or else only glass or quartz optical fibers and ball lenses.
7. The simplest approach to minimizing the problem is to use a short capillary and to keep the buffer ionic strength as high as practical. An effective way to minimize waveform distortion is to use a driven shield. Heiger and coworkers *(51)* found that a 2.54-cm id Plexiglas tube with an outer wall covered with aluminum foil was satisfactory. The foil must be connected to the high-voltage supply so that it is maintained at the same driving voltage as the capillary.
8. In dilute polymer solutions, DNA may aggregate during electrophoresis. Aggregation occurs as a result of local electroosmotic effects caused by the large and slow-moving nucleic acid. Large aggregates can form. If a detector of high spatial resolution is used, such as a LIF detector, characteristic artifacts are observed. A series of sharp spikes is seen as the aggregates pass through the detector window.
9. Aggregation is exacerbated by oscillating electric fields because the fields lead to unstable hydrodynamic patterns *(52)*. In semi-dilute polymer solutions or in gels, hydrodynamic instabilities are suppressed because electroosmosis is suppressed. Aggregation is therefore absent or minor under these conditions.
10. There is a threshold DNA concentration for aggregate formation. Under typical dilute polymer solution FICE conditions, the threshold is about 3–4 ng/µL. It is best to work at concentrations below the aggregation threshold, although size separation is possible in the presence of some aggregation, because aggregation does not begin immediately at the start of electrophoresis. In dilute polymer solution, the onset time is about 30 s. During high-speed separations, the effect of aggregation is often to cause each band to appear as a series of sharp spikes because aggregates form after the DNA fragments have separated.

References

1. Kim, Y. and Morris, M. D. (1996) Ultrafast high resolution separation of large DNA fragments by pulsed-field capillary electrophoresis. *Electrophoresis* **17,** 152–160.
2. Heller, C., Magnúsdóttir, S., and Viovy, J.-L. (2001) Robust field inversion capillary electrophoretic separation of long DNA fragments, in *Capillary Electrophoresis of Nucleic Acids,* Vol. 1 (Mitchelson, K. R. and Cheng, J., eds.), Humana Press, Totowa, NJ, pp. 293–305.
3. Magnúsdóttir, S., Heller, C., Sergot, P., and Viovy, J.-L. (2001) Collection of capillary electrophoresis fractions on a moving membrane, in *Capillary Electrophoresis of Nucleic Acids,* Vol. 1 (Mitchelson, K. R. and Cheng, J., eds.), Humana Press, Totowa, NJ, pp. 323–331.
4. Ulfelder, K. J. and McCord, B. R. (1997) Separation of DNA by capillary electrophoresis, in *Handbook of Capillary Electrophoresis, 2nd ed.* (Landers, J. P., ed.), CRC Press, Boca Raton, pp. 347–378.
5. Grossman, P. D. (1992) Factors affecting the performance of capillary electrophoresis separations: Joule heating, electroosmosis, and zone dispersion, in *Capillary Electrophoresis: Theory and Practice* (Grossman, P. D. and Colburn, J. C., ed.), Academic Press, San Diego, pp. 3–43.
6. Viovy, J.-L. and Duke, T. (1993) DNA electrophoresis in polymer solutions: Ogston sieving, reptation, and constraint release. *Biopolymers* **24,** 1573–1593.
7. Barron, A. E., Soane, D. S., and Blanch, H. W. (1993) Capillary electrophoresis of DNA in uncross-linked polymer solutions. *J. Chromatogr. A* **652,** 3–16.
8. Barron, A. E., Blanch, H. W., and Soane, D. S. (1994) A transient entanglement coupling mechanism for DNA separation by capillary electrophoresis in ultradilute polymer solutions. *Electrophoresis* **15,** 597–615.
9. Barron, A. E., Sunada, W. M., and Blanch, H. W. (1996) The effects of polymer properties on DNA separations by capillary electrophoresis in uncross-linked polymer solutions. *Electrophoresis* **17,** 744–757.
10. Hubert, S. J., Slater, G. W., and Viovy, J.-L. (1996) Theory of capillary electrophoretic separation of DNA using ultradilute polymer solutions. *Macromolecules* **29,** 1006–1009.
11. Lumpkin, O. J., Dejardin, P., and Zimm, B. H. (1985) Theory of gel electrophoresis of DNA. *Biopolymers* **24,** 1573–1593.
12. Slater, G. W. and Noolandi, J. (1985) New biased-reptation model for charged polymers. *Phys. Rev. Lett.* **55,** 1579–1583.
13. Slater, G. W. and Noolandi, J. (1986) On the reptation theory of gel electrophoresis. *Biopolymers* **25,** 431–454.
14. Duke, T. (1993) Molecular mechanisms of DNA electrophoresis. *Int. J. Genome Res.* **1,** 227–247.
15. Smith, S. B., Aldridge, P. K., and Callis, J. B. (1989) Observation of individual DNA molecules undergoing gel electrophoresis. *Science* **243,** 203–206.
16. Oana, H., Masubuchi, Y., Matsumoto, M., Doi, M., Matsuzawa, Y., and Yoshikawa, K. (1994) Periodic motion of large DNA molecules during steady field gel electrophoresis. *Macromolecules* **27,** 6061–6067.
17. Shi, X., Hammond, R. W., and Morris, M. D. (1995) DNA conformational dynamics in polymer solutions above and below the entanglement limit. *Anal. Chem.* **67,** 1132–1138.
18. Carlsson, C., Larsson, A., Jonsson, M., and Nordén, B. (1995) Dancing DNA in capillary solution electrophoresis. *J. Am. Chem. Soc.* **117,** 3871–3872.
19. Shi, X., Hammond, R. W., and Morris, M. D. (1995) Dynamics of DNA during pulsed field electrophoresis in entangled and dilute polymer solutions. *Anal. Chem.* **67,** 3219–3222.

20. Hammond, R. W., Shi, X., and Morris, M. D. (1996) Dynamics of T2 DNA during capillary electrophoresis in entangled and ultradilute hydroxyethyl cellulose solutions. *J. Microcolumn Sep.* **8,** 201–210.
21. Masubuchi, Y., Oana, H., Matsumoto, M., Doi, M., and Yoshikawa, K. (1996) Conformational dynamics of DNA during biased sinusoidal field gel electrophoresis. *Electrophoresis* **17,** 1065–1074.
22. Duke, T. A. J. (1989) Tube model of field-inversion electrophoresis. *Phys. Rev. Lett.* **62,** 2877–2880.
23. Deutsch, J. M. (1990) Theoretical aspects of electrophoresis, in *Electrophoresis of Large DNA Molecules: Theory and Applications* (Lai, E. and Birren, B. W., eds.), Cold Spring Harbor Laboratory Press, Plainview, NY, pp. 81–99.
24. Schwartz, D. C. and Cantor, C. R. (1984) Separation of yeast chromosome-size DNAs by pulsed field gradient gel electrophoresis. *Cell* **37,** 67–75.
25. Sudor, J. and Novotny, M. V. (1994) Separation of large DNA fragments by capillary electrophoresis under pulsed-field conditions. *Anal. Chem.* **66,** 2446–2450.
26. Sudor, J. and Novotny, M. V. (1995) The mobility minima in pulsed-field capillary electrophoresis of large DNA. *Nucleic Acids Res.* **23,** 2538–2543.
27. Viovy, J.-L. (1987) Pulsed electrophoresis: some implications of reptation theories. *Biopolymers* **26,** 1929–1940.
28. Viovy, J.-L. (1989) Reptation-breathing theory of pulsed electrophoresis: Dynamic regimes, antiresonance and symmetry breakdown effects. *Electrophoresis* **10,** 429–441.
29. Schwinefus, J. J. and Morris, M. D. (1998) Periodicity of λ DNA motions during field inversion electrophoresis in dilute hydroxyethyl cellulose visualized by high-speed video fluorescence microscopy. *Macromolecules* **32,** 3678–3684.
30. Weiss, G. H., Sokoloff, H., Zakharov, S. F., and Chrambach, A. (1996) Interpretation of electrophoretic band shapes by a partition chromatographic model. *Electrophoresis* **17,** 1325–1332.
31. Yarmola, E., Calabrese, P. P., Chrambach, A., and Weiss, G. H. (1997) Interaction with the matrix: the dominant factor in macromolecular band spreading in gel electrophoresis. *J. Phys. Chem. B* **101,** 2381–2387.
32. Starkweather, M. E., Muthukumar, M., and Hoagland, D. A. (1998) Single chain entanglement: a monte carlo simulation of dilute solution capillary electrophoresis. *Macromolecules* **31,** 5495–5501.
33. Kim, Y. and Morris, M. D. (1994) Pulsed field capillary electrophoresis of multikilobase length nucleic acids in dilute methyl cellulose solutions. *Anal. Chem.* **66,** 3081–3085.
34. Kim, Y. and Morris, M. D. (1995) Rapid pulsed field capillary electrophoretic separation of megabase nucleic acids. *Anal. Chem.* **67,** 784–786.
35. Schwinefus, J. J. and Morris, M. D. (1998) Band broadening during DC and field inversion capillary electrophoresis of λ DNA in dilute hydroxyethyl cellulose solutions. *Analyst* **123,** 1481–1485.
36. Kim, Y. S. and Yeung, E. S. (1997) Separation of DNA sequencing fragments up to 1000 bases by using poly(ethylene oxide)-filled capillary electrophoresis. *J. Chromatogr. A* **781,** 315–325.
37. Kim, Y. S. and Yeung, E. S. (1997) DNA sequencing with pulsed-field capillary electrophoresis in poly(ethylene oxide) matrix. *Electrophoresis* **18,** 2901–2908.
38. Navin, M. J., Rapp, T. L., and Morris, M. D. (1994) Variable frequency modulation in DNA separations using field inversion capillary gel electrophoresis. *Anal. Chem.* **66,** 1179–1182.

39. Heiger, D. N., Cohen, A. S., and Karger, B. L. (1990) Separation of DNA restriction fragments by high performance capillary electrophoresis with low and zero crosslinked polyacrylamide using continuous and pulsed electric fields. *J. Chromatogr.* **516,** 33–48.
40. Demana, T., Lanan, M., and. Morris, M. D. (1991) Improved separation of nucleic acids with analyte velocity modulation capillary electrophoresis. *Anal. Chem.* **63,** 2795–2797.
41. Chen, N.,Wu, L., Palm, A., Srichaiyo, T., and Hjertén, S. (1996) High-performance field inversion capillary electrophoresis of 0.1-23 kbp DNA fragments with low-gelling, replaceable agarose gels. *Electrophoresis* **17,** 1443–1450.
42. Heller, C., Pakleza, C., and Viovy, J.-L. (1995) DNA separation with field inversion capillary electrophoresis. *Electrophoresis* **16,** 1423–1428.
43. Sudor, J. and Novotny, M. (1993) Electromigration behavior of polysaccharides in capillary electrophoresis under pulsed-field conditions. *Proc. Natl. Acad. Sci. USA* **90,** 9451–9455.
44. Oda, R. P. and Landers, J. P. (1997) Introduction to capillary electrophoresis, in *Handbook of Capillary Electrophoresis, 2nd ed.* (Landers, J.P., ed.), CRC Press, Boca Raton, pp. 1–47.
45. Morris, M. D., Kim, Y., and Hammond, R. W. (1996) Pulsed field electrophoresis of nucleic acids: ultrafast separations in ultrashort capillaries. *Proc. SPIE* **2680,** 219–224.
46. Gobie, W. A. and Ivory, C. F. (1990) Thermal model of capillary electrophoresis and a method for counteracting band broadening. *J. Chromatogr.* **516,** 191–210.
47. Davis, K. L., Liu, K.-L., Lanan, M., and Morris, M. D. (1993) Spatially resolved temperature measurements in electrophoresis capillaries by Raman thermometry. *Anal. Chem.* **65,** 293–298.
48. Liu, K.-L., Davis, K. L., and Morris, M. D. (1994) Raman spectroscopic measurement of spatial and temporal gradients in operating electrophoresis capillaries. *Anal. Chem.* **66,** 3744–3750.
49. Nelson, R. J., Paulus, A., Cohen, A. S., Guttman, A., and Karger, B. L. (1989) Use of Peltier thermoelectric devices to control column temperature in high performance capillary electrophoresis. *J. Chromatogr.* **480,** 111–127.
50. Rapp, T. L. and Morris, M. D. (1996) An aluminum heat sink and radiator for electrophoresis capillaries. *Anal. Chem.* **68,** 4446–4450.
51. Heiger, D. N., Carson, S. M., Cohen, A. S., and Karger, B. L. (1992) Wave form fidelity in pulsed field capillary electrophoresis. *Anal. Chem.* **64,** 192–199.
52. Mitnik, L., Heller, C., Prost, J., and Viovy, J.-L. (1995) Segregation in DNA solutions induced by electric fields. *Science* **267,** 219–222.

21

Collection of Capillary Electrophoresis Fractions on a Moving Membrane

Soffia Magnúsdóttir, Christoph Heller, Philippe Sergot, and Jean-Louis Viovy

1. Introduction

During the last 10 years, capillary electrophoresis (CE) has developed into a powerful analytical method for a wide variety of applications. It tends to replace traditional slab-gel electrophoresis for the separation of complex biological samples (DNA, proteins, carbohydrates) *(1)*. The ability to further analyze the separated components using specific secondary reactions, analogous to well-established methods such as the Southern blot *(2)* and direct blotting *(3)* is important for the full development of this rapidly growing method. The total amount of sample separated in CE is very small when compared to slab gels, but thanks to the sensitivity and miniaturization of modern detection methods, it may still be sufficient for many postseparation analysis applications, such as hybridization, sequencing, enzyme digestion, reamplification by polymerase chain reaction (PCR), and so on.

1.1. Nano-Preparative Devices

Several designs for "nano-preparative" CE devices have already been described. One of the approaches consists of collecting the CE fractions into microvials containing a small amount of buffer and an electrode. The fractions are then eluted into the vial by pressure, or by electric field whenever a separated zone is calculated to exit from end of the capillary *(4–10)*. This procedure, which is easily adapted to commercial CE apparatus, can also be used to collect fractions from a "microchip" CE device into a microtiter plate *(11)*.

The collection procedure, however, has several limitations. It is difficult to implement the collection of pure fractions if the mobilities of the fractions are too close together. Moreover, the method inevitably leads to a strong dilution of the sample, because producing the electric connection between the capillary and the electrode in the outlet vial requires several microliters of buffer. Another limitation of this approach

From: *Methods in Molecular Biology, Vol. 162:*
Capillary Electrophoresis of Nucleic Acids, Vol. 1: Introduction to the Capillary Electrophoresis of Nucleic Acids
Edited by: K. R. Mitchelson and J. Cheng © Humana Press Inc., Totowa, NJ

is that the interruption of the electric field during the collection process is inevitable, and introduces extra diffusion and temporal uncertainity in the migration of zones. Interesting variants of this mode of collection avoid some of the mentioned problems (e.g., the dilution problem) and involve a selective isolation of fraction zones after separation on a CE-chip ***(12)***, or the collection of fractions into a series of microwells precast in an agarose gel ***(13)***.

1.2. Porous Glass Joint

In another family of methods, the collection end of the capillary is electrically isolated by closing the electric circuit at a point situated a short distance from the capillary outlet. Examples of this approach are based on the porous glass joint arrangement originally introduced by Wallingford and Ewing ***(14)***. This junction is prepared by cutting the capillary into two segments and placing both into a porous glass sleeve such that the capillary outlet is not part of the electrical circuit. Using a strong electroosmotic flow (EOF), the analytes are transported across the electrically isolated part, where they can be collected into a dry vial (thus with no dilution of the sample) without interruption of the separation potential ***(15,16)***. An alternative to the same principle is presented by Huang and Zare ***(17)*** by placing an on-column frit close to the capillary outlet. Another approach consists of stepwise increase in the EOF, which is accomplished by switching between several coupled capillary systems having different EOF characteristics which enables collection of the fractions into capillary tubes ***(18)***. A major disadvantage of these approaches is that they require significant electroosmosis, a situation not always achievable nor desirable, especially for the separation of biomolecules (in particular, the vaste majority of DNA separation protocols involve viscous polymer solutions which strongly reduce the EOF). Müller et al. ***(19,20)*** have presented another collection method, in which a sheath liquid arrangement is selected to establish an electrical connection (continuous collection). A coaxial capillary flow cell is also proposed as another way to establish a complete circuit ***(21)***.

1.3. Collection onto a Surface

Finally, several micropreparative methods have been proposed that involve the direct collection of eluted fractions onto a surface that is able to adsorb the analytes. Sample collection may be accomplished onto a flat surface ***(20,22)***, or onto a continuously rotating membrane ***(23–29)***. The membrane has to be kept wet, in order to ensure electric continuity. In this case, the major difficulty is to avoid the spreading of the sample on the surface, driven by hydrodynamics of the carrying fluid (owing to several factors such as capillarity, motion of the membrane, or evaporation). If these problems are avoided, a very high concentration of the analyte can be achieved on the membrane site, and the "micromembrane" format is actually very convenient for numerous further detection methods (hybridization, mass spectroscopy, and so on). In this chapter, we describe how to collect DNA fragments onto a moving membrane from the outlet of a capillary with minimal dispersion. We also provide an example of a procedure for the analytical development of the spots using dioxigenin. The possibility for further analysis of the immobilized DNA is also discussed.

2. Materials

2.1. Separation Unit

1. The DNA separation is performed on a P/ACE Beckman 2100 CE device, modified for fraction collection. A schematic overview of the micropreparative system is given in **Fig. 1**. The commercial apparatus is (reversibly) modified by connecting it to a bipolar power supply grounded at the outlet of the capillary, and to a negative high voltage at the inlet. This latter power supply can be obtained as part of the MS coupling kit from Beckman.
2. An appropriate length of the capillary is 60 cm with the detection window at 20 cm from the injection end. The 40-cm long outlet can be connected to the collection system, located out of the apparatus on its right side. A notch present in the instrument lid allows for an easy passage of the capillary.

2.2. Micropreparative System

1. The collection system (detailed scheme in **Fig. 2**) is the key feature for high-resolution collection fraction. A guide and spring arrangement gently presses the capillary end onto the moving membrane, which is guided along a slit in the holder. Considering the small outer diameter of the capillary, a very soft spring and a fine adjustment screw will be necessary to avoid scratching of the membrane by the capillary. The presence of a very shallow and uniform trace on the membrane, without scratching, is the signature of the correct pressure (*see* **Note 1**).
2. It is also mandatory that the capillary end be both flat and smooth. This can be achieved using a tool for optical fiber fracture (preferably of the type involving axial extention). Best results are obtained when the polyimide coating is removed from the capillary at cutting location, before fracture.
3. Across the membrane, a 300-µm pinhole, connected to a buffer vial by tygon tubing, is located coaxially with the capillary outlet. The outlet electrode is located in the buffer vial, so that field lines are focused across the membrane into the pinhole. This attachment reduces band spreading and favors strong adsorption of DNA onto the membrane upon exit from the capillary.
4. Besides this requirement for the pinhole, and a correct low-friction guiding of the capillary and of the membrane, the dimensions of the system are not critical and can be adapted to available material. The body of the collection device can be made of various insulating polymers, such as poly(methyl methacrylate) (PMMA). It should be located so that the outlet of the capillary is at the same level as fluid in the inlet vial during regular operation (unlike the schematic view given in **Fig. 1**).
5. The outlet buffer vial is attached to a vertical translator, in order to adapt the relative level of the fluid for different steps of the protocol.
6. Membrane strips, cut at a width adapted to the guide in the capillary/membrane holder, are attached to a cylinder driven at constant rotation rate by a synchronous or stepper motor.
7. Other aspects of the separation method are quite conventional, and can be adapted to specific needs. In the separation presented in this chapter, we have used the following conditions:
   a. Capillary: DB-17, 100 µm id from J&W Scientific (Folsom CA).
   b. Separation matrix: Hydroxypropyl cellulose (Mw 10^6) from Sigma Aldrich (Dorset, UK), at 0.2% or 0.4% depending on the application, diluted in 1X TBE (89 m*M* Tris-HCl, pH 8.3, 89 m*M* boric acid, 2.5 m*M* EDTA).

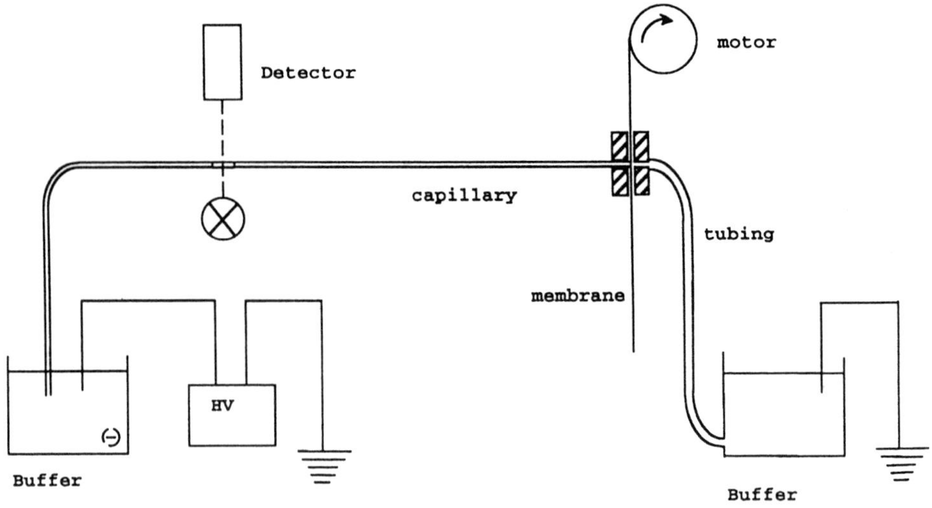

Fig. 1. Schematic representation of the arrangement of the micro-preparation device (not drawn to scale).

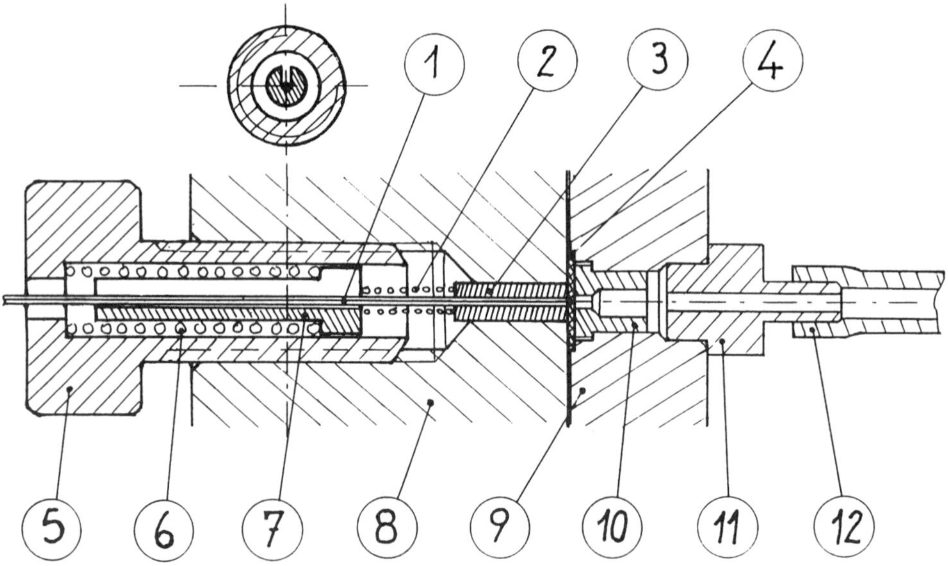

Fig. 2. Detail of the capillary holder/pinhole attachment. (1) capillary, (2) fine-tuning spring, (3) elastomer gasket such as a piece of Tygon® tubing, (4) membrane, (5) pressure-adjusting screw, (6) primary spring, (7) capillary holder, (8) a "body," which is attached to the CE machine, (9) membrane presser and pinhole holder, (10) pinhole, (11) tubing adaptor, and (12) tubing to outlet buffer vial (tygon).

c. DNA samples: pBR328/*Bg1*I-*Hinf*I (marker VI), Lambda/*Hind*III (marker II) labeled with dioxigenin, from Boehringer Mannheim GmbH (Mannheim, Germany). pBR328 /*Bgl*I-*Hinf*I contains the following fragments (base pair): 154(n1), 220(n2), 234(n3), 298(n4), 394(n5), 453(n6), 517(n7), 653(n8), 1033(n9), 1230(n10), 1766(n11), 2176(n12).
d. Collection is performed on a positively charged nylon membrane Hybond®-N+ from Amersham International Plc. (Buckinghamshire, UK).
e. All the material needed for the development of the DNA spots on the membrane is provided in the "DIG Nucleic Acids Detection" Kit from Boehringer Mannheim GmbH (Mannheim, Germany). The development procedure is described under **Subheading 3.3.**

3. Methods

3.1. Preparation

1. The aim of this step is to achieve a continuous buffer contact from the end of the capillary to the outlet buffer vial, without bubbles. This is obtained by hydrodynamic pressure, raising the outlet buffer vial above the pinhole level (*see* **Note 2**).
2. If the polymer material used for making the cell does not seem hydrophilic enough, its wetting can be improved by reducing the surface tension of the buffer, e.g., by addition of a few percent of ethyl alcohol, or of nonionic surfactant (e.g., Pluronic® F127, BASF).
3. After rinsing the buffer vial with separation medium, the vial is lowered to 0–2 mm above the pinhole level, the membrane is located in its guide and the cell is secured.
4. The capillary is rinsed with the separation medium by increasing the hydrostatic pressure in the inlet vial. This can be performed before the membrane is introduced into the cell. Ensure a fast rinsing operation to avoid drying of the capillary outlet (*see* **Note 3**).
5. The capillary may also be rinsed with the separation medium after insertion of the membrane into the cell. In this case, the membrane should be driven during the capillary rinsing to avoid accumulation of separation medium in the cell.

3.2. Separation and Fraction Collection

1. After securing the cell and starting the membrane rotatory motion, sample injection must be done electrokinetically. Use the same injection and separation conditions as for conventional separations.
2. The results shown in **Fig. 3** for example, are obtained with the following conditions. A mixture of DNA fragments from pBR328/*Bgl*I-*Hinf*I is labeled with digoxigenin, and is diluted to 10 µg/mL and injected electrokinetically for 50 s at 10kV.
3. The fragments are separated at constant field 83 V/cm, and are detected by UV absorption at 254 nm.
4. Allow for a run time sufficient to let the slowest analytes reach the membrane (e.g., about 4× the position of the last peak at the detector, for 20/60 cm capillaries). We observed a good linearity between the position of the peaks on the electrophoregram and the positions on the membrane, making the peak/spot correlation easy.
5. After the run, dismount the membrane, gently rinse it in pure water and store it dry for further detection, or run it through the detection procedure immediately.
6. Sequential runs can also be performed. In this case, handwritten marks, or more sophisticated labeling procedures should be made to identify the relevant sections on the membrane and to facilitate the following band allocation step.

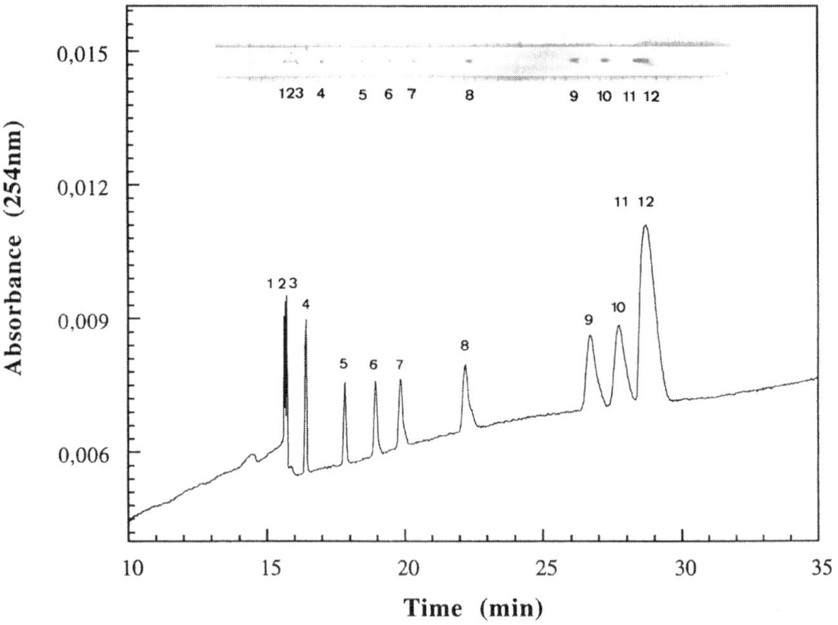

Fig. 3. Electropherogram and photograph of the corresponding blotting membrane, for the separation of MW-marker VI. The peaks and the dots correspond to the following sizes in base pairs: (1) 154, (2) 220, (3) 234, (4) 298, (5) 394, (6) 453, (7) 517, (8) 653, (9) 1033, (10) 1230, (11) 1766 and (12) 2176. Sample concentration: 10 ng/µL, HPC concentration: 0.4% in 1X TBE buffer. Separation field strength: 83 V/cm. Membrane speed: 1.57 mm/min. One division on the photograph corresponds to 2 mm on the membrane.

7. Various methods can be used for the detection of the DNA spots. These are prelabeling of the fragments or probe hybridization using colorimetry, radioactive labeling, fluorescence, luminescence or MS. We preferred to perform a colorimetric detection of dioxigenin prelabeled DNA fragments.

3.3. Immunological Detection of DNA After Collection

1. DNA fragments are prelabeled with dioxigenin.
2. All the incubations are performed at room temperature except for the color reaction accompanied with shaking or mixing.
3. The volumes of the solutions are calculated for a filter size of 100 cm^2 and should be adjusted appropriate to other sizes of filter membrane.
4. Wash the membrane briefly (1 min) in the DIG kit Buffer-1 plus 0.3% (w/v) Tween 20.
5. Incubate the membrane for 30 min with about 100 mL of DIG kit Buffer-2.
6. Dilute antidioxigenine-AP conjugate to 150 mU/mL (1:5000) in Buffer-2. Note that diluted antibody-conjugate solution are stable only at 4°C for about 12 h.
7. Incubate the membrane for 30 min with 20 mL of diluted antibody-conjugate solution.
8. Remove the unbound antibody-conjugate by washing the membrane twice for 15 min with 100 mL of kit Buffer-1.

9. Equilibrate the membrane for 2 min with 20 mL of DIG kit Buffer-3.
10. Incubate the membrane with ~10 mL of freshly prepared color-substrate solution sealed in a plastic bag, or in a suitable box in the dark. The color precipitate starts to form within a few minutes, and the reaction is usually complete after 16 h.
11. Do not shake or mix the bag while the color is developing.
12. When the desired spots are detected, stop the reaction by washing the membrane for 5 min with 50 mL of kit Buffer-4.
13. The results can be documented by photographing the membrane.
14. The membrane may then be dried at room temperature, or by baking at 80°C and stored, however the color fades upon drying. Membranes must not be allowed to dry if they are to be reprobed.
15. The color can be revitalized by wetting the membrane with kit Buffer-4. The membrane can also be stored in sealed bags containing Buffer-4. The color remains unchanged in this case.

3.4. Results

1. The electropherogram resulting from the separation of MW-marker VI and a photograph of the corresponding membrane on which the DNA fragments are collected are shown in **Fig. 3**. The first three fragments are well separated on the membrane, whereas on the electropherogram they are barely resolved. This difference can be explained by the longer distance of travel to the membrane, than the distance to the detector.
2. Note that minor "tailing" or background effects are introduced by the collection process. The DNA on the membrane is concentrated into very small, quasi-circular spots (typically 0.5–1 mm in diameter). The small spots allow for the detection of the small amounts of injected sample (*see* **Note 4**).
3. To evaluate the amount of DNA collected on the membrane, we compared the color intensity of the CE deposited microspots to that of a standard membrane, and then scaled the area of the standard spots to the area of the microspots. The DNA content of the first microspot on the membrane (154-bp fragment) is estimated as 35 pg. The detection limit in colorimetry should be of order 10 pg or below, and if higher sensitivity is required it can certainly be achieved with radioactivity or luminescence.
4. An important improvement to this CE-blotting setup, compared to other direct blotting systems, is the refocusing of the electric field lines across the membrane. This idea was proposed in Eriksson et al. *(25)*. Combining this method with a cationic membrane ensures the strong driving of the analytes into the membrane, since electrophoretic mobility and EOF both act in the same direction inside the membrane.
5. Also, the use of a linear strip of membrane, which can be of arbitrary length, allows for the collection of several sequential CE separations in a row, without further manual intervention. This would provide scope for automation of fraction collection by CE. The use of membranes in a "cassette" format, similar to magnetic tapes, could be considered. However, this suggested modification has not yet been implemented, and the setup described here is still a manual laboratory tool.

4. Notes

1. Scratching of the membrane: This can be because of an uneven capillary end (check with a microscope or binocular), or of a "too high" capillary pressure onto the membrane. Adjust the pressure screw, or the spring strength to lessen the contact. Note that the capillary itself is quite rigid, and if the body of the collection device *(8)* is not securely

attached to the CE machine it can lead to a considerable pressure on the membrane. Excess pressure can also occur if the capillary is accidently touched or displaced after adjustment of the pressure-controlling screw.

2. Maintainance of the electrical continuity of the system is the most common problem encountered during the implementation step: this may usually be due to bubbles in the tubing. Bubbles can be avoided in the following ways: by the careful degassing of the buffer solutions; by an increase in the time and rate of rinsing of the collection device (increase the altitude of the outlet buffer vial during rinsing); and by an increase the surfactant content of the outlet buffer; by avoiding use of "too hydrophobic" materials (e.g., fluorocarbons) for the construction of the cell.
3. The loss of electric continuity can also be caused by the drying of the membrane. This can be avoided by: increasing the altitude of the outlet buffer vial during the run; prewetting the membrane before a run; and placing a vial containing buffer under the device, in which the lower part of the membrane is wetted and stored during the run.
4. Spreading, or disappearance of DNA spots: If the peaks appear to be uniform and unperturbed on the electropherogram, but not so on the membrane, this may be because of current breakdown between passage of the analytes in front of the detector and collection (check for current stability during the whole run), or of dispersion at the outlet of the capillary due to excess flow of liquid. In the latter case, this may be owing to excessive prewetting of the membrane, to excessive height of the outlet buffer vial, or to a "too loose fitting" of the membrane in its guide. Excessive quantity of separation medium in the collection cell may also be the cause, which can be remedied by performing capillary rinsing steps and then removing excess polymer solution before attaching the membrane.

References

1. Righetti, P. G. (1996) *Capillary Electrophoresis in Analytical Biotechnology*. CRC Press, Boca Raton, FL.
2. Southern, E. M. (1975) Detection of specific sequences among DNA fragments separated by gel electrophoresis. *J. Mol. Biol.* **98,** 503–517.
3. Beck, S. (1988) Protein blotting with direct blotting electrophoresis. *Anal. Biochem.* **170,** 361–366.
4. Rose, D. J. and Jorgenson, J. W. (1988) Fraction collector for capillary zone electrophoresis. *J. Chromatogr.* **438,** 23–34.
5. Cohen, A. S., Njarian, D. R., Paulus, A., Guttman, A., Smith, J. A., and Karger, B. L. (1988) Rapid separation and purification of oligonucleotides by high-performance capillary gel electrophoresis. *Proc. Natl. Acad. Sci. USA* **85,** 9660–9663.
6. Guttman, A., Cohen, A. S., Heiger, D. N., and Karger, B. L (1990) Analytical and micropreparative ultra-high resolution for oligonucleotides by polyacrylamide gel high-performance capillary electrophoresis. *Anal. Chem.* **62,** 137–141.
7. Camilleri, P., Okafo, G. N., and Southan, C. (1991) Separation by capillary electrophoresis followed by dynamic elution. *Anal. Biochem.* **196,** 178–182.
8. Bergman, T., Agerbeth, B., and Jornvall, H. (1991) Direct analysis of peptides and amino acids from capillary electrophoresis. *FEBS* **283,** 100–103.
9. Albin, M., Chen, S.-M., Andrea, L., Pairaud, C., Colburn, J., and Wiktorowich, J. (1992) The use of capillary electrophoresis in a micropreparative mode: Methods and application. *Anal. Biochem.* **206,** 382–388.
10. Altria, K. D. and Dave, Y. K. (1993) Peak homogeneity determination and micropreparative fraction collection by capillary electrophoresis for pharmaceutical analysis. *J. Chromatogr.* **633,** 221–225.

11. Karger, B. L., Foret, F., Zavracky, P., McGruger, E. N., Xue, Q., and Dunayevskiy, Y. M. (1995) Microscale fluid handling system, US patent 60/001,349.
12. Effenhauser, C. S., Manz, A., and Widmer, M. (1995) Manipulation of sample fractions on a capillary electrophoresis chip. *Anal. Chem.* **67,** 2284–2287.
13. Minarik, M., Gilar, M., Klépárník, K., Foret, F., and Karger, B. L. (1999) A multicapillary array fraction collector, in *HPCE '99*, Palms Springs CA, Jan 23–28.
14. Wallingford, R. A. and Ewing, A. G. (1987) Capillary zone electrophoresis with chemical detection. *Anal. Chem.* **59,** 1762–1766.
15. Guzman, N. A., Trebilcock, M. A., and Advis, J. P. (1991) Affinity capillary electrophoresis: two semi-preparative approaches to concentrate samples on the capillary column and to recover microgram quantities of material, in *The 42nd Pittsburgh Conf. Expo. Anal. Chem. Appl. Spectroscopy*, Chicago, Illinois, March 4–7, pp. 160.
16. Guzman, N. A., Trebilcock, M. A., and Advis, J. P. (1991) The use of a concentration step to collect urinary components separated by capillary electrophoresis and further characterization of collected analytes by mass spectrometry. *J. Liq. Chromatogr.* **14,** 997–1015.
17. Huang, X. and Zare, R. N. (1990) Use of an on-column frit in capillary zone electrophoresis: Sample collection. *Anal. Chem.* **62,** 443–446.
18. Nashabeh, W., Smith, J. T., and El Rassi, Z. (1993) Studies in capillary zone electrophoresis with a post-column multiple capillary device for fraction collection and stepwise increase in electroosmotic flow during analysis. *Electrophoresis* **14,** 407–416.
19. Müller, O., Foret, F., and Karger, B. L. (1995) Design of a high-precision fraction collector for capillary electrophoresis. *Anal. Chem.* **67,** 2974–2980.
20. Muth, J., Müller, O., Berka, J., Leonard, J. T., and Karger, B. L (1996) DNA sequence analysis of Prinker-modified restriction fragments after collection from capillary electrophoresis with replaceble matrices. *J. Chromatogr. A* **744,** 303–310.
21. Chiu, R. W., Walker, K. L., Hagen, J. J., Monning, C. A., and Wilkins, C. L. (1995) Coaxial capillary and conductive capillary interfaces for collection of fractions isolated by capillary electrophoresis. *Anal. Chem.* **67,** 4190–4196.
22. Konse, T., Takahashi, T., Nagashima, H., and Iwaoka, T. (1993) Blotting membrane micropreparation in capillary electrophoresis to evaluate enzyme purity and activity. *Anal. Biochem.* **214,** 179–181.
23. Huang, X. and Zare, R. N. (1990) Continuous sample collection in capillary zone electrophoresis by coupling the outlet of a capillary to a moving surface. *J. Chromatogr.* **516,** 185–189.
24. Cheng, Y.-F., Fuchs, M., and Carson, W. (1993) Post-capillary Immobilon™ –P membrane fraction collection for capillary electrophoresis. *BioFeedback* **14,** 51–55.
25. Eriksson, K. O., Palm, A., and Hjertén, S. (1992) Preparative capillary electrophoresis base on adsorption of the solutes (proteins) onto a moving blotting membrane as they migrate out of the capillary. *Anal. Biochem.* **201,** 211–215.
26. Murata, H., Toshifumi, T., Anahara, S., and Shimonishi, Y. (1993) On-line membrane blotting of peptides and proteins from capillary outlet. *Anal. Biochem.* **210,** 206–208.
27. Anon (1994) Capillary electrophoresis using new membrane fraction collector technology. *WatersNews* **5**/94 1.
28. Liao, J.-L., Zeng, C.-M., Palm, A., and Hjertén, S. (1995) Continuous beds for microchromatography: Detection of proteins by a blotting membrane technique. *Anal. Biochem.* **241,** 195–198.
29. Magnúsdóttir, S., Heller, C., Sergot, P., and Viovy, J.-L. (1997) Micropreparative capillary electrophoresis of DNA by direct transfer onto a membrane. *Electrophoresis* **18,** 1990–1993.

22

Separation of Supercoiled DNA Using Capillary Electrophoresis

David T. Mao and Roy M. A. Lautamo

1. Introduction

DNA supercoiling is a general phenomenon of almost all DNA in vivo. Plasmids, bacterial chromosomes, the mitochondrial genome and many viral genomes occur as closed circular DNA. Even some linear DNAs, such as eukaryotic chromosomes and yeast plasmids, behave like circular DNA because part of the DNA is anchored to the nuclear matrix creating closed-circular loops. Many DNA-associated processes are affected by changes in DNA supercoiling. These processes include the genome organization, transcription, gene expression, DNA replication, and recombination events *(1)*. Plasmids are double-stranded circular DNAs which are capable of replicating independent of the chromosome. This property allows plasmids to be carriers of desired DNA sequences and to be replicated in large quantity. Plasmids were actually the first successful DNA vectors used for molecular cloning in bacteria. Recently, the potential of plasmids in gene therapy has been explored extensively. The rising interest in gene therapy for treatment of human diseases, such as cancer and genetic disorders, is reflected in the growing number of clinical trials *(2,3)*.

In the early 1960s, studies of viral DNA revealed three DNA forms of structure. These three forms were labeled I, II, and III in order of their decreasing sedimentation velocity. Form III was clearly linear, but forms I and II were both double-stranded circular DNA of same molecular weight as the linear form III. Vinograd *(4)* suggested that form I was a twisted circular DNA compared to the open circular form II. Subsequently, viewing of DNA isolated from polyoma virus by means of high-power electron microscopy supported this interpretation.

1.1. Electrophoretic Analysis of Plasmids

The most common method for the analysis of supercoiled DNAs is agarose gel electrophoresis. Electrophoresis separates DNA molecules on the basis of size and compactness: The smaller and more compact molecules will migrate more rapidly

From: *Methods in Molecular Biology, Vol. 162:*
Capillary Electrophoresis of Nucleic Acids, Vol. 1: Introduction to the Capillary Electrophoresis of Nucleic Acids
Edited by: K. R. Mitchelson and J. Cheng © Humana Press Inc., Totowa, NJ

through the matrix of the gel than open, linear molecules, or less compact molecules. Capillary electrophoresis (CE), in which separation is carried out in tubing of small internal diameter, is a radically improved extension of traditional slab-gel electrophoresis. Because of its improved speed, automation and data archiving ability, CE has emerged as a new analytical technique for DNA, including synthetic oligonucleotides *(5)*, restriction enzyme fragments *(6)*, and plasmid DNAs *(7–11)*. In this chapter, methods are described using capillary electrophoresis for separation of plasmid DNAs of different conformations which include linear, open circular, and supercoiled topoisomers. A method for analysis of a viral DNA, MP13mp18 (+), which is mainly a single-stranded open circular DNA, is also included for comparison.

CE is a versatile tool for analysis of plasmid DNA samples. The lower mobility and greater peak width allow easy identification of supercoiled plasmids in the presence of both linear and open circular DNA. Topoisomers of supercoiled DNAs can be separated, and the monitoring of conformational change of a plasmid under different environments is possible. The fast-analysis time, quantitation, and simplicity of this methodology are well suited for modern molecular biology analysis.

2. Materials

2.1. CE Instrumentation

1. Power supply: Model 230, 0–30 kV (Bertan).
2. Detector: UV-VIS detector model 200 (Linear).
3. Capillary: μSil-FC coated capillary, 50 μm id (J&W Scientific) (*see* **Note 1**).

2.2. Separation Buffer

1. Hydroxypropyl methyl cellulose (HPMC): viscosity 4000 cP at 2% aqueous solution (Sigma).
2. 10X TBE: 890 m*M* Tris-base, 890 m*M* boric acid, and 20 m*M* EDTA, pH 8.4.
3. Separation buffer is 0.1 *M* HPMC in 1X TBE (*see* **Note 2**).

2.3. Membrane Dialysis

1. Membrane: MF-Millipore membrane filter, 0.025 μm pore, 25 mm diameter (Millipore, cat. no. VSWP 025 00).
2. Deionized water: from Barnstead NANO Pure water system.

2.4. Capillary Column Preconditioning Solution

1. Methyl cellulose: 1500 cP at 2% aqueous solution (Sigma).
2. Filter unit: sterile filter unit with 0.45 μm pore.

2.5. Supercoiled DNA and Linear DNA Ladders

1. Supercoiled DNA ladder; 0.25 μg/μL in TE buffer: 10 m*M* Tris-HCl, pH 7.6, 1 m*M* EDTA (Life Technologies).
2. Linear DNA ladder: 1 kb ladder, 0.8 μg/μL in TE (Life Technologies).

2.6. Linearized, Nicked Circular, and Native pBR322 Plasmid

1. pBR322 plasmid: 1.34 μg/μL in TE buffer (Sigma).
2. *Ava*I: 5 U/μL and 10X REact 2 buffer: 500 m*M* Tris-HCl, pH 8.0, 100 m*M* MgCl, 500 m*M* NaCl (Life Technologies).

3. DNase I: 1 U/μL and 10X DNase I reaction buffer: 200 m*M* Tris-HCl, pH 8.4, 20 m*M* $MgCl_2$, 500 m*M* KCl (Life Technologies).

2.7. CE Separation of Supercoiled DNA Ladder in the Presence of Ethidium Bromide

1. Supercoiled DNA ladder: 0.25 μg/mL (Life Technologies).
2. 10 m*M* ethidium bromide in deionized water: freshly prepared in a colored centrifuge tube.

2.8. M13mp18 RFI Plasmid DNA and M13mp18(+) Viral DNA

1. M13mp18 RFI plasmid DNA: 7249 bp, 0.295 μg/μL in TE buffer (Sigma).
2. M13mp18 (+) viral DNA: 1.05 μg/μL in TE buffer (Sigma).

2.9. pBR322 Topoisomerase Reaction

1. Topoisomerase I, 10 U/μL (Life Technologies).
2. 2X topoisomerase I reaction buffer: 100 m*M* Tris-HCl, pH 7.5, 100 m*M* KCl, 20 m*M* $MgCl_2$, 0.2 m*M* EDTA, 1 m*M* DTT, and 60 μg/mL BSA.

3. Methods

3.1. Instrument Setup

1. Capillary: μSil-FC coated capillary, 50 μm id × 50 cm total length, window is placed 20 cm from injection inlet to detector. Separation buffer is 0.1% HPMC in 1X TBE.
2. Injection of sample is electrokinetic at –7.5 kV for 5 s. Applied voltageis –6.0 kV.
3. Capillary run temperature is at ambient (around 22°C).
4. Between runs, flush the capillary with separation buffer for 5 min at 30 psi vacuum.

3.2. Preparation of 2% HPMC Stock Solution

1. Add 2 g of HPMC powder to 100 mL of 1X TBE in 250-mL bottle.
2. Stir the mixture vigorously for 2 h over a water bath preheated to 70°C.
3. Turn off the heat and stir the mixture continuously overnight or until a clear solution reached. The clear solution is kept undisturbed for at least 24 h before use.

3.3. Preparation of Separation Buffer

1. Dilute the 2% HPMC solution to 0.1% with 1X TBE, then rock the mixture gently until it is homogeneous over several hours.

3.4. Preparation of Capillary Preconditioning Solution (see Note 3)

1. Add 0.5 g of methyl cellulose to 100 mL of deionized water to 0.5% methyl cellulose.
2. Stir the mixture vigorously for 2 h over a waterbath preheated to 70°C.
3. Turn off the heat and the stirring is continued overnight. A slight cloudy solution is kept in a refrigerator overnight to allow further dissolution.
4. The clear solution is filtered through a 0.45 μm membrane.

3.5. New Capillary Preconditioning (see Note 4)

1. Fill the capillary with the preconditioning solution prepared in **Subheading 3.4.**
2. Let it sit for 2 h before use.

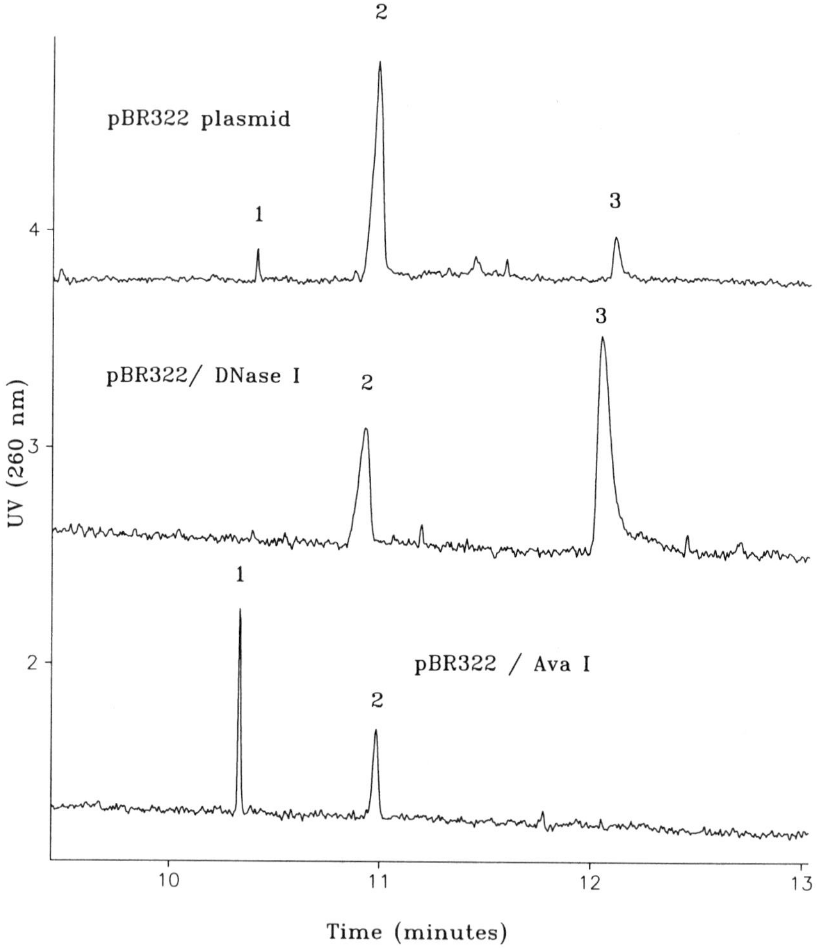

Fig. 1. CE separation of pBR322 plasmid sample. 1, linear; 2, supercoiled; 3, nicked open circular DNA. *Top*, pBR322 plasmid sample as supplied; *middle*, DNAse I treated pBR322; *bottom*, *Ava*I treated pBR322.

3.6. Membrane Dialysis Procedure

1. Float an MF-Millipore membrane filter shiny side up on deionized water.
2. Apply a drop of sample (5–25 µL) to the top of the filter.
3. Keep the sample in a refrigerator for 15–20 min.
4. Transfer the prepared sample to a sample vial for CE injection.

3.7. CE of Native pBR322 Plasmid (Fig. 1, top)

1. pBR322 plasmid sample (Sigma) is dialyzed as described in **Subheading 3.6.**
2. Perform CE analysis as described in **Subheading 3.1.**

3.8. Nicked Circular pBR322 Plasmid (Fig. 1, middle) (see Note 5)

1. 2.5 µL of pBR322 plasmid is dialyzed as described in **Subheading 3.6.**
2. Make 1X DNase I reaction solution by diluting the 10X DNase I solution 10-fold in deionized water. Dilute the DNase I enzyme with 1X DNase I reaction solution to approx 500 pg/µL.
3. Add the dialyzed pBR322 to a mixture of 1.5 µL of the dilute DNase I, 1.5 µL of 10X DNase I reaction buffer and 10 µL of deionized water.
4. Incubate at room temperature for 20 min (*see* **Note 5**).
5. Stop the reaction by adding 1 µL of 0.5 *M* EDTA solution.
6. Dialyzed the reaction mixture as described in **Subheading 3.6.**
7. Perform CE analysis as described in **Subheading 3.1.**

3.9. Linearized pBR322 Plasmid (Fig. 1, bottom)

1. Add 1 µL of *Ava*I enzyme to a mixture of 1 µL of pBR322 plasmid, 1 µL of 10X REact 2 buffer and 10 µL of deionized water.
2. Incubate at 37°C for 30 min. Chill the sample in ice-water to stop the reaction.
3. Dialyze the sample as in **Subheading 3.6**
4. Perform CE analysis as described in **Subheading 3.1.**, except that the total length of the capillary is 60 cm.

3.10. CE of 1 kb DNA Ladder and Supercoiled DNA Ladder (Fig. 2)

1. Both DNA sample are dialyzed as described in **Subheading 3.6.**
2. Dilute DNA samples to approx 0.08 µg/µL with deionized water.
3. Perform CE analysis as described in **Subheading 3.1.**

3.11. High-Resolution CE Separation of a Supercoiled DNA Ladder (Fig. 3)

1. Supercoiled DNA ladder is dialyzed as described in **Subheading 3.6.**
2. Adjust the sample concentration to about 0.08 µg/µL with deionized water.
3. CE conditions are the same as describe in **Subheading 3.1.** except that the applied voltage is set to 9.0 kV.

3.12. CE of Topoisomers of pBR322 Plasmid (Fig. 4)

1. 3 µL of pBR322 is dialyzed as described in **Subheading 3.6.**
2. Pipet the dialyzed sample into a vial.
3. Add 0.6 µL of Topoisomerase I, 7 µL of 2X reaction buffer, and 5 µL of deionized water to the vial.
4. Incubate at 37°C for 30 min. Then, chill the sample in ice-water to stop the reaction.
5. Dialyze the sample as described in **Subheading 3.6.**
6. CE as described in **Subheading 3.1.**

3.13. Separation of Supercoiled DNA Standards in Separation Buffer with or Without Ethidium Bromide (Fig. 5)

1. Dilute sample in four parts of deionized water (no dialysis).
2. For separation of the sample in a solution containing no EtBr, CE as described in **Subheading 3.1.** (*see* **Fig. 5**, top).
3. For separation of the sample in a solution containing EtBr, a separation solution containing 10 µm EtBr is prepared by adding 10 mL of 10 µ*M* ethidium solution (prepared as in

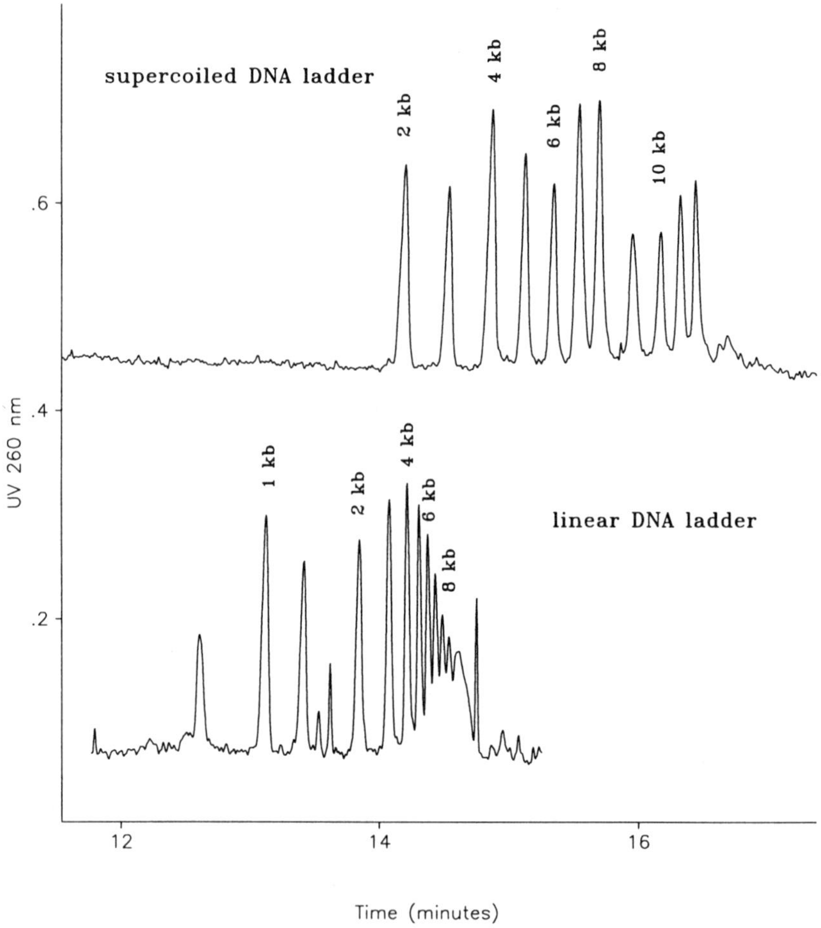

Fig. 2. Relative migration speeds of linear and supercoiled DNAs. *Top*, supercoiled DNA standards (sizes from 2–12 kb); *bottom*, linear DNA standards (1 kb DNA ladder, sizes from 1–11 kb).

Subheading 2.8.) to 10 mL of the separation buffer (0.1% HPMC in TBE). Electrophoresis conditions are same as described in **Subheading 3.1.** (*see* **Fig. 5**, bottom).

3.14. M13m18 RFI and M13mp18 (Fig. 6)

1. Samples are dialyzed as described in **Subheading 3.6.**
2. Dilute the dialyzed M13mp18RFI sample threefold in deionized water.
3. Dilute the dialyzed M13mp18 (+) sample 10-fold in deionized water.
4. CE as described in **Subheading 3.1.**

3.15. Results of CE Separations

1. In CE, a typical plasmid DNA sample, such as *Escherichia coli* plasmid pBR322 (4361 bp), separates into three peaks as shown in **Fig. 1**, top. The slowest migrating peak (peak 3)

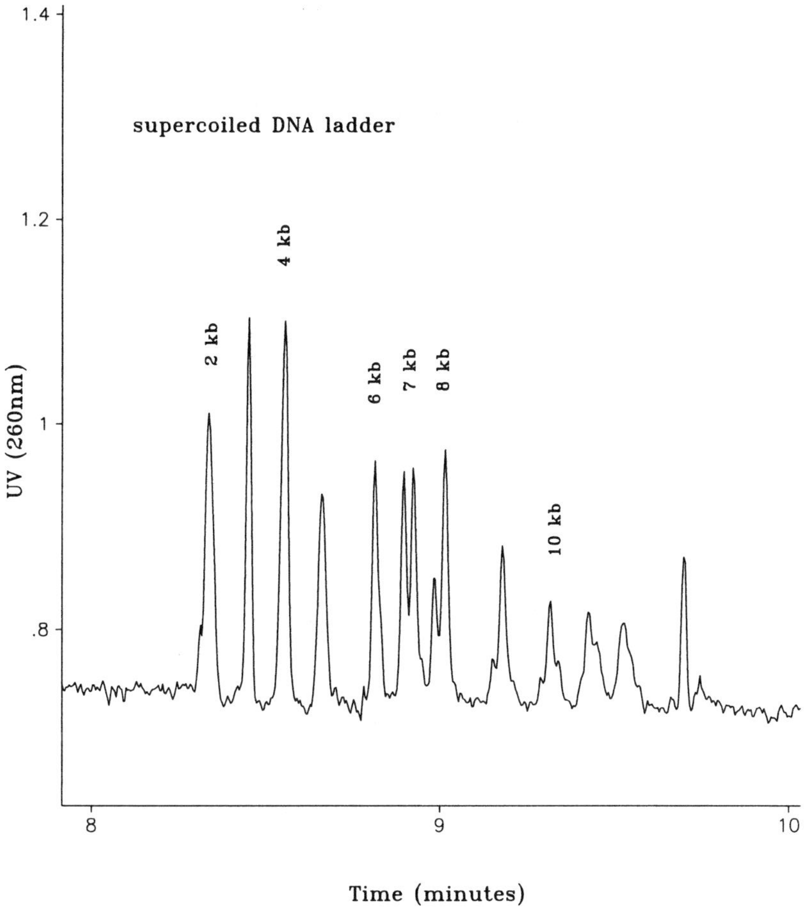

Fig. 3. Separation of supercoiled DNA at high electric field strength.

corresponds to the nicked, circular DNA (form II) which may be formed by breakage of one strand during the plasmid preparation. The largest peak represents the native supercoiled pBR322 (form I) which migrates faster than the nicked circular form II because of its more compact, twisted conformation. Contrary to the order of electrophoretic mobility seen in agarose gel electrophoresis, a linear DNA migrates faster than either of the supercoiled or the nicked circular forms of the same size during CE.

2. The identification of the linear and nicked circular forms, which are present in a pBR322 plasmid sample in small amounts, can be confirmed by comigration of the corresponding enzymatically prepared forms. The open circular form (peak 3) is generated by nicking the plasmid with dilute DNase I, and the supercoiled DNA is completely unwound (**Fig. 1**, middle). The linear form (peak 1) is generated by a single site cleavage using the restriction enzyme, *Ava*I (**Fig. 1**, bottom). In general, the peaks of native plasmid DNAs are broad because the native plasmids consist of a distribution of topoisomers which cannot be completely resolved.

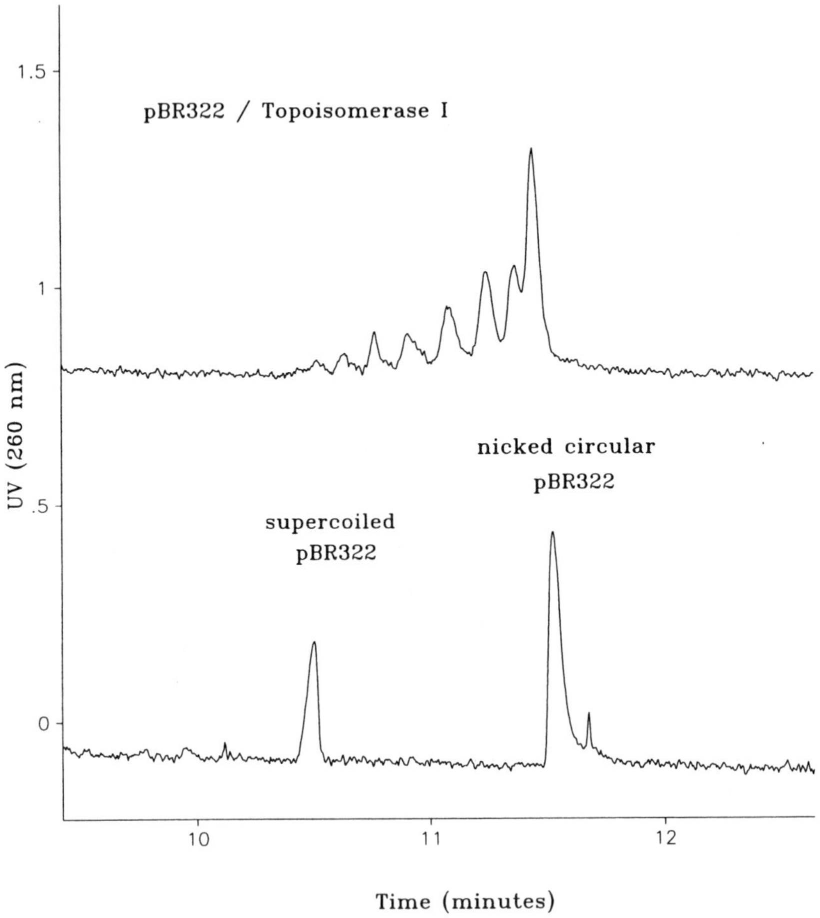

Fig. 4. Separation of topoisomers of pBR322 plasmid. *Top*, topoisomerse I treated pBR322 plasmid; *bottom*, DNAse I treated pBR322 plasmid.

3. **Figure 2** shows the relative migration speeds of a range of different linear and supercoiled DNAs. Typically, linear DNAs migrate faster than the supercoiled DNAs of same size using the method described in this chapter. The identification of the linear and nicked circular forms, which are present in a pBR322 plasmid sample in small quantity, can be confirmed by comigration of their enzymatically prepared corresponding forms.
4. Topoisomers of a plasmid are circular DNAs of the same molecular weight, but are different topological conformations, i.e., different in the numbers of twists. The topoisomers present in preparations of native plasmids are not able to be separated by slab-gel electrophoresis. However, they can be partially resolved in CE under a high applied voltage (**Fig. 3**), because the separation efficiency of DNA during CE is dependent on the applied electric field strength ***(12)***. The higher the field strength, the better the resolution.

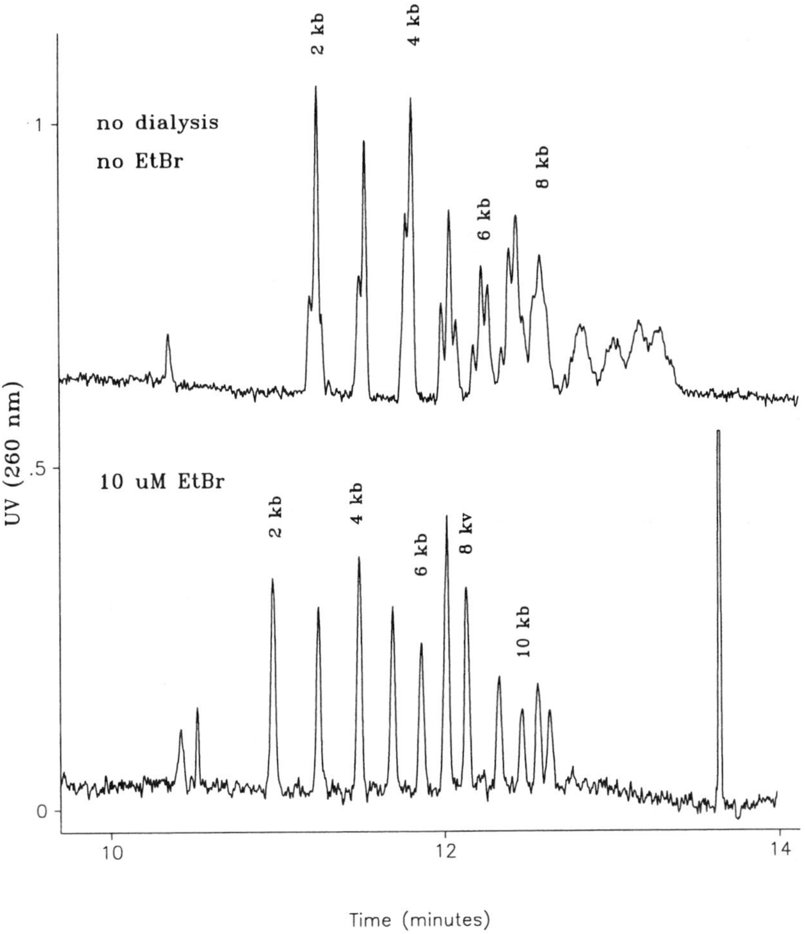

Fig. 5. Influence of ethidium bromide in supercoiled DNA separation. *Top*, sample is diluted fourfold, no sample dialysis, no EtBr in separation buffer; *bottom*, same sample as above, except the separation buffer contains 10 μ*M* EtBr.

5. **Figure 4** shows that the products of plasmid pBR322 catalysed by Topoisomerase I, which forms a series of topological isomers that differ in the number of imposed (superhelical) twists, migrate as a series of peaks between the native and nicked circular forms. By breaking and rejoining the closed circular DNA chain, Topoisomerase I, originally isolated from *E. coli*, catalyzes the formation of a distribution of partially relaxed circular DNAs (topoisomers) ***(13)***.
6. The analysis of supercoiled DNAs by CE can be complicated by factors such as temperature, ionic strength, or the presence of intercalating agents ***(14)*** as these factors are known to change the conformation of DNA. To minimize the effects of ions and to eliminate sample to sample variation, the samples are always dialyzed against deionized water using a membrane dialysis procedure, as shown in **Fig. 2** (top) and **Fig. 6** (top). If the sample is

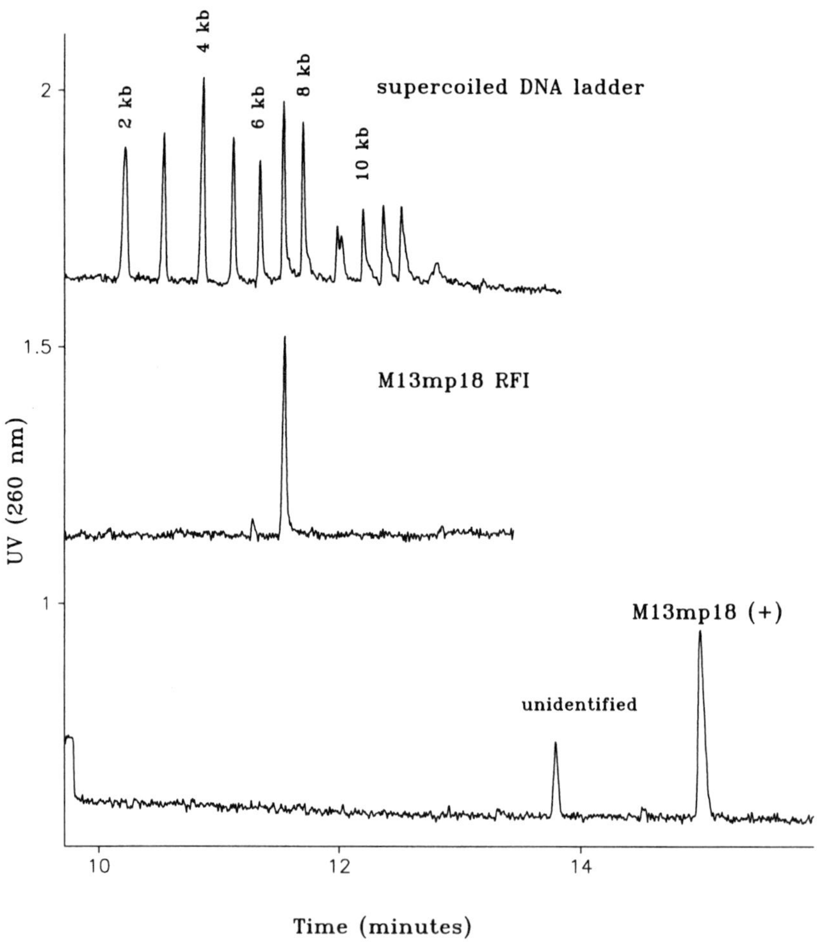

Fig. 6. Separation of viral DNA (ss) and plasmid DNA (ds). *Top*, supercoiled DNA ladder; *middle*, M13mp18 RFI plasmid (ds); *bottom*, M13mp18 (+) viral DNA (ss).

not dialyzed to remove salts, the supercoiled plasmid DNAs display a complex series of peaks (**Fig. 5**, top), representative of the distribution of topoisomers present in isolates of native plasmid. Intercalators such as EtBr are commonly used in DNA analyses for enhancing the detection of DNA bands. Because it inserts itself between two base pairs of dsDNA, intercalators unwind the DNA helix, resulting in a more relaxed structure. At an EtBr concentration in separation buffer greater than 1 µg/mL, supercoiled DNAs are saturated with the dye and assume a uniform constrained form, in which the topoisomers are unresolvable and appear as sharper peaks (**Fig. 5**, bottom).

7. Viral DNA, such as the M13mp18(+), is mainly single-stranded circular DNA with a hairpin (region of self-complementary helix). As shown in **Fig. 6**, the migration speed of M13mp18(+) is significantly slower than its supercoiled counterpart, M13mp18 RFI. This

is totally expected since the supercoiled structure is more compact than the circular structure, thereby migrating faster.

8. CE is thus shown to be a versatile tool for analysis of the topological state of plasmid DNA samples. The lower mobility and greater peak width of supercoiled plasmids allow easy identification although both linear and open circular DNA are also present. Using CE, it is possible to fractionate the topoisomers of supercoiled DNAs, permitting the monitoring of conformational change of a plasmid under different environments.

4. Notes

1. Alternative capillaries including polyacrylamide coated, poly(ethylene glycol) (PEG) coated capillary (such as DB-wax coated) can be used. The μSil-FC coated capillary is found to be most stable and reproducible.
2. Alternative separation buffers reported for similar analyses are the diluted Sigma buffer (1:10 dilution in 1X TBE), 0.125% PEG in TBE, and dilute HEC solution.
3. Alternative conditioning solutions are 0.5% HPMC, or 0.5% HEC.
4. For a new capillary being used for the first time, preconditioning is required to wet the surface, otherwise bubbles could form inside the capillary during electrophoresis which could be a serious problem.
5. Since DNAse I cleaves phosphodiester linkages, a digestion time longer than necessary could result in complete degradation of the DNA. We suggest monitoring the reaction periodically using CE to determine optimal conditions.

References

1. Gellert, M. (1981) DNA Topoisomerase. *Ann. Rev. Biochem.* **50,** 879–910.
2. Sobol, R. E. and Scanlon, K. J. (eds.) (1995) Clinical protocols list. *Cancer Gene Therapy* **2,** 67–74.
3. Culver, K. M. (1994) *Gene Therapy: A Handbook for Physicians.* Mary Ann Liebert, Inc. New York.
4. Vinograd, J., Lebowitz, J., Radloff, R., Watson, R., and Laipis, P. (1965) The twisted circular form of polyoma viral DNA. *Proc. Natl. Acad. Sci. USA* **53,** 1104–1111.
5. Cohen, A. S., Najarian, D. R., Paulus, A., Guttman, A., Smith, J. A., and Karger B. L. (1988) Rapid separation and purification of oligonucleotides by high-performance capillary gel electrophoresis. *Proc. Natl. Acad. Sci. USA* **85,** 9660–9663.
6. Maschke, H. E., Frenz, J., Belenkii, A., Karger, B. L., and Hancock, W. S. (1993) Ultrasensitive plasmid mapping by high performance capillary electrophoresis. *Electrophoresis* **14,** 509–514.
7. Hammond, W. R., Oana, H., Schwinefus, J. J., Bonadio, J., Levy, R. L., and Morris, M. D. (1997) Capillary electrophoresis of supercoiled and linear DNA in dilute hydroxyethyl cellulose solution. *Anal. Chem.* **69,** 1192–1196.
8. Hebenbrok, K., Schiigerl, K., and Freitag, R. (1993) Analysis of plasmid-DNA and cell protein of recombinant *Escherichia coli* using capillary gel electrophoresis. *Electrophoresis* **14,** 753–758.
9. Nackerdien, Z., Morris, S., Choquette, S., Ramos, B., and Atha, D. (1996) Analysis of laser-induced plasmid DNA photolysis by capillary electrophoresis. *J. Chromatogr. B* **683,** 91–96.
10. Mao, D. T., Levin, J. D., Yu, L., and Lautamo, R. M. A. (1998) High-resolution capillary electrophoretic separation of supercoiled plasmid DNAs and their conformers in dilute

hydroxypropylmethyl cellulose solutions containing no intercalating agent. *J. Chromatogr. B* **714,** 21–27.

11. Courtney, B. C., Williams, K. C., Bing, Q. A., and Schlager, J. J. (1995) Capillary gel electrophoresis as a method to determine ligation efficiency. *Anal. Biochem.* **228,** 281–286.
12. Slater, G. W., Mayer, P., and Grossman, P. D. (1995) Diffusion, Joule heating, and band broadening in capillary gel electrophoresis of DNA. *Electrophoresis* **16,** 75–84.
13. Wang, J. C. (1980) Superhelical DNA. *Trends Biochem. Sci.* **5,** 219–221.
14. Pulleyblank, D. E. and Morgan, A. R. (1975) The sense of naturally occurring superhelices and the unwinding angle of intercalated ethidium *J. Mol. Biol.* **91,** 1–15.

V

Practical Applications for Oligonucleotide and Nucleotide Analysis

23

Quality Control of Nucleotides and Primers for PCR

Martin J. Pearce and Nigel D. Watson

1. Introduction

The polymerase chain reaction (PCR) has become an essential tool in molecular biology for the rapid amplification and analysis of DNA *(1)*. The technique requires the use of oligonucleotide primers that bind to the DNA at a specific region of interest. The region is subsequently copied by enzymatic polymerization of deoxynucleotides. Among its many applications, PCR has found use in fields where the quality of results is of primary importance, such as in disease diagnosis *(2)* or forensic DNA analysis *(3)*. In these cases, PCR amplification from very small quantities of DNA is often required *(4,5)*, requiring high-reaction efficiency. Additionally, the consequences of ambiguous or false results in assays are far-reaching, and it is therefore essential that the PCR is highly specific. In order to obtain such high-performance standards the components of the PCR must be carefully controlled, particularly with respect to deoxynucleoside triphosphates (dNTPs) and oligonucleotide primers.

The quality of dNTPs and primers can have significant consequences for PCR performance. The dNTPs, which are polymerized into DNA with the release of pyrophosphate, can undergo successive dephosphorylation with extended storage to form the deoxynucleoside diphosphate (dNDP) and monophosphate (dNMP) compounds. Neither of these forms is a suitable substrate for polymerization during PCR and their presence will consequently reduce reaction efficiency. Primer quality has a similarly key role in PCR performance, the specificity of the reaction being dependent on the specificity of primer binding to target DNA. Other than primer design, the main determinant in specificity of binding is primer purity and this can display considerable variation between batches. Methods for assessing quality of dNTPs and primers are therefore highly desirable, particularly where unequivocal PCR results are required.

A number of methods have been reported for the separation of either nucleotides or primers by capillary electrophoresis (CE) *(6–13)* but none are suitable for the detection of both species. This is due to the fact that nucleotides are best separated at acidic pH in free solution *(6,7)* whereas oligonucleotides are best separated at basic pH in gel

From: *Methods in Molecular Biology, Vol. 162:*
Capillary Electrophoresis of Nucleic Acids, Vol. 1: Introduction to the Capillary Electrophoresis of Nucleic Acids
Edited by: K. R. Mitchelson and J. Cheng © Humana Press Inc., Totowa, NJ

or linear polymer filled capillaries ***(10,11)***. However, where separations are to be performed on high concentration samples to assess purity only, it is possible to design compromise methods that can be used for either sample type. This we have achieved with a pH 5.7 separation medium containing linear polymers. In addition to uniquely providing fast, high-resolution quality control of both nucleotides and primers in a single protocol, this method has the distinct advantage in comparison to alkaline oligonucleotide separation systems of operating at acidic pH, thereby causing less degradation of internal capillary coatings. This results in greatly increased capillary life-span. As an additional benefit, we have found that the system is also capable of extremely high-resolution separations of dsDNA, such as restriction digests or PCR amplification products. The system is therefore ideal for all CE separations associated with a PCR analysis, including both pre-PCR quality control and post-PCR product detection.

2. Materials

1. Separation medium: 0.1 *M* potassium phosphate, pH 5.7, 1.25% hydroxypropyl methyl cellulose (HPMC) (viscosity 50 cP at 2%), stored at 4°C (*see* **Notes 1–3**).
2. Capillary wash solution: 0.1 *M* phosphoric acid, stored at 4°C (*see* **Note 1**).
3. Capillary of 50 cm length and 50 μm id, internally coated with linear polyacrylamide (*see* **Note 4**). The detection window is formed 3 cm from the outlet.

3. Methods

1. Assemble the capillary within the instrument and purge with at least 10 vol (40 μL) of deionized water followed by 10 vol of capillary wash solution and then with a further 10 vol of deionized water. Fill the capillary with separation medium. Due to the viscosity of the medium, care must be taken not to over pressurize the system.
2. Dilute the dNTPs in deionized water to a final concentration of 200 μ*M* of each nucleotide. Dilute the PCR primers in deionized water to a final concentration of 50 ng/μL of each oligonucleotide primer.
3. Place 10 μL of the nucleotide or the primer sample to be analyzed at the cathode. With separation medium at the anode, load the sample onto the capillary by applying 8 kV for 6 s with negative to positive polarity. Rinse the cathodic end of the capillary by dipping it into deionized water and place it in the separation medium. Separate at 8 kV with detection by UV absorbance at 260 nm, 3 cm from the anode (*see* **Note 5**).
4. High purity dNTP solutions separate into four individual species at between 15 and 20 min migration time (**Fig. 1**). The separation order is according to charge to mass ratio, i.e., dCTP, dTTP, dATP, and dGTP. Degradation products of dNTPs (diphosphates and monophosphates) display increased migration time due to decreased charge to mass ratio (**Fig. 2**).
5. High purity PCR primer separates as a single species, with a 20-mer oligonucleotide having a migration time of between 15 and 20 min depending on the DNA sequence (**Fig. 3**). Low purity primers, which have either degraded or contain synthesis failure sequences, display additional peaks at varying migration times (**Fig. 4**).
6. Following each separation, purge the capillary with deionized water and capillary wash solution as described in **step 1**. If the capillary is not to be reused immediately, purge with nitrogen for 2 min to dry.

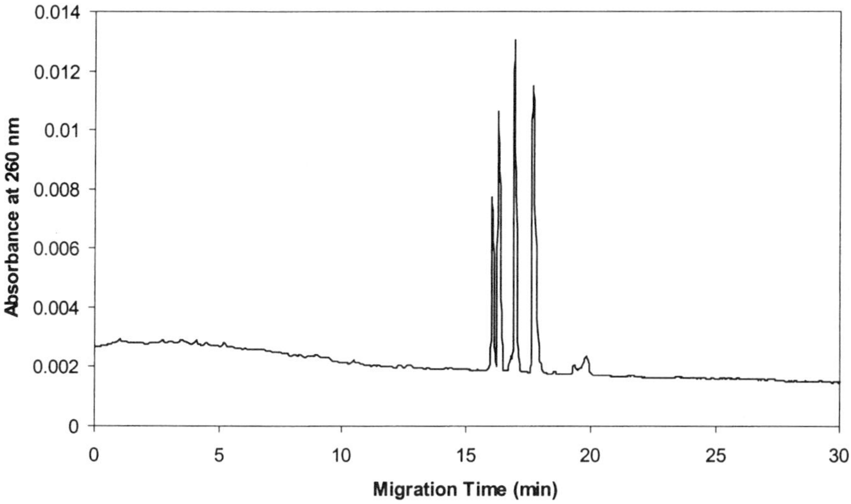

Fig. 1. Separation of high purity dNTP mixture in 0.1 *M* phosphate, pH 5.7, 1.25% HPMC. The separation order is dCTP, dTTP, dATP, and dGTP.

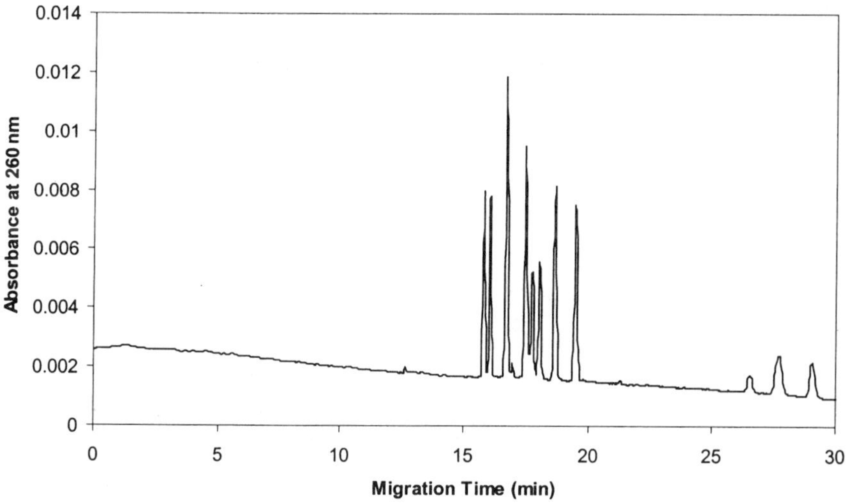

Fig. 2. Separation of a low purity dNTP mixture, which contains diphosphate and monophosphate degradation products. The separation order is 4× dNTPs, 4× dNDPs, and 3× dNMPs. The use of the dNTP mixture in a PCR could reduce amplification efficiency.

4. Notes

1. Despite the relatively large diameter capillary used in the method, any capillary is prone to blocking. We therefore filter all separation and wash solutions through a 0.45-µm mem-

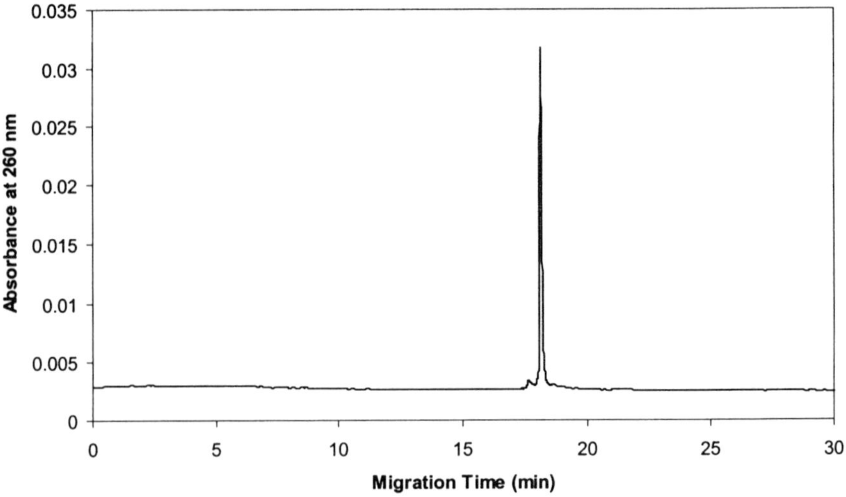

Fig. 3. Separation of high purity 20-mer PCR primer in 0.1 *M* phosphate, pH 5.7, 1.25% HPMC.

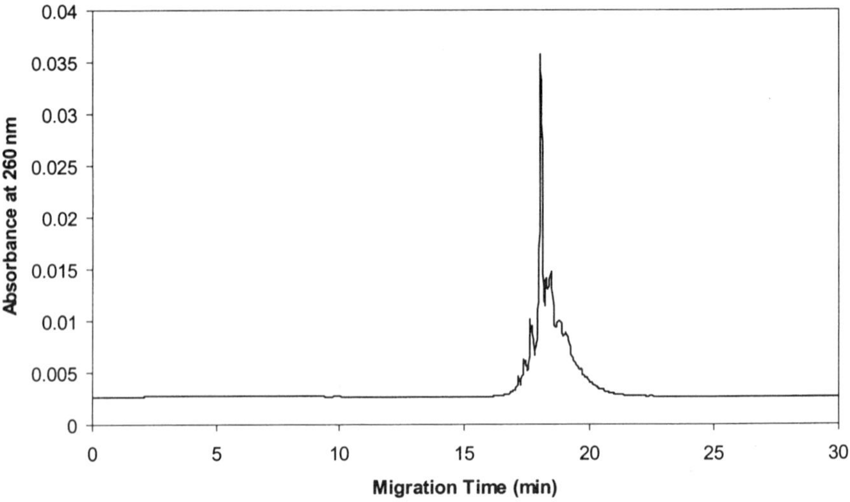

Fig. 4. Separation of low purity 20-mer PCR primer containing synthesis failure sequences. The use of this incomplete primer in a PCR could compromise both the efficiency and the specificity of amplification.

brane to remove any particulate matter, followed by vacuum degassing to remove dissolved gases. In the case of separation medium, filtration is performed prior to addition of HPMC.

2. As with all cellulose polymers, HPMC can be difficult to solubilize. We dissolve the polymer overnight at 4°C with constant gentle agitation. Other researchers have reported heating the solution to be beneficial.

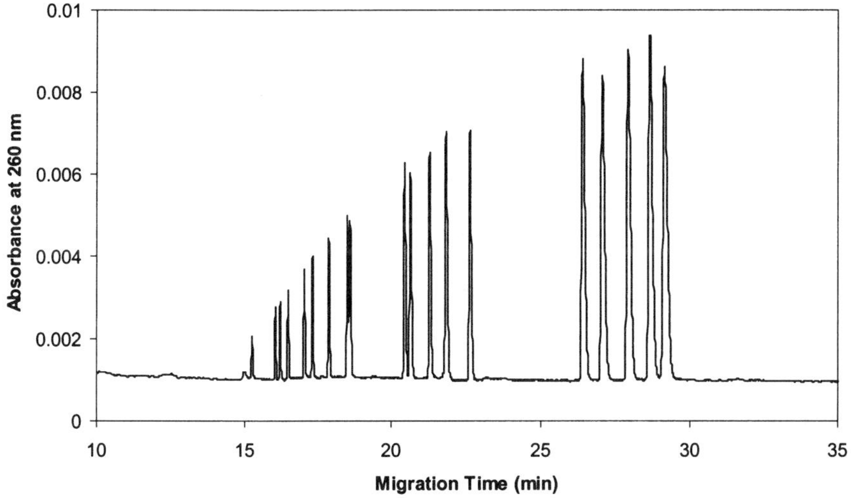

Fig. 5. Separation of dsDNA (100 ng/µL *Hae*III digest of pBR322 DNA) in 0.1 *M* phosphate, pH 5.7, 1.25% HPMC. The fragment sizes are 21, 51, 57, 64, 80, 89, 104, 123, 124, 184, 192, 213, 234, 267, 434, 458, 504, 540, and 587 bp.

3. We routinely use a low viscosity (shorter chain length) HPMC polymer in preference to the more widely used high viscosity forms (up to 4000 cP at 2% [w/v]). We have found that the low viscosity polymer gives equivalent resolution of nucleotides, primers and dsDNA to the resolution achieved with high viscosity polymer. The use of shorter HPMC polymers has the benefit of easier solubilisation, faster capillary purging, and the ability to use higher concentrations, thereby increasing the range of oligonucleotides that may be separated and analyzed.
4. The role of the internal linear polyacrylamide coating of the capillary is to reduce the electroendosmotic flow. It should therefore be possible to substitute the specified capillary for others, utilizing internal coating chemistries to the same effect.
5. The separation medium and protocol can also be used for highly efficient separation of dsDNA in the size range 50–1500 bp (**Fig. 5**), with single base resolution achieved between fragments up to at least 125 bp. A review of other methods for the purification of short PCR products is given in **ref. *14***.

References

1. Saiki, R. K., Gelfand, D. H., Stoffel, S., Scharf, S. J., Higuchi, R., Horn, G. T., Mullis, K. B., and Erlich, H. A. (1988) Primer-directed enzymatic amplification of DNA with a thermostable DNA polymerase. *Science* **239,** 487–491.
2. Schochetman, G., Ou, C.-Y., and Jones, W. K. (1988) Polymerase chain reaction. *J. Infect. Dis.* **158,** 1154–1157.
3. Reynolds, R., Sensabaugh, G., and Blake, E. (1991) Analysis of genetic markers in forensic DNA samples using the polymerase chain reaction. *Anal. Chem.* **63,** 2–15.
4. Higuchi, R., von Beroldingen, C. H., Sensabaugh, G. F., and Erlich, H. A. (1988) DNA typing from single hairs. *Nature* **332,** 543–545.

5. Jeffreys, A. J., Wilson, V., Newman, R., and Keyte, J. (1988) Amplification of human minisatellites: Towards DNA fingerprinting of single cells. *Nucleic Acids Res.* **16,** 10,953–10,971.
6. Takigiku, R. and Schneider, R. E. (1991) Reproducibility and quantitation of separation for ribonucleoside triphosphates and deoxyribonucleoside triphosphates by capillary zone electrophoresis. *J. Chromatogr. A* **559,** 247–256.
7. O'Neill, K., Shao, X., Zhao, Z., Malik, A., and Lee, M. L. (1994) Capillary electrophoresis of nucleotides on ucon-coated fused silica columns. *Anal. Biochem.* **222,** 185–189.
8. Uhrova, M., Deyl, M., and Suchanek, M. (1996) Separation of common nucleotides, mono-, di- and triphosphates by capillary electrophoresis. *J. Chromatogr. B-Biomedical Applications* **681,** 99–105.
9. Geldart, S. E. and Brown, P. R. (1998) Analysis of nucleotides by capillary electrophoresis. *J. Chromatogr. A* **828,** 317–336.
10. Cohen, A. S., Najarian, D. R., Paulus, A., Guttman, A., Smith, J. A., and Karger, B. L. (1988) Rapid separation and purification of oligonucleotides by high-performance capillary gel electrophoresis. *Proc. Natl. Acad. Sci. USA* **85,** 9660–9663.
11. Manabe, T., Chen, N., Terabe, S., Yohda, M., and Endo, I. (1994) Effects of linear polyacrylamide concentrations and applied voltages on the separation of oligonucleotides and DNA sequencing fragments by capillary electrophoresis. *Anal. Chem.* **66,** 4243–4252.
12. Khan, K., Van Schepdael, A., and Hoogmartens, J. (1996) Capillary electrophoresis of oligonucleotides using a replaceable sieving buffer with low viscosity-grade hydroxyethyl cellulose. *J. Chromatogr. A.* **742,** 267–274.
13. Chiari, M., Damin, F., Melis, A., and Consonni, R. (1998) Separation of oligonucleotides and DNA fragments by capillary electrophoresis in dynamically and permanently coated capillaries, using a copolymer of acrylamide and beta-D-glucopyranoside as a new low viscosity matrix with a high sieving capacity. *Electrophoresis* **19,** 3154–3159.
14. Devaney, J. M. and Marino, M. A. (2001) Purification methods for preparing polymerase chain reaction products for capillary electrophoresis analysis, in *Capillary Electrophoresis of Nucleic Acids,* Vol. 1 (Mitchelson, K. R. and Cheng, J., eds.), Humana Press, Totowa, NJ, pp. 43–49.

24

Analysis of Modified Oligonucleotides with Capillary Gel Electrophoresis

Lawrence A. DeDionisio

1. Introduction

Short segments of DNA, which have been synthesized chemically, are commonly known as synthetic oligonucleotides (ODNs). The length of the ODNs typically utilized in the lab varies but is on average between five and forty nucleotides. Chemical modifications may be selectively placed at various locations within the molecule including the backbone, the heterocyclic base, or the sugar moiety *(1,2)*. For the work discussed in this chapter, chemical alterations were placed in three locations: the backbone, the 3'-bridging position between the backbone and the sugar moiety, and at the 5' end.

1.1. Examples of Modified Oligonucleotides

1.1.1. Phosphorothioate Containing ODNs

One important and often utilized backbone modification involves the substitution of a nonbridging oxygen atom with a sulfur atom. This modification is known as a phosphorothioate linkage (**Fig. 1**). The phosphorothioate linkage was first incorporated into DNA by enzymatic polymerization of dNTPαS, where N represents one of four nucleotide bases, adenine (A), guanine (G), cytidine (C), or thymine (T), to elucidate enzyme functions and mechanisms *(3,4)*. It was also found that polyribonucleotides with phosphorothioate backbones possess antiviral properties *(5)*. This effect was attributed to the ability of the phosphorothioate to protect the poyribonucleotide against premature enzymatic degradation by ribonuclease within the cell. After the phosphorothioate modification was shown to be useful for studies involving nucleic acids, the modification was incorporated into ODNs via chemical synthesis methods, which served to improve the efficiency of chain assembly and enable larger quantities of phosphorothioate ODNs *(2)*.

From: *Methods in Molecular Biology, Vol. 162:*
Capillary Electrophoresis of Nucleic Acids, Vol. 1: Introduction to the Capillary Electrophoresis of Nucleic Acids
Edited by: K. R. Mitchelson and J. Cheng © Humana Press Inc., Totowa, NJ

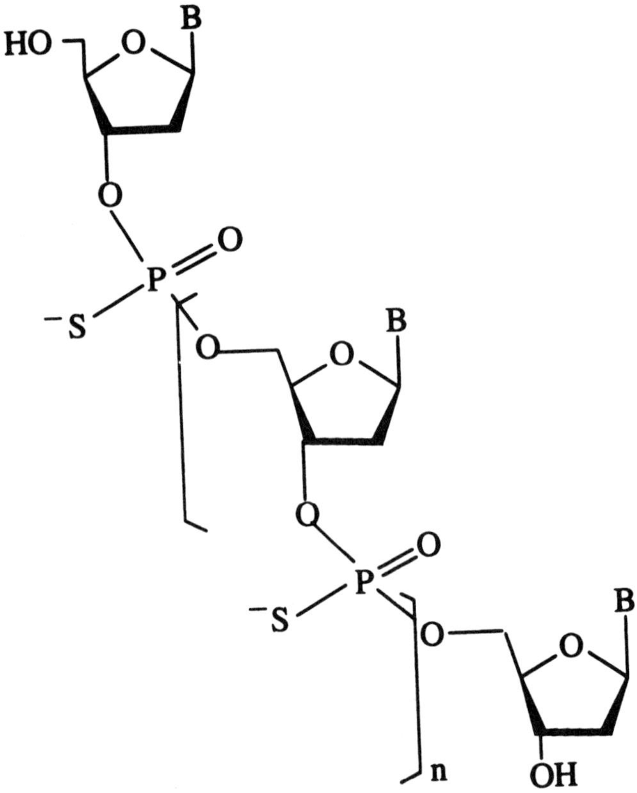

Fig. 1. Molecular structure of the phosphorothioate linkage in an ODN. A phosphodiester linkage possesses an O^- in place of the S^- atom found in the phosphorothioate linkage. The length of the ODN = $n + 2$. A chiral center exists in the phosphorothioate linkage at the P atom. B represents one of the four nucleotide bases, adenine (A), guanine (G), cytidine (C), and thymine (T).

1.1.2. N3'→ P5' Phosphoramidate Linkage and Mixed-Backbone ODNs

Another backbone modification, known as a N3'→ P5' phosphoramidate (amidate) linkage, involves the substitution of the 3'-bridging oxygen atom with an amino group (**Fig. 2**). For this discussion, amidate linkages are used in conjunction with phosphorothioate linkages within the same ODN. ODNs, which contain two or more different backbone modifications, are referred to by various names; here, we will refer to them as mixed backbone ODNs. Mixed backbone ODNs are also referred to as second generation, in contrast to first generation or uniformly modified ODNs *(6,7)*. They are useful when the properties of more than one backbone modification is needed to illicit a chemical or biological response that one modification alone could provide. Two mixed backbone ODNs that will be covered are the phosphorothioate-amidate and the

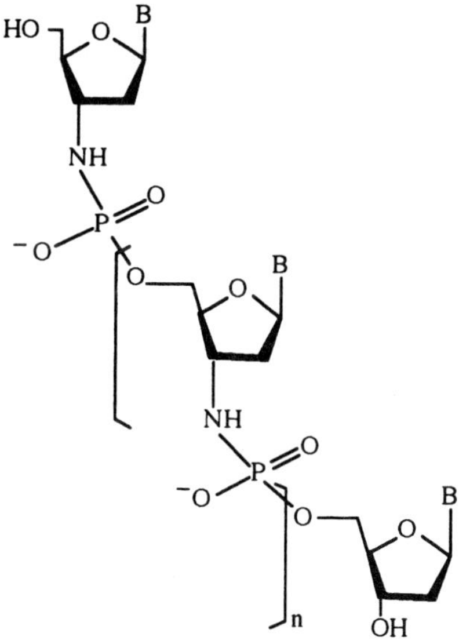

Fig. 2. Molecular structure of the N3'→P5' phosphoramidate (amidate) linkage. The –NH- group in the amidate linkage is substituted with an -O- atom in the phosphodiester linkage. As in **Fig. 1**, the length of the ODN = n + 2, and B represents one of four bases (A, G, C, or T).

phosphorothioate-phosphodiester chimeras. The phosphodiester linkage is the naturally occuring backbone found in DNA. There are a variety of other mixed backbone combinations apart from these two ***(6–9)***, which are not applicable to our discussion.

1.1.3. 5'-Fluorescein-Labeled ODNs

Another type of synthetically modified ODN discussed herein is the DNA-fluorescein conjugate. These conjugates are either automatically synthesized on a DNA synthesizer, or the ODN is first modified during its synthesis at the 5'-end with a terminal amine (-NH_2), and then this modified ODN is reacted with NHS-fluorescein ***(10)*** in a separate reaction. Reacting NHS-fluorescein with the ODN in this fashion insures a higher purity because the conjugation occurs apart from the automated synthesis, after the ODN has been purified. Alternatively, commercially available FAM or fluorescein-on phosphoramidite can be used during the automated synthesis, and then the fluorescein conjugate is purified from the crude ODN mixture. Fluorescein-labeled ODNs are utilized as probes for hybridization experiments ***(11)*** or as fluorescently labeled primers for polymerase chain reaction (PCR) ***(12)***.

1.2. Reasons for Modifying an ODN

1.2.1. Modified ODNs and Antisense Technology

Chemically modified ODNs have potential utility in various applications. For our work, backbone modifications have served to increase stability toward nuclease degradation within a cell. However, the effects of modified ODNs within a cell are not completely understood, but much has been accomplished within the field to correlate the interactions of modified ODNs with cellular mechanisms *(2)*. Within the field of antisense technology, backbone modifications inhibit naturally occuring nucleases that serve to break down foreign DNA within the cell *(1,2)*. Some of the first antisense experiments were conducted with unmodified phosphodiester ODNs in the inhibition of *Rous sarcoma* virus *(13)*. In principal, an antisense ODN is designed to hybridize with mRNA via Watson/Crick base pairing and block the translation machinery to inhibit protein translation. This mechanism came to be known as steric blocking *(14)*. Earlier work within the field had provided evidence that the results of an antisense hybridization also served to degrade the RNA strand by RNase H *(15)*. RNase H acts by facilitating the hydrolysis of the RNA sense strand in the RNA/DNA duplex. In this case, the DNA strand is the antisense ODN. The phosphorothioate linkage was one of the first modifications utilized to protect the antisense ODN from enzymatic degradation. It was later discovered that phosphorothioate ODNs exhibit both specific and non-specific protein binding within the cell *(16,17)*.

In an effort to develop better antisense ODNs, other backbone modifications were soon investigated. The amidate linkage (**Fig. 2**) was one result of these investigations. This backbone modification was first considered valuable for antisense technology because, unlike the phosphorothioate backbone, it is achiral *(18–20)*. Subsequent studies proved that uniformly modified amidate ODNs form duplexes with RNA with high affinity (i.e., possess relatively high melting temperatures, Tms) *(21)*; higher Tms were considered advantageous. After it was determined that this amidate linkage was a viable modification, studies involving amidate ODNs and RNA duplexes revealed that ODNs with uniformly modified amidate backbones do not illicit the RNase H response described above *(22)*. Instead, amidate ODNs appear to work solely through steric blocking of the translation machinery. Furthermore, amidate ODNs were found to have less nonspecific protein binding than phosphorothioate ODNs *(23,24)*, and could be designed to specifically bind ribonucleo-proteins. This discovery was exploited to create a structural RNA mimetic *(24)*. Thus, benefits of the uniformly modified amidate ODN outweighed the fact that the existence of this linkage within an ODN/RNA duplex does not activate an RNase H response. Consequently, much effort was applied toward optimizing the synthesis and purification of amidate containing ODNs *(25–27)*.

1.2.2. Modified ODNs as PCR Primers

Modified ODNs as PCR primers have also been used to block exonuclease digestion for the detection of known point mutations *(28)*. For this application, three or four of the nucleotide linkages at the 5'-end are synthesized as phosphorothioate linkages. When this partially modified ODN is used as a primer for PCR, the phosphorothioate containing strand of the PCR product will be protected from subsequent 5'-exonuclease digestion. Exonuclease digestion of the unmodified strand in the dsPCR product gives ssDNA, which is then probed for point mutations. A modified ODN PCR primer will be used here to demonstrate the utility of CGE for the analysis of ODNs that contain the phosphorothioate linkage.

1.3. Advantages of Capillary Gel Electrophoresis as an Analytical Tool

1.3.1. ODNs that Contain Phosphorothioate Linkages

When phosphorothioate linkages are introduced into an ODN, a chiral center is created at the phosphorous atom during synthesis (**Fig. 1**), resulting in a mixture of 2^n diastereomers, where n = the number of phosphorothioate linkages *(2)*. This often causes peak broadening and subsequent loss of resolution during high performance liquid chromatography (HPLC) analyses of phosphorothioate containing ODNs *(29,30)*. A loss of resolution is thus manifested between the product ODN and the primary impurities, known as failure sequences. Failure sequences result during the automated synthesis of the ODN when stepwise coupling to the nascent strand fails. The largest failure sequence is the full-length ODN minus one nucleotide base. For example, if the full-length ODN is 21 nucleotide bases long, the largest failure sequence would be a mixture of sequences 20 nucleotide bases in length *(31)*. This mixture is collectively known as the $N-1$ mer, where N is the full-length or product ODN. For an analysis of the length integrity of an ODN, it is crucial that the $N-1$ mer is fully resolved from the product peak. Consequently, for the analysis of phosphorothioate containing ODNs, most existing HPLC methods cannot separate the $N-1$ mer from the product with baseline resolution. A likely reason for this is that slight differences in the hydrophobicity between the diastereomers results in their partial separation on an HPLC column which manifests as broader peaks within resultant chromatograms (**Fig. 3**). In fact, reverse phase (RP)-HPLC is capable of separating various diastereomers of small ODNs that contain two or three phosphorothioate linkages *(32)*. Such ODNs would have $2^2 = 4$ and $2^3 = 8$ diastereomers, respectively. If more than three phosphorothioate linkages are present and if the ODN is longer than five nucleotide bases, then RP-HPLC cannot fully resolve these diastereomers. To reiterate, in the interest of determining full-length integrity of the product ODN, separation of diastereomers introduced by the chiral center present in the phosphorothioate linkage is unwanted.

In addition to differences in hydrophobicity, slight differences also exist in the charge distribution between the diastereomers *(32)*. Consequently, for electrophoretic methods of analysis, peak or band broadening of a phosphorothioate ODN is also observed; however, with the proper running conditions, CGE can resolve phosphorothioate ODNs that differ in length by one nucleotide base (**Fig. 4**). This is also true when polyacrylamide gel electrophoresis (PAGE) is used to analyze phosphorothioate ODNs, but capillary gel electrophoresis (CGE) is considered superior to PAGE for these types of analyses *(33)*. Thus, one advantage of electrophoresis relative to HPLC, is the ability of eletrophoresis to resolve the critical $N-1$ mer impurity from the product peak of any ODN that contains the phosphorothioate linkage.

1.3.2. Analysis of 5' Fluorescein-Labeled ODNs

Similarly, HPLC methods are often unable to resolve the $N-1$ mer of flourescein-labeled ODNs that have been synthesized with the FAM or fluorescein-on phosphoramidite reagent (*see* **Subheading 1.1.3.**). The fluorescein molecule is similar in size to a nucleotide base; however, its molecular structure *(10)* imparts a more hydrophobic

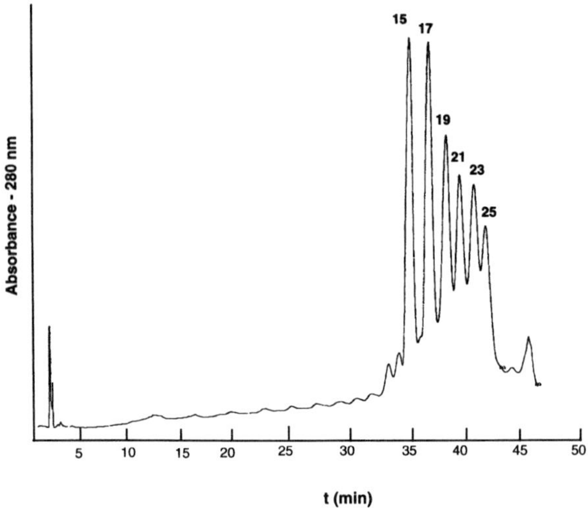

Fig. 3. Ion exchange HPLC of a mixture of six ODNs (15–25 mers), differing in length by two nucleotides, that were uniformily modified throughout their backbones with the phosphorothioate linkage. The profile illustrates typical peak broadening that occurs with HPLC analyses of phosphorothioate containing ODNs. Chromatographic conditions are as follows: buffer A is 0.05 *M* Na_3PO_4 (pH 8.1), buffer B is 1.0 *M* NaSCN, buffer C is CH_3CN; the flow rate is 1 mL/min, and a linear gradient of 0.8%/min of buffer B is applied over 50 min. Buffer C is kept constant at 30%. A Dionex (Sunnyvale, CA) Nucleopac PA-100 (250 × 4 mm) column is used, and the detector is set at 280 nm.

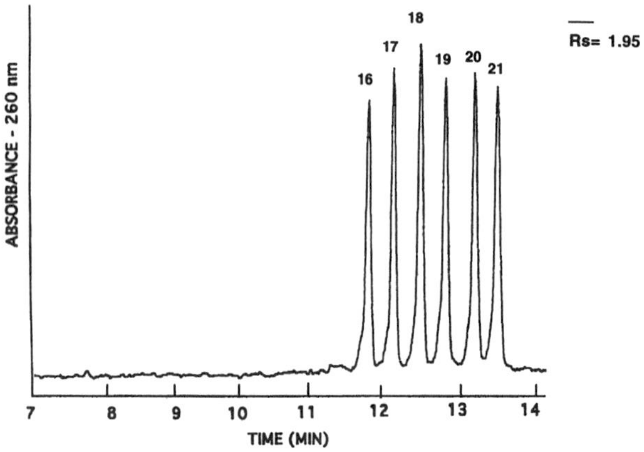

Fig. 4. Electropherogram of a mixture of six uniformily modified phosphorothioate ODNs (16–21 mers) that differ in length by one nucleotide. Sample concentration is 0.5 AU_{260}/mL. Electrophoresis is conducted with an electrokinetic injection at –8 kV for 5 s, and a constant running voltage of –22 kV is used. Gel matrix and running buffer consist of 10% hydroxyethyl cellulose (HEC) in 35 m*M* Tris-HCl, 5.6 m*M* H_3BO_3, and 15% ethylene glycol (pH 9.0) with 50 cm × 100 μm id capillaries. The average resolution (R_s) between the six major peaks is 1.95. R_s is defined here as $2(ty - tx)/(Wy + Wx)$, where *y* and *x* refer to the peaks of interest, *t* is migration time in decimal min and *W* is the peak width in decimal min at the base.

character than the four nucleotide bases that are normally incorporated during ODN synthesis. Therefore, ion exchange (IE) HPLC methods that are used for the analysis of ODNs cannot easily resolve the $N - 1$ mer of these 5' labeled ODNs since IE-HPLC methods rely on charge differences between the full-length ODN and the $N - 1$ mer. The hydrophobic nature of fluorescein can, theoretically, affect the HPLC elution profile of fluorescein-labeled ODNs. This would serve to counter-act IE-HPLC separations based on charge density. On the other hand, gel eletrophoresis separates on the principles of both charge to mass (q/m) and size or molecular weight. The molecular weight of the $N - 1$ mer and full-length ODN is sufficiently different to give good resolution with electrophoretic techniques regardless of the hydrophobic nature of a modification. This principle also holds true for phosphorothioate containing ODNs, which are difficult to separate with HPLC methods because of slight differences in hyrophobicity between diastereomers (*see* **Subheading 1.3.1.**). Although we are not including data here regarding the analysis of fluorescein-labeled ODNs, their purification and analysis follows similar methods as described in **Subheadings 2.** and **3.**

The automated synthesis of each modified ODN requires a unique set of reagents and synthesis protocols. Although methods of synthesis are important, in the discussion of an analytical technique for determining the purity of modified ODNs is the focus, a discussion of the reagents and protocols for their synthesis is, for the most part, not applicable. For those interested, there is literature available that describes the synthesis protocols for the type of mixed backbone ODNs that we are describing here *(27,34,35)*. Even though we feel a discussion of the synthesis protocols is unnecessary, the purification of the mixed backbone ODNs is germane to a discussion of their analysis. Therefore, the reagents, instruments, and protocols used for the purification as well as for the analysis of specific modified ODNs will be covered in detail below.

2. Materials

2.1. Reverse-Phase High Performance Liquid Chromatography

1. RP-HPLC purifications are conducted with a Rainin Dynamax PC™ liquid chromatography system equipped with a Rainin Dynamax absorbance detector (model UV-C) from Varian (Palo Alto, CA), and a 100 Å, 8 µm PLRP-S column (300 × 7.5 mm id) from Polymer Laboratories (Amherst, MA).
2. Alternatively, for ODNs that contained amidate linkages, RP-HPLC purifications are conducted with a Perkin-Elmer-Applied Biosystems (PE-ABI, Foster City, CA) series 410 LC and a 757 absorbance detector (PE-ABI) with the same PLRP column described above. For the sake of utility, a separate LC system (the PE-ABI series 410 LC) is used for ODNs that contain amidate linkages because a different set of buffers are utilized for their purification.
3. Buffers used for purifications were as follows. For mixed backbone ODNs with phosphodiester and phosphorothioate backbones, buffer A is 2% acetonitrile (ACN) in triethylammonium acetate (0.1 *M* TEAA, pH 7.0); buffer B is 100% ACN. Buffer A is formulated by diluting 2 *M* TEAA with deionized water and adding UV-grade ACN such that the final concentration of ACN is 2% v/v. Final volumes vary depending upon how the buffer is formulated at any one time. Any brand of UV-grade ACN is acceptable for buffers A and B (*see* **Note 1**).

4. For mixed backbone ODNs with amidate linkages, buffer A is 2% ACN in triethylammonium bicarbonate (0.1 *M* TEAB, pH 8), and buffer B is 100% ACN. For buffer A, a stock solution of 1 *M* TEAB must first be formulated. This is done by mixing 139 mL of triethylamine (Aldrich Milwaukee, WI), on a stir plate with approx 800 mL of deionized water and then slowly adding dry ice (CO_2) with constant stirring until the pH measures 8.0. The volume must then be brought to 1 L to give a 1 *M* solution. From this 1 *M* stock solution, a 0.1 *M* running buffer is formulated. ACN is also added to this buffer such that its final concentration is 2% (v/v) (*see* **Note 1**).

2.2. Capillary Gel Electrophoresis (see Note 2)

1. CGE analysis is conducted with a Beckman (Fullerton, CA) P/ACE 5510. The capillaries, the gel used, and the running buffer are all purchased from Beckman. The capillaries are 65 cm long and have an id of 100 μm. The capillary detector window lies 20 cm from the origin or sample injection end of the capillary.
2. The gel matrix is eCAP® ssDNA 100-R gel (Beckman), and the running buffer is 44% Tris, 56% boric acid, and 7 *M* urea. Beckman provides reagents and protocols for buffer formulation and gel hydration. The buffer is formulated by adding 135 mL deionized water to a bottle of dry Tris-borate reagent. After stirring for 20–30 min, dry urea is added slowly to this bottle with constant stirring. The reagent becomes very cold because the dissolution of urea is endothermic. Thus, the buffer must be allowed to come to ambient temperature (approx 1.5–2 h) before it can be used, and it is stable for 30 d if stored at 4°C. The magnetic stir bar can be kept with the buffer because each time the buffer is used it must be allowed to come to room temperature with constant stirring.
3. Each time the buffer is used, aliquot only what is needed for the day and filter this through a 0.2-μm filter. We use a disposable syringe in combination with an *Acrodisc*® (Gelman Sciences Ann Arbor, MI) filter. To hydrate the lyophilized gel, filter 5.0 mL of the Tris-borate-urea buffer directly into the bottle containing the gel material and stir with a magnetic stir bar until the gel is dissolved (approx 5 h). This gel solution is stable for up to 30 d if stored at 4°C.
4. We often utilize an internal standard for CGE analysis. For the work presented here, a polythymidine ODN consisting of 21 thymidine bases with phosphodiester linkages [(dT_{21})] is the internal standard (*see* **Note 3**).

3. Methods

3.1. RP-HPLC Purifications and Final Product Workup (see Note 1)

1. The running conditions for each modified ODN are all the same, apart from the differences in the buffer system described in **Subheading 2.1.** After a synthesis reaction, the volume of the crude modified ODN product is adjusted to approx 1 mL and it is buffered with 2 *M* TEAA (0.15 mL) immediately prior to injection onto the RP-HPLC column.
2. Initial running conditions are 95% buffer A and 5% buffer B, and the detector is set at 292 nm. After the crude sample is injected, A linear gradient is run, ramping 1%/min of buffer B over 40 min and fractions are collected between 30 and 38 min. Fractions are collected every 30 s so that each fraction contains approx 0.5 mL, and the number of fractions vary depending upon the amount of crude material injected. Thus, the product ODN is collected when the ACN concentration is between 30 and 38%. After fractions are collected, the post-RP-HPLC purification protocol varies depending upon whether or not the amidate linkage is present within the ODN.

3. Fractions from purifications that did not contain amidate linkages are concentrated to approx 0.2 mL *in vacuo* with a Speed-Vac, and 0.8 mL of glacial acetic acid is added to each fraction to cleave the dimethoxytrityl (DMT) group from the 5'-end of the ODN. The fractions are kept in 80% aq. acetic acid at ambient temperature for approx 1 h and then evaporated to dryness *in vacuo* with a Speed-Vac.
4. Each fraction is then dissolved in 0.32 mL of sterile water and transferred into 1.5-mL Eppendorf tube. At this point, 0.08 mL of aq. 5 *M* NaCl solution is added. The modified ODN is precipitated from solution with 1.0 mL of ethanol and allowed to sit at 0°C for one h after which the precipitate is pelleted out at ~20,000*g* (14,000 rpm) for 10 min in an Eppendorf centrifuge.
5. The precipitates are then dissolved in 0.02 mL of sterile water, and aliquots (0.005 mL) are taken from each fraction and are analyzed by CGE (*see* **Subheading 3.2.3.**) to determine their purity.
6. After this analysis, those fractions which contain the product are combined and concentrated *in vacuo* to approx 0.05 mL. This solution is then desalted over a Sephadex G-25 (NAP-5) column (Pharmacia, Piscataway, NJ). A 1:100 dilution is made on the desalted product and the concentration (absorbance units at 260 nm (AU_{260}/mL) is measured with a Hewlett Packard (Palo Alto, CA) 8452A diode array spectrophotometer.
7. RP-HPLC purifications of ODNs that contain the amidate linkage are worked up differently because of a 3'-3' conjugation of a dimethoxytrityl-uridine nucleotide (DMT-U) (*see* **Note 1**). Fractions from these purifications are always collected directly into 1.5-mL Eppendorf tubes and then concentrated *in vacuo* to approx 0.2 mL.
8. Aliquots (0.005 mL) are taken from each of these fractions and analyzed with CGE (*see* **Subheading 3.2.3.**). The fractions, which contain the product are combined into a 4.0-mL glass screw-cap vial, are concentrated to dryness *in vacuo*. The resulting residue is dissolved in a mixture of 1 *M* NaF (0.2 mL) and concentrated NH_4OH (0.2 mL) and is then incubated at 58°C for approx 16 h.
9. This solution is then concentrated to 0.2 mL *in vacuo,* and the DNA is precipitated with 0.6 mL of ethanol and allowed to sit at 0°C for 1 h. The precipitate is pelleted via centrifugation, and after the supernatant is removed the product is dissolved in 0.5-mL sterile water and desalted over a Sephadex G-25 (NAP-5) column. A 1:100 dilution is conducted on the final product and the concentration is determined spectrophotometrically (*see* **step 6**).

3.2. Capillary Gel Electrophoresis

3.2.1. RP-HPLC Fraction and Final Product Analysis with CGE

1. Beckman provides a protocol with their eCAP® ssDNA 100-R kit that describes in detail the conditions necessary for installing, filling, and maintaining their capillaries. This protocol also describes conditions for preliminary test runs and explains how to change the polarity of the instrument, which is necessary when analyzing DNA samples. We follow these protocols with some variations.
2. The following paragraphs describe our method for filling the capillary with the hydrated eCAP® ssDNA 100-R gel and our protocols for analyzing fractions and the final product from a typical RP-HPLC purification of a modified ODN. With the use of an internal standard, relative mobility (RM) is used in place of real time migration values. RM is defined as t_S/t_{IS} = RM, where t_S is the migration time of the sample, and t_{IS} is the migration time of the internal standard.
3. The following conditions are constant for all CGE runs on the P/ACE 5510. All reagents, including Eppendorf tubes that contain either reagents or samples, are placed into spe-

cialized 4.5-mL P/ACE vials. UV detection is set at 254 nm and the running temperature is kept at 30°C. The electrical connections on the instrument must be configured in the following manner. The inlet connector should be connected to the positive (+) terminal, and the outlet connector should be connected to the negative (–) terminal.

3.2.2. Filling the Capillary and the Equilibration Step

1. After a capillary has been installed on the P/ACE 5510, a high-pressure rinse is used to fill it with the hydrated eCAP® ssDNA 100-R gel. Approximately 0.1 mL of hydrated gel is first pipeted into a 0.5-mL Eppendorf tube and is centrifuged for 30 s to remove any trapped air bubbles. The top of this tube is cut off and the tube is placed in position 33 on the instrument (*see* **Note 2**).
2. A 10 min high-pressure rinse is conducted with position 33 defined as the inlet, and position 10, which contains an empty vial, defined as the outlet. Any gel on the outside of the capillary is then cleaned off it by dipping the column ends in sterile water. We do this by placing vials of sterile water in positions 32 and 2, moving the autosampler so that the capillary ends are above these positions, and lowering the ends into the water for a few seconds.
3. A manual run is then conducted to equilibrate the capillary in the following manner. Two vials of running buffer (3 mL in each) are first filtered through a 0.2-µm *Acrodisc®* filter and degassed *in vacuo* for about 2 min. These are placed in position 11, which is defined as the inlet, and position 1, which is defined as the outlet. The voltage is then ramped up over 3 min to 8.1 kV and then the run is continued at this voltage for 10 min. If the baseline remains relatively flat and the current does not fluctuate significantly, then the capillary is equilibrated and ready for routine analyses.
4. The above procedures are conducted at least once a day, usually in the morning, and repeated during the day if the gel within the capillary showed signs of failure. At the end of each day, or in the morning if the instrument is utilized for overnight runs, the gel is flushed out of the capillary with sterile water. This is done with a high-pressure rinse from inlet position 32, which contains a vial of sterile water, to outlet position 10, which holds an empty vial.

3.2.3. Analytical Runs for Mixed Backbone ODNs

1. After the equilibration step, the gel-filled capillary is used to analyze both the fractions from RP-HPLC purifications and the final products. The running conditions described here are utilized for mixed-backbone ODNs containing phosphodiester and phosphorothioate linkages. Slight variations are necessary if an amidate linkage is present. These differences in protocol are noted below.
2. Sterile water is always used for sample dilutions and the final volume for each sample vial is kept at 0.04 mL. Each sample vial contains 0.002 mL of sample, 0.020 mL of an internal standard with a concentration of 0.528 AU_{260}/mL, and 0.018 mL of sterile water. Thus, the final concentration of the internal standard is 0.264 AU_{260}/mL, while the concentration of the sample is varied.
3. Sample injections are electrokinetic for 3.0 s at 10 kV, and run times are 35 min at a constant voltage of 8.1 kV. Positions 11 and 1 are defined as the inlet and the outlet respectively, and the vials put into these positions contain the running buffer, which is used to equilibrate the capillary. In general, these buffers are changed daily, and a gel is reliable for approx 12 runs of 35 min. If amidate linkages are present, then the sample aliquots taken from the RP-HPLC fractions are increased to 0.005 mL, and injections are conducted for 20 s. The total volume for these samples is still 0.04 mL and the injection voltage is 10 kV.

3.3. Interpretation of Data

1. Although these purification and analytical methods are designed for automation and routine analysis, there are many nuances about these procedures which are best explained with a "hands on" approach within the laboratory. With this said, we will try to explain here within the context of our discussion how we handle the specific details and lab technique that come with daily experience. While CGE is an efficient method for the analysis of an ODN's length relative to failure sequences like the $N - 1$ mer, it is not able to give an accurate measurement of an ODN's molecular weight. Determining the molecular weight of an ODN with electrophoresis is difficult because of issues concerning secondary structures, tertiary aggregates, and whether the 7 *M* urea can disrupt these secondary and tertiary phenomena *(29)*.
2. The CGE method described here is used to systematically monitor an RP-HPLC purification of a modified ODN. It is designed as an alternative to the more traditional polyacrylamide gel (PAGE). It has the potential to provide a fully automated method for routine analysis of any modified ODN. However, each modified ODN is unique by nature of its particular modification and therefore, some level of data interpretation, specific to the modification, becomes necessary. For this discussion, we have chosen a mixed backbone ODN, which contains three phosphorothioate linkages on the 5' end followed by 18 phosphodiester linkages. Its nucleotide sequence is 5'-CsTsCs GCA TTC TCT CTT CCT TCT T-3' [1] (s = phosphorothioate), and it is utilized as a PCR primer for the detection of known point mutations *(22)*. The molecular structure of the phosphorothioate linkage (**Fig. 1**) gives rise to diastereomers *(2)* that partially separate during the RP-HPLC purification (*see* **Subheading 1.3.1.**) (*see* **Note 3**).

3.3.1. Correlating the Analysis of a Phosphorothioate Containing ODN with its Purification

1. The presence of diastereomers within phosphorothioate containing ODNs *(2)* allows us to illustrate the benefits of CGE, which primarily separates impurities from the full-length ODN on the basis of size. The 5'-end of ODN [1] has been modified with three phosphorothioate linkages (**Fig. 1**). The diastereomers that are created because of these phosphorothioate linkages are partially separated on our RP-HPLC column.
2. **Figure 5** illustrates two chromatograms obtained during the purification of this ODN [1]. **Figure 5B** is a chromatogram of the actual purification, and the main peak, which is collected in nine fractions, is partially split. **Figure 5A** is an analytical run conducted on a 1:100 dilution of the crude mixed backbone ODN [1]. Running conditions for analytical RP-HPLC runs are exactly the same as preparative RP-HPLC except that the UV detection is set at 260 nm instead of 292 nm. RP-HPLC analytical runs like this are only conducted when we begin work with an ODN that contains modifications that are novel or troublesome in some way. In this case, we had never before purified a phosphodiester-phosphorothioate mixed backbone ODN of this type, and all subsequent work with similar mixed backbone ODNs contained this characteristic peak split within the product peak. After the fraction analysis with CGE, it is apparent that the peak splitting of the product peak in the chromatograms of **Fig. 5** is not caused by failure sequences.
3. **Figure 6** illustrates the fraction analysis after the purification of our mixed backbone ODN [1]. Based on the quantitative agreement of RM values and a qualitative evaluation of each electropherogram, fractions 2–8 are combined, and a final product workup is conducted (*see* **Subheading 3.1.** and **Note 3**).

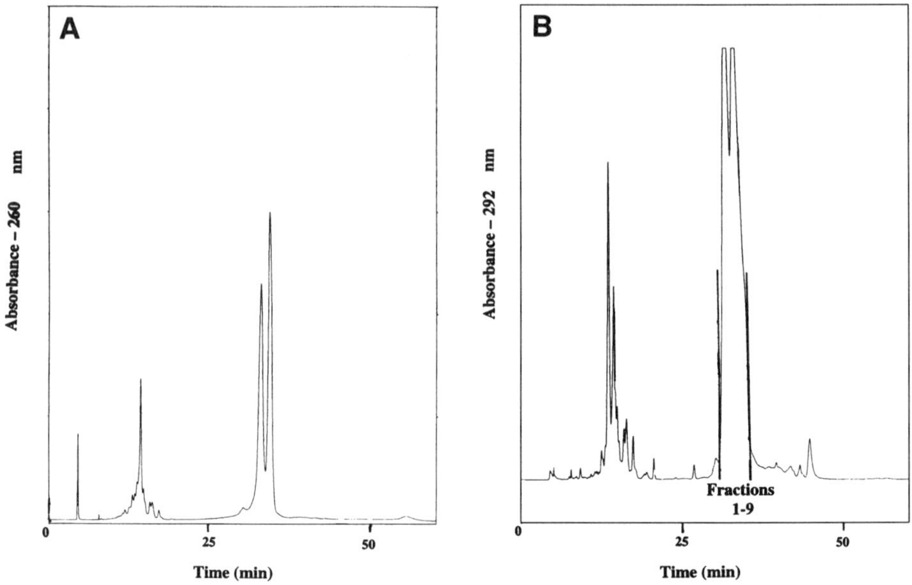

Fig. 5. Analytical **(A)** and preparative **(B)** RP-HPLC profiles of a mixed-backbone ODN, 22 nt long, that contained three phosphorothioate linkages at the 5'-end and 19 phosphodiester linkages. Chromatographic conditions are as follows: Buffer A is 2% acetonitrile (ACN) in 0.1 *M* triethylammonium acetate (TEAA, pH 7.0); Buffer B is 100% ACN. The initial running conditions are 95% buffer A and 5% buffer B, and a linear gradient is applied of 1%/min of buffer B over 40 min. For the analytical run (A), the detector is set at 260 nm and the sample (1 mL) is a 1:100 dilution of the crude mixed-backbone ODN. For the preparative run (B), the detector is set at 292 nm, and the crude sample mixture (approx 1 mL) is buffered with 0.15 mL of 2 *M* TEAA. The product ODN is collected in nine fractions between 30 and 38 min in the preparative run (B).

4. **Figure 7B** shows an electropherogram of the final product. This electropherogram verified that the split peak of our RP-HPLC purification is caused by something other than impurities in the form of failure sequences, since the fractions that are combined contain the split (**Fig. 5B**). If our product did contain a significant amount of failure sequences, CGE would resolve these from the main peak because the principle of separation for electrophoresis is based on differences in molecular weight as well as charge. Failure sequences are smaller in length and thus have smaller molecular weights than the full-length product. The split peak seen in **Fig. 5** is most likely due to the phenomena caused by the presence of diastereomers *(32)* described in **Subheading 1.3.1.**

3.3.2. Conclusions

In conclusion, it is important that the analyst understands how a modification of an ODN affects the chemical and physical structure of the ODN before trying to interpret results either from an HPLC profile, or a CGE electropherogram. The split in the

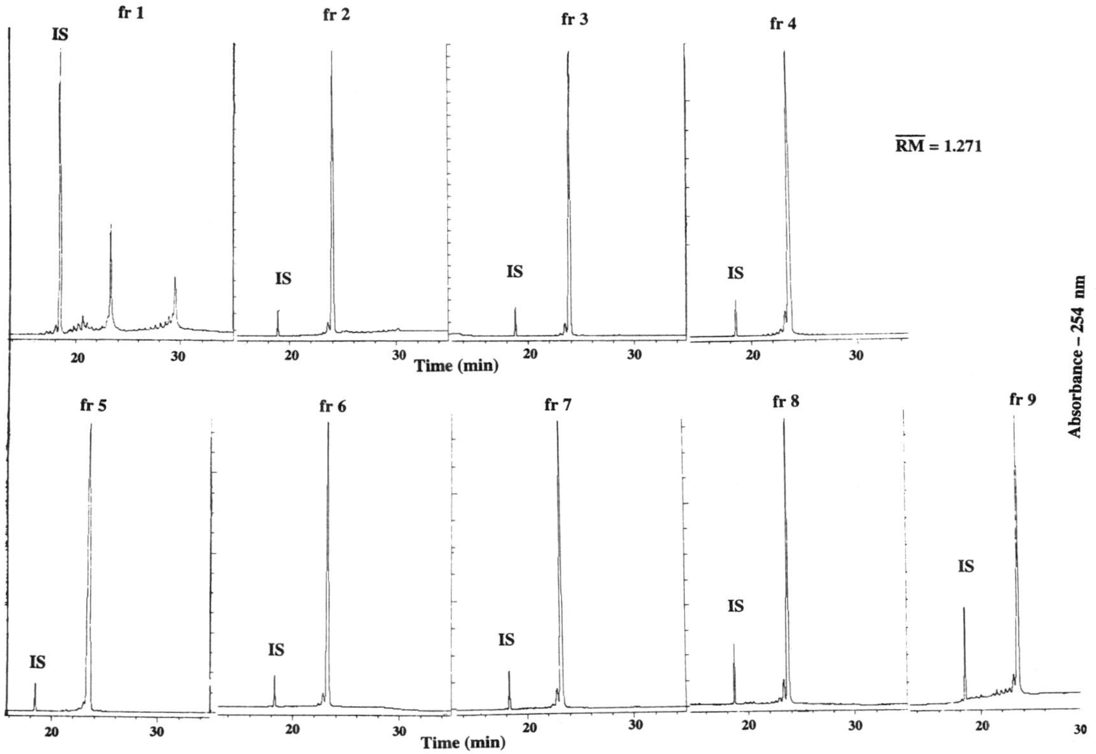

Fig. 6. CGE analysis of fractions 1–9 from the RP-HPLC purification of a mixed-backbone ODN. Electrophoresis is conducted with an electrokinetic injection of –10 kV for 3.0 s and a running voltage of –8.19 kV. The running buffer consists of 44% Tris-HCl, 56% boric acid, and 7 *M* urea. The capillary (100 μm id; 20 cm to detector window) is filled with eCAP® ssDNA 100-R gel and the detector is set a 254 nm. The internal standard (IS) is a polythmidine ODN (dT_{21}), and its concentration in each of the nine samples is 9.264 AU_{260}/mL. Aliquots (0.002 mL) are taken from each fraction and combined with IS and diluted to give a sample volume of 0.04 mL. The average relative mobility (RM) for this mixed-backbone ODN is 1.271.

product peak of **Fig. 5** is a good illustration of an effect of an ODN modification. An awareness of the chiral nature of the phosphorothioate linkage *(2)* is important in interpreting the HPLC profiles shown in **Fig. 5**. Consequently, the results from our CGE analyses (**Figs. 6** and **7**) give us confidence in the purity of our final product. A modification on an ODN does not inherently cause anomalies during its purification or analysis. Ideally, the analyst should be able to predict the effects that a modification of an ODN might have on HPLC profiles or CGE electropherograms.

4. Notes

1. The following precautions and protocol notes are applicable to our RP-HPLC method. Aqueous stock solutions that are used to make buffers (2 *M* TEAA and 1 *M* TEAB) must be stored at 4°C. For the following reasons, formulation of the TEAB stock solution takes approx 8 h to complete. First, when the triethylamine is combined with the water it is imiscible, but with good stirring and upon the addition of CO_2 it slowly dissolves. Second, dissolving the dry ice into the triethylamine and water solution is a very slow process that must be monitored continually with a pH meter, or with narrow range pH paper within a fume hood. It is important to use TEAB instead of the TEAA buffer for the purification of ODNs that contain the amidate linkage (**Fig. 2**) because these ODNs are subject to acid-mediated degradation *(27)*. The TEAB buffer has a higher pH than the TEAA buffer (*see* **Subheading 2.1.**). In general, the running buffers remain stable on the chromatography systems (Rainin and PE-ABI) for at least 2 mo. The final product workup for amidate containing ODNs requires an incubation step at 58°C in 1 *M* NaF and concentrated NH_4OH, shown in **Subheading 3.1.** This must be done to remove a DMT-U from the 3'-end of the ODN. This nucleotide base is attached to the 3'-end of the amidate ODN during its automated synthesis. The purpose of this 3'-3' attachment of a DMT-U is to facilitate the RP-HPLC purification *(27,34)* which requires the presence of a DMT group. After purification, the DMT group is normally removed by treating the crude ODN mixture with glacial acetic acid (*see* **Subheading 3.1.**); however, the amidate linkage is not stable to these acidic conditions. Therefore, after the purification and fraction pooling of amidate containing ODNs, the DMT-U group is cleaved with the nonacidic NaF/NH_4OH incubation. It is very important that amidate containing ODNs are never exposed to acidic conditions (pHs less than 7.0) or they will degrade ***(19,20)***.
2. Whereas the actual procedure for filling the capillary and conducting routine analyses with the Beckman P/ACE is fairly straightforward, working with any CE system takes some time to get used to. For instance, concluding whether or not a gel-filled capillary is equilibrated requires experience. Current and baseline are the best indicators of a gel's status. If the current is radically fluctuating, this may be an indication of an air bubble within the gel, and the best solution for this is to clear the gel out with a high-pressure rinse and reload the capillary with fresh gel. Beause 0.1 mL of gel solution is enough gel for several capillary loads, gel replacement is easily accommodated by simply repeating the high-pressure rinse from a gel fill step (*see* **Subheading 3.2.2.**). If the baseline is unstable, this could also be an indication of gel failure; however, a noisy baseline may result from detector lamp failure or from some other instrument problem. The autosampler within the P/ACE requires a specialized type of tube for the small volumes used for sample injection. In our case, this is a 0.040-mL sample volume. In order to keep our cost down, we construct our own tubes from 0.5-mL Eppendorf tubes. We cut the tops of these Eppendorf tubes off with veterinary nail clippers that are normally used to trim a house

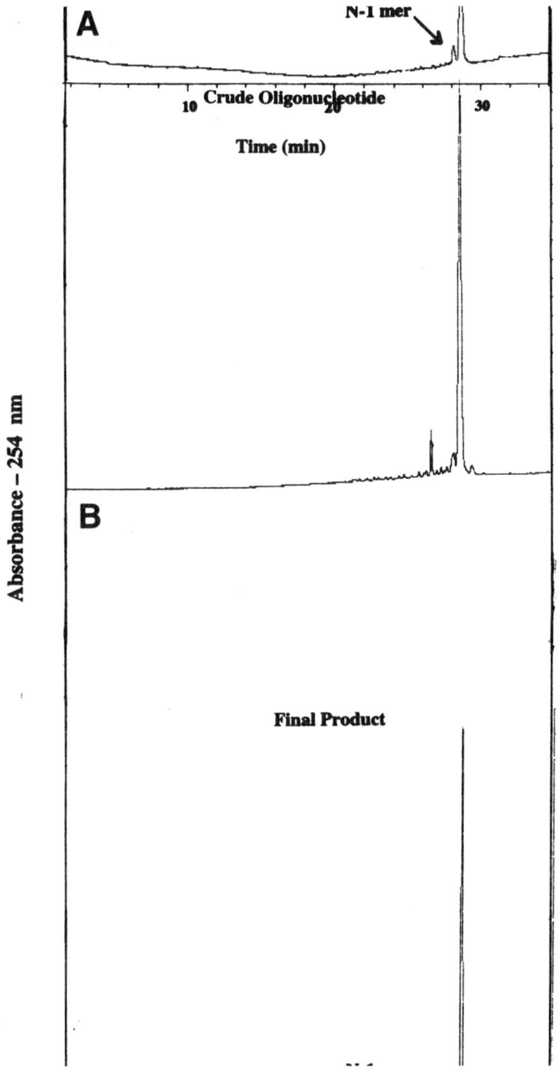

Fig. 7. CGE analysis of **(A)** the crude mixed-backbone ODN prior to its RP-HPLC purification and **(B)** the final product after the RP-HPLC purification illustrating an $N - 1$ mer that is baseline resolved from the main peak. Running conditions are the same as in **Fig. 5**. Sample aliquots (0.005 mL) are taken from a 1:100 dilution of the (A) the crude sample mixture and (B) the final product and combined with 0.035 mL of sterile water to give sample volumes of 0.04 mL.

pet's claws. One other potential problem that we encounter with this capillary gel system has to do with the urea within the buffer system. Since urea is used as a denaturant in this system, its concentration within the buffer is relatively high (approx 7 *M*). As buffer condenses within the buffer vials, urea precipitates as a salt on the vials' septa as well as on and around the electrodes. This must be constantly monitored and any precipitate of urea should be cleaned away from the electrode area to avoid arcing.

3. The use of an internal standard is not absolutely necessary with CGE. **Figure 7** illustrates two electropherograms that do not utilize an internal standard. Here, we characterize our mixed backbone ODN [1] before and after its purification. The percent purity of the main peak increases from 86.3% in the crude sample (**Fig. 7A**) to 96.2% in the purified sample (**Fig. 7B**). **Figure 7B** also demonstrates the resolution of the $N-1$ mer from the product peak. The $N-1$ mer is often the main impurity after the automated synthesis of an ODN (*see* **Subheading 1.3.**). Consequently, we found Beckman's eCAP® ssDNA 100-R gel fairly reliable, and the migration times are often reproducible which would preclude the need for an internal standard ***(29,33)***. However, in our case, using the internal standard serves to insure that the peak of interest is indeed our product. The RM value for the product of our purification is obtained during the fraction analysis (**Fig. 6**). It is measured to be 1.271, and the standard deviation, based on the nine fractions from the purification is 0.00298 giving a relative standard deviation of 0.234%. This approach is by no means the standard way to determine an RM value. RM values can be determined by conducting a CGE analysis of the crude ODN prior to its purification ***(34)***; however, this requires the preparation of a small aliquot of the crude material that would mimic the final product workup (*see* **Subheading 3.1.**). The rule of thumb for a crude analysis is to treat the ODN in the same manner as if it had been through the RP-HPLC purification step. This approach is often impractical if the amount of crude material is scant. Furthermore, the decision to use an internal standard is dependant on many other factors as well. Familiarity with the sample that is being analyzed plays a large role in deciding if an internal standard is necessary. Routine use of an internal standard can also be cumbersome in that it has the potential to introduce contaminants into the sample vial and provides an opportunity for errors during pipeting.

Acknowledgment

Jeff Nelson is acknowledged for helpful discussions and for critically evaluating this manuscript.

References

1. Matteucci, M. (1996) Structural modifications toward improved antisense oligonucleotides. *Perspectives Drug Discovery Design* **4,** 1–16.
2. Uhlmann, E. and Peyman, A. (1990) Antisense oligonucleotides: A new therapeutic principle. *Chem. Rev.* **90,** 544–584.
3. Eckstein, F. (1985) Nucleoside phosphorothioates. *Ann. Rev. Biochem.* **54,** 367–402.
4. Eckstein, F. (1983) Phosphorothioate analogues of nucleotides – tools for the investigation of biochemical processes. *Angew. Chem.* **6,** 423–439.
5. De Clerq, E., Eckstein, F., Sternbach, H., and Merigan, C. (1970) The antiviral activity of thiophosphate-substituted polyribonucleotides in vitro and in vivo. *Virology* **42,** 421–428.
6. Altmann, K. H., Fabbro, D., Dean, N. M., Geiger, T., Monia, B. P., Muller, M., and Nicklin, P. (1996) Second generation antisense oligonucleotides: Structure-activity relationships and the design of improved signal-transduction inhibitors. *Biochem. Soc. Trans.* **24,** 630–637.

7. Agrawal, S., Jiang, Z., Zhao, Q., Shaw, P., Cai, Q., Roskey, A., Channavajjala, L., Saxinger, C., and Zhang, R. (1997) Mixed-backbone oligonucleotides as second generation antisense oligonucleotides: In vitro and in vivo studies. *Proc. Natl. Acad. Sci. USA* **94,** 2620–2625.
8. Heidenreich, O., Gryaznov, S., and Nerenberg, M. (1997) RNase H-independent antisense activity of oligonucleotide N3'→ P5' phosphoramidates. *Nucleic Acids Res.* **25,** 776–780.
9. Giles, R. V. and Tidd, D.M. (1992) Enhanced RNase H activity with methylphosphodiester/phosphodiester chimeric antisense oligonucleotides. *Anti-Cancer Drug Design* **7,** 37–48.
10. Hermanson, G. T. (1996) Tags and probes, in *Bioconjugate Techniques*, Academic Press, San Diego, pp. 297–416.
11. Murakami, A., Nakaura, M., Nakatsuji, Y., Nagahara, S., Tran-Cong, Q., and Makino, K. (1991) Fluorescent-labeled oligonucleotide probes: detection of hybrid formation in solution by fluorescence polarization spectroscopy. *Nucleic Acids Res.* **19,** 4097–4102.
12. Landgraf, A., Reckmann, B., and Pingoud, A. (1991) Quantitative analysis of polymerase chain reaction (PCR) products using primers labeled with biotin and a fluorescent dye. *Anal. Biochem.* **193,** 231–235.
13. Zamecnik, P. C. and Stephenson, M. L. (1978) Inhibition of *Rous sarcoma virus* replication and cell transformation by a specific oligonucleotide. *Proc. Natl. Acad. Sci. USA* **75,** 280–284.
14. Bonham, M. A., Brown, S., Boyd, A. L., Brown, P. H., Bruckenstein, D. A., Hanvey, J. C., et al. (1995) An assessment of the antisense properties of RNase H-competent and steric-blocking oligomers. *Nucleic Acids Res.* **23,** 1197–1203.
15. Furdon, P. J., Dominski, Z., and Kole, R. (1989) RNase H cleavage of RNA hybridized to oligonucleotides containing methylphosphonates, phosphorothioate and phosphodiester bonds. *Nucleic Acids Res.* **17,** 9193–9204.
16. Brown, D. A., Kang, S., Gryaznov, S. M., DeDionisio, L., Heidenreich, O., Sullivan, S., Xu, X., and Nerenberg, M. I. (1994) Effect of phosphorothioate modification of oligodeoxynucleotides on specific protein binding. *J. Biol. Chem.* **269,** 26,801–26,805.
17. Beltinger, C., Saragovi, H. U., Smith, R. M., LeSauteur, L., Shah, N., DeDionisio, L., Christensen, L., Raible, A., Jarett, L., and Gerwitz, A. M. (1995) Binding, uptake, and intracellular trafficking of phosphorothioate-modified oligodeoxynucleotides. *J. Clin. Invest.* **95,** 1814–1823.
18. Bannwarth, W. (1988) Solid-phase synthesis of oligonucleotides containing phosphoramidate internucleotide linkages and their specific chemical cleavage. *Helvetica Chimica Acta* **71,** 1517–1527.
19. Gryaznov, S. M. and Letsinger, R. L. (1992) Synthesis and properties of oligonucleotides containing aminodeoxythymidine units. *Nucleic Acids Res.* **20,** 3403–3409.
20. Mag, M., Schmidt, R., and Engels, J. W. (1992) Synthesis and selective cleavage of an oligonucleotide containing a bridged non-chiral internucleotide 3'-phosphoramidate linkage. *Tetrahedron Lett.* **48,** 7319–7322.
21. Gryaznov, S. M., Lloyd, D. H., Chen, J., Schultz, R. G., DeDionisio, L. A., Ratmeyer, L., and Wilson, W. D. (1995) Oligonucleotide N3'→P5' phosphoramidates. *Proc. Natl. Acad. Sci. USA* **92,** 5798–5802.
22. DeDionisio, L. and Gryaznov, S. M. (1995) Analysis of a ribonuclease H digestion of N3'→P5' phosphoramidates-RNA duplexes by capillary gel electrophoresis. *J. Chromatogr. B* **669,** 125–131.

23. Hawley, P., Nelson, J. S., Fearon, K. L., Zon, G., and Gibson, I. (1999) Comparison of binding of N3'→P5' phosphoramidate and phosphorothioate oligonucleotides to cell surface proteins of cultured cells. *Antisense Nucleic Acid Drug Dev.* **9,** 61–69.
24. Rigl, C. T., Lloyd, D. H., Tsou, D. S., Gryaznov, S. G., and Wilson, W. D. (1997) Structural RNA mimetics:: N3'→P5' phosphoramidate DNA analogs of HIV-1 RRE and TAR helices that bind specifically to rev and tat-related peptides. *Biochemistry* **36,** 650–659.
25. McCurdy, S. N., Nelson, J. S., Hirschbein, B. L., and Fearon, K. L. (1997) An improved method for the synthesis of N3'→P5' phosphoramidate oligonucleotides. *Tetrahedron Letts.* **38,** 207–210.
26. Nelson, J. S., Fearon, K. L., Nguyen, M. Q., McCurdy, S. N., Frediani, J. E., Foy, M. F., and Hirschbein, B. L. (1997) N3'→P5' oligodeoxyribonucleotide phosphoramidates: a new method of synthesis based on a phosphoramidite amine-exchange reaction. *J. Org. Chem.* **62,** 7278–7287.
27. Fearon, K. L., Nelson, J. S., Hirschbein, B. L., Foy, M. F., Nguyen, M. Q., Okruszek, A., McCurdy, S. N., Frediani, J. E., DeDionisio, L. A., Raible, A. M., and Boyd, V. (1998) An improved synthesis of oligonucleotide N3'→P5' phosphoramidates and their chimera using hindered phosphoramidite monomers and a novel handle for reverse phase purification. *Nucleic Acids Res.* **26,** 3813–3824.
28. Somers, V., Moerkerk, P., Murtagh, J., and Thunnissen, F. (1994) A rapid, reliable method for detection of known point mutations: Point-EXACCT. *Nucleic Acids Res.* **22,** 4840–4841.
29. DeDionisio, L. A. and Lloyd, D. H. (1996) Capillary gel electrophoresis and antisense therapeutics: Analysis of DNA analogs. *J. Chromatogr. A* **735,** 191–208.
30. Cohen, A. S., Vilenchik, M., Dudley, J. L., Gembroys, M. W., and Bourque, A. J. (1993) High-performance liquid chromatography and capillary gel electrophoresis as applied to antisense DNA. *J. Chromatogr.* **638,** 293–301.
31. Fearon, K. L., Stults, J. T., Bergot, B. J., Christensen L. M., and Raible, A. M. (1995) Investigation of the 'n-1' impurity in phosphorothioate oligodeoxynucleotides synthesized by the solid-phase β-cyanoethyl phosphoramidite method using stepwise sulfurization. *Nucleic Acids Res.* **23,** 2754–2761.
32. Wilk, A. and Stec, W. J. (1995) Analysis of oligo(deoxynucleoside phosphorothioate)s and their diastereomeric composition. *Nucleic Acid Res.* **23,** 530–534.
33. Srivatsa, G. S., Batt, M., Schuette, J., Carlson, R. H., Fitchett, J., Lee, C., and Cole, P. L. (1994) Quantitative capillary gel electrophoresis assay of phosphorothioate oligonucleotides in pharmaceutical formulations. *J. Chromatogr. A* **680,** 469–477.
34. DeDionisio, L. A., Raible, A. M., and Nelson, J. S. (1998) Analysis of an oligonucleotide N3'→P5' phosphoramidate/phosphorothioate chimera with capillary gel electrophoresis. *Electrophoresis* **19,** 2935–2938.

25

Use of Capillary Electrophoresis for Concentration Analysis of Phosphorothioate Oligonucleotides

G. Susan Srivatsa, Roya Pourmand, and Steven Winters

1. Introduction

The rapid advance of phosphorothioate oligonucleotides in clinical development has accelerated the need for accurate quantitation of these compounds for quality control and bioanalytical applications *(1)*. Whereas slab-gel polyacrylamide gel electrophoresis has been in routine use for the analysis of oligonucleotides, it has been replaced by capillary gel electrophoresis (CGE) as the technique of choice for length-based separation of synthetic oligonucleotides *(2–10)*. This can be attributed to many factors including, superior resolution as a result of 10–100-fold higher electric fields, on line detection, and automation.

The major goals in quantitative analysis of drug products are for the profiling of impurities and for the assay of drug concentration. Impurity profiling, the quantitation of synthesis-related deletion sequence impurities *(11)* is straightforward because it involves a relative area-% determination. As such, CGE results are independent of applied sample concentration, so long as the concentration of the injected analyte is within the optimum range for electrophoretic resolution. The assay of drug concentration, however, is a w/w measurement, requiring assay of the drug product sample against a reference standard of known concentration and purity. This poses a challenge for quantitative capillary gel electrophoresis (QCGE) analysis as it employs, in general, an electrokinetic mode of introducing the sample onto the capillary column. Electrokinetic loading places stringent requirements on the sample matrix since the quantity of analyte loaded onto the capillary is a function of the electrophoretic mobilities of the solutes present in the sample. For oligonucleotide drug formulations, the buffer salts will be preferentially loaded onto the gel capillary due to their relatively high charge-to-mass ratios. Thus, relatively small differences in the amounts of buffer salts present in the sample and the external standard can lead to dramatic differences in the amount of oligonucleotide loaded onto the column and the observed

From: *Methods in Molecular Biology, Vol. 162:*
Capillary Electrophoresis of Nucleic Acids, Vol. 1: Introduction to the Capillary Electrophoresis of Nucleic Acids
Edited by: K. R. Mitchelson and J. Cheng © Humana Press Inc., Totowa, NJ

detector response *(10)*. Most pharmaceutical products intended for intravenous or ophthalmic use are formulated in isotonic salt solutions. It is crucial to the direct, accurate quantitation of oligonucleotide concentrations in these pharmaceutical dosage forms, therefore to formulate the external reference standard in an identical sample matrix to that of the drug product being assayed. Alternatively, it is possible to desalt the sample prior to analysis. However, this has not proven as effective. In addition, use of an internal standard has been shown to be necessary to correct for migration time and irreproducibility of peak area *(12)*.

The procedure described in this chapter is based on the original publication which described the use of QCGE for the analysis and stability monitoring of phosphorothioate oligonucleotides in pharmaceutical preparations *(12)*. This approach, referred to as QCGE, has been demonstrated to have the linearity, accuracy, selectivity, precision, and ruggedness required for routine drug product analysis and has been successfully transferred to a number of analytical laboratories.

2. Materials

All reagents and chemicals used must be reagent grade or of higher purity. The water to prepare solutions should be purified MilliQ water, or water of equivalent purity.

1. 10% polyacrylamide, preformed static gel capillary, effective length 40 cm (J & W Scientific μ-PAGE 10, cat. no. 195-1311, or equivalent). Cut the column to 47 cm length, of which 40 cm spans between the injector to the detector (*see* **Notes 1** and **2**).
2. CE System (CE), e.g., Beckman P/ACE 5000 or equivalent with UV detector, auto-injector and digital peak area integration system.
3. Tris-base and boric acid (dry).
4. Urea.
5. 0.2-μm filter (Nalgene, cat. no. 190-2520 or equivalent).
6. Sonicator for degassing solutions.
7. Stirring plate mixer.
8. Migration Time Marker (MTM), an oligonucleotide marker whose migration time is greater than that of the oligonucleotide being investigated.
9. Dibasic sodium phosphate, heptahydrate.
10. Monobasic sodium phosphate, monohydrate.
11. Sodium chloride and sodium hydroxide.
12. Standard oligonucleotide (with the same sequence as the sample, known concentration and impurity profile), if available.

3. Methods

3.1. Buffers

1. Prepare a 100 m*M* Tris-borate, 7 *M* urea buffer from the individual reagents as follows: For 500 mL of Tris-borate, urea buffer, as an example, add 295 mL of purified water to 6.1 g of Tris-base and 3.1 g of boric acid. To dissolve the buffer, stir for 20–30 min using a magnetic stir bar and stirring mixer and ensure the boric acid is completely dissolved.
2. Slowly add 210.2 g of dry urea to the fully dissolved Tris-borate buffer. The dissolution of urea is endothermic, so the buffer solution will get very cold. Quantity Sufficient (Q.S.)

to 500 mL, if necessary. Before use, the buffer solution must come to ambient temperature. **Do not heat.** The buffer solution should be completely clear and all salts completely dissolved. Degas the buffer by ultrasound or by vacuum and store it in a stoppered bottle at 2–8°C until use. Filter the required volume of buffer for the day's use through a 0.2-μm filter. Place the buffer at both the inlet and the outlet vials in the CE instrument.

3. To use stored buffer solution, bring entire container to ambient temperature before use while stirring slowly with a clean stir bar to ensure complete dissolution of urea. Remove the necessary volume, sonicate the buffer to degas and then carefully filter through a 0.2-μm filter.

3.2. Appropriate Diluent Buffer

1. Prepare an appropriate diluent as follows: for a one liter buffer preparation, combine 14.33 g of dibasic sodium phosphate, 1.73 g of monobasic sodium phosphate, and 4.4 grams of sodium chloride in an appropriate beaker containing approx 600 mL of water. After mixing thoroughly, adjust pH to 7.3 with NaOH, if necessary. Transfer to a 1-L volumetric flask and adjust the volume to 1 L with purified water.
2. Preparation of MTM: Prepare a 1 mg/mL stock solution of MTM in water. Store at –15°C. Prepare the "working" MTM by transferring 250 μL of stock solution to a clean 5-mL volumetric flask and Q.S. with purified water to 5 mL. Mix well by vortexing. Store at 2–8°C, when not in use. The expiration date is 1 mo from the date of preparation.

3.3. Sample Preparation

1. Sample preparation: Dilute the sample using at least 50 μL of sample to a final concentration of 1 mg/mL in the appropriate diluent.
2. Using fixed volume pipets dilute each sample by 1:100 in a volumetric flask. Examples of typical dilution sequences are as follows: 50 μL of 1 mg/mL sample, 250 μL of "working" MTM.
3. Q.S. to 5 mL with purified water. Transfer the sample to CE sample vials and place in the auto-injector tray.
4. Standard Preparation: prepare the reference standard solution (in triplicate) by diluting a reference standard using the "suggested dilution scheme" shown above for sample preparation (*see* **Note 3**).

3.4. Capillary Electrophoresis

1. Instrument parameters: The instrument should be run at reversed electrical polarity for oligonucleotides due to their negative charge. Set the detector wavelength to the maximum absorbing wavelength for the specific oligonucleotide being tested or, if using bandpass filters, use the filter closest to the required wavelength. Before commencing the column equilibration run, the column temperature should be set at 30°C (*see* **Note 4**).
2. Equilibrate column as follows: Filter and degas the 100 m*M* Tris-borate, 7 *M* urea buffer prior to use. Fill the buffer solutions into vials for both the inlet and outlet ends of the column (*see* **Note 5**).
3. Run the following voltage gradient:

Applied voltage (kV)	Time (min)
5	10
10	10
14.1	10

4. System suitability: Run a standard oligonucleotide at a constant applied voltage of 14.1 kV for at least 33 min. Use the following table to set the voltage and appropriate injection time for the sample (*see* **Note 4**).

Concentration	Injection time (s)	Voltage
1 mg/mL	20	7.5 kV

5. **Important**: If the amount of oligonucleotide loaded onto the column is low, as seen from the absorbance values less than 0.01 AU, increase the injection time appropriately to obtain an absorbance between 0.01 and 0.02 AU for the main oligonucleotide peak. The current should be stable. For a voltage of 14.1 kV the current should be >7 μA (*see* **Note 6**). Ensure that the duration of the run is adequate to include the MTM (typically, 32–40 min for a 27-mer oligonucleotide such as MTM). Calculate the relative standard deviation (RSD) of the concentration factor for all the reference standard normalized peak areas. Typical suitable systems produce RSD of about 3%. *See* **Fig. 1** for an example of a typical electropherogram (*see* **Note 7**).

3.5. Calculations of Analyte Concentration

1. Correct the standard peak areas of both the sample and the internal reference. In CE, the observed peak areas are influenced by the difference in migration velocities of the analytes and consequently the length of time that the analytes reside at the detector.
2. To correct for this phenomenon, divide the observed areas of the peaks of interest by their corresponding migration times:

$$\text{Corrected area of peak}_x = \text{Observed peak arear of peak}_x/\text{Migration time of peak}_x$$

3. Then calculate the total corrected area of detected "nonbackground peaks" in each sample run. Baseline shifts and electronic spikes are typical examples of the types of background peaks encountered. Express the peak area for each oligonucleotide, and the peak area of all the nonbackground peak areas as a percent of the total corrected peak area: This is the area-% result of the method.

$$\text{Area \%} = (\text{Corrected individual peak area/corrected total peak area}) \times 100$$

4. Calculation of the assay: Normalize the corrected peak areas by dividing by the area of the MTM peak:

$$\text{Normalized area of peak}_x = \text{Corrected peak of area of peak}_x/\text{Corrected peak area of MTM}$$

5. Calculate the concentration factor for the formulated reference standard as follows:

$$\text{Concentration factor } (K_f) = \text{Normalized peal area of sample/C}$$

in which, C = Concentration of formulated reference standard in mg/mL.

6. Calculate the concentration factor for the active pharmaceutical ingredient (API) samples by the following formula:

$$K_f = \text{Normalized peak area of sample}/(1 - W/100) \times C$$

in which, W= the moisture content determined by Karl Fischer expressed as a percent, and C = the concentration of the sample in mg/mL.

7. Calculate the %-purity of API samples by use of the following formula:

$$\text{\% Purity} = (K_f \text{ of sample}/K_f \text{ of reference}) \times P \times 100$$

for which, P = Purity of the Reference expressed as a decimal, K_f = The average K_f.

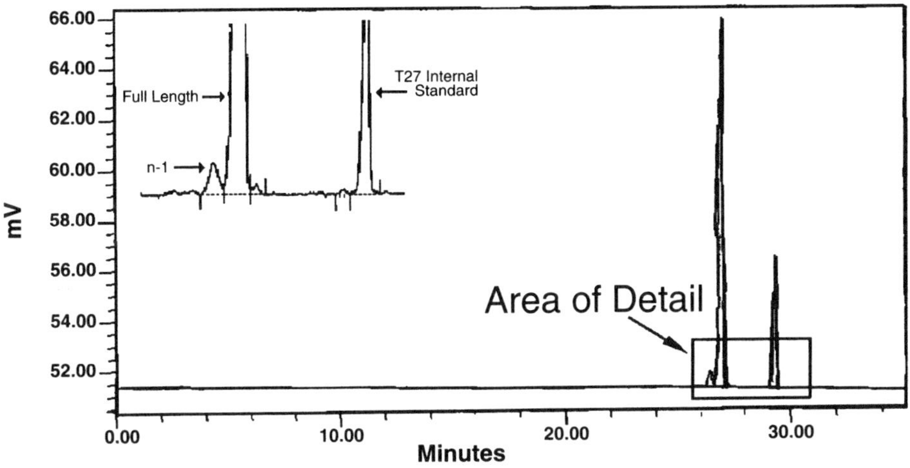

Fig. 1. A typical electropherogram of phosphorothioate oligonucleotides.

8. Calculate the concentration for drug products using the following formula:

$$\text{Concentration} = (\text{Normalized sample area}/\text{Kf of reference}) \times P \times D$$

for which, P = Purity of the Reference Standard expressed as a decimal, D = Dilution factor to get to 1 mg/mL or 0.33 mg/mL, K_f = The average Kf.

4. Notes

1. Special care should be given to preparation of the capillary columns after they have been cut to the appropriate size. The ends of these columns can dry very quickly and they should be put in the appropriate running CGE buffers as soon as they are cut. Low currents or loading can result if the ends are exposed to air for more than a couple of minutes.
2. The window of the column should be clean and scratch free otherwise one might observe low absorbances.
3. It is important that the initial sample and reference standard dilutions are made to 1 mg/mL with diluent buffer. It is also important that the second 1:100 dilution of the MTM is made with water, in order to achieve proper sample loading onto the capillary. Because this method uses electrokinetic sample injection, sample loading is highly affected by differences in sample ionic strength.
4. Some phosphorothioate oligonucleotides require a capillary temperature of 40°C in order to achieve sufficient resolution between the full-length peak and the *n*–1 species. The temperature setting needs to be optimized for each particular oligonucleotide being analyzed.
5. Inspect the columns for tiny bubbles, which can also cause an unstable current or spikes.
6. Occasionally an "overloading" or "underloading" of a sample onto the column may be detected in the middle of a series of sequential assays. This phenomenon can be caused by current fluctuations—hence, current should be monitored for each run. One of the likely causes could be the accumulation of salt precipitate around the electrodes if the instrument is not cleaned regularly and thoroughly in-between runs. Simple cleaning usually

remedies the problem (refer to the manual for the maintenance and upkeep of the instrument used).

7. An alternative buffer system for the separation of small oligonucleotides and single nucleotides is provided in **ref. *13***.

References

1. Uhlmann, E. and Peyman, A. (1990) Antisense oligonucleotides: a new therapeutic principle. *Chem. Rev.* **90,** 543.
2. Paulus, A. and Ohms, J. I. (1990) Analysis of oligonucleotides by capillary gel electrophoresis. *J. Chromatogr.* **507,** 113–123.
3. Guttman, A., Cohen, A. S., Heiger D. N., and Karger, B. L. (1990) Analytical and micropreparative ultrahigh resolution of oligonucleotides by polyacrlamide gel high-performance capillary electrophoresis. *Anal. Chem.* **62,** 137–141.
4. Heiger, D. N., Cohen A. S., and Karger, B. L. (1990) Separation of DNA restriction fragments by high performance capillary electrophoresis with low and zero crosslinked polyacrylamide using continuous and pulsed electric fields. *J. Chromatogr.* **516,** 33–48.
5. Cohen, A. S., Najarian D. R., and Karger, B. L. (1990) Separation and analysis of DNA sequence reaction products by capillary gel electrophoresis. *J. Chromatogr.* **516,** 49–60.
6. Cohen, A. S., Najarian, D. R., Paulus, A., Guttman, A., Smith J. A., and Karger, B. L. (1988) Rapid separation and purification of oligonucleotides by high-performance capillary gel electrophoresis. *Proc. Natl. Acad. Sci. USA* **85,** 9660–9663.
7. Cohen, A.S., Vilenchik, M., Dudley, J. L., Gemborys, M. W., and Bourque, A. J. (1993) High-performance liquid chromatography and capillary gel electrophoresis as applied to antisense DNA. *J. Chromatogr.* **638,** 293–301.
8. Warren, W. J. and Vella, G. (1993) Analysis of synthetic oligodeoxyribonucleotides by capillary gel electrophoresis and anion-exchange HPLC. *BioTechniques* **14,** 598–606.
9. United States Pharmacopoeia, XXII (1990) 1711.
10. Huang, X., Gordon, M. J., and Zare, R. N. (1988) Bias in quantitative capillary zone electrophoresis caused by electrokinetic sample injection. *Anal. Chem.* **60,** 375–377.
11. Srivatsa, G. S., Klopchin, P., Batt, M., Feldman, M., Carlson, R. H., and Cole, D. L. (1997) Selectivity of anion exchange chromatography and capillary gel electrophoresis for the analysis of phosphorothioate oligonucleotides. *J. Pharm. Biomed. Anal.* **16,** 619–603.
12. Srivatsa, G. S., Batt, M., Schuette, J., Carlson, R., Fitchett, J., Lee, C., and Cole, D. L. (1994) Quantitative capillary gel electrophoresis assay of phosphorothioate oligonucleotides in pharmaceutical formulations. *J. Chromatogr.* **680,** 469–477.
13. Pearce, M. and Watson, N. (2001) Quality control of nucleotides and primers for PCR, in *Capillary Electrophoresis of Nucleic Acids,* Vol. 1 (Mitchelson, K. R. and Cheng, J., eds.), Humana Press, Totowa, NJ, pp. 347–352.

26

Capillary Electrophoresis Separation of Ribonucleosides

Dan-Ke Xu and Hong-Yuan Chen

1. Introduction

Electrochemical detection (ED), primarily because of the extraordinary sensitivity and unusual selectivity, has established itself based on the characteristics of microelectrodes as one of important detection methods for capillary electrophoresis (CE) *(1)*. Compared with the popular spectrometric detection techniques, the advantage of ED derives from the fact that the sensitivity of analyte detection is not hampered by the small volumes of the sample zone which limit the signal strength. This problem is frequently encountered in CE using spectrometric detection *(2)*. In order to realize this significant potential of ED in practice, two principal difficulties must be overcome. First, there is the need to diminish the influence of the high voltage employed in the CE separation on the electrochemical detection potential. This problem was solved initially by placing a small fracture near the downstream end of the separation capillary, and covering the crack with a porous glass *(3)*, or Nafion® *(4)* coating. This procedure permits this point of the capillary to be held at ground, but maintains the electroosmotic flow (EOF) out of the end of the capillary, without any detrimental effects to the electrophoresis current. It has been shown *(5)* that when the smaller diameter capillary columns (i.e., less than 25 μm) are employed, the electrophoresis current and associated *iR* drop in the detection cell are both small. The effects are small enough in most cases that the exit end of the capillary column can simply be placed into the electrochemical cell, very near the working electrode surface. Secondly, microelectrodes comprising 5–50-μm wires, or fibers were initially employed in order to match the size of the separation capillary. This fabrication has to be accomplished under a microscope. This requirement greatly complicates electrode fabrication, as it makes the alignment of the electrode and capillary difficult to maintain during an electrophoresis run, and during subsequent runs. Experimental reproducibility is thus difficult to maintain with such fine electrodes.

However, two recent developments have overcome these problems. Both working electrodes of more than 100 μm and the wall-jet electrochemical detection approach

From: *Methods in Molecular Biology, Vol. 162:*
Capillary Electrophoresis of Nucleic Acids, Vol. 1: Introduction to the Capillary Electrophoresis of Nucleic Acids
Edited by: K. R. Mitchelson and J. Cheng © Humana Press Inc., Totowa, NJ

have both successfully been developed for use in CE *(6)*. In these approaches, a disk-shaped electrode with only its tip cross-section exposed is positioned immediately in front of the much smaller capillary opening. Detection is performed on the solution as it exits the capillary, flowing radially across the face of the disk electrode. The main advantages of these developments are more rugged electrode construction and the easier handling of the modified electrode, which results in a more reproducible alignment of the capillary and electrode. More recently, the vertical electrochemical detection cell has been developed in our laboratory *(7)*. In our proposed CE-ED system, the manipulation of the working electrode is performed without the aid of a microscope. In addition, the configuration will be easier to replace the working electrode as well as the running buffer. These characteristics would routinely promote the development of ED for use in CE.

This chapter describes a simple method for the simultaneous determination of purine bases, ribonucleosides, and ribonucleotides by CE-ED, based on a vertical wall-jet electrochemical detector. In this case, a copper (Cu) electrode is employed as the working electrode, and in combination with strongly alkaline condition offers the direct electrocatalytic oxidation for the compounds of interest. The method has a wide linear response range and the calibration curves are linear over 2–3 orders of magnitude. Moreover, the sensitivity is high and average limit of detection approaches to the fmol level. In addition, this technique is robust and specific and could also be applied to analyze these compounds in biological samples containing mixtures, such as plasma or serum.

2. Materials

2.1. Apparatus

1. The CE system described is assembled in-house. This requires a high-voltage power supply, an uncoated fused silica capillary of 25 μm id, an electrochemical detector including a detection cell, and a strip chart recorder or an integrator.
2. A schematic diagram of the detailed construction for the separation and detection system is shown in **Fig. 1**. The high-voltage end of the capillary and anode buffer reservoir are contained in a plexiglass cabinet equipped with an interlock for operator safety. The outlet of the capillary is always maintained at ground.
3. The solution reservoir is made of plexiglass and it serves both as the cathodic buffer reservoir for the CE system, and as the electrochemical cell for the ED (**Fig. 2A**). The end of the capillary is vertically inserted into the buffer reservoir and is fixed at the bottom of the cell through the fitting (*see* **Note 1**).
4. A simple, home-constructed manipulator is mounted on the cell as a cover (**Fig. 2B**). The working electrode fixed in the manipulator can be positioned centrically, directly against the exit of the capillary column. A saturated calomel reference electrode (SCE), a platinum (Pt) wire counter electrode and the grounding electrode for the CE are also placed into the cell besides the working electrode (*see* **Notes 2** and **3**).
5. Preparation of the Cu electrode: Insert a 200-μm diameter Cu wire through a glass capillary tube until the wire protrudes about 3–4 mm from the tip. At the junction of the capillary and the Cu wire, apply a small amount of epoxy glue to seal the Cu wire to the glass. Cure the epoxy resin at 40–60°C overnight. Once the resin is cured, cut one end of the Cu

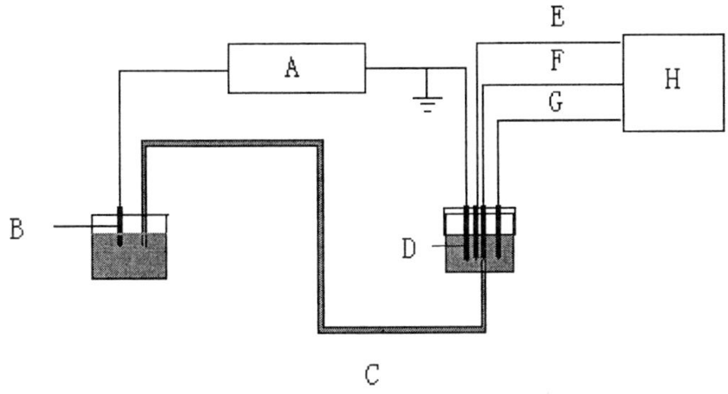

Fig. 1. Diagram of the ED-CE system. Schematic: (A), high-voltage power; (B), anode; (C), capillary column; (D), cathode (grounding electrode); (E), reference electrode; (F), working electrode; (G), counter electrode; and (H), electrochemical detector.

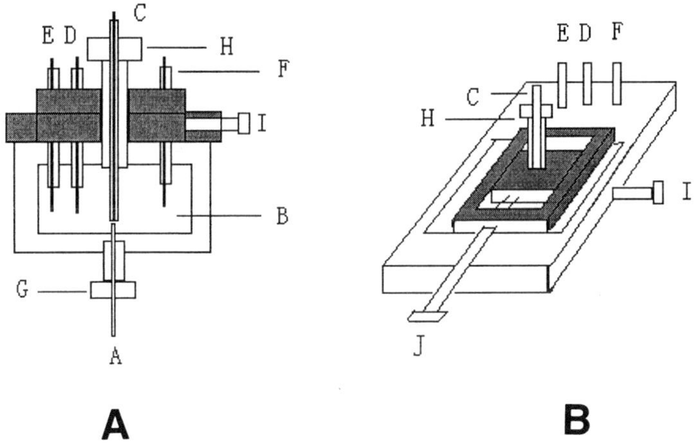

Fig. 2. Electrochemical detection cell: **(A)**, cross-sectional view; and **(B)**, vertical view of the vertical wall-jet electrochemical detection cell. Parts are: (A) capillary column; (B) electrolyte solution; (C) working electrode; (D) reference electrode; (E) cathode (grounding electrode); (F) counter electrode; (G) fitting; and (H–J) adjustable screws.

wire with surgical scissors and "wet-polish" the tip with 0.05-μm alumina powder abrasive until the Cu surface is flush with the epoxy seal. This end of the wire is used for detection as a working electrode and the other end is used to support the electrical connection (*see* **Note 4**).

6. A high-sensitivity amperometric detector should be employed because only a very small amount of the analyte exits from the capillary column, and the current range of amperometric detection based on constant-potential is usually in a range from pA to nA. The types of detector used ordinarily for high performance liquid chromatography are sufficiently sensitive to be employed for EC-CE separation technique.

2.2. Chemicals

1. Separation electrolyte: 40 m*M* sodium hydroxide (NaOH).
2. The purine, ribonucleoside, and ribonucleotide standards (Sigma, St Louis, MO).

3. Methods

3.1. Sample Preparation

1. The standard sample solutions are prepared by serial dilution in 40 m*M* NaOH in order to obtain appropriate concentrations (*see* **Note 5**).
2. The serum or plasma sample (e.g., 100 μL vol) is prepared through the addition of an equal amount of 10% (v/v) trichloroacetic acid and then centrifuged for 5 min at 3000*g*. The supernatant is neutralized with 100 m*M* NaOH. This solution is further diluted with 200 μL of 40 m*M* NaOH and could be injected onto the CE system.

3.2. CE and Electrochemical Detection

1. Freshly diluted running buffer is added to the electrochemical cell until it is 2–3 mm above the top of the capillary outlet within the cell (*see* **Note 1**). Meanwhile, replace the electrolyte solution in the anode reservoir with the running buffer. Fill the separation capillary column with the running buffer through a vacuum pump.
2. Fix the Cu electrode in the manipulator, and then position it directly against the exit of the capillary column. Connect the Cu electrode, the reference electrode and the counter electrode with the relevant leads of the electrochemical detector, respectively. Set a +0.65 V constant voltage on the detector, and then turn on the cell (*see* **Note 2**). The amperometric detection current signal will slowly decrease until a stable Cu(III) oxidation layer is formed at the Cu electrode surface. This usually requires about 30 min of stabilization time (*see* **Notes 5** and **6**).
3. Apply a separation voltage of 12.5 kV between the inlet and outlet of the capillary column through the high-voltage power supply. When the high voltage is turned on the detection current will show a pulse signal. As the run progresses, the separation current is about 50 μA (*see* **Note 3**).
4. Prior to injection of the sample, switch off the high voltage of the electrophoresis system and put both the inlet of the capillary and the anode Pt wire of the electrophoresis system into the sample reservoir. Apply a high voltage of +12.5 kV for 3 s, the sample will be introduced into the capillary through "electromigration injection."
5. After the loading of the sample, remove the capillary and the anode from the sample reservoir to the anode buffer reservoir. Switch on the high voltage again and the electrophoresis separation commences. Check the separation current to insure that there are no bubbles within the capillary, as low current will indicate impeded electrophoresis. The amperometric detection signal will be shown at the recorder. The separation of purine bases, ribonucleosides, and ribonucleotides should be completed in 15 min, as shown in **Fig. 3** (*see* **Note 7**).
6. To ensure reproducible separation and to avoid the decrease in separation current resulting from the presence of low concentrations of carbon dioxide and the Joule heating, replace the electrolyte solution in the capillary with the freshly prepared solution before each run (*see* **Note 8**).
7. After the run, the purine, ribonucleoside, and ribonucleotide compositions of the unknown samples are characterized by comparison with the migration times given by the standard compounds. The concentrations of each of the compounds are calculated through the linear regression equations (*see* **Note 7**).

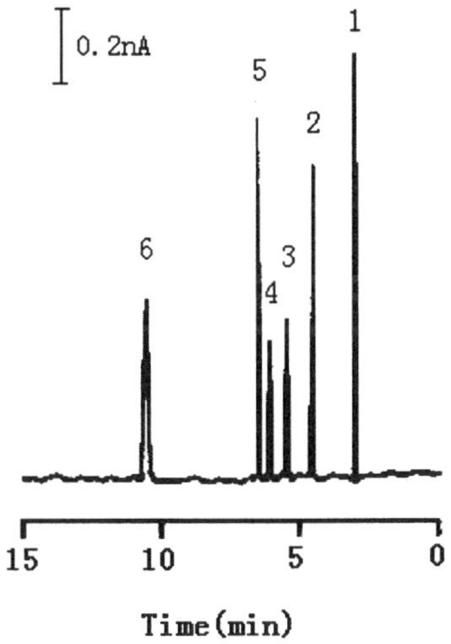

Fig. 3. Electropherogram of a mixture of nucleotides, which are detected electrochemically: (1) adenosine; (2) guanosine; (3) adenine; (4) adenosine-5'-monophosphate; (5) guanine; and (6) guanosine-5'-monophosphate (*see* **Note 7**).

4. Notes

1. In this wall-jet electrochemical detection approach, the cell has a dimension of 4 cm (L) × 4 cm (W) × 1.5 cm (H). A volume of 10–15 mL electrolyte solution should be sufficient to fill the cell.
2. **Safety Warning**: This electrochemical detection scheme is based on the wall-jet approach, in which there is no interface to isolate between the high voltage employed in the CE separation and the detection system. **Thus, before switching on the high voltage, it is absolutely necessary to check that the electrochemical detection cell has been held at ground voltage through the grounding electrode**. The high voltage from the CE power source would cause damage to the electrochemical detector.
3. Use of a 25-μm id capillary column is highly recommended to decrease the separation current. This safeguard will avoid the interference of the amperometric detection current by the CE current, and result in higher sensitivity for the detector.
4. The diameter of the Cu electrode will influence both the separation efficiency and the detection sensitivity. An electrode of 200 μm id is highly recommended. Although the smaller diameter of electrode will produce a better resolution efficiency, it is more difficult to fix the electrode opposite the capillary outlet, and thus more difficult to maintain the reproducibility of analyte detection.
5. In order to keep the catalytic oxidation activity of the Cu electrode, a basic running buffer solution (NaOH, at a concentration > 5 m*M*) must be present, in which a Cu(III) oxidation layer can be formed at +0.65 V potential such that it can oxide the purine, ribonucleoside, and ribonucleotide compounds. As the hydroxide ion concentration is increased, the migration time of the analytes accordingly becomes longer. On the other hand, the resolution between these compounds improves. However, at buffer concentrations above 40 m*M*, the migration time becomes too long, and the separation will not improve further, as diffusion becomes a significant factor. The elevated buffer concentration will also

increase the separation current, which results in interference with the electrochemical detection system. Thus, the optimum concentration should be 40 m*M* NaOH.

6. The activity of fresh Cu electrodes will decrease slightly within the first two injections. After the subsequent injections, the electrode response tends to stabilization. The same electrode can be used without any further treatment at least 2–3 d.
7. The linear response ranges of the compounds of interest are as follows: adenine ($5 \times 10^{-6} - 5 \times 10^{-4}$ *M*), adenosine ($1 \times 10^{-6} - 2 \times 10^{-3}$ *M*), adenosine-5'-monophosphate ($5 \times 10^{-6} - 5 \times 10^{-4}$ *M*), guanine ($4 \times 10^{-7} - 5 \times 10^{-4}$ *M*), guanosine ($1 \times 10^{-6} - 2 \times 10^{-3}$ *M*), guanosine-5'-monophosphate ($2 \times 10^{-6} - 5 \times 10^{-4}$ *M*). The limits of detection should be below 9 fmol (S/N =3).
8. Increasing the applied voltage of the electrophoresis system can reduce the migration time, but cannot improve the resolution. On the other hand, the higher voltage will produce larger separation current and cause more Joule heating.

Acknowledgments

The financial support of this work by the National Natural Science Foundation of China is gratefully acknowledged. We thank Mr. Lin Hua for his assistance in developing this system. We also thank Dr. Guo-Rong Fan for his helpful correspondence.

References

1. Ewing, A. G., Mesaros, J. M., and Gavin, P. F. (1994) Electrochemical detection in microcolumn separation. *Anal. Chem.* **66,** 527A–537A.
2. Matysik, F.-M., Meister A., and Werner, G. (1995) Electrochemical detection with microelectrodes in capillary flow systems. *Anal. Chim. Acta* **305,** 114–120.
3. Wallingford, R. A. and Ewing, A. G. (1988) Amperometric detection of catechols in capillary zone electrophoresis with normal and micellar solutions. *Anal. Chem.* **60,** 258–263.
4. O'Shea, T. J., Greenhagen, R. D., Lunte, S. M., Lunte, C. E., Smyth, M. R., Radzik, D. M., and Watanabe N. (1992) Capillary electrophoresis with electrochemical detection employing an on-column Nafion joint. *J. Chromatogr.* **593,** 305–312.
5. Sloss, S. and Ewing, A. G. (1993) Improved method for end-column amperometric detection for capillary electrophoresis. *Anal. Chem.* **65,** 577–581.
6. Ye, J. and Baldwin, R. P. (1994) Determination of amino acids and peptides by capillary electrophoresis and electrochemical detection at a copper electrode. *Anal. Chem.* **66,** 2669–2674.
7. Xu, D.-K., Hua, L., and Chen, H.-Y. (1996) Determination of purine bases by capillary zone electrophoresis with wall-jet amperometric detection. *Anal. Chim. Acta* **335,** 95–101.

27

On-Line Preconcentration and Separation of Antisense Oligonucleotides by ITP-CE in Dextran-Filled Capillaries

Iris Barmé and Gerard J. M. Bruin

1. Introduction

The therapeutic use of "antisense oligonucleotides" is a new approach to rational drug design *(1)*. Antisense oligonucleotides are short, single-stranded oligonucleotides, which will bind to a complementary sequence of a cellular messenger RNA due to the high specificity of nucleic acid base pairing. Most of the antisense oligonucleotides, currently in drug development programs, are phosphorothioates. In drug development, both the purity of the active compound as well as possible metabolites need to be quantitatively analyzed and characterized *(2,3)*. Capillary electrophoresis (CE) has proven to be a separation method with higher resolving power than traditional separation methods for oligonucleotides.

To detect and quantify the parent compound and possible metabolites, concentration detection limits in the nanomolar range are necessary. Although UV detection has one of the highest detection limits of all detection limits in CE, typically only in the submicromolar range, it is still the most popular method because of simplicity and versatility. To allow the detection of nanomolar concentrations with UV detection, sample preconcentration is an absolute necessity.

Isotachophoresis (ITP) is an efficient on-line sample pretreatment method because sample concentration enhancement is an intrinsic part of ITP *(4)*. This chapter describes an ITP-CE procedure for the preconcentration and separation of antisense phosphorothioate oligonucleotides using a dextran solution in a single capillary. The aim of this work is to develop a sensitive method for tissue extracts and serum with commercially existing instrumentation. The influence of leading and terminating electrolyte, (LE and TE, respectively), the choice of the capillary coating, the composition of the separation media, and the effect of matrix constituents in serum extracts on the preconcentration and separation performance are described *(5)*.

From: *Methods in Molecular Biology, Vol. 162:*
Capillary Electrophoresis of Nucleic Acids, Vol. 1: Introduction to the Capillary Electrophoresis of Nucleic Acids
Edited by: K. R. Mitchelson and J. Cheng © Humana Press Inc., Totowa, NJ

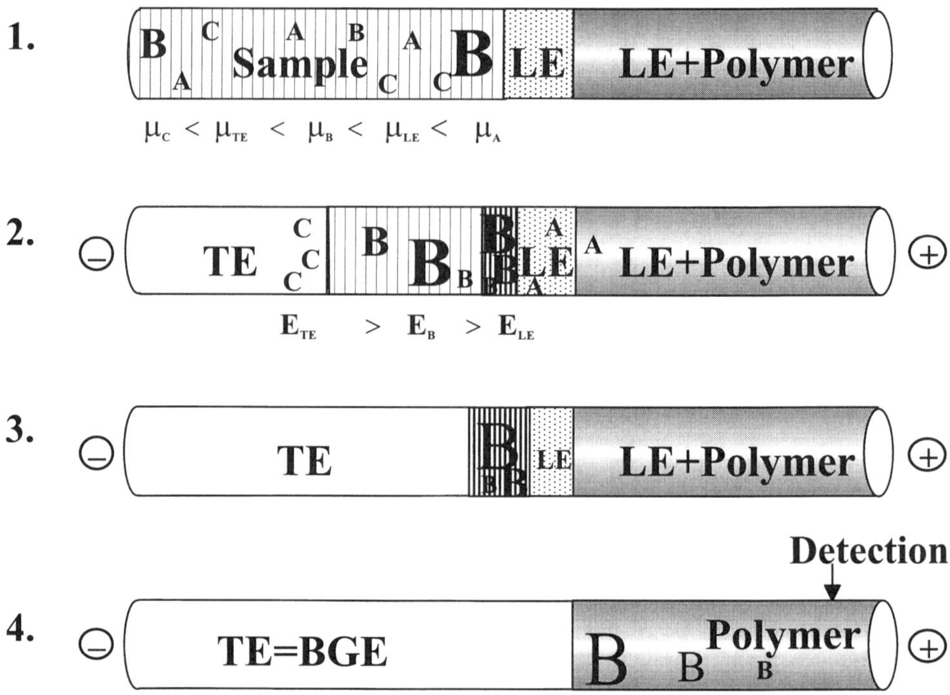

Fig. 1. Schematic representation showing the ITP-CE procedure. For a detailed description, *see* text. Reproduced from **ref. 5**.

The principle of ITP-CE coupling in our setup is similar to the one, described by van der Schans et al. *(6)*. The general procedure consists of the following steps (*see* **Fig. 1**):

1. Filling of the capillary with polymer solution, leading electrolyte and a sample consisting of a mixture of oligonucleotides B, and fast and slowly moving matrix constituents A and C, respectively. The inlet vial is filled with terminating electrolyte.
2. The focusing step to concentrate the sample plug is performed between the leading and terminating electrolyte. After applying the electric field under constant current or constant voltage conditions, terminating electrolyte with a lower conductivity than the leading electrolyte enters the capillary. Formation of an adjusted zone with increased concentration of oligonucleotides takes place. Until the leading electrolyte has left the capillary, the voltage or current increases. Then the voltage or current will be constant.
3. Steady-state in the ITP step. The different zones move at constant velocity through the capillary until the sample zone reaches the polymer solution.
4. Separation of oligonucleotides B starts when they enter the polymer solution zone. The separation takes place in the terminating electrolyte as background electrolyte (BGE).

2. Materials

2.1. Chemicals

2.1.1. Isotachophoresis-CE

1. Tris(hydroxymethyl)-aminomethane (Tris), butyric acid, dextran 200,000 and urea were purchased from Fluka (Buchs, Switzerland). Care should be taken when working with urea. It is a denaturing agent.

2.1.2. Sample Treatment

1. Phosphodiesters $p(dA)_{25-30}$ are obtained from Pharmacia Biotech AG (Dübendorf, Switzerland).
2. The internal standard is the 25-mer phosphorothioate sequence: 5'-TCC CGC CTG TGA CAT GCA TTG GTG A-3').
3. The 20-mer phosphorothioate (sequence: 5'-TCC CGC CTG TGA CAT GCA TT -3', ISIS5132), the 15, 16, 17, 18, and 19-mer phosphorothioates (sequence: as 20-mer cut from the 3' end site) are each synthesized at MWG Biotech GmbH (Ebersberg, Germany).
4. The phenol/chloroform/isoamylalcohol (25:24:1 v/v/v) solution was purchased from Sigma (Buchs, Switzerland). **NOTE:** This extraction reagent is toxic. The safety sheet from the supplier *should be read carefully* before the use of the phenol/chloroform/isoamylalcohol solution.

2.2. Instrumentation

2.2.1. CE Instrumentation

1. The experiments are performed on a P/ACE 5000 at 25°C (Beckman Instruments, Fullerton, CA) (*see* **Note 1**).
2. To allow filling of the 50- or 100-µm id capillary with the highly viscous dextran solution in a reasonable time frame (mins), the settings for rinsing pressure are modified from 20 to 50 psi (≈ 3.5 bar) and checked with an external manometer (*see* **Notes 2–5**).
3. The detection wavelength was set to 260 nm and the capillary temperature to 25°C.

2.2.2. Instrumentation for Sample Cleanup

1. Eppendorf centrifuge, run at top speed (20,800*g*).
2. "Speed-Vac" vacuum centrifuge.
3. Ultrasonic bath.

2.2.3. Coated Capillaries

1. In order to work without electroosmotic flow (EOF), the separations were performed in poly(vinyl alcohol) (PVA)-coated capillaries (Hewlett Packard, Waldbronn, Germany) (*see* **Note 6**).
2. The overall length of the capillary was 66.7 cm and the length from injection end to detection window was 60.0 cm (*see* **Notes 7–9**).

3. Methods

3.1. Choice of Leading and Terminating Electrolyte

1. To create suitable conditions for ITP-CE, the choice of LE and TE is very important. The electrophoretic mobility of the analyte has to be sandwiched between the electrophoretic mobilities of the LE and TE.

2. The anions of the TE have to have a lower electrophoretic mobility than the sample analytes, whereas the anions of the LE have to have a higher mobility.
3. Due to the different conductivities in the TE, sample and LE-zone and the condition of constant current through the capillary, the electric fields across the various zones will adapt according to Ohm's law *(4–6)*.
4. First, the electrophoretic mobility of the analytes of interest in free solution has to be determined. Using a BGE that consists of 50 m*M* chloride, 100 m*M* Tris-base, oligonucleotides exhibit an electrophoretic mobility (μ_{el}) of -39×10^{-5} cm^2/V/s (in free solution, not in the sieving polymer solution!).
5. Chloride ($\mu_{el} = -68.5 \times 10^{-5}$ cm^2/V/s), which is often used as LE in ITP experiments, was chosen in this work as LE anion. The choice of the anion in the TE is more critical. In this ITP-CE system, it was found that the electrophoretic mobility should be smaller (less negative) than -34×10^{-5} cm^2/V/s.
6. The preconcentration effect was smaller, when a TE anion with a μ_{el} close to the μ_{el} of the oligonucleotides was used. For instance, with lactic acid ($\mu_{el} = -35.8 \times 10^{-5}$ cm^2/V/s) as TE, a fourfold decrease in sensitivity was observed. All ITP-CE-experiments were performed with butyrate ($\mu_{el} = -33.7 \times 10^{-5}$ cm^2/V/s) as TE anion.

3.2. Preparation of Buffers and Polymer Solution

1. The final pH of all buffers is between 8.0 and 8.5. The pH is not adjusted following addition of the components, in order to avoid a change in the ionic strength.
2. LE (100 m*M* Tris-base, 50 m*M* Cl): dissolve 1.21 g Tris-base and 2.5 mL 2.0 *M* HCl in water in a 100-mL volumetric flask and fill it up to 100 mL.
3. TE (100 m*M* Tris-base, 50 m*M* butyric acid): dissolve 1.21 g Tris-base and 0.460 mL butyric acid in water in a 100-mL volumetric flask and fill it up to 100 mL.
4. The dextran polymer solution (28% dextran, 6 *M* urea, 100 m*M* Tris-base, 50 m*M* Cl) (*see* **Note 10**). Dissolve 14 g dextran 200,000, 36.04 g urea, 0.605 g Tris-base and 1.25 mL of 2 *M* HCl in approx 20 mL water in a 50-mL volumetric flask, shake and then degas in an ultrasonic bath for 5 min. Degas the solution *in vacuo* in a vacuum dessicator, placed on a magnetic stirrer, and increase the vacuum slowly to approx 100 mbar. If too much foam develops, the vacuum has to be decreased for a short time and again slowly increased. After about 30 min, a homogeneous, transparent polymer solution without air bubbles should be obtained. Fill up to a final volume of 50 mL with water, mix and degas once more for approx 30 min.
5. Storage: Buffers and polymer solutions could be stored at 4°C up to 2 wk without decreasing separation performance.

3.3. On-Column ITP-CE of Antisense Phosphorothioates in Biological Fluids

1. It is possible to separate phosphodiesters from serum constituents such as bovine serum albumin (BSA) in the ITP step without any sample pretreatment *(5)*. This reduces the time taken for sample cleanup steps considerably, in case of serum and other biological fluids.
2. Unlike phosphodiesters, phosphorothioates exhibit a strong tendency to adsorb onto proteins. Detecting phosphorothioates after direct injection of crude, spiked serum samples into an on-column ITP-CE system is found to be impossible. Therefore, the on-column ITP-CE approach fails when using phosphorothioates in biological fluids without an additional sample cleanup.

3. A phenol-chloroform extraction (cleanup) step for protein removal with subsequent drop dialysis for desalting is described below. Desalting is also necessary, because of the very high salt concentration (close to 1 *M*), after sample preconcentration in the Speed-Vac.

3.4. Protein Removal Procedure and Desalting (see Note 11)

1. Dilute 100 µL rat serum with 300 µL LE.
2. Extract three times with 400 µL phenol-chloroform-isoamyl alcohol to remove matrix proteins. A centrifugation step is recommended for good phase separation.
3. Concentrate the aqueous phase to dryness in a Speed-Vac for about 1 h at high temperature (about 60°C).
4. Dissolve the residue in 30 µL water using an ultrasonic bath.
5. Desalting by drop dialysis: Place the 30–µL droplet on a 0.025-µm cellulose triacetate membrane (Millipore) floating on 25 mL water and leave it for around 1h.
6. Pipet the droplet into a microvial and analyze by ITP-CE.

3.5. ITP-CE Procedure

1. For separation of phosphorothioates, a sieving polymer solution consisting of 28% dextran 200,000 and 6 *M* urea in 100 m*M* Tris-HCl was used. Because of the excellent resolution obtained with the dextran solution, short zones of sieving medium and thus long sample plugs can be injected, resulting in high sensitivity (*see* **Note 12**).
2. The following procedures are used for pretreated samples with high salt content, such as serum extracts. Procedure I: Use a 100-µm id, PVA-coated capillaries, 60.0/66.7 cm.
3. Rinse the capillary with LE: 3.5 min at 50 psi. Fill with sample: 0.2 min at 50 psi.
4. Fill with polymer from the outlet vial: 2.5 min, 50 psi, (backflush mode).
5. Place the inlet end of the capillary in TE and the outlet end in LE and apply constant current 40 µA. (reversed polarity, injection at the anodic site).
6. Procedure II: For 50-µm id, PVA-coated capillaries, 60.0/66.7 cm, rinse with LE: 5 min at 50 psi, fill with sample: 0.2 min at 50 psi, fill with polymer from the outlet vial for 4.0 min, 50 psi, (backflush mode).
7. Place the inlet end of the capillary in TE and the outlet end in LE and apply constant voltage –28 kV.

3.5.1. ITP-CE Run

1. An example of the analysis of serum spiked with ISIS 5132, a 20-mer heterogeneous phosphorothioate, and 5 shortmers at concentrations of 4×10^{-8} *M* and 8×10^{-9} *M*, respectively, after sample cleanup is shown in **Fig. 2A**. The separation was performed in a 100-µm id capillary. This electropherogram shows that the concentration detection limit is in the low nanomolar range. Additional peaks between 12 and 23 min could be observed in the electropherogram. These signals, which result from serum constituents, did not interfere with the phosphorothioates and allow a reliable identification of the phosphorothioates.
2. **Figure 2B** shows an electropherogram, resulting from the ITP-CE separation of 50 n*M* ISIS 5132, and five of its possible metabolites at 10 n*M* concentration in a 50-µm id capillary. The sample was dissolved in 50 m*M* chloride and 100 m*M* Tris-HCl. In this case, sample clean up was not necessary. A phosphorothioate of 25-base length is added to the sample as an internal standard to determine normalized peak areas. The relative standard deviations for migration times and normalized peak areas of ISIS 5132 were 2.2 and 5.0%, respectively ($n = 6$). The calibration curve is found to be linear over two orders

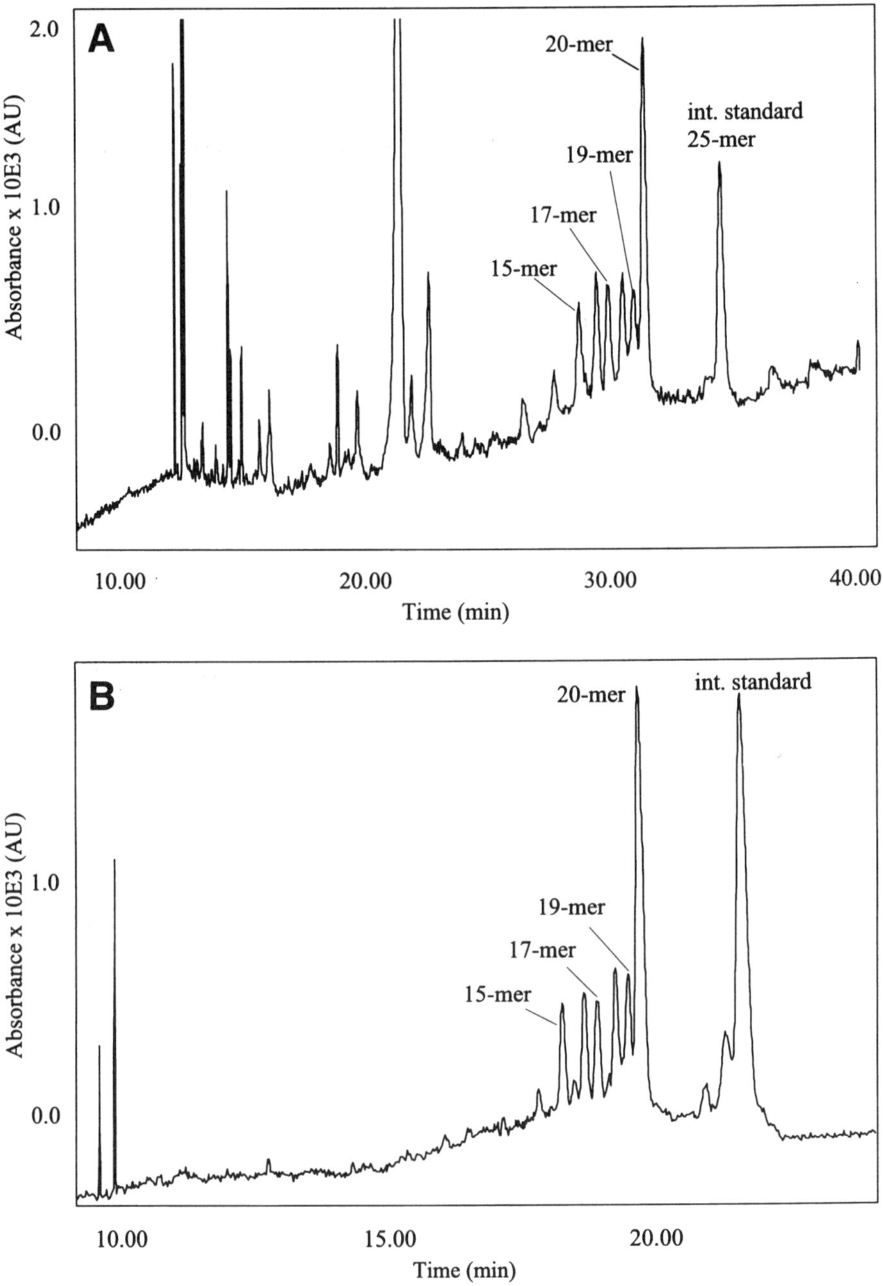

Fig. 2. Separation of ISIS5132 and five shortmers by ITP-CE, **(A)** 100-µm id capillary with procedure I (*see* **Subheading 3.**). Sample: serum, spiked with ISIS 5132, a 20-mer heterogeneous phosphorothioate, and 5 shortmers at concentrations of 4×10^{-8} *M* and 8×10^{-9} *M*, respectively, after sample cleanup. **(B)** 50-µm id capillary using procedure II (*see* **Subheading 3.5.1.**). Sample: 5×10^{-8} *M* ISIS5132, 1×10^{-8} *M* of the shortmers and 5×10^{-8} *M* internal standard dissolved in 50 m*M* chloride, 50 m*M* Tris-base. Reproduced from **ref. 5**.

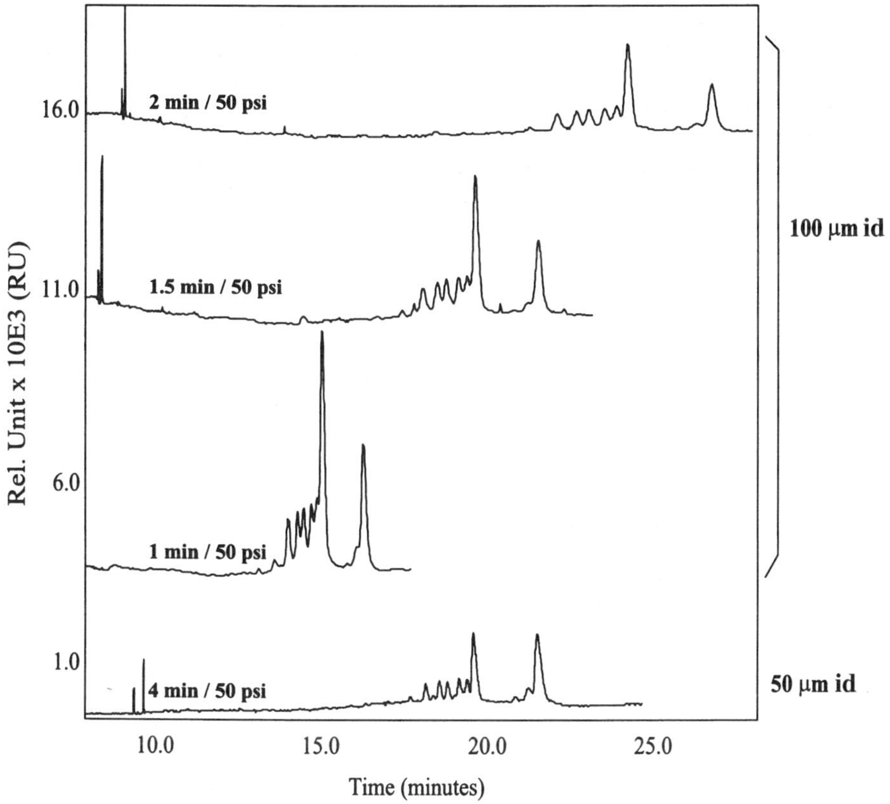

Fig. 3. Effect of polymer filling time on sensitivity and resolution of phosphorothioates. LE: 50 m*M* chloride, 100 m*M* Tris-base at pH 8.4. TE: 50 m*M* butyrate, 100 m*M* Tris-base at pH 8.4. Sieving medium is 28% dextran 200,000, 6 *M* urea, 50 m*M* chloride, 100 m*M* Tris-base, pH 8.4. Sample: 5×10^{-8} *M* ISIS5132, 1×10^{-8} *M* of the shortmers and 5×10^{-8} *M* internal standard dissolved in 50 m*M* chloride, 50 m*M* Tris-base, pH 8.4; constant current 40 μA; Capillaries: 100-μm id, PVA-coated, 60/66.7 cm; 50-μm id, PVA-coated, 60/66.7 cm.

of magnitude between 4 and 500 n*M* ($r = 0.999$). The concentration detection limit (S/N = 3) for these phosphorothioates is 2 n*M* in buffer solution.

3. Comparing **Fig. 2A**,**B**, it can be seen that a better resolution for the separation of the phosphorothioates is obtained in the 50-μm id capillary than in the 100-μm id capillary.

3.6. Choice of Polymer Plug Length

1. Any change of capillary length, inner diameter, polymer viscosity, or pressure applied will influence the time needed to fill the capillary with a zone of polymer solution. The optimal zone length has to be found out experimentally. As can be seen in **Fig. 3**, the peak height and resolution are strongly influenced by the length of the injected polymer plug.
2. Longer polymer plugs automatically decrease the length of the sample zone and thus the sample amount injected. Shortening the polymer plug still allows a good separation performance with higher sensitivity.

3. When a too short polymer plug is injected, resolution diminishes as can be seen in the electropherogram resulting from a 1 min filling time in the 100 µm id capillary (**Fig. 2**). Possibly, the steady-state in the ITP step is not reached, resulting in a contribution from the width of the sample plug to the total band broadening. Also, the polymer plug length could become too small to obtain adequate resolution.
4. As seen for the serum sample, a desalting step is still mandatory to reduce the time needed for the ITP step. Other desalting techniques instead of drop dialysis, such as reversed phase cartridges, can be used as well.

4. Notes

1. To enable the filling of the highly viscous polymer solution within a reasonable time frame (minutes) into the capillary, it is recommended to use CE-instrumentation that allows pressure application of at least 3 bar across the capillary. Other instruments than the Beckman P/ACE can be used as well. For instance, the Beckman P/ACE-MDQ allows around 7 bar (100 psi) pressure drop across the capillary. The ^{3D}CE instrument from Hewlett Packard (Waldbronn, Germany) has a maximum applicable pressure of 12 bar.
2. The capillaries are held in cartridges in both the Beckman and HP instruments. The different capillary housings result in somewhat different lengths to the detection window and total capillary lengths that can be used. For instance, in the Beckman P/ACE and the HP ^{3D}CE instrument, the lengths from detection window to outlet end are 6.7 and 8.5 cm, respectively. Therefore, the capillary lengths given above for the Beckman P/ACE have to be adapted when experiments will be performed with other instrumentation. Also, the filling times for the dextran solution zones have to be adapted for each machine.
3. When using the Beckman P/ACE 5000 instrument with a maximum applied pressure of 50 psi, pressure leakage at the sample vial occurs occasionally. This can be prevented by fixing the rubber cap of the sample vial in place with a thin, plastic ring. The inner diameter of the ring is the same as the outer diameter of the rubber cap. Alternatively, the rubber vial cap can also be hold in place by wrapping it with a piece of plastic tape. Pressure leakage can also occur when small crystals of buffer salts or urea (6 *M*) hinder the formation of an airtight seal between the injection device and the rubber cap. Check this regularly and clean when necessary.
4. Working with larger id capillaries, 100 instead of 50 µm, shortens the filling time considerably, because the filling time is inversely proportional with the square of the capillary diameter. A constant current of only 40 µA is applied with 100-µm id capillaries as the current increases quadratically with increasing id. Maximum voltage (30 kV) is not used with 100-µm id capillaries, as the Joule heating could disturb both the ITP preconcentration and the CE separation processes (air bubbles).
5. Because of the lower current in a 50-µm id capillary, higher electric fields can be applied and thus faster analysis times than in 100-µm id capillaries are achieved. Although a lower sensitivity in 50-µm id capillaries is expected due to the smaller injected sample volume and the shorter path length for detection light, this is compensated by a reduced band broadening of the peaks (compare **Fig. 2A,B**). Therefore, a longer sample plug can be injected into 50-µm id capillaries, resulting in very comparable sensitivities for both the 50- and the 100-µm id capillaries.
6. In order to operate a stable ITP-CE system, it is very important to minimize the EOF. Neutral coated capillaries are optimal for zero EOF conditions. Of the several commercial coatings tested, the PVA-coating capillaries proved to be satisfactory with respect to both capillary life time and separation performance. Other coated capillaries with strongly

reduced EOF, such as CElect N capillaries from Supelco (Bellefonte, PA) or DB17 from J&W (Folsom, CA) can also be used.

7. For short-term storage, do not leave the capillary filled with polymer solution. The capillary has to be flushed with water at the end of the day, which will prevent capillary clogging. For long-term storage, the capillary is also flushed with water for about 3-5 min and dried with air using empty, capped vials.
8. It should be noted here that flushing coated capillaries with strongly acidic or basic solutions should be avoided, because the coating could be destroyed by such a rinsing treatment.
9. It is not recommended to remove the polyimide coating by heat to create a detection window, because this will damage the capillary inner coating. A better way is to remove the outer polyimide coating by gently scraping the window area with a scalpel blade over a length of approx 3 mm. Alternatively, hot, concentrated sulphuric acid can be dropped onto the coating until the coating can be easily removed with paper tissue. Before installing the capillary into the CE-instrument, the detection window has to be carefully wiped with a tissue with methanol to remove any remaining polyimide or grease.
10. Several sieving polymers other than dextran solutions may be used when separating modified phosphodiester oligonucleotides. For instance, poly(ethylene glycol) solutions, which have a lower viscosity, have already been described for ITP-CE of phosphodiesters *(5,7)*. The polymer solution always has to contain the leading electrolyte anion. Dextran sieving polymer solutions are recommended when separating phosphorothioates. In case of problems with air bubbles within the capillary during operation, degas the polymer solution and, when necessary, also the buffer solutions, once more for about one hour.
11. Other methods can also be used for protein removal. Protein precipitation with acetonitrile or methanol are very easy and fast ways to remove proteins. However, because of the strong adsorption of phosphorothioates onto proteins the recovery will be low *(5)*. Ion-exchange chromatography is another effective way to remove proteins from biological samples, using small commercially available cartridges. However, this method has the disadvantage that two desalting steps of the eluent are needed *(2)* because of the high salt concentration, which makes this procedure quite labor-intensive.
12. The side from which the capillary is filled with polymer solution is very important! When the sieving polymer is pushed from the inlet site through the capillary, separation performance diminishes after only 10–15 runs. Most likely, the polymer changes the characteristics of the capillary wall, thereby disturbing the ITP process that takes place at the inlet site in a discontinuous buffer system. When the polymer is injected from the outlet site, the capillary lifetime exceeds 50 runs!

References

1. Crooke, S. T. (1992) Therapeutic applications of oligonucleotides. *Annu. Rev. Pharmacol. Toxicol.* **32,** 329–376.
2. Leeds, J. M., Graham, M. J., Truong, L. A., and Cummins, L. L. (1996) Quantitation of phosphorothioate oligonucleotides in human plasma. *Anal. Biochem.* **235,** 36–43.
3. Bruin, G. J. M., Börnsen, K. O., Hüsken, D., Gassmann, E., Widmer, H. M., and Paulus, A. (1995) Stability measurements of antisense oligonucleotides by capillary gel electrophoresis. *J. Chromatogr.* **709,** 181–195.
4. Everaerts, F. M., Beckers, J. L., and Verheggen, Th. P. E. M. (1976) Isotachophoresis. Theory, Instrumentation and Applications. *J. Chromatogr. Library*, Vol. 5, Elsevier, Amsterdam.

5. Barmé, I., Bruin, G. J. M., Paulus, A., and Ehrat, M. (1998) Preconcentration and separation of antisense oligonucleotides by on-column isotachophoresis and capillary electrophoresis in polymer-filled capillaries. *Electrophoresis* **19,** 1445–1451.
6. van der Schans, M. J., Beckers, J. L., Molling, M. C., and Everaerts, F. M. (1995) Intrinsic isotachophoretic preconcentration in capillary gel electrophoresis of DNA restriction fragments. *J. Chromatogr.* **717,** 139–147.
7. Auriola, S., Jääskeläinen, I., Regina, M., and Urtti, A. (1996) Analysis of oligonucleotides by on-column transient capillary isotachophoresis and capillary electrophoresis in poly(ethylene glycol) filled columns. *Anal. Chem.* **68,** 3907–3911.

VI

Practical Applications for Analysis of Environmental Damage and Cellular Metabolism of DNA

28

Capillary Electrophoresis–Mass Spectrometric Analysis of DNA Adducts

Lisa A. Marzilli, Cheryl Koertje, and Paul Vouros

1. Introduction

It is believed that the majority of cancers are caused by contact with naturally occurring or synthetic chemicals present in the environment. Hence, in theory, if contact with these carcinogens is minimized or eliminated, most cancers could be preventable *(1)*. Cancer is a multistage process in which the first stage is called initiation. It is during the initiation stage that chemical damage or modification occurs to DNA *(2)*. The compounds, which result from a covalent attachment of the chemical with DNA, are called "DNA adducts." Because of their potential for initiating cancer, the characterization and quantitation of DNA adducts are important. The specific identification of the adduct compound, the position of the adduct group on the base and the location of the modified base in the DNA sequence are significant factors. Carcinogens vary from the highly reactive to relatively inert, however they are typically electrophilic in nature or become electrophilic after being metabolized in vivo *(1,2)*. The most reactive sites are the purine nitrogens of guanine and adenine *(1)*. These nucleophilic sites are ideal for covalent reactions with electrophilic carcinogens. DNA adduct formation is complex, but there is site selectivity or specificity *(2)*.

The ability to detect DNA adducts at levels present in humans is crucial because two facts are validated: exposure to a carcinogen and risk of cancer development. Typically, one adducted base is present in 10^7–10^8 normal bases *(2)*. Although high performance liquid chromatography (HPLC) has been effective in the analysis of DNA adducts *(3,4)*, this review will focus on the technique of capillary electrophoresis (CE) coupled to mass spectrometry (CE-MS). Compared to HPLC, CE offers the advantages of higher column efficiencies, rapid analysis time, low sample consumption, and minimal operational costs. From a practical viewpoint, CE also consumes less buffer and solvent reagents and produces less waste than HPLC. A major drawback of CE for biological samples is the limited volume of the capillary. Small injection volumes, typically 1–10 nL, can make CE impractical for real life applications due to poor

From: *Methods in Molecular Biology, Vol. 162:*
Capillary Electrophoresis of Nucleic Acids, Vol. 1: Introduction to the Capillary Electrophoresis of Nucleic Acids
Edited by: K. R. Mitchelson and J. Cheng © Humana Press Inc., Totowa, NJ

concentration detection limits. However, preconcentration techniques, such as sample-stacking, have been devised to increase the loadability of the capillary and improve detection limits for CE-MS.

Several reviews have been written on the subject of CE and CE-MS for biological samples ***(5–8)***. Although CE can be performed in several modes when coupled to MS, the most common method for DNA adduct analysis has been capillary zone electrophoresis (CZE). Several types of detectors have been coupled with CZE for adduct detection including UV absorption ***(9,10)*** and fluorescence ***(11)***. However, neither UV nor fluorescent detection provides any intrinsic information (e.g., molecular weight, structure), and the latter technique usually requires a fluorescent moiety to permit detection. Thus, use of MS detection warrants an added level of selectivity and specificity for an analysis. This chapter focuses on the application of CZE-MS for the study of DNA adducts. In particular, the nucleotide adducts resulting from exposure to the polycyclic aromatic hydrocarbon (PAH), benzo[a]pyrene, are discussed. These particular DNA-adducts result from the in vitro reaction of BaP diol epoxide and calf thymus DNA. Using preconcentration techniques, it is possible to extend the dynamic range for oligonucleotide BaP adducts detection. Separation techniques based on the use of water-soluble polymer and MS are explored for the analysis of oligonucleotide adducts. A model system based on the analysis of (*N*-acetoxy-*N*-acetylamino) fluorene or AAF-modified oligonucleotides is briefly reviewed.

2. Materials

2.1. Apparatus

1. In the modified nucleotide experiments, the CE instrument was built in-house using a Spellman high-voltage power supply (Plainview, NY), connected to the electrolyte cell by a platinum wire electrode.
2. The silica capillary (Polymicro Technologies Inc., Phoenix, AZ) with a 30-cm effective length and 100 μm id had a constant 5 kV voltage applied during the electrophoretic run. This generated a current of approx 20 μA.
3. A Finnigan MAT TSQ 700 triple quadrupole mass spectrometer (Finnigan, San Jose, CA) equipped with an electrospray source was used. The electrospray needle was maintained at –3 kV for negative ion detection.
4. In the modified oligonucleotide experiments, the capillary utilized was a polyvinylalcohol (PVA)-coated capillary column with a 100 μm id (Beckman Instruments, Fullerton, CA) and an effective length ranging from 36–44 cm.
5. The capillary was interfaced with a Finnigan LCQ ion trap mass spectrometer (Finnigan MAT, San Jose, CA) using an electrospray ion source. The electrospray needle was set at –3.7 kV for negative ion detection.
6. Analytes were introduced into the capillary column electrokinetically (–10 to –16 kV, 3–6 s) and a constant voltage of 16 kV was applied across the polymer-filled capillary. This voltage produced currents between 35 and 40 μA. All analyses were carried out at room temperature (25°C).

2.2. Chemicals

2.2.1. Anti-BPDE Modification of DNA

1. All buffers and sheath liquids are made fresh each day using only HPLC grade solvents purchased from Fisher Scientific Inc. (Pittsburgh, PA).

2. A buffer solution of 20 m*M* NH_4OAc, pH 9.3 is used for the CZE experiments.
3. Calf thymus DNA and digestion enzymes were purchased from Sigma Chemical Co. (St. Louis, MO).
4. *Trans*-7,8,9,10-tetrahydrobenzo[a]pyrene-7,8-diol 9,10-epoxide (anti-BPDE) purchased from National Cancer Institute Repository (Kansas City, MO). **Caution**: **This is an active carcinogen and should be handled with latex gloves and in a fume hood!**
5. Approximately 11 mg of calf thymus DNA is dissolved in 10 mL of 0.05 *M* Tris-HCl, pH 6.8.
6. Approximately 1 mg of benzol[a]pyrene diol epoxide (BPDE) is dissolved in 0.5 mL of THF and added to the DNA solution. The resultant mixture was incubated at 50°C for ~8 h. The unreacted BPDE is extracted from the mixture with ethyl acetate (5X) and diethyl ether (2X). The sample was then centrifuged at –10°C for 20 min at 5600*g* (12,000 rpm) after which the excess liquid is siphoned from the DNA pellet.
7. DNA is hydrolyzed to deoxynucleotides with DNAase 1 and snake venom phosphodiesterase 1 at pH 8.7 in the presence of Ca^{2+} (0.5 *M*) and Mg^{2+} (0.4 *M*) at 37°C for approx 6 h.

2.2.2. The 2-(N-Acetoxy-N-Acetylamino) Fluorene Modification of Oligonucleotides

1. In the modified oligonucleotide experiments, a 15 m*M* ammonium acetate buffer (Sigma Chemical Corp., St. Louis, MO) solution is prepared in water from an ultrapure filtration system (Millipore, Bedford, MA). This buffer pH is adjusted to 9.5 with ammonium hydroxide (Fisher Scientific Inc.).
2. Unmodified oligonucleotides are from Amitof Biotech (Boston, MA) and typically stored at –20°C until analysis is performed. Modification with a 7.8 m*M* stock solution of 2-(*N*-acetoxy-*N*-acetylamino) fluorene (AAAF) in acetonitrile is utilized. **Caution: Extreme care should be used when working with carcinogenic compounds such as AAAF. Use facilities mentioned previously for work with BPDE.**
3. For each reaction, 4 µL of the AAAF stock solution is added to 10 µL of a 2 m*M* oligonucleotide solution. The reaction is allowed to stand at room temperature overnight (~12 h) and then analyzed immediately by CE-MS to avoid thermal degradation.

3. Methods

3.1. Preconcentration Techniques

3.1.1. Purification of Adduct Nucleotides

1. In order to concentrate the amount of modified nucleotides in a sample, it may be necessary to remove the normal nucleotides and impurities from the sample. In the case of B[a]P nucleotide adducts, this is accomplished by hydrophobic interaction liquid chromatography. This technique involves:
   a. Use of a solid-phase interaction column consisting of a very small volume bed of poly-(styrene divinylbenzene) beads sandwiched between two frits. The column has a 7-µm id and contains a 2-mm thick bed of stationary phase.
   b. The DNA hydrolysate mixture is added to the column and the unmodified nucleotides are removed by repeated washings with high-purity water. The nucleotide adducts are retained on the hydrophobic column.
   c. The adducted nucleotides are eluted by rinsing the column with a 50:50 (v/v) methanol: water solution, with a recovery of approx 75% of the adducts.

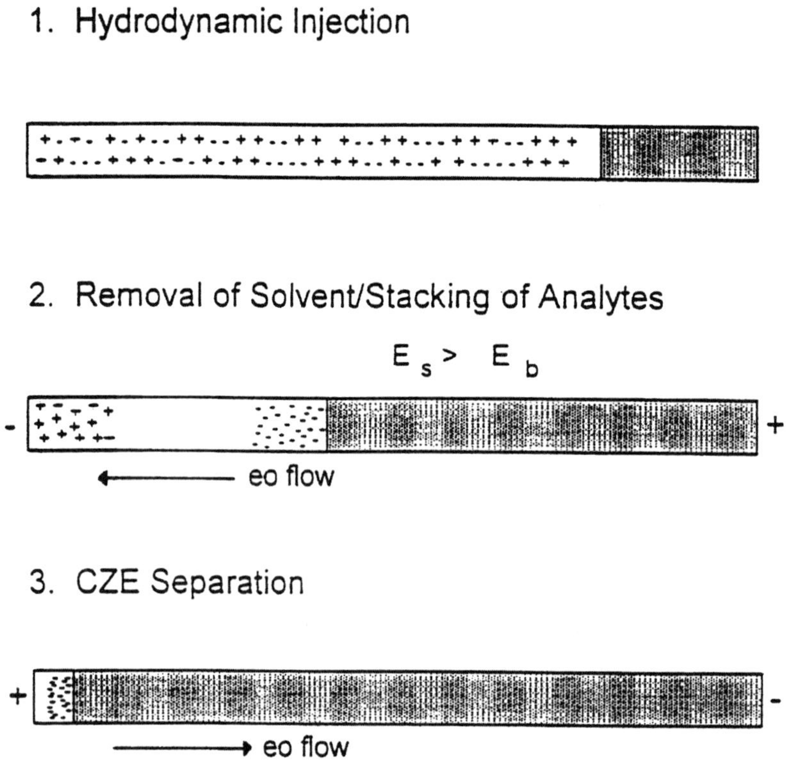

Fig. 1. Steps involved in the preconcentration technique of sample stacking.

3.1.2. Sample-Stacking

1. Following the above cleanup, the B[a]P nucleotide adducts can be concentrated in the capillary column by a method called "sample stacking" (*see* **Note 1**).
2. Compared to analytical techniques such as LC-MS, CE-MS generally suffers from poor concentration detection limits. However, this potential weakness may be ameliorated with use of sample stacking techniques. Sample stacking, which was first proposed by Chien and Burgi ***(12)***, involves the sequence of steps depicted in **Fig. 1**:
   a. A large volume of the sample is dissolved in a solvent of lower conductivity than the running buffer and hydrodynamically injected on the column.
   b. High voltage (~5kV) with a polarity opposite to that used for the electrophoretic separation is applied. This causes the sample to experience a higher electric field than the running buffer ($E_s > E_b$) and therefore, negatively charged species will rapidly stack at the boundary between the sample solvent and the CZE buffer. The positive ions and the sample solvent are then pumped out at the anode (injection) end by electroosmotic flow (EOF).
   c. The current is monitored and when it reaches about 95% of its initial (buffer only) value, the voltage is reversed. This increase in current is a result of fresh buffer enter-

ing the capillary by EOF. Thus, the stacking process is stopped (and polarities switched) when the current has reached a level comparable to that observed when the capillary is filled with buffer.

3. Results indicate that without sample stacking, significant changes in migration times occur with different size sample injection plugs. However, when solvent removal is employed, various injection volumes result in reproducible migration times with similar analyte peak shapes, and no significant loss of resolution. Therefore, depending on the length of column used, a significant increase in sensitivity can be achieved with the aid of sample stacking.
4. The use of this technique has a notable effect on the detection limits of CE-MS ***(13–15)***. In fact, sample stacking with solvent removal can effectively preconcentrate adducted nucleotides by as much as 1000-fold without significant loss in resolution ***(13)***.
5. Since the B[a]P nucleotide adducts are negatively charged, sample stacking is used to concentrate the sample. This technique affords detection of B[a]P adducts at a concentration detection limit of approx 10^{-8} *M* ***(15)***. Deforce and coworkers ***(14)*** have also used sample stacking for DNA adducts. They obtained signal enhancement by a factor of several hundred when sample stacking was employed in combination with CZE/ESI-MS for the analysis of in vitro adducted phenyl glycidyl ethers (PGEs) on 2'-deoxynucleotides. Phenyl glycidyl ethers are found in the paint and resin industries and belong to the PAH group of suspected carcinogens.

3.2. Interfacing CE and MS

Since the first report of coupling MS detection to CZE ***(16)***, a significant amount of research has occurred in this area ***(6,7,17)***. Although fast atom bombardment (FAB) ionization set the stage for DNA adduct work by CZE-MS ***(18)***, limitations such as matrix interference, poor sensitivity, and difficulty in analyzing nucleotides larger than 10 mers have made this ionization technique obsolete. Thus, the current ionization method of choice for the determination and identification of DNA adducts by CZE-MS is electrospray ionization (ESI) ***(19,20)***.

1. In ESI, a liquid sample is nebulized to produce charged droplets, a few μm in diameter ***(21)***. As the solvent evaporates during transport to the mass spectrometer, the droplets become progressively smaller in size. ESI has a high ionization efficiency and can be used for analysis of large nonvolatile molecular species such as DNA adducts ***(19,20)***.
2. In addition, the formation of multiply charged ions facilitates the use of a low-mass range instrument for high molecular weight compounds. A schematic representation of a CE-ESI-MS setup is depicted in **Fig. 2**.
3. Two critical factors in combining CZE with MS are the buffer system and the interface (*see* **Notes 2** and **3**). One needs to consider a buffer system appropriate for both the CE and MS components. The buffer should not only be capable of high-resolution separations but also be sufficiently volatile in order to guarantee a stable ESI signal. Therefore, criteria to consider for the buffer include volatility, salt concentration, and pH.
4. The buffer system used for the following DNA adduct analyses by CE-MS was an ammonium acetate solution. This buffer is capable of separating adducted from non-adducted species, yet, is volatile enough to produce a stable electrospray signal. The buffer solution is pH adjusted to an alkaline pH to aid in the formation of negatively charged oligonucleotide species. These charges result from deprotonation of the phosphate backbone.

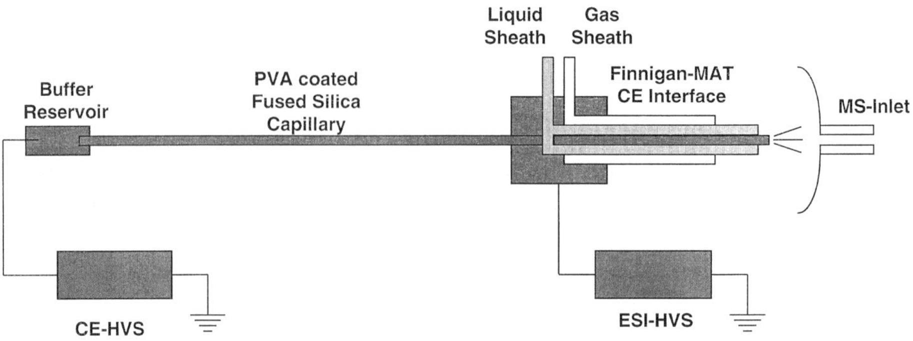

Fig. 2. Schematic representation of a CE-ESI-MS instrument.

5. Another important consideration in coupling these two techniques is the interface between the end of the CE capillary and the mass spectrometer for sample transfer (*see* **Note 3**). The interface should provide electrical contact and minimize band broadening.
6. Several types of interfaces have been incorporated for CE-MS, including the sheathless interface ***(22)***, the liquid–junction ***(23)***, and the liquid-sheath interfaces ***(24)***.
7. The liquid-sheath is most commonly used interface, due to its simple construction and reproducibility. It relies on the coaxial introduction of sheath liquid at a higher flow rate than the CE system (3–4 µL/min). The sheath liquid effectively serves as the terminal buffer reservoir for CE and aids in producing a stable ESI current. It typically consists of an organic solvent (e.g., 2-propanol, methanol, or acetonitrile) combined with a percentage (5–60%) of the running buffer.
8. In the case of B[a]P adducts, a sheath liquid of 75:25 (v/v) (2-propanol: 20 m*M* ammonium acetate, pH 9.3) was used. This combination of organic solvent and buffer constituents ensures electrical contact and spray stability. The coupling of CZE and ESI-MS allows the detection and identification of the BaP-dGMP nucleotide adduct formed from the in vitro reaction of the active intermediate of BaP, i.e., anti-BPDE and DNA.
9. However, short, undigested oligonucleotide adducts are also produced in this in vitro reaction mixture. **Figure 3** shows that several hydrolysis products are formed in addition to the expected mononucleotide adduct (BaP-dGMP). The presence of these modified di- and trinucleotides (TG-BaP, ATG-BaP, GTG-BaP) can result in very complicated electrospray mass spectra, due to multiple charge states and metal cation adduction (Na^+, K^+). Thus, both sieving and partitioning polymer matrices are employed to improve the CZE separation of the modified oligonucleotides ***(25)***.

3.3. Buffer Additives

1. Buffer additives such as polyvinylpyrrolidone (PVP) can be used to provide added selectivity in CE-MS experiments for modified oligonucleotides ***(25)***. The following steps are employed to prepare the capillary:
   a. A 5% (w/v) polymer solution is made using the 15 m*M* NH_4OAc, pH 9.5 running buffer.
   b. This solution is stirred at room temperature for 6 h and subsequently sonicated for approx 15 min to remove dissolved gases.

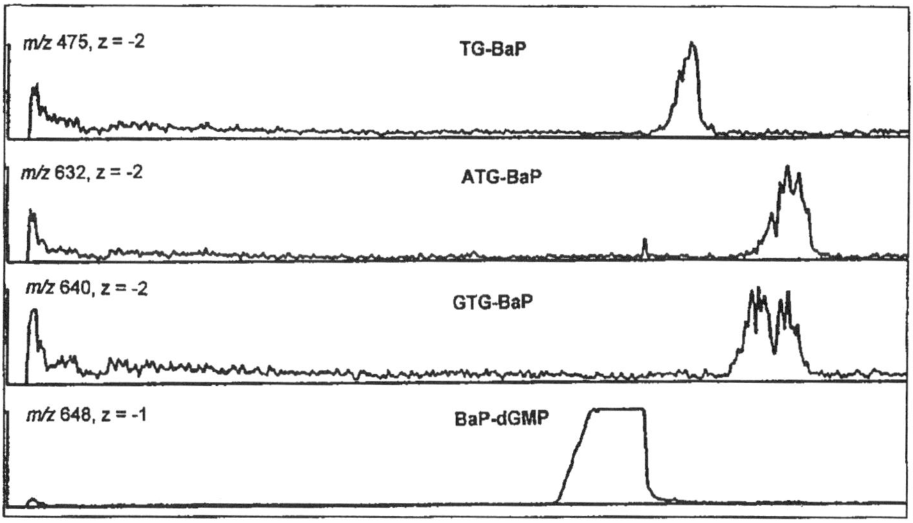

Fig. 3. Mass electropherogram of an in vitro reaction mixture. Besides the expected modified nucleotide (BaP-dGMP), several short, modified di-mers, and tri-mers are observed (TG-BaP, ATG-BaP, GTG-BaP). Reprinted with permission from **ref. *15***, Copyright (1996), American Chemical Society.

c. The viscous polymer solution is loaded into the PVA-coated silica capillary using a syringe.

2. Recently, Harsch et al. *(26)* published data which clearly demonstrates the separation of oligonucleotides and their biologically relevant AAF-modified analogs. **Figure 4** shows that these adducts are formed when AAAF reacts at the C8 position of guanine residues. AAAF serves as a valuable model system for other C8-binding mutagens, such as dietary carcinogens *(27)*. AAF-modified oligonucleotides, designated with an asterisk (*), range in length from 3–7 nt (ATG*, ATG*C, ATG*CT, ATG*CTA, and ATG*CTAT) and are analyzed by CZE-MS.
3. Using CE with a PVP solution, all analyte peaks can be separated to baseline in 35 min, and are detected by ion trap mass spectrometry (**Fig. 5**). The longer analogs elute prior to the shorter analogs, and the unmodified nucleotides before the modified species. The addition of AAAF to an oligonucleotide increases the hydrophobicity of the molecule, and thus contributes to its greater interaction with the PVP additive. Thus, it is concluded that PVP played a prominent role in the resolution of the DNA adducts.
4. To further prove the usefulness of PVP as a buffer additive, the separation of four AAF-modified structural isomers was investigated. These isomers have the same base composition but different sequences. **Figure 6** shows that all four peaks are observed in the extracted mass electropherogram and are resolved to baseline level. This work underscores the power of the PVP additive to the CE running buffer to improve the separation of DNA adducts. Although the mass limits were low, the concentration injected was about 10^{-5} *M*.
5. In other CZE-MS work *(28)*, a strategy for gaining nucleobase sequence information of in vitro DNA modifications is described. In this approach, DNA is enzymatically digested with benzonase and alkaline phosphatase into 2–6 nt lengths, and the analysis of the

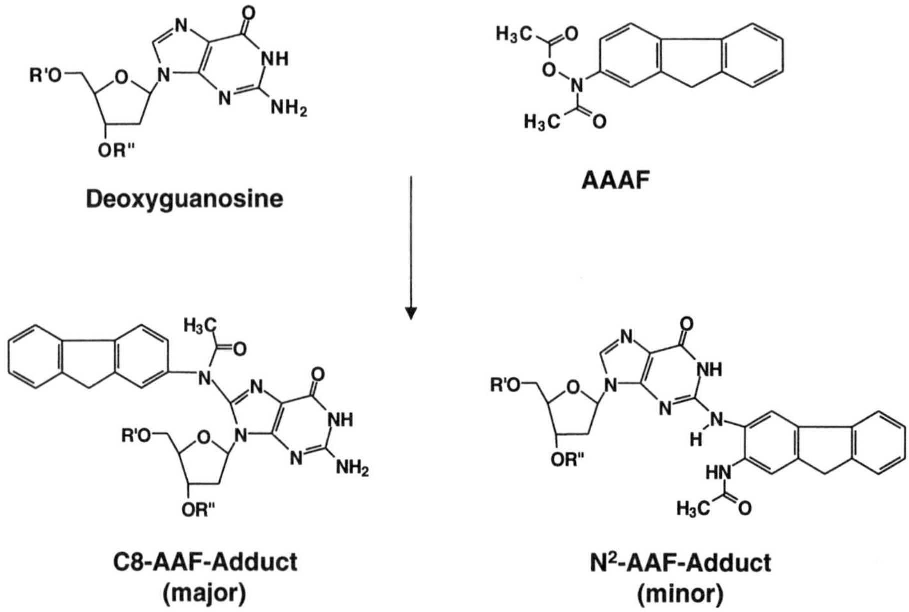

Fig. 4. Modification of deoxyguanosine units with AAAF. Reprinted with permission from **ref. *26***, Copyright (1998), American Chemical Society.

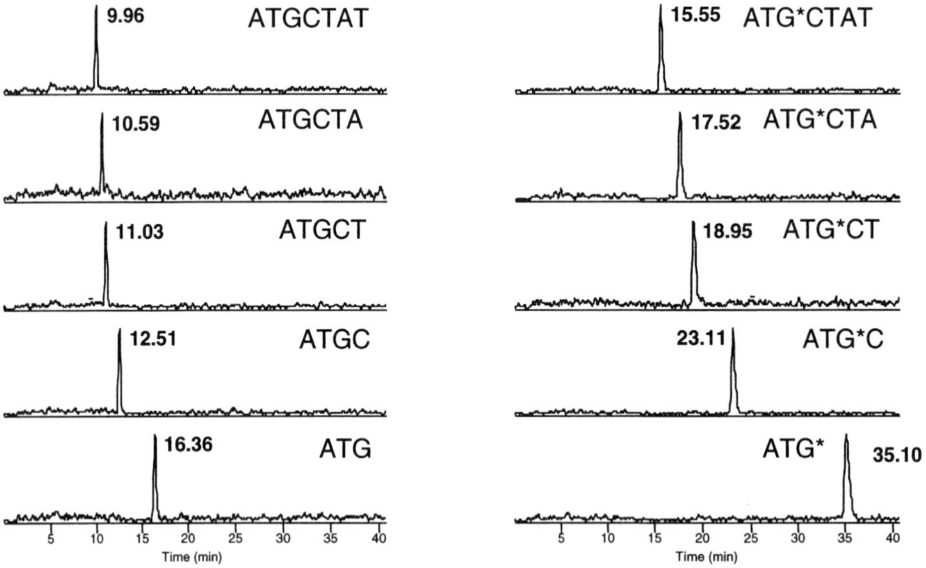

Fig. 5. Separation of analogs using CE with a 5% (w/v) aqueous PVP solution, and using on-line MS detection. Extracted mass electropherograms show that the longer analogs elute prior to the shorter analogs. Reprinted with permission from **ref. *26***, Copyright (1998), American Chemical Society.

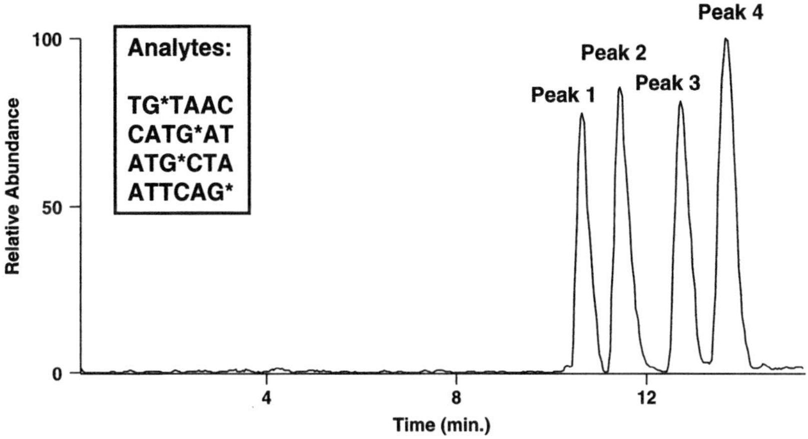

Fig. 6. CE-MS in a PVP matrix. Reprinted with permission from **ref.** ***26***, Copyright (1998), American Chemical Society.

resulting mixture of alkylated and nonalkylated nucleotides is accomplished either by differences in electrophoretic mobility using CZE, or by the addition of adduct mass (MS).

6. The same group also examined styrene oxide, a compound which possesses both mutagenic and carcinogenic effects. They found that purine bases are more likely than pyrimidine bases to be alkylated by styrene oxide ***(29)***. Previously unknown adenine adducts could also be identified. However, MS/MS experiments are not possible with these experiments due to limiting concentrations of adduct products.
7. A solid phase extraction (SPE) step can be employed to obtain a higher sample concentration. Deforce et al. ***(30,31)*** demonstrated that an additional sample cleanup step using polystyrene divinylbenzene copolymer results in a more concentrated sample of PGE-modified nucleotides. The concentration of modified adducts is then sufficient for MS/MS experiments after the removal of unmodified species and with use of sample-stacking (*see* **Note 1**). The resulting PGE-modified nucleotides are separated by CZE and show evidence both of base alkylation, and of phosphate alkylated products by collision induced dissociation (CID) analysis ***(30,31)***.

3.4. Improved Detection and Quantitation of Nucleotide Adducts

1. DNA adducts are important molecular biomarkers that indicate damage to DNA and possible risk of cancer. Because of their significance, the identification, structural characterization, and quantitation of DNA adducts are important.
2. CE is one analytical technique that has been shown to be well suited for this purpose because of its simplicity, high efficiency, excellent selectivity, and low operational costs.
3. Combined with the selectivity and specificity of MS detection, structural information of DNA adducts is possible.
4. Sample stacking techniques are used to preconcentrate the sample on the capillary column and result in an improvement in the lower concentration limits of detection of adducts by several orders of magnitude.
5. With the aid of buffer additives such as PVP, the separation of various modified nucleotides and oligonucleotide structural isomers can be achieved.

6. It is likely that further improvements to detector capabilities and to other aspects of CE-MS instrumentation will also play an important role in increasing the detection sensitivity for low concentrations of modified DNA adducts. This will have important consequences for the quantitative analysis of DNA damage in both environmental and industrial situations.

4. Notes

1. Because this preconcentration technique is performed in the off-line mode with the mass spectrometer, the execution of the stacking process can be mechanically challenging. For example, following the sample stacking procedure, the capillary must be repositioned into the ESI source without disturbing the stacked sample plug or unintentionally removing the capillary from the anode buffer reservoir.
2. The polymer solution is replaced after each run by pressure displacement. This takes several minutes and involves moving the ESI source away from the instrument and refilling the PVP solution with a syringe. The capillary is not removed when the fresh polymer solution is added. As mentioned above, the spray stability is very sensitive to capillary positioning.
3. The stability of the spray into the mass spectrometer can be difficult to maintain because of spray blockage. Such instability is evidenced by a sudden decrease in current. When this occurs, the experiment is stopped, and by simply cutting off the end of the capillary (~ 0.25 cm), the problem can often be alleviated. In addition, repositioning the capillary in the electrospray source can also be beneficial. Moving the capillary less than 1 mm in either direction can sometimes dramatically improve spray stability. Lifting the inlet of the capillary (~1 cm) was also a useful maneuver in minimizing resolution loss.

References

1. Calladine, C. R. and Drew, H. R. (1997) *Understanding DNA: The Molecule and How It Works.* 2nd ed., Academic Press, San Diego.
2. Farmer, P. B. and Sweetman, G. M. (1995) Mass spectrometric detection of carcinogen adducts. *J. Mass Spectrom.* **30,** 1369–1379.
3. Apruzzese, W. A. and Vouros, P. (1998) Analysis of DNA adducts by capillary methods coupled to mass spectrometry: a perspective. *J. Chromatogr.* **794,** 97–108.
4. Esmans, E. L., Broes, D., Hoes, I., Lemiere, F., and Vanhoutte, K. (1998) LC-MS in nucleoside, nucleotide and modified nucleotide characterization. *J. Chromatogr. A* **794,** 109–127.
5. Karger, B. L., Cohen, A. S., and Guttman, A. (1989) High-performance capillary electrophoresis in the biological sciences. *J. Chromatogr.* **492,** 585–614.
6. Smith, R. D., Wahl, J. H., Goodlett, D. R., and Hofstadler, S. A. (1993) Capillary electrophoresis–mass spectrometry. *Anal. Chem.* **65,** 574A-584A.
7. Cai, J. and Henion, J. (1995) Review: Capillary electrophoresis–mass spectrometry. *J. Chromatogr. A* **703,** 667–692.
8. Andrews, C. L., Vouros, P., and Harsch, A. (1999) Analysis of DNA adducts using high-performance separation techniques coupled to electrospray ionization mass spectrometry. *J. Chromatogr. A* **856**, 515–526.
9. Jackim, E. and Norwood, C. (1990) Separation and detection of a benzo[a]pyrene \deoxyguanosyl-5-monophosphate adduct by capillary zone electrophoresis. *J. High Res. Chromatogr.* **13,** 195–196.
10. Norwood, C. B., Jackim, E., and Cheer, S. (1993) DNA adduct research with capillary electrophoresis. *Anal. Biochem.* **213,** 194–199.

11. Li, W., Moussa, A., and Giese, R. W. (1993) Capillary electrophoresis with laser fluorescence detection for profiling damage to fluorescein-labeled deoxyadenylic acid by background, ionizing radiation and hydrogen peroxide. *J. Chromatogr.* **633,** 315–319.
12. Chien, R.-L. and Burgi, D. S. (1992) Sample stacking of an extremely large injection volume in high-performance capillary electrophoresis. *Anal. Chem.* **64,** 1046–1050.
13. Wolf, S. M. and Vouros, P. (1995) Incorporation of sample stacking techniques into capillary electrophoresis CF-FAB mass spectrometric analysis of DNA adducts. *Anal. Chem.* **67,** 891–900.
14. Deforce, D. L., Ryniers, F. P., Van den Eeckhout, E. G., Lemiere, F., and Esmans, E. L. (1996) Analysis of DNA adducts in DNA hydrolysates by capillary zone electrophoresis and capillary zone electrophoresis - electrospray mass spectrometry. *Anal. Chem.* **68,** 3575–3584.
15. Barry, J. P., Norwood, C., and Vouros, P. (1996) Detection and identification of benzo[a]pyrene diol epoxide adducts to DNA utilizing capillary electrophoresis-electrospray mass spectrometry. *Anal. Chem.* **68,** 1432–1438.
16. Olivares, J. A., Nguyen, N. T., Yonker, C. R., and Smith, R. D. (1987) On-line mass spectrometric detection for capillary zone electrophoresis. *Anal. Chem.* **59,** 1230–1232.
17. Banks, J. F. (1997) Recent advances in CE-ESI-MS. Review. *Electrophoresis* **18,** 2255–2266.
18. Wolf, S. M., Vouros, P., Norwood, C., and Jackim, E. (1992) Identification of deoxynucleoside-polyaromatic hydrocarbon adducts by capillary zone electrophoresis-continuous flow-fast atom bombardment mass spectrometry. *J. Am. Soc. Mass Spectrom.* **3,** 757–761.
19. Fenn, J. B., Mann, M., Meng, C. K., Wong, S. F., and Whitehouse, C. M. (1989) Electrospray ionization for mass spectrometry of large biomolecules. *Science* **246,** 64–71.
20. Fenn, J. B., Mann, M., Meng, C. K., Wong, S. F., and Whitehouse, C. M. (1990) Electrospray ionization-principles and practice. *Mass Spectrom. Rev.* **9,** 37–70.
21. Yamashita, M. and Fenn, J. B. (1984) Electrospray ion source. Another variation on the free-jet theme. *J. Chem. Phys.* **80,** 4451–4459.
22. Fang, L., Zhang, R., Williams, E. R., and Zare, R. N. (1994) On-line time-of-flight mass spectrometer analysis of peptides separated by capillary electrophoresis. *Anal. Chem.* **66,** 3696–3701.
23. Lee, E. D., Muck, W., Henion, J. D., Covey, T. R. (1989) Liquid junction coupling for capillary zone electrophoresis/ion spray mass spectrometry. *Biomed. Environ. Mass Spectrom.* **18,** 844–850.
24. Smith, R. D., Barinaga, C. J., and Udseth, H. R. (1988) Capillary zone electrophoresis-MS. *Nature* **331,** 639–640.
25. Barry, J. P., Muth, J., Law, S.-J., Karger, B. L., and Vouros, P. (1996) Analysis of modified oligonucleotides by capillary electrophoresis in a polyvinylpyrrolidone matrix coupled with electrospray mass spectrometry. *J. Chromatogr.* **732,** 159–166.
26. Harsch, A. and Vouros, P. (1998) Interfacing of CE in a PVP matrix to ion trap mass spectrometry: Analysis of isomeric and structurally related (N-Acetylamino)fluorene-modified oligonucleotides. *Anal. Chem.* **70,** 3021–3027.
27. Hoffmann, G. R. and Fuchs, R. P. (1997) Mechanisms of frameshift mutations: Insight from aromatic amines. *Chem. Res. Toxicol.* **10,** 347–359.
28. Janning, P., Schrader, W., and Linscheid, M. (1994) A new mass spectrometric approach to detect modifications in DNA. *Rapid Commun. Mass Spectrom.* **8,** 1035–1040.
29. Linscheid, M. and Schrader, W. (1997) Styrene oxide DNA adducts: *in vitro* reaction and sensitive detection of modified oligonucleotides using capillary zone electrophoresis interfaced to electrospray mass spectrometry. *Arch. Toxicol.* **71,** 588–595.

30. Deforce, D. L., Lemiere, F., Hoes, I., Millecamps, R. E., Esmans, E. L., De Leenheer, A., and Van den Eeckhout, E. G. (1998) Analysis of the DNA adducts of phenyl glycidyl ether in a calf thymus DNA hydrolysate by capillary zone electrophoresis-electrospray mass spectrometry: evidence for phosphate alkylation. *Carcinogenesis* **19,** 1077–1086.
31. Deforce, D. L. and Van den Eeckhout, E. G. (2001) Analysis of DNA damage using capillary zone electrophoresis and electrospray mass spectrometry, in *Capillary Electrophoresis of Nucleic Acids,* Vol. 1 (Mitchelson, K. R. and Cheng, J., eds.), Humana Press, Totowa, NJ, pp. 429–441.

29

Analysis of Environment-Induced DNA Damage by Capillary Electrophoresis

Zeena E. Nackerdien, Barbara A. Siles, Stacey A. Nevins, and Donald H. Atha

1. Introduction

Chemical pollutants, ionizing radiation, the products of aerobic metabolism and UV light are important metabolic and environmental factors contributing to DNA damage. Ionizing radiation, in particular, induces damage in the form of single-strand breaks, double-strand breaks, crosslinks, and associated conformational changes to the higher-order structure of DNA inside the cell. Several techniques, such as the single cell gel assay are used to analyze strand breakage in subpopulations of cells. Capillary electrophoresis (CE), by virtue of its high-separation efficiency, throughput, and automation has become recognized as a powerful tool to detect base, sequence, and structural changes in nanogram quantities of different DNA substrates. CE can also be combined with mass spectrometry (MS) to detect precursors to DNA strand breaks, i.e., DNA base modifications in human cancer cells (*see* **refs.** ***1*** and ***2***). In this chapter, we describe methods for the analysis of environment-induced structural changes in linear, plasmid, and genomic DNA. We describe the response of DNA to graded doses of laser or ionizing radiation as examples of how these methods can be used in a practical manner. Using CE to detect small radiation-induced fragments of DNA and conformational changes in large DNA fragments and plasmids provides a faster and more efficient alternative to other techniques such as standard and pulsed-field agarose gel electrophoresis.

CE can also serve as a quality control tool to rapidly monitor the effects of environment-induced DNA damage ***(2–10)***. This is particularly useful for large plasmids (≥27,000 bp), which are routinely used in molecular biology and DNA repair assays. Environment-induced DNA damage leads to programmed cell death, otherwise known as apoptosis. CE can be used to monitor the signature pattern of apoptosis, nucleosomal laddering of DNA. The methods described in this chapter are designed to help the

From: *Methods in Molecular Biology, Vol. 162:*
Capillary Electrophoresis of Nucleic Acids, Vol. 1: Introduction to the Capillary Electrophoresis of Nucleic Acids
Edited by: K. R. Mitchelson and J. Cheng © Humana Press Inc., Totowa, NJ

development of more efficient strategies for studying mutagenesis or apoptosis. These improved strategies will shed more light on the contributions and levels required of various types of DNA structural damage to apoptosis.

2. Materials and Methods

2.1. Electrophoresis Equipment (see Disclaimer)

1. Biofocus 3000 Electrophoresis system (Bio-Rad, Hercules, CA) with UV absorbance detector set at 260 nm. Coated capillary (Bio-Rad, cat. no. 148-3033) of 50 μm id × 50 cm (45-cm effective length). Capillaries are conditioned with Capillary Wash Solution pH 2.5 (Bio-Rad) between analyses.
2. P/ACE 2200 system with laser-induced fluorescence (LIF) detection (Beckman Instruments, Fullerton, CA) with an argon-ion laser (488 nm) set at 3 mW. Coated capillary (Beckman, cat. no. 47412) of 100 μm id × 37 cm (30.2-cm effective length). Capillaries are conditioned with Bio-Rad Capillary Wash Solution pH 2.5 between analyses.
3. Hewlett-Packard^{3D} CE system. The UV absorbance detector is set at 260 nm. Capillary (Supelco, Bellefonte, PA) Supelco CElect-H75 of 75 μm id × 35.5 cm (23-cm effective length) and 50 μm id × 47 cm (38.5-cm effective length). Capillaries are conditioned with Bio-Rad Capillary Wash Solution, pH 2.5 between analyses.
4. A home-built CE system with a Spellman (Hauppauge, NY) High Voltage CE power supply and a Unicam (Philips Scientific, Inc.) 4225 UV detector set at 260 nm. Data points are recorded every 100 ms using Axxiom Chromatography (Moorpark, CA), Model 737 data acquisition software. Uncoated Capillaries (Polymicro Technologies Inc., Phoenix, AZ) of 50 μm id × 40 cm (27-cm effective length). Capillaries are initially pretreated with a rinse for 4 h with 0.1 *N* HCL at 274 kPa (40 psi), followed by a rinse for 30 min with high performance liquid chromatography (HPLC) grade water. Capillaries are conditioned between analyses with a rinse for 10–20 min with 0.1 *M* HCL at 685 kPa (100 psi), followed by a rinse for 5 min with HPLC grade water ***(11)***.

2.2. Dynamic Size-Sieving Matrices

1. Bio-Rad Biofocus 3000 system: Beckman eCap® ds 1000 polyacrylamide mixture (Beckman, cat. no. 477411) containing 100 μmol/L ethidium bromide.
2. Beckman P/ACE system: Beckman eCap® ds 1000 polyacrylamide mixture (Beckman, cat. no. 477411) containing 10 μmol/L thiazole orange.
3. Hewlett Packard ^{3D} CE System®: 0.20% TreviSol® (TS-CE) solution (a blend of polysaccharides in Tris-phosphate-EDTA, pH 7.0) (Patent pending, Trevigen Inc., Gaithersburg, MD). Prior to each run, the capillary is flushed with TS-CE solution for 4 min at the maximum pressure of the instrument, 1 bar. The capillary is postconditioned with aqueous Bio-Rad Capillary Wash Solution, pH 2.5 for one min and then with HPLC-grade water for 2 min.
4. Home-built system: Hydroxyethyl cellulose (mol wt: 90,000–105,000 g/mol), HEC-90K, Polysciences, Inc. Polymer solutions are prepared by adding the dry powders to THE Buffer (89 m*M* Tris-base, 89 m*M* HCl, 2 m*M* Na_2 EDTA, pH 7.0) in the appropriate % w/w concentration ***(9)***. Polymer/buffer solutions are heated in a microwave oven and shaken mechanically until fully dissolved. The water lost during heating is replaced gravimetrically with ultrapure water. Polymer solutions are loaded into the capillary for 4 min at 685 kPa (100 psi), and are then preelectrophoresed at –550 V/cm for 1 min. This initial electrophoresis step enhances the reproducibility of analysis.

2.3. Electrophoresis Conditions

1. Bio-Rad Biofocus 3000 system: The plasmid samples (0.1 mg/mL) are pressure injected for 1 s at 137 kPa (20 psi), resulting in the loading of 0.2 ng of DNA. Runs are conducted in reversed polarity (cathode—negatively charged electrode at the capillary inlet and anode—positively charged electrode at the capillary outlet) at 10 kV for 90 min at a temperature of 25⁰C.
2. Beckman P/ACE 2200 system: The plasmid samples (0.1 mg/mL) are pressure injected for 20 s at 3.4×10^3 Pa resulting in the loading of 2 ng of DNA. The runs are conducted in reversed polarity mode at 7.4 kV for 30 min at 20°C.
3. Hewlett-Packard ^{3D} CE system: The DNA samples are injected (at least in duplicate) electrokinetically at –2.5 kV for 5 s unless otherwise noted. The runs are conducted in reversed polarity at 5.0 kV at 30°C.
4. Home-built system: The plasmid samples are injected electrokinetically at –5 kV for 5 s. The runs are conducted in reversed polarity at 15 kV for 10 min at 21°C.

2.4. Reagents

1. Ethidium bromide (Aldrich, Milwaukee, WI, cat. no. 16,053-9), 15 m*M* stock solution in distilled water.
2. Thiazole orange (Beckman, cat. no. 477409).
3. Fluorescent dyes are kept in the dark to prevent photo-bleaching.
4. Stock solutions and buffers are prepared from analytical grade reagents and are stored at 4°C unless otherwise indicated.
5. DNA substrates are desalted using nitrocellulose filters prior to CE.

2.5. DNA Substrates

2.5.1. Linear DNA Substrates

1. Several double-stranded, linear DNA substrates are used to develop the system. These include the 1-kbp DNA ladder and a ϕx174-*Hae*III DNA standard (Life Technologies, Inc. Gaithersburg, MD).
2. The 1-kbp DNA ladder consisted of 23 blunt-ended DNA fragments ranging in length from 75–12,000 bp. However, only the 10 *Hinf*I DNA fragments of the 1-kb DNA ladder standard (75–517 bp) are monitored in this study. The manufacturer has increased the amounts of the 1000- and 3000-bp fragments in the mixture, to ensure their clear visualization after separation by agarose slab-gel electrophoresis. The radiation damage, which can be observed in the smaller DNA fragments is readily resolved by CE.
3. The ϕx174-*Hae*III DNA standard contains 11 fragments, ranging in size from 72–1353 bp. The 1-kbp DNA ladder and the ϕx174-*Hae*III DNA standards were diluted prior to irradiation to final concentrations of 0.09 and 0.10 μg/μL, respectively.

2.5.2. Supercoiled DNA Substrates

1. The supercoiled pKOL8UV5 plasmid (3890 bp) and its *Hind*III-digested conformer were gifts from Dr. K. McKenney (George Mason University, VA). The plasmids are diluted to 1 mg/mL in 10 m*M* Tris/EDTA, pH 8.0 and stored at –20°C until use (*see* **Note 1**).
2. Samples of the small plasmids (pKOL8UV5) are diluted to 0.1 mg/mL prior to irradiation. Fluorescent dyes such as ethidium bromide and thiazole orange are added to CE runs to improve the detection of individual DNA peaks. However, the binding of these dyes to the DNA is dependent on DNA structure and the amount of dye bound can influence the migration times and relative peak heights. Appropriate DNA size standards must be

included in runs to verify the size of damaged DNA products. Comparison of CE and slab-gel results will aid in the identification of different plasmid conformers *(7)*.

3. *Deinococcus radiodurans* plasmid (pMD68) is used to examine the CE separations of a large plasmid with UV on-line detection. The 27-kb plasmid (gift from Dr. M. Daly, Uniformed Armed Services, Bethesda, MD) is derived from the *Eschericia coli* plasmid pBR322 with *Deinococcal* promoting sequence insertions derived from pS18 and pUC13. This plasmid is diluted in THE buffer (89 m*M* Tris-base, 89 m*M* HCl, 2 m*M* Na_2EDTA pH 7) prior to irradiation. HEC-90K matrix is used for the CE separation of this plasmid.

2.5.3. Genomic DNA Substrates

1. Human myeloblastic leukemia cells (GM03798) from NIGMS Human Genetic Mutant Cell Repository (Camden, NJ) serve as a source of genomic DNA. These cells are grown in suspension at 37°C and 5% CO_2 in RPMI media, supplemented with 15% fetal bovine serum (Sigma), penicillin (100 U/mL) and streptomycin (100 μg/mL).
2. ML-1 human leukemia cells (American Type Culture Collection) are cultured under the same conditions, except for the addition of 50 μg/mL gentamycin to the RPMI medium.
3. The progressive degradation of genomic DNA by nucleases constitutes another form of environment-induced DNA damage. In this model case, nuclease-released DNA fragments are produced by first isolating nuclei from human leukemia cells, and by digestion with 15,000 U of micrococcal nuclease (MN) (Boehringer Mannheim, Indianapolis, IN). Cells are harvested and incubated for 15 min in the presence of 10 m*M* Tris-HCl, pH 7.4, 10 m*M* NaCl, 5 m*M* $MgCl_2$. The swollen cells are lysed by dropwise addition of 10% NP-40 to a final concentration of 0.5% and the released nuclei are sedimented by centrifugation at 300*g* for 5 min.
4. Nuclei are resuspended in 10 m*M* Tris-HCl, pH 7.4, 50 m*M* NaCl, 24 m*M* KCl, 0.75 m*M* $CaCl_2$. The concentration of nuclei is determined by optical absorbance at 260 nm. MN is then added at 10 U/A_{260} and the nuclei are incubated at 37°C for periods of 1, 2, and 5 min, respectively. The digestions are terminated by the addition of 0.1 mL of 0.1 *M* EDTA, pH 8.0. DNA from the treated cells is isolated using the TACS Apoptotic DNA laddering kit (Trevigen Inc., Gaithersburg, MD), and the DNA is redissolved in 10 m*M* Tris-EDTA and the DNA: protein ratio as monitored by the UV absorbance of DNA and protein respectivly at 260 nm and 280 nm. The absorbance ratio is typically 1.8:1 (A_{260nm}:A_{280nm}).
5. Apoptotic DNA fragments are generated from ML-1 cells by exposure to 20 Gy of ionizing radiation. Approximately 10^6 cells/mL are incubated for 48 h at 37°C, 5% CO_2 (volume fraction in air) following irradiation, and prior to the isolation of DNA using the "TACS-DNA® apoptotic DNA laddering kit." The DNA: protein ratio is monitored by optical absorbance and is typically 1.77:1 (A_{260nm}:A_{280nm}) (*see* **Note 2**).

2.6. Ionizing Radiation

1. Absorbed dose to water calibrations of a ^{60}Co-γ-source are made using a thermoluminescent dosimetry device, or by use of a carbon-walled cavity ion chamber. Calibration services were available at the NIST Ionizing radiation Division and the dose rates are calculated after correcting for scatter and attenuation.
2. The 1-kb DNA ladder was exposed to ^{60}Gy γ-radiation for periods of 4.1 and 15 min *(6)*.
3. A ϕX174-*Hae*III DNA standard is also exposed to 0–100 Gy of γ-radiation.
4. The large pMD68 plasmid is irradiated at a dose rate of 58.2 Gy/min (*see* **Note 1**).
5. Samples of the different plasmid DNAs are irradiated in both the absence or presence of 110 μ*M* ethidium bromide, and the resulting DNA fragments are sized by CE and detected in combination with UV absorbance, or laser-induced fluorescence (LIF).

6. In the latter case, thiazole orange serves as the detecting dye with an excitation wavelength of 509 nm and an emission wavelength of 533 nm.

2.7. Laser Radiation

1. The sample holder and optical beam are constructed so that an argon-ion laser beam (Coherent Innova 200, Sunnyvale, CA) illuminates the sample from the top and down through the center of the open Eppendorf tube *(7)*.
2. Focusing optics are not necessary, and the beam diameter at the sample surface is 2.1 mm.
3. Approximately 50% of the sample volume is illuminated by the beam for a total of 250 s. The total exposure time was divided into three cycles of 60 s, allowing a 60-s cooling period between each illumination, and a final 70-s exposure. The measured intensity of the argon-ion laser (488 nm) is 0.44 W (approx 110 J) at the sample tube.

2.8. Capillary Electrophoresis

1. Two types of on-line CE, CE-UV and LIF-CE are used to observe structural and conformational changes in the different irradiated DNA substrates. A home-built CE system and three commercially available CE instruments (Hewlett-Packard, Bio-Rad Biofocus 3000, Beckman P/ACE system 2200) have been used in our studies to measure damage to the irradiated DNA substrates.
2. The Hewlett Packard and Bio-Rad CE instruments were used in combination with UV-absorbance to detect damage to linear DNA, to plasmids and to apoptotic genomic DNA. The 3890-bp pKOL8UV5 plasmid was also analyzed with the Beckman P/ACE system, in combination with the fluorescent dye, thiazole orange (*see* **Note 3**).
3. Electrophoretic separations of DNA, on both types of instrument, depend on a number of factors including the electric field strength, the column length, the mode of injection (hydrodynamic, or electrokinetic). The viscosity of polymer solution, column temperature, pH and ionic strength of the running buffer, as well as the dielectric constant, the net charge, and the size and the shape of irradiated DNA substrates also effect separation.
4. Dynamic size-sieving CE with detection by UV absorption of the DNA bases can provide direct detection of high concentrations of irradiated DNA.
5. LIF-CE provides a further increase in sensitivity (femtomole detection and quantitation of nanoliter sample volumes), when used in combination with the appropriate fluorescent dyes.

3. Results

3.1. ^{60}Co-γ-Ray-Induced Damage to Linear DNA

1. ^{60}Co-γ-ray induced damage to linear DNA is detected using CE with UV absorption detection. Both the ϕX 174-*Hae*III and the 1-kbp DNA ladder serve as model substrates (*see* **Fig. 1**). The single-strand breaks, double-strand breaks, strand-to-strand crosslinks, and other associated conformational changes are observable as band broadening, and are produced by high doses of ionizing radiation (*see* **Note 4**). These DNA structural changes may be quantified by changes in electrophoretic peak-width-at-half-height.
2. The choice of buffer system and polymer matrix is particularly influential in the separation of damaged DNA. When using HEC as the size-sieving polymer solution, only damaged DNA fragments larger than 1000 bp can be observed clearly *(6)*. However, using the Trevisol polymer solution, as shown in **Fig. 1**, two discrete subpopulations of DNA fragments (200 and 240 bp) can be observed in the irradiated ϕX174-*Hae*III digest *(2,3)* (**Note 5**).
3. The reproducibility and distinctness of the new subpopulations of DNA after exposure to 20–60 Gy radiation is most likely to be because of energetically favorable cleavage, as opposed to sequence-specific radiation-induced DNA fragmentation *(3)*.

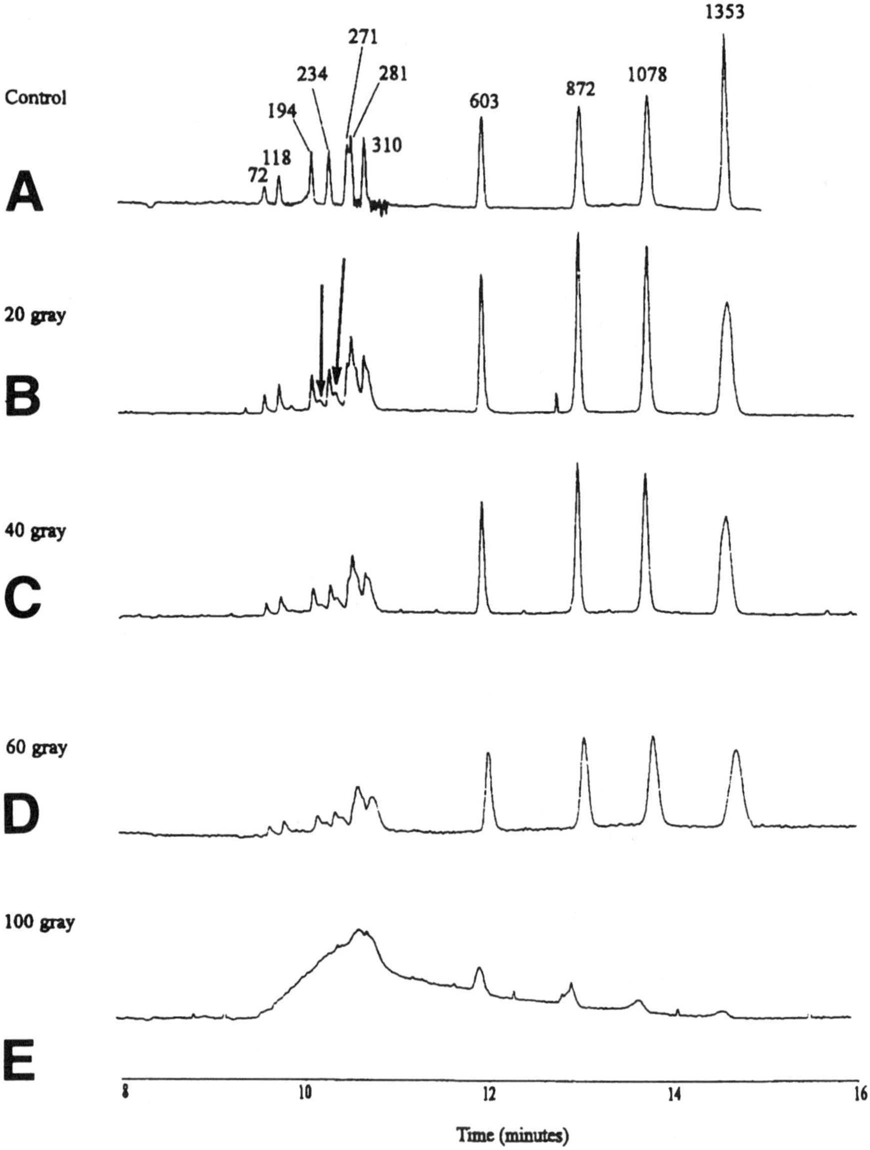

Fig. 1. The effect of γ-radiation on the ϕX174-*Hae*III DNA standard. Reprinted with permission from **ref. *3***.

3.2. Laser and Ionizing Radiation-Induced Damage to Plasmid DNA

1. Environment-induced DNA damage to small plasmids can be detected by CE, using either native UV absorbance at 260 nm, or by LIF in combination with a fluorescent dye.

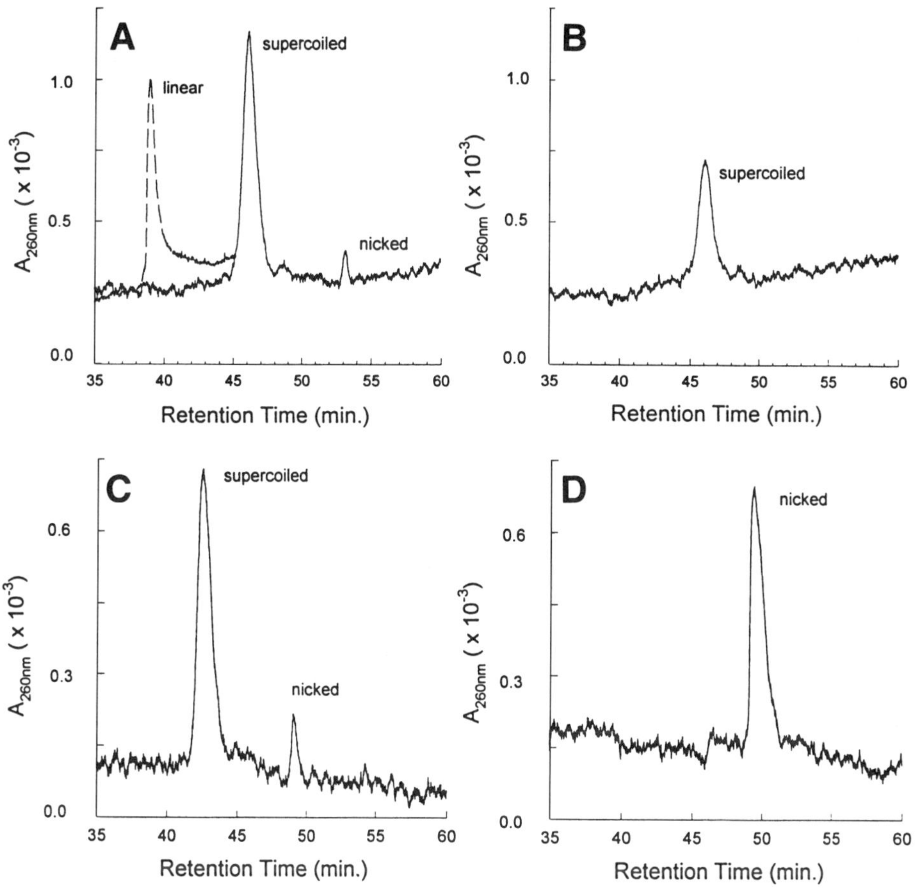

Fig. 2. The CE separation of pKOL8UV5 plasmid DNA detected by UV absorption. Samples were laser-irradiated in the absence and presence of ethidium bromide. Panels: **(A)** control DNA; **(B)** irradiated DNA; **(C)** control DNA +110 μ*M* ethidium bromide; **(D)** irradiated DNA in the presence of 110 μ*M* ethidium bromide. Reprinted with permission from **ref.** *7*.

2. The Bio-Rad Biofocus 3000 Electrophoresis system with UV detection is used to perform the first series of analyses (*see* **Fig. 2**).
3. Alternatively, the P/ACE system 2200 LIF system is used to monitor the separation of plasmid pKOL8UV5 conformers in the presence of a fluorescent dye thiazole orange as shown in **Fig. 3** (*see* **Note 6**).
4. Both techniques indicate that exposure of the plasmid sample to laser radiation in the presence of ethidium bromide causes conversion of the predominantly supercoiled conformer to the nicked conformer via photo-scission, in agreement with observations on molecules separated by agarose gel electrophoresis (*see* **Note 4**).
5. However, large differences are observed in the electropherograms using the two methods of detection. The UV detection system yields CE peaks that are indicative of the relative

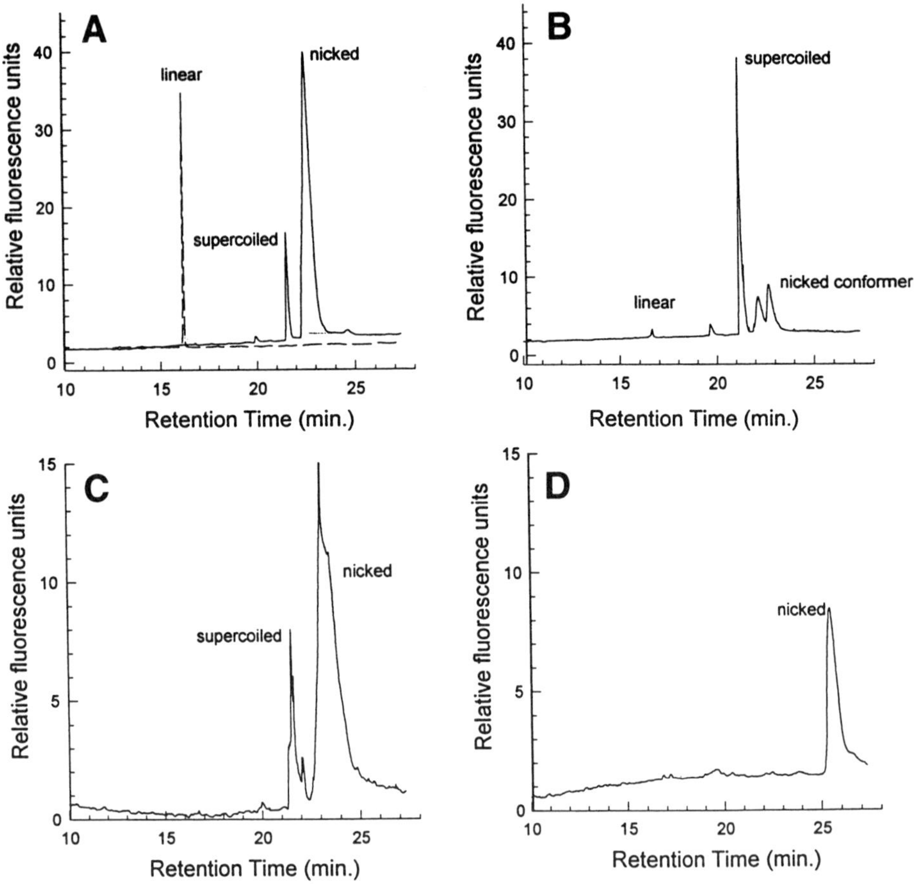

Fig. 3. CE separations of control (unirradiated) and laser-irradiated pKOL8UV5 plasmid DNA detected by LIF-fluorescence. Panels: **(A)** control DNA; **(B)** irradiated DNA; **(C)** control DNA + 110 μ*M* ethidium bromide; **(D)** irradiated DNA in the presence of 110 μ*M* ethidium bromide. Reprinted with permission from **ref.** *7*.

molar proportions of the various conformers. The sensitivity of LIF-fluorescent dye detection is dependent on the amount of dye that may intercalate the different plasmid conformers, with linear and nicked conformers binding much more dye than supercoiled plasmid. Thus, although the LIF detection is approx 100 times more sensitive than UV detection in revealing the presence of plasmid conformers, the relative concentration of each species can not be determined (*see* **Note 7**).

6. We are also able to detect supercoiled conformers of the large pMD68 plasmid with a home-built CE system. The multiple supercoiled conformers are represented by a partially resolved, highly reproducible distribution of peaks in the capillary electropherogram (*see* **Note 7**). Exposure of the plasmid to ionizing radiation of 200 Gy or less predominantly causes an increase in this peak distribution. This indicates the conversion

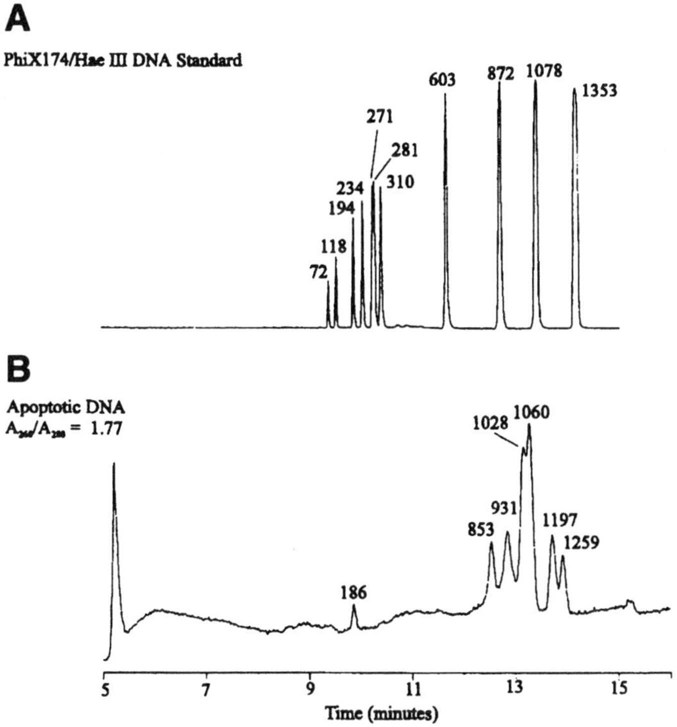

Fig. 4. Electrophoretic separation of various DNA conformers: **(A)** linear ϕX174-*Hae*III DNA standard; **(B)** apoptotic DNA fragments isolated from a 20 Gy γ-irradiated ML-1 cell line. Reprinted with permission from **ref.** *3*.

of discrete supercoiled domains into a wider distribution of levels of supercoiling. A decrease in this peak height distribution is observed upon exposure to 400 Gy of radiation, possibly due to the generation of the fully linear form or the fully open circular form, which would both only be observed with pulsed-field CE.

3.3. Micrococcal Nuclease Digested Genomic DNA

1. Micrococcal nuclease degrades genomic DNA into a discrete ladder of nucleosomal multimers of approx 100–240 bp. Similarly, apoptosis or programmed cell death may cause a signature pattern of DNA laddering which is usually detected by agarose slab-gel electrophoresis (*see* **Note 4**).
2. DNA fragments generated by both forms of environmental stress are monitored using the Hewlett-Packard CE system (*see* **Fig. 4**). The enzymatically or apoptotically digested genomic DNA fragments (from ML-1 cells) can be sized by comparison with the ϕX 174-*Hae*III DNA reference standard. The DNA sizes are obtained by a calibration plot of DNA sizes (bp) for the ϕX 174/*Hae*III DNA standard vs the migration time (min).
3. Regular peaks corresponding to multiples of 140–250 bp are indicative of nucleosomal laddering which is a hallmark of apoptosis. Sometimes distinct subpopulations of peaks can be observed, indicative of necrosis or the presence of protein contaminants.

4. A broad spectrum of DNA peaks corresponding to nucleosomes and random DNA fragments can also be generated by CE separation of MN-digested genomic DNA under the same electrophoretic conditions. The mono-nucleosomes generated by MN-digestion of lympho-blastoid cell DNA vary in size from 140–205 bp. The TS-CE matrix is able to resolve DNA sizes in increments of 10 bp, which means that it is able to detect the presence of endogenous nucleases, in addition to the 200-bp multimer cutting pattern normally generated by MN.

4. Notes

1. Both linear and dsDNA are highly sensitive to storage procedures employed following radiation exposure. Plasmids are particularly prone to nicking after repeated freezing and thawing, resulting in the generation of several of the conformer types. The spurious degradation of linear DNA can be avoided by immediate analysis, or by storing sample aliquots at –20°C.
2. Genomic DNA can be isolated and analyzed 24–48 h after radiation exposure depending on the radiation dose employed. Contaminating RNA and protein can interfere with CE separations, hence a brief digestion of any RNA using RNAse A at 37°C for 30 min is recommended prior to DNA isolation. The separation of apoptotic-nicked/enzymatically-digested genomic DNA by CE can be problematic, if the protein or the salt content of the DNA mixture is very high. Dialysis of the sample prior to electrophoresis is recommended.
3. Fluorescent dyes such as ethidium bromide can improve the separation of linear DNA. The ethidium interacts with the duplex DNA to intercalate between stacked base pairs, making the duplex more rigid and hence altering the DNA retention by the seiving matrix.
4. When irradiated DNAs are analyzed by agarose slab-gel electrophoresis the conformationally diverse DNA molecules appear as diffuse bands, or as smears and are difficult to quantify.
5. The choice of running buffer has influence on the separation of both linear and dsDNAs by CE. The buffering capacity of TAE (0.04 *M* Tris-acetate, 0.001*M* EDTA, pH 8.0) is lower than the buffer capacity of both TPE (0.09 *M* Tris-phosphate, 0.002 *M* EDTA, pH 8.0) and TBE (0.045 *M* Tris-borate, 0.001 *M* EDTA, pH 8.0). However, TPE buffer used in conjunction with Trevesol-CE separation solution provides greater separation efficiency than either of the more commonly employed TAE, or TBE buffers. The size-sieving polymer chosen should be of the appropriate average molecular weight and stiffness relative to the flexibility of the DNA molecules to be separated; the appropriate polymer concentration should also be chosen with consideration of the size of the DNA fragments being separated ***(4)***.
6. A polyacrylamide polymer matrix is used here with dynamic size-sieving CE to separate laser-irradiated small plasmids. However, it must be noted that mobility through this matrix can also be affected by the DNA base composition and the sequence, presumably due to kinks in the DNA structure that form at specific sequences. Under the buffer solution and size-sieving polymer employed, the migration order of the plasmid conformers is consistent from the fastest to slowest: linear, supercoiled, and nicked conformers.
7. Fluorescent dyes can be used in combination with LIF CE to detect different conformers of irradiated plasmids. This method is about 100 times more sensitive compared to UV-CE. However, care must be taken in the interpretation of results, as the dye can affect both the peak intensities and the separation order of intact plasmids and plasmid conformers.

Disclaimer: Electrophoresis Equipment and Associated Materials

Certain commercial electrophoresis instruments and CE materials are identified in this paper in order to specify an experimental procedure as completely as possible. In no case does this identification of particular equipment or materials imply a recommendation or endorsement by the National Institute of Standards and Technology, nor does it imply that the material, instrument, or equipment is necessarily the best available for the purpose.

References

1. Le, X. C., Xing, J. Z., Lee, J., Leadon, S. A., Weinfeld, S. A., and Weinfeld, M. (1999) Inducible repair of thymine glycol detected by an ultrasensitive assay for DNA damage. *Science* **280,** 1066–1069.
2. Xing, J. Z., Carnelley, T., Lee, J., Watson, W. P., Weinfeld, M., and Le, X. C. (2001) Assay for DNA damage using immunochemical recogition and capillary electrophoresis, in *Capillary Electrophoresis of Nucleic Acids,* Vol. 1 (Mitchelson, K. R. and Cheng, J., eds.), Humana Press, Totowa, NJ, pp. 419–428.
3. Siles, B. A., Nackerdien, Z. E., and Collier, G. B. (1997) Analysis of DNA fragmentation using a dynamic size-sieving polymer solution in capillary electrophoresis. *J. Chromatogr. A* **771,** 319–329.
4. Hammond, R. W., Oana, H., Schwinefuss, J. J., Bonadio, J., Levy, R. J., and Morris, M. D. (1997) Capillary electrophoresis of supercoiled and linear DNA in dilute hydroxyethylcellulose solution. *Anal. Chem.* **69,** 1192–1196.
5. Nevins, S. A., Siles, B. A., and Nackerdien, Z. E. (2000) Analysis of gamma radiation-induced damage to plasmid DNA using dynamic size-sieving capillary electrophoresis. *J. Chromatogr. B.* **741,** 243–255.
6. Nackerdien, Z. E., and Atha, D. (1996) Measurement of ^{60}Co-ray induced DNA damage by capillary electrophoresis. *J. Chromatogr. B* **683**, 85-89.
7. Nackerdien, Z., Morris, S., Choquette, S., Ramos, B., and Atha, D. (1996) Analysis of laser-induced plasmid DNA photolysis by capillary electrophoresis. *J. Chromatogr. B* **683,** 91–96.
8. Daly, M. J., Ouyang, L., Fuchs, P., and Minton, K. (1994) *In vivo* damage and recA-dependent repair of plasmid and chromosomal DNA in the radiation-resistant bacterium *Deinococcus radiodurans. J. Bacteriol.* **176,** 3508–3517.
9. Siles, B. A., Anderson, D. A., Buchanan, N. S., and Warder, M. F. (1997) The characterization of composite agarose/hydroxyethylcellulose matrices for the separation of DNA fragments during capillary electrophoresis. *Electrophoresis* **18,** 1980–1989.
10. Deforce, D. L. and van den Eeckhout, E. E. (2001) Analysis of DNA damage using capillary zone electrophoresis and electrospray mass spectrometry, in *Capillary Electrophoresis of Nucleic Acids,* Vol. 1 (Mitchelson, K. R. and Cheng, J., eds.), Humana Press, Totowa, NJ, pp. 429–441.
11. Fung, E. N. and Yeung, E. S. (1995) High-speed DNA-sequencing by using mixed poly(ethylene oxide) solutions in uncoated capillary columns. *Anal. Chem.* **67,** 1913–1919.
12. Kim, Y. and Yeung, E. S. (2001) Capillary electrophoresis of DNA fragments using poly(ethylene oxide) as a sieving material, in *Capillary Electrophoresis of Nucleic Acids,* Vol. 1 (Mitchelson, K. R. and Cheng, J., eds.), Humana Press, Totowa, NJ, pp. 215–223.

30

Assay for DNA Damage Using Immunochemical Recognition and Capillary Electrophoresis

James Z. Xing, Trevor Carnelley, Jane Lee, William P. Watson, Ewan Booth, Michael Weinfeld, and X. Chris Le

1. Introduction

Damage to DNA by environmental and endogenous agents is considered to be a critical step leading to mutation and subsequent neoplastic transformation and cancer induction. It is also a key event in the cytotoxicity induced by several anti-cancer agents such as ionizing radiation and many of the commonly used chemotherapeutic agents. To counter insult to their DNA by genotoxic agents, cells have recourse to a variety of DNA repair pathways that each deal with a different class of lesions *(1)*. Because of the pivotal role played by DNA damage and repair in carcinogenesis and cancer therapy, there is a clear need to be able to detect and quantify the damage. In addition to their use in laboratory studies of the biochemistry and genetics of mutagenesis and DNA repair, such assays present the possibility of using DNA damage as an endpoint in monitoring environmental and occupational exposure to carcinogens and in clinical dosimetry of cellular exposure to anti-cancer agents.

Many assays have been developed to measure DNA damage. Currently, the most widely used approaches include ^{32}P-postlabeling, gas chromatography/mass spectrometry, polymerase chain reaction (PCR)-based assays, strand break-based assays, such as the single-cell gel electrophoresis (comet) assay and pulsed-field gel electrophoresis, and immunoassays *(2)*. Each of these has its advantages and disadvantages and work better with some types of lesions than with others. For the purposes of environmental and clinical monitoring, several factors are of crucial importance. These include sensitivity, selectivity, and complexity of manipulation. For example, ^{32}P-postlabeling assays can be applied to many different types of DNA adducts. Depending on the base modification and the amount of DNA available, these can be detected at levels of 1 in 10^5–10^9 nt by this method; the absolute level of detection being approx 10^{-15}–10^{-17}mol. However, the general protocol is quite protracted, requiring complete digestion of the

From: *Methods in Molecular Biology, Vol. 162: Capillary Electrophoresis of Nucleic Acids, Vol. 1: Introduction to the Capillary Electrophoresis of Nucleic Acids*
Edited by: K. R. Mitchelson and J. Cheng © Humana Press Inc., Totowa, NJ

DNA to mono or dinucleotides, followed by enzymatic labeling, and separation of the radiolabeled adducts by multidimensional thin layer chromatography (TLC) ***(3,4)***. A further complication is, of course, the usage of large quantities of radioactivity, although fluorescent protocols have been devised ***(5,6)***. A limitation of postlabeling approaches, whether based on radioactivity or fluorescence, is the need to carry out chemical or biochemical reactions with minute quantities of analyte.

This particular limitation is reduced in immunoassays where binding to the sites of DNA damage is dependent solely on the affinity of the antibodies. Immunoassays also offer the advantages of specificity and simplicity, because there is no need to digest the DNA or perform any other biochemical reaction. Until recently, the limitation of sensitivity of immunoassays, particularly enzyme-linked immunosorbent assay (ELISA)-based assays, has been set by the background arising from nonspecific binding of antibody to the solid-phase matrix. However, by combining the use of fluorescent antibodies with capillary electrophoresis (CE) and laser-induced fluorescence (LIF) detection, we have been able to develop an assay capable of detecting DNA base damage at the zeptomole (10^{-21} mol) level *(7)*.

We have previously shown that thymine glycol, a radiation-induced DNA lesion, is amenable to detection by a capillary electrophoresis-laser induced fluorescence based immunoassay *(7)*. This chapter describes the modification of this protocol for the detection of adducts produced by the well-known carcinogen benzo[a]pyrene, which is metabolized in cells to its active form, benzo[a]pyrene diol epoxide (BPDE). Benzo[a]pyrene belongs to an important class of environmental genotoxins, polycyclic aromatic hydrocarbons, generated by incomplete combustion of fossil fuels and other organic material such as tobacco. Aside from the use of different primary antibodies, there are some other differences between the methodology for measuring thymine glycols and for BPDE-DNA adducts.

2. Materials

2.1. Biologicals

1. BPDE: (±) r-7,t-8-dihydroxy-t-9,10-epoxy-7,8,9,10-tetrahydrobenzo[a]pyrene (anti), is obtained from the Chemical Carcinogen Reference Standard Repository of the National Cancer Institute, Midwest Research Institute, Kansas City, MO.
2. Mouse monoclonal anti-BPDE-DNA antibody E5 is prepared as described by Booth et al. ***(8)***.
3. The tetramethylrhodamine conjugated anti-mouse IgG (Calbiochem, La Jolla, CA) is used as the secondary antibody.
4. The anti-mouse IgG is purified by chromatography on a HiTrap® protein G column (Amersham-Pharmacia Biotech, Baie d'Urfe, PQ, Canada) prior to use.
5. Mouse monoclonal anti-BrdU antibody ***(9)*** is kindly provided by Dr. S.A. Leadon of the University of North Carolina School of Medicine.
6. DNA polymerase I (Klenow fragment, Gibco-BRL, Burlington, ON, Canada).
7. T4 DNA ligase (New England Biolabs, Beverly, MA).

2.2. Equipment

1. Fused silica capillaries of 20 μm id, 150 μm od and of 40-cm length are from Polymicro Technologies (Phoenix, AZ).

2. A CZE1000R high-voltage power supply: (Spellman High Voltage Electronics, Plainview, NY).
3. The LIF detector is constructed on an optical table using commercially available optics and laboratory-built accessories.
4. A 1.0-mW green helium-neon laser (Melles Griot, Irvine, CA) with a wavelength of 543.5 nm is used as the excitation source.
5. A 6.3X microscope objective (Melles Griot).
6. A sheath-flow cuvet (NSG Precision Cells, Farmingdale, NY).
7. A high numerical aperture microscope objective (60X, 0.7 NA, Universe Kogaku, Oyster Bay, NY).

3. Methods

3.1. Preparation of BrdU Standards

1. Plasmid pUC 18 (100 μg) is cleaved by incubation with 300 U of *Sal*I endonuclease in 100 μL of buffer (20 m*M* Tris-acetate, pH 7.5, 20 m*M* magnesium acetate, 100 m*M* potassium acetate) at 37°C for 4 h. The restriction endonuclease is removed by phenol extraction and the DNA is precipitated by addition of sodium acetate and ethanol ***(10)***.
2. For BrdU incorporation, the DNA is resuspended in a 300-μL reaction mixture containing 50 m*M* Tris-HCl, pH 8.0, 10 m*M* $MgCl_2$, 50 m*M* NaCl, 0.2 m*M* dATP, 0.2 m*M* dGTP, 0.2 m*M* dCTP, 0.2 m*M* BrdUTP and 30 U of DNA polymerase I (Klenow fragment, Gibco-BRL). The mixture is incubated at room temperature for 10 min. The DNA polymerase is then removed by phenol/chloroform extraction, and the DNA is precipitated by addition of sodium acetate and ethanol.
3. For DNA ligation, the DNA is resuspended in a 200-μL reaction mixture containing 50 m*M* Tris-HCl, pH 7.8, 10 m*M* $MgCl_2$, 1 m*M* dithiothreitol, 5% (w/v) poly(ethylene glycol)-8000, 1 m*M* ATP and 400 U of T4 DNA ligase (New England Biolabs) and incubated overnight at 16°C. One unit of ligase is defined as the amount of enzyme required to give 50% ligation of *Hin*dIII fragments of λ phage DNA in 30 min at 16°C in 20 μL with concentration of 5'-DNA termini of 0.12 μ*M*. This is approximately equivalent to 0.015 Weiss units. Protein is again removed by phenol/chloroform extraction.
4. The ligated DNA is precipitated by addition of sodium acetate and ethanol, then is resuspended in TE (10 m*M* Tris-HCl, pH 8.0, 0.1 m*M* EDTA), and quantified by UV spectrophotometry (1 AU_{260}/μL = 50 mg/mL). This DNA is mixed with known amounts of linearized plasmid DNA or calf thymus DNA to generate a calibration curve.

3.2. Treatment of Cells with BPDE and Recovery of Cellular DNA

1. BPDE is made up and stored in the dark under argon at –20°C as a 3 m*M* solution in dimethyl sulfoxide (DMSO) and serially diluted in DMSO as required.
2. A549 human lung adenocarcinoma cells are obtained from the American Type Culture Collection (Rockville, MD). Cells are cultured as monolayers in Dulbecco's modified Eagle's medium/Ham's F12 medium (1:1 v/v) supplemented with 10% fetal bovine serum at 37°C in a humidified (75–85%) atmosphere of 5% CO_2-95% air.
3. About 7.5×10^5 cells are seeded in plastic Petri dishes (diameter 60 mm; Becton Dickinson Labware, Franklin Lakes, NJ) and are incubated overnight in 3 mL of media per dish. The media is then replaced by fresh media containing BPDE (1 μL of DMSO solution of BPDE per 3-mL media). The cells are then incubated for 4 h, after which the media is discarded and the cells rinsed twice with ice-cold PBS.
4. The cells in each dish are lysed by addition of 1 mL DNAzol (Sigma) and the contents transferred to a 10-mL tube. The sample in each tube is gently mixed with 0.5 mL 100%

ethanol to precipitate the DNA, which is then spooled onto a pipet tip and transferred to an Eppendorf tube containing 0.5 mL 70% ethanol.

5. The DNA is collected by centrifugation and washed twice with 95% ethanol before being dissolved in 600 mL of 8 m*M* NaOH. The final concentration of this DNA solution is determined spectrophotometrically and is typically found to be approx 0.1 mg/mL.

3.3. Incubation of DNA with Antibodies

1. A BPDE-DNA sample (5–20 µL) is heat denatured at 100°C for 5 min followed by cooling and quenching on ice-water (*see* **Note 1**).
2. 2.5 µL of 200 µg/mL anti-BPDE-DNA antibody is added to the denatured DNA sample. The mixture is incubated at 4°C overnight, then 20 µL of 100 µg/mL of fluorescently labeled anti-mouse IgG is added to this mixture and it is incubated at 4°C for a further 2 h.
3. 1X TBE buffer (89 m*M* Tris-borate, pH 8.3, 2 m*M* EDTA) is added to make the final volume 50 µL.
4. The BPDE-DNA solution is subjected to CE analysis. Control mixtures containing all of the above reagents except the BPDE-DNA are prepared and analyzed similarly by CE.

3.4. CE with LIF Detection

1. A laboratory-built CE-LIF system is used as previously described *(11)*. Briefly, a high-voltage power supply (CZE1000R, Spellman High Voltage Electronics) is used to drive the electrophoresis.
2. The separation voltage, injection voltage, and injection time are controlled by a Macintosh computer employing a program written in LabVIEW (National Instruments, Austin, TX).
3. Separation is typically performed in a 40-cm long, 20 µm id, 150 µm od, fused silica capillary at an electric field of 500 V/cm.
4. The high-voltage injection end of the capillary, along with a platinum electrode, is inserted into a sample solution (when injecting sample) or running buffer (when performing separation) and housed in a Plexiglas box equipped with a safety interlock. The other end of the capillary is grounded and inserted into a sheath-flow cuvet detection cell.

3.4.1. LIF Detection

1. The LIF detector (**Fig. 1**) is constructed on an optical table using commercially available optics and laboratory-built accessories.
2. A 1.0-mW green helium-neon laser (Melles Griot) with a wavelength of 543.5 nm is used as the excitation source. The laser beam is focused with a 6.3X microscope objective (Melles Griot) into a sheath-flow cuvet (NSG Precision Cells, Farmingdale, NY). The sheath-flow cuvet has been previously used as the fluorescence detection cell, as developed by Cheng and Dovichi *(12)*.
3. Fluorescence is detected at 90° with respect to both the laser beam and the sample stream by using a high numerical aperture microscope objective (60X, 0.7 NA, Universe Kogaku, Oyster Bay, NY), which has the required working distance of greater than 2 mm. The fluorescence is then spectrally filtered with a bandpass filter (580DF40, Melles Griot) to reject scattered laser light. A 1-mm diameter pinhole is placed in the reticle position of the microscope objective to reduce background light.
4. The fluorescent light is detected with a photomultiplier tube (PMT, R1477, Hamamatsu Photonics, Japan). The output from the PMT is digitized by a PCI data acquisition board and by the LabVIEW software in a Power Macintosh computer.
5. An auxiliary microscope, placed opposite the fluorescence collection optic, is used to assist the alignment of the tightly focused laser beam with respect to a small-diameter

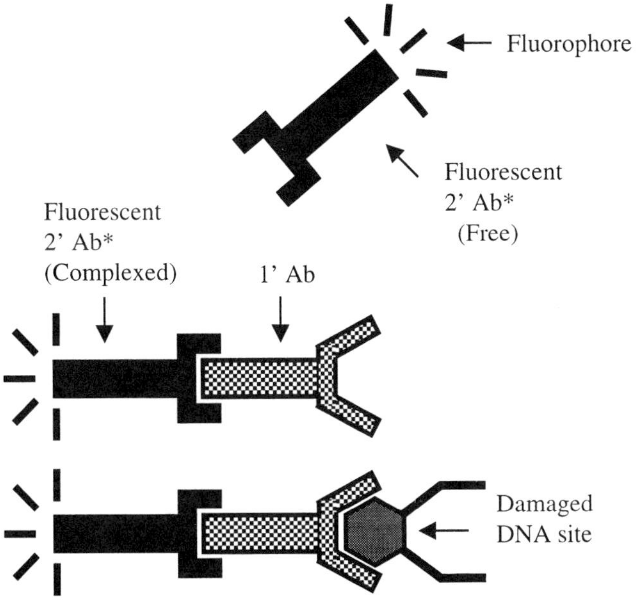

Fig. 1. A schematic representation of immunochemical recognition for the CE assay of DNA damage.

sample stream. To align the detection system, we fix the location of the PMTs, the collection optics, and the limiting aperture. All other components are aligned with respect to the collection optics. The sheath-flow cuvet and the laser-beam focusing objective are each mounted on a set of three-axis translation stages, and thus their positions could be adjusted with ease and precision.

3.5. Electrophoretic Analysis

1. Samples are injected into the separation capillary either electrokinetically by applying a 5–10 kV potential for 5 s, or hydrodynamically by raising the injection end 10 cm higher than the detection end of the capillary for 20 s.
2. The electrophoretic separation is performed at room temperature. The running buffer contains 1X TBE (pH 8.3).
3. Following each sample analysis, the capillary is flushed sequentially, by passing through solutions of:
   a. Approx 250 nL of 1X TBE buffer, pH 7.8, pumped electroosmatically for 5 min,
   b. 2 μL of 0.1 *M* NaOH applied by a syringe, and finally,
   c. Approx 500 nL of 1X TBE running buffer, pH 8.3, pumped electroosmatically for 10 min.

3.6. Results

3.6.1. Immunochemical Recognition

1. Quantifying a few damaged bases in a DNA molecule of several million to a billion base pairs represents a major analytical challenge with a need for high selectivity and sensitivity. Excellent selectivity can be obtained by using antibodies that recognize specific DNA

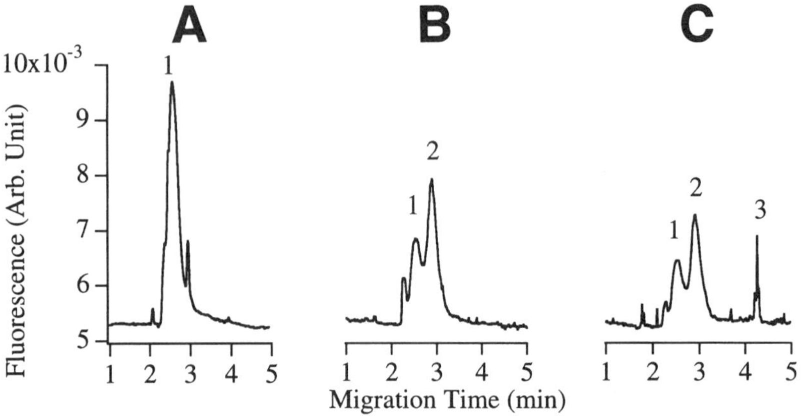

Fig. 2. Electropherograms showing the analysis of tetramethylrhodamine labeled anti-mouse IgG (**A**), a mixture of (A) with mouse anti-BPDE-DNA antibody (**B**), and a mixture of (B) with DNA from A549 cells incubated with 0.1 μ*M* BPDE for 4 h (**C**).

modifications. High sensitivity can be achieved by combining the use of a fluorescent probe with LIF detection. Thus, for the DNA damage assay we employ a mouse monoclonal primary antibody (1' Ab) raised against a specific DNA damage that recognizes and binds to the DNA at these damaged sites, and a fluorescently-labeled secondary anti-mouse antibody (2' Ab*) that recognizes the 1' Ab (**Fig. 1**). We chose tetramethylrhodamine-labeled secondary anti-mouse antibodies because of the convenient fluorescence wavelengths (λ_{ex} = 555 nm and λ_{em} = 580 nm).

2. To maximize formation of the fluorescent complex so that the DNA damage can be quantified, we use excess amounts of the 2' Ab* to saturate the 1' Ab and excess 1' Ab to saturate the DNA damage. We then use CE to separate the complex and LIF to detect the fluorescent components.
3. **Figure 2** demonstrates the general approach of the assay. It shows electropherograms from the analyses of 2' Ab* alone (**Fig. 2A**), a mixture of 2' Ab* and 1' Ab (**Fig. 2B**), and a mixture of BPDE-DNA with the two antibodies (**Fig. 2C**). The amount of DNA damage can be quantified from peak 3, the complex of the DNA damage with the antibodies. Using different primary antibodies to specific DNA lesions, assays for other DNA lesions can be developed similarly.

3.6.2. Assay for Bromodeoxyuridine (BrdU)

1. We used DNA antigens containing a specified quantity of BrdU as a standard model DNA base modification. For many types of DNA damage, there are no straightforward methods to reproducibly introduce a known number of DNA lesions into high molecular weight DNA, other than through the incorporation of an oligonucleotide bearing a particular lesion into a larger DNA molecule such as a plasmid. However, the synthesis and purification of oligonucleotides with specifically damaged moieties is not an easy undertaking.
2. We have, therefore, used an alternative approach to develop calibration standards. This involves the incorporation of a known number of BrdU bases into plasmid DNA, which are then detected by anti-BrdU monoclonal antibodies, which in turn, are detected by the same fluorescently-labeled secondary antibodies as used for the detection of antibodies

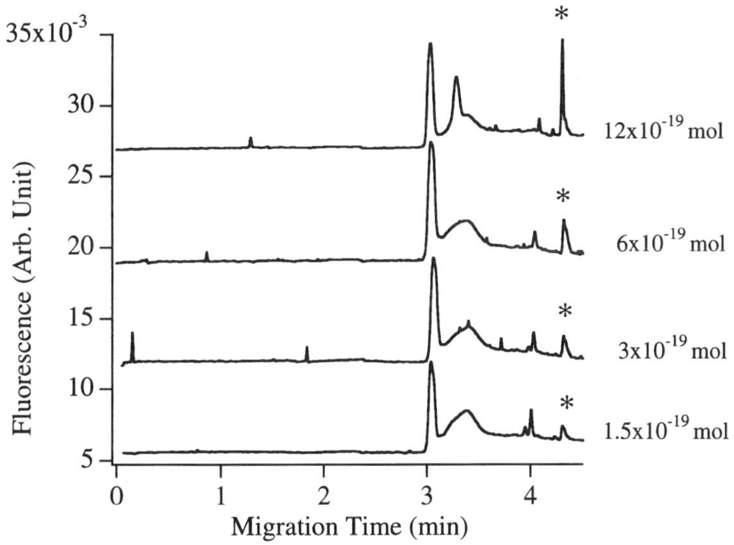

Fig. 3. Electropherograms obtained from the CE/LIF analysis of plasmid DNA (0.12–1 μg/mL) containing BrdU (1.4×10^{-10}–1.2×10^{-9} *M*). Tetramethylrhodamine conjugated anti-mouse IgG and mouse monoclonal anti-BrdU antibodies are used to react with the BrdU-containing DNA. The complex of modified DNA bound to the antibodies is marked with an asterisk.

that recognize DNA lesions. Because the primary antibody is saturated with the secondary antibody, and because the fluorescence from the same secondary antibody is measured, the relative fluorescence intensity from one mole of a specific DNA adduct should be equal to that from one mole of BrdU. Thus, other DNA lesions can be quantified from measured fluorescence intensities using calibration against BrdU-DNA standards. With our protocol, we incorporate 2 mol of BrdU per molecule of pUC18 (2690 bp).

3. **Figure 3** shows a series of electropherograms obtained with increasing amounts of BrdU. The concentration of plasmid DNA in the reaction mixture with the two antibodies ranged from 0.12 μg/mL to 1 μg/mL. Because two BrdU nucleosides are present in each plasmid DNA molecule (pUC18, 2690 base pairs, ~1.6×10^6 Dalton), the concentration of BrdU in the mixture ranged from 1.4×10^{-10} *M* to 3×10^{-9} *M*. For a typical CE/LIF run, the injection volume for each analysis is approx 1 nL (10^{-9} L). Therefore, the signals in peak 3 (**Fig. 3**) are produced by 1.4×10^{-19} mol to 1.2×10^{-18} mol of BrdU. The detection limit, based on a signal-to-noise ratio of three, is on the order of 10^{-20} mol.
4. Results in **Fig. 3** show a linear increase in fluorescence intensity with the increase of BrdU concentration. A calibration curve is constructed and the regression coefficient (r^2) is found to be 0.988. The BrdU is then used as a standard against which other DNA damage measurements can be calibrated.

3.6.3. Assay for BPDE-DNA in A549 Cells Incubated with BPDE

1. **Figure 4** shows electropherograms obtained from the analyses of a series of DNA samples from A549 cells incubated with various concentrations of BPDE for 4 h. Little background is observed from the control cells that are incubated in the medium without BPDE.

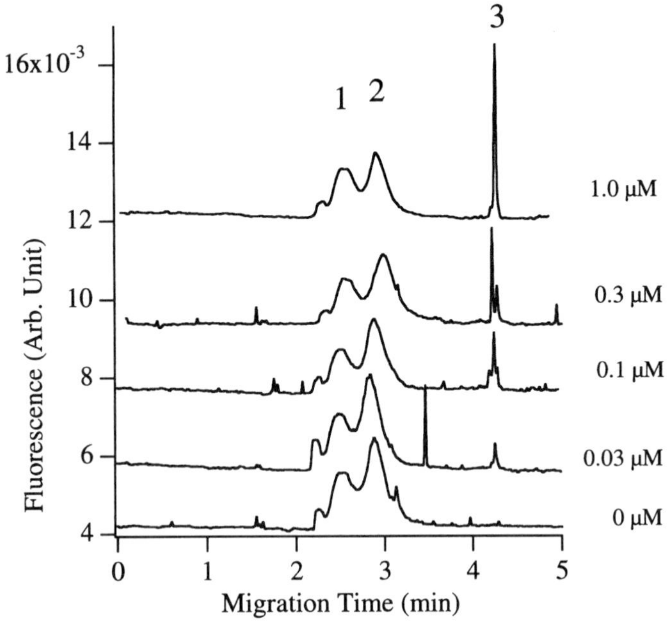

Fig. 4. Electropherograms obtained from the CE/LIF analysis of DNA from A549 cells incubated with 0.03–1 μ*M* BPDE.

With the increase of BPDE in the incubation medium, we observe increasing intensity of the signal corresponding to the BPDE-DNA complex with the antibodies (peak 3).

2. However, integration of the peak area appears to increase linearly with the log of the concentration of BPDE. Shinozaki et al. ***(13)*** reported a similar dose-response for BPDE and BPDE-DNA adduct levels, although the concentration range they used is approx 10-fold higher (0.3–13.2 μ*M*). The nonlinear relationship probably reflects competing parameters including DNA damage by BPDE, DNA repair, and the hydrolysis of BPDE.
3. The quantitation of BPDE-DNA levels is achieved by calibrating the peak intensity of BPDE-DNA adduct against those of BrdU. We find approx 1.7 BPDE adducts per 10^6 nt when the A549 cells are incubated with 1 μ*M* BPDE for 4 h. This is in comparison with 2.5 BPDE-DNA adducts per 10^6 nt when human fibroblasts are incubated with 1 μ*M* BPDE for 2 h ***(14)***. Exposure to the lowest concentration of BPDE, 0.03 μ*M*, generates an adduct level of approx 0.2 BPDE adducts per 10^6 nt.
4. As an alternative method of quantitation, we compare the level of BPDE-DNA adducts in A549 cells to that in a sample of DNA isolated from benzo[a]pyrene-treated mouse skin DNA, for which the BPDE-DNA adduct level has been previously determined by the ^{32}P-postlabeling assay, using the BPDE-mononucleotide from NCI Chemical Carcinogen Repository as the primary standard, to be 1 adduct in 10^5 nt (*see* **Fig. 5**). This gave a result of 1.1 adducts per 10^6 nt in A549 cells exposed to 1 μ*M* BPDE.

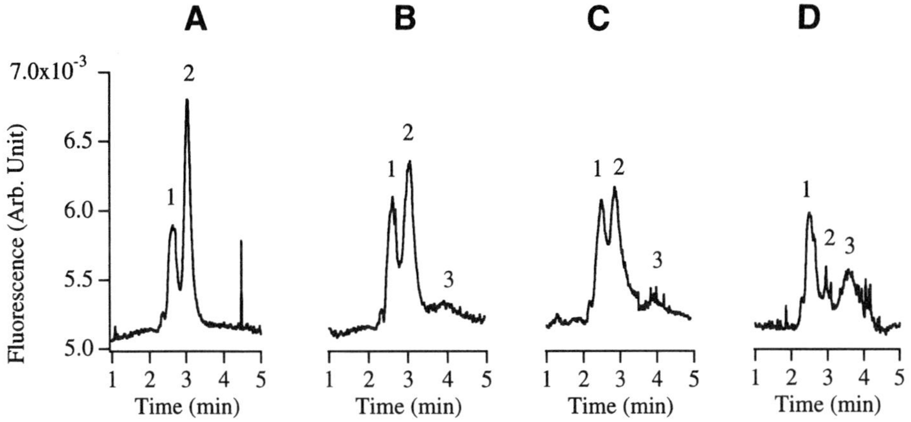

Fig. 5. Electropherograms showing the analysis of mixtures containing varying amounts of DNA from benzo[a]pyrene-treated mouse skin. **(A)** 0, **(B)** 0.6 µg/mL, **(C)** 1.2 µg/mL, and **(D)** 5 µg/mL. Based on a determination by postlabeling assay of 1 BPDE-DNA adduct per 10^5 nt and an injection volume of approx 1 nL, the absolute quantity of adduct visualized is (A) 0, (B) 1.8×10^{-20} mol, (C), 3.6×10^{-20} mol, and (D) 1.5×10^{-19} mol.

3.6.4. BPDE Adducts in A549 Cellular DNA and in Mouse Skin DNA

1. **Figure 5** shows electropherograms from the analysis of mouse skin DNA. The BPDE-DNA complex with the antibodies is a much broader peak compared to the DNA isolated from A549 cells. The mouse skin is treated with benzo[a]pyrene as previously described *(8)*. Vigorous procedures are used to treat the mouse skin samples and to extract the DNA. Agarose gel electrophoresis of the DNA isolated from A549 cells and mouse skin confirmed that the mouse skin DNA is markedly more fragmented, particularly after heat denaturation.
2. In addition, the levels of BPDE adducts in the two samples being examined differed greatly; the mouse skin DNA has approx 10 times more adducts than the A549 DNA from cells treated with the highest dose of BPDE. As a result, fewer antibody molecules are bound to each DNA molecule from A549 cells, and the electrophoretic mobility of the complex is more like high molecular weight DNA.
3. In contrast, the adduct level in the mouse skin DNA is more extensive and each DNA fragment is bound to more antibody molecules. Because of the smaller size of the DNA fragments and more antibodies bound to the fragment the electrophoretic mobility of the complex is more like that of the 2' Ab*–1'Ab complex (i.e., peak 2).

4. Notes

1. For measurement of the BPDE-DNA adducts, it is found that heat-denatured DNA gives a higher signal-to-noise ratio than native DNA. Secondly, the pH of the running buffer is reduced from 10.5 to 8.3. We are able to detect BPDE adducts at levels below 2×10^{-20} mol both in DNA from human cells treated with BPDE and from mouse skin treated with the parent compound.

Acknowledgments

The work was supported by grants from the Natural Sciences and Engineering Research Council of Canada, the National Cancer Institute of Canada, the Alberta Heritage Foundation for Medical Research through the Eco-Research Chair, and the Alberta Cancer Board. The authors thank Dr. S.A. Leadon of University of North Carolina School of Medicine for providing the mouse monoclonal anti-BrdU antibody and Dr. Aghdass Rasouli-Nia for technical assistance.

References

1. Friedberg, E. C., Walker, G. C., and Siede, W. (1995) *DNA Repair and Mutagenesis*, ASM Press, Washington, DC.
2. Pfeifer, G. P., ed. (1996) *Technologies for Detection of DNA Damage and Mutations.* Plenum Press, New York, NY.
3. Gupta, R. C. (1996) ^{32}P-Postlabeling for detection of DNA adducts, in *Technologies for Detection of DNA Damage and Mutations* (Pfeifer, G. P., ed.), Plenum Press, New York, NY, pp. 45–61.
4. Weinfeld, M. and Buchko, G. W. (1993) Postlabelling methods for the detection of apurinic sites and radiation-induced DNA damage, in *Postlabelling Methods for the Detection of DNA Damage* (Phillips, D. H., Castegnaro, M., and Bartsch, H., eds.), IARC Scientific Publications No. 124, Lyon, France, pp. 95–103.
5. Kelman, D. J., Lilga, K. T., and Sharma, M. (1988) Synthesis and application of fluorescent labeled nucleotides to assay DNA damage. *Chem. Biol. Interact.* **66,** 85–100.
6. Sharma, M., Jain, R., Ionescu, E., and Slocum, H. K. (1995) Capillary electrophoretic separation and laser-induced fluorescence detection of the major DNA adducts of cisplatin and carboplatin. *Anal. Biochem.* **228,** 307–311.
7. Le, X. C., Xing, J. Z., Lee, J., Leadon, S. A., and Weinfeld, M. (1998) Inducible repair of thymine glycol detected by an ultrasensitive assay for DNA damage. *Science* **280,** 1066–1069.
8. Booth, E. D., Aston, J. P., van den Berg, P. T., Baan, R. A., Riddick, D. A., Wade, L. T., Wright, A. S., and Watson, W. P. (1994) Class-specific immunoadsorption purification for polycyclic aromatic hydrocarbon-DNA adducts. *Carcinogenesis* **15,** 2099–2106.
9. Leadon, S. A. (1986) Differential repair of DNA damage in specific nucleotide sequences in monkey cells. *Nucleic Acids Res.* **14,** 8979–8995.
10. Sambrook, J., Fritsch, E. F., and Maniatis, T. (1989) *Molecular Cloning: A Laboratory Manual.* Cold Spring Harbor Laboratory Press, Cold Spring Harbor, NY, pp. E3–E4.
11. Le, X., Scaman, C., Zhang, Y., Zhang, J., Dovichi, N. J., Hindsgaul, O., and Palcic, M. M. (1995) Analysis by capillary electrophoresis laser-induced fluorescence detection of oligosaccharides produced from enzyme reactions. *J. Chromatogr. A* **716,** 215–220.
12. Cheng, Y. F. and Dovichi, N. J. (1988) Subattomole amino acid analysis by capillary zone electrophoresis and laser-induced fluorescence. *Science* **242,** 562–564.
13. Shinozaki, R., Inoue, S., and Choi, K.-S. (1998) Flow cytometric measurement of Benzo[a]pyrene-diol-epoxide-DNA adducts in normal human peripheral lymphocytes and cultured human lung cancer cells. *Cytometry* **31,** 300–306.
14. Hanelt, S., Helbig, R., Hartmann, A., Lang, M., Seidel, A., and Speit, G. (1997) A comparative investigation of DNA adducts, DNA strand breaks and gene mutations induced by benzo[a]pyrene and (±)-anti-benzo[*a*]pyrene-7,8-diol 9,10-oxide in cultured human cells. *Mutat. Res.* **390,** 179–188.

31

Analysis of DNA Damage Using Capillary Zone Electrophoresis and Electrospray Mass Spectrometry

Dieter L. Deforce and Elfride G. Van den Eeckhout

1. Introduction

The methodology described in this chapter was developed to perform DNA adduct research. Chemicals causing DNA adducts can lead to DNA damage, such as strand breaks, by hydrolysis of the DNA adducts. A preliminary screening procedure to detect DNA strand breaks using agarose gel electrophoresis is described. However, the focus of this chapter is on the analysis of the products resulting from this DNA damage and the DNA adducts by capillary zone electrophoresis coupled on-line to electrospray mass spectrometry (CZE-ESI-MS). The sample preparation, the coupling of CZE to ESI-MS and the use of CZE-ESI-MS/MS are discussed in detail. A sample stacking procedure, used to concentrate the samples up to a factor of 200 on the CZE capillary, is described.

Chemicals interacting with DNA, forming so-called DNA adducts have been studied extensively *(1,2)*. The ^{32}P postlabeling of DNA technique is very sensitive *(3–5)* and is a frequently used technique to study the interaction of xenobiotics with DNA, but in the absence of suitable reference compounds it cannot provide structural information. Moreover, it cannot be used to study DNA damage such as the liberation of low molecular weight adducts by DNA solvo-lysis, which is observed using the methodology described here. Therefore our preference is to use mass spectral techniques, which give both structural information and are applicable to the field of adduct research *(6–11)*.

One approach that is already proven in the field of DNA adduct research *(12)* is the coupling of CZE, as a highly efficient separation technique to an electrospray ionization interface (ESI). In order to be able to load enough sample on the capillary and to obtain the desired sensitivity, a sample stacking technique is applied, as described here. In general, in the field of DNA adduct research, most of the attention goes to the adducts formed by interaction of the alkylating agent with the purine or pyrimidine

From: *Methods in Molecular Biology, Vol. 162:*
Capillary Electrophoresis of Nucleic Acids, Vol. 1: Introduction to the Capillary Electrophoresis of Nucleic Acids
Edited by: K. R. Mitchelson and J. Cheng © Humana Press Inc., Totowa, NJ

moiety of DNA *(13)*. In those cases where enzymatic hydrolysis gives rise to the formation of nucleosides, phosphate alkylated adducts are likely to be lost for further analysis. However, during previous experiments with phenyl glycidyl ether (PGE) and 5'-deoxynucleotides it was observed that PGE extensively reacted with the 5'-phosphate group *(10)*. Therefore, an MS analysis for the detection of liberated nucleoside-adducts is performed on the supernatant, after precipitation of the DNA. This methodology is used to investigate specific types of DNA damage, for instance the hydrolysis of the phosphate backbone of DNA upon formation of phosphotriesters by the chemical under investigation. The CZE-ESI-MS methodology, especially in conjunction with online sample stacking, can also be used for the analysis of DNA adducts present in enzymatic hydrolysates of precipitated DNA pellets.

Phosphate alkylation has been observed earlier, particularly in the interaction of mustards with DNA *(14–16)*. The reaction of nitrosourea with RNA has been well investigated and the phosphotriester formation has been held responsible for the degradation of RNA *(17)*. Phosphotriester hydrolysis is also reported to be responsible for DNA degradation upon reaction with pyrrolo[1,2-a]benzimidazole-based aziridinyl quinones *(18)*.

2. Materials

2.1. Sample Preparation

1. All products and solvents used are of analytical grade and are used without further purification. The calf thymus DNA used is highly polymerized DNA type I from Sigma (St. Louis, MO).
2. High performance liquid chromatogrphy (HPLC) grade water is obtained from a Milli-Q Reagent Grade water purification system (Millipore, Bedford, MA).
3. The Chromabond HR-P solid phase extraction (SPE) columns containing a polystyrene-di-vinylbenzene solid phase are from Machery-Nagel, Düren, Germany. This matrix retains molecules with phenolic residues based on Van der Waals interactions.
4. Freshly prepared calf thymus DNA solution in HPLC grade water (at 10 mg/mL) (*see* **Note 1**).
5. Methanolic solution of 0.5 *M* PGE.

2.2. Agarose Gel Electrophoresis

1. Agarose gel electrophoresis is performed with an EPS 600 (Pharmacia Biotech, Uppssala, Sweden) power supply and a GNA 100 agarose submersing electrophoresis apparatus (Pharmacia Biotech).
2. 1% agarose gel: Prepare by heating agarose NA (Pharmacia Biotech) in 1X Tris-borate-EDTA (TBE) buffer containing 0.5 μg/mL ethidium bromide. **Caution**: Ethidium bromide is highly toxic and can cause genetic damage!
3. Loading buffer: 0.25% bromophenol blue, 0.25% xylene cyanol, and 30% glycerol in water.

2.3. CZE-ESI-MS, On-line Sample Stacking, and CZE-ESI-MS/MS

1. The CZE-instrument used is a Lauerlabs PRINCE system (Lauerlabs, Emmen, The Netherlands) with autosampler and control software version 4.201.
2. In theory, any other CZE-instrument should also allow the coupling to ESI-MS. Important features are that the CZE-instrument can measure the current leaving its high-voltage

power supply, instead of the current going through the capillary. This is a prerequisite to be able to perform the sample stacking when coupled to ESI-MS because there is no outlet vial in that configuration.

3. The methods are applied to two different mass analyzers both from Micromass (Manchester, England). The first is a triple quadrupole mass spectrometer (VG Quattro II triple-quadrupole) and the second is a quadrupole time-of-flight (Q-TOF®) mass spectrometer.
4. CZE-ESI ionization is performed using the triaxial CZE-ESI probe (Micromass, Manchester, England). Data is collected and is analyzed using the MassLynx software (Micromass, Manchester, England).
5. In order to deliver the sheath liquids, two HPLC pumps are necessary, one to deliver the normal sheath flow and one to deliver the buffer necessary to perform sample stacking. Flow splitters are used (LC packings, Amsterdam, The Netherlands) to be able to deliver the low volume flows (7 μL/min) at a constant flow rate.
6.. Sheath liquid: 80% (v/v) isopropanol, 20% (v/v) of 2.5-m*M* ammonium carbonate buffer (pH 9.68). The buffer stock is 100-m*M* ammonium carbonate buffer, which is brought to pH 9.68 with ammonium hydroxide. Both sheath liquids are thoroughly degassed by sonication for 5 min.

3. Methods

3.1. Sample Preparation

1. The sample preparation is optimized for the analysis of the DNA damage induced by PGE. This method should be easily applicable to all chemicals containing an aromatic group (such as polyaromatic hydrocarbons). This principle of action also holds for all chemicals suspected to induce strand breaks in DNA.
2. Take great care during the sample preparation to avoid bacterial contamination. Perform all procedures in Laminar Air Flow (LAF) cabinets and use only autoclaved plasticware recipients.
3. Incubate a mixture of 1 mL of freshly prepared calf thymus DNA (at 10 mg/mL) (*see* **Note 1**) with the desired test chemical at 37°C for several different periods of time (e.g., 15 min., 5, 10, 24, and 48 h). In our experiments: 2 mL of a 0.5 *M* methanolic solution of PGE.
4. The reaction is stopped by removing the chemical from the above incubation mixture. Unreacted PGE is removed by extraction of the aqueous phase three times with 3 mL of chloroform. Take 1-μL aliquots of the resulting DNA solutions for subsequent analysis on agarose gel (*see* **Subheading 3.2.**).
5. Precipitate the DNA from the reaction mixtures by adding absolute ethanol to a final concentration of 70% and sodium acetate to a final concentration of 0.3 *M*, incubate at –20°C for 30 min, and then centrifuge the DNA pellets down at 15,000*g* for 10 min.
6. In order to identify possible products resulting from the breaking of the DNA strands, the ethanolic supernatant is dried under a stream of nitrogen gas and redissolved in 500-μL HPLC water (*see* **Note 2**).
7. Reduce the solid phase of the Chromabond HR-P SPE columns to 125 mg. Precondition these columns by sequential rinses with 3 mL methanol, then with 3 mL of 50/50 (v/v) methanol/HPLC water and finally with 9 mL of HPLC water.
8. Apply 200 μL of the sample to the column and rinse with 6 mL HPLC water to remove the salts from these samples (*see* **Note 3**). During this washing procedure, the analytes bearing an aromatic group (in our case the PGE-group) and thus being possible adducts, are retained on the solid phase by Van der Waals interactions.

9. The compounds of interest are then eluted by rinsing the column sequentially with 2 mL of 50/50 (v/v) of methanol/HPLC water, followed by 5 mL of methanol. The collected fractions are then dried under a stream of nitrogen gas. The residue of all the separated fractions are then combined by dissolving in the same 100 µL of HPLC water.
10. The sample is now free of any electrolytes and is ready for analysis by CZE-ESI-MS.

3.2. Agarose Gel Electrophoresis

1. Agarose gel electrophoresis is performed to examine the integrity of the DNA molecule upon incubation with the chemical of interest (in our experiments PGE) in order to evaluate whether the chemical under investigation causes strand breaks in the DNA (or other DNA damage). It should be kept in mind that this experiment does not detect if the chemical causes DNA adducts, because DNA adducts can be formed without causing DNA strand breaks.
2. In order to gain convincing evidence of DNA strand cleavage, an incubation time experiment needs to be performed. We test incubation periods of 15 min, 5, 10, 24, and 48 h using PGE.
3. Add 9-µL HPLC water to the 1-µL aliquots of the DNA solutions obtained in **step 4** of **Subheading 3.1.**
4. After careful mixing, take 1 µL of these diluted samples and add 1 µL of loading buffer and 9 µL of water. Load these samples onto the agarose gel.
5. Run the samples in the gel by submersing the gel in 1X TBE running buffer and applying an electric potential of 10 V/cm over the gel for 1 h.
6. Visualize the ethidium bromide stained DNA bands using an UV transilluminator at 312 nm.
7. Inspect the visualized DNA for degradation. **Figure 1** shows an agarose gel that is indicative of DNA strand breaking damage. DNA strand breaking is likely to be present when the high molecular weight DNA band decreases in intensity, shifts to lower molecular weights or becomes smeared over a range of lower molecular weights.

3.3. CZE-ESI-MS

1. This procedure is optimized for the analysis of negatively charged products resulting from DNA damage or DNA adducts.
2. Prepare a fused silica capillary, dimensions: 75 µm id and 365 µm od by cutting it to the desired length (1 m) using a ceramic cutter (*see* **Note 4**).
3. The polyimide coating needs to be removed both from the injection end (± 5 cm) and over the whole length of the capillary end which will be inside the inner sheath capillary of the ESI-probe (± 30 cm). This is necessary because the polyimide coating loosens during operation and may cause malfunctions and clog the sample orifice.
4. Insert the capillary in the CZE-instrument and in the inner sheath capillary of the triaxial ESI-probe and let it protrude about 200 µm from this sheath capillary, which in turn is positioned to protrude approx 500 µm from the nebulizing capillary (*see* **Fig. 2**). The positioning of the capillaries needs slight optimization (*see* **Notes 5** and **6**).
5. Connect the HPLC pump to the sheath capillary using a T-piece and then pump the sheath-liquid (80% (v/v) isopropanol, 20% (v/v) of 2.5-m*M* ammonium carbonate buffer, pH 9.68) at a flow rate of 7 µL/min (*see* **Note 7**). A nitrogen gas flow of 30 L/h is delivered to the ESI tip through the nebulizing capillary to aid the ESI process. A drying gas flow of 100 L/h nitrogen is delivered to the source. Set the temperature in the source to 75°C.
6. Make the liquid surface in the anode buffer reservoir level with the ESI probe tip when inserted in the source of the mass spectrometer. This is necessary to prevent hydrodynamic flow effects during electrophoresis.

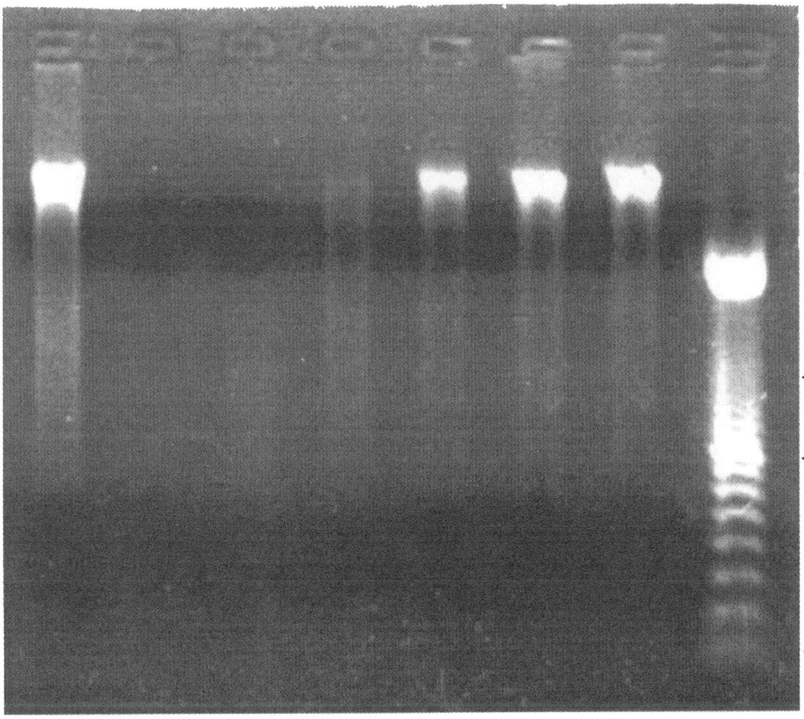

Fig. 1. Ethidium bromide stained agarose gel containing in the lanes: 1, Calf thymus DNA solution freshly prepared; 2–6, Calf thymus DNA is incubated for various periods with PGE in a methanolic solution: 2, 48 h: 3, 24 h; 4, 10 h; 5, 5 h; 6, 15 min; 7, control lane, calf thymus DNA is incubated 48 h in methanol; 8, 100-bp ladder molecular weight marker. The (h) denotes high molecular weight DNA (1.6, 0.8, 0.1) denote the 1600-bp fragment, the 800- and the 100-bp fragment markers.

7. Set the current control of the CZE instrument to the inlet electrode. This electrode measures the amount of current going out of the CZE high-voltage source. It is necessary to be able to monitor the current during electrophoresis (*see* **Note 8**).
8. Fill the capillary with a solution of 0.1 mg/mL dAMP/TMP in 0.1 *M* ammonium carbonate buffer, pH 9.68 by applying a pressure of 200 mbar on the vial at the inlet of the capillary. Position the ESI probe in the source of the mass spectrometer, apply an electrospray voltage of -4.5 kV to the ESI probe and a cone voltage of +30 V.
9. Wait for about 20 min to allow the temperature of the ESI probe to reach 75°C.
10. Apply a pressure of 10 mbar and 13 kV over the electrophoresis capillary and monitor the signal at m/z 321 (TMP) and 330 (dAMP) on the mass spectrometer for tuning purposes.
11. Optimize the *X*-, *Y*-, and *Z*-axis position of the ESI probe in the MS source to obtain an optimal signal for TMP and dAMP. This step needs to be done with great care as slight deviations from the optimal position result in a dramatic decrease in sensitivity. Tune the mass spectrometer to attain optimal sensitivity.

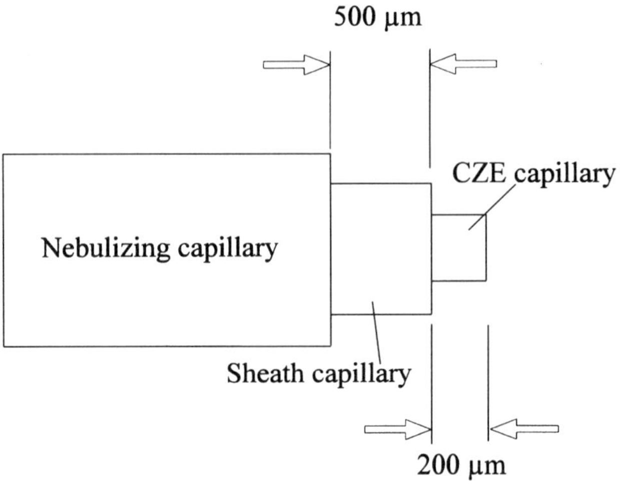

Fig. 2. Schematic representation of the positioning of the three capillaries at the triaxial ESI probe tip.

12. Position a vial containing the electrophoresis buffer (0.1 *M* ammonium carbonate, pH 9.68) at the capillary inlet and rinse the CZE capillary with electrophoresis buffer by applying a pressure of 200 mbar (*see* **Note 9**).
13. Fill the CZE autosampler tray with the properly prepared samples (*see* **Subheading 3.1.**).
14. Prepare the data collection settings on the mass spectrometer, be sure to set the mass range to be detected to include the anticipated molecular weights. In our application a mass range between 160 and 800 Dalton was scanned at a scan speed of 320 Dalton/s.
15. Before injecting a sample on the CZE capillary switch off the electrospray voltage in the ESI probe. Inject the sample on the CZE capillary by applying a pressure of 22 mbar for 0.2 min.
16. Move the autosampler back to the electrophoresis buffer and start the electrophoresis by applying a voltage of 13 kV and a pressure of 10 mbar over the CZE capillary.
17. Switch the electrospray voltage back on and start the data collection on the mass spectrometer (*see* **Note 10**). Monitor an extracted ion electropherogram of the m/z value of a product known to be present in the sample (e.g.m an unmodified nucleotide) on the mass spectrometer to follow the electrophoresis process. A typical reconstructed mass electropherogram is shown in **Fig. 3**.
18. When the analysis is completed, switch of the electrophoresis voltage and apply a pressure of 200 mbar during 5 min to rinse the CZE capillary with electrophoresis buffer. The system is now ready for the analysis of a new sample (repeat from **step 14**) (*see* **Note 11**).
19. Inspect the CZE-ESI-MS results for unknown products resulting from DNA damage or unknown DNA adducts. This can best be performed in the map view (MassLynx software) or by taking a summation of all m/z values present in the electrophoretic zone of interest.

3.4. On-Line Sample Stacking

1. The sample stacking procedure described here is only suited for the concentration of negatively charged analytes (such as DNA adducts). In order to be able to perform this

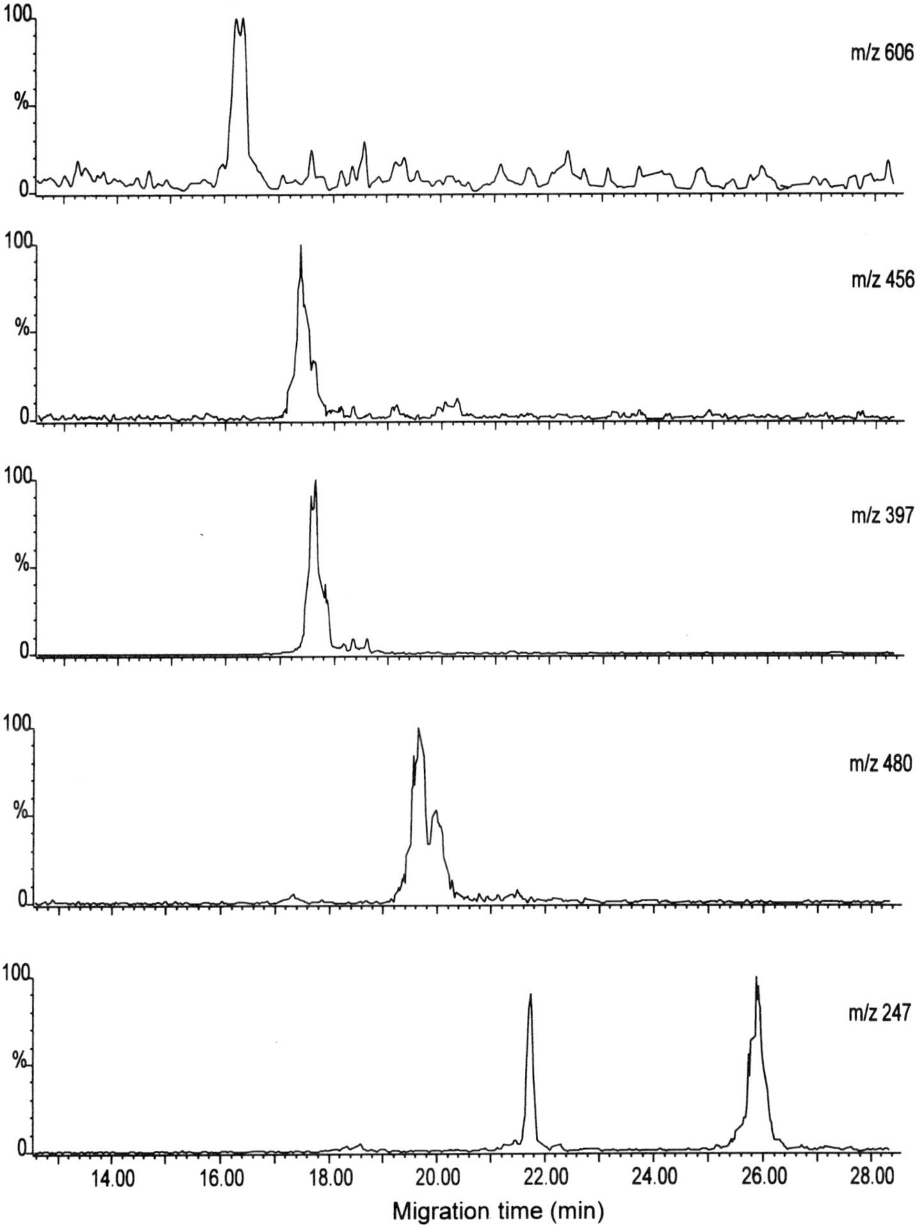

Fig. 3. Reconstructed mass electropherogram of the calf thymus DNA sample following 48-h incubation with PGE in a methanolic solution, and then subjected to the sample preparation procedure using full-scan conditions (75–800 Dalton, at a scan speed of 725 Dalton/s). (CZE-ESI-MS conditions, *see* **Subheading 3.3.**). $[M-H]^-$ in order of increasing t_M: m/z 606, *bis*-PGE-dCMP adduct; m/z 456, PGE-dCMP alkylated on the phosphate moiety; m/z 397: PGE phosphodiester; m/z 480: PGE-dAMP adduct alkylated on the base moiety; m/z 247: PGE monophosphate.

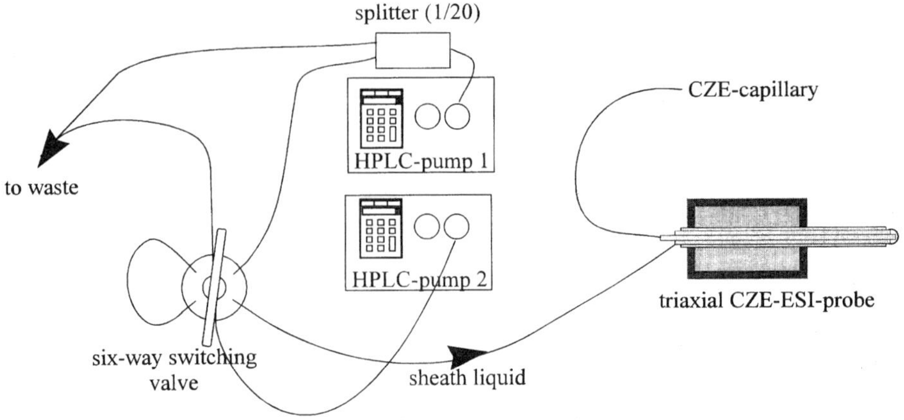

Fig. 4. Schematic representation of the setup used to perform on-line sample stacking in a CZE-ESI-MS setup.

sample stacking procedure, it is a prerequisite that the samples are present in a solution with a lower electrolyte concentration (or ideally in a solution containing no electrolytes) than the electrophoresis buffer.

2. Perform the **steps 1–14** described in **Subheading 3.3.**, if you have not yet done so.
3. Connect a second HPLC pump to the sheath capillary using a Valco six-way switching valve, according to **Fig. 4**, to select the flow from the first or the second HPLC pump to enter the sheath capillary. This second HPLC pump is used to deliver the electrophoresis buffer (0.1 *M* ammonium carbonate, pH 9.68) at a flow rate of 10 µL/min to the CZE capillary end in the electrospray probe (*see* **Note 12**).
4. Monitor the m/z value of 59, this is the $[M\text{-}H]^-$ ion of isopropanol, on the mass spectrometers tuning page. Apply a constant pressure of 20 mbar on the buffer vial at the CZE inlet to deliver a constant flow of electrophoresis buffer through the CZE capillary. Then switch the Valco six-way valve to deliver the flow from the second HPLC pump to the ESI probe. After some time the signal at m/z 59 will decrease and eventually drop down to zero (about 10 min after switching the valve, *see* **Note 13**).
5. Set the pressure on the CZE instrument to zero and apply a voltage of –9 kV to the CZE capillary (relative to the –4.5 kV at the ESI tip this results in a voltage over the capillary of –4.5 kV). Monitor the current and write it down as the stack limit current (–48 to –54 µA). Program the CZE instrument to start the normal electrophoresis when the stack limit current +0.5 µA is reached.
6. Set the voltage over the CZE capillary to zero and rinse the capillary with CZE buffer by applying a pressure of 200 mbar for 2 min.
7. Switch of the electrospray voltage in the ESI probe and inject the sample on the CZE capillary by applying a suitable pressure, e.g., 100 mbar for 0.2 min (*see* **Note 14**). Move the autosampler back to the electrophoresis buffer and start the sample stacking process by applying a voltage of –9 kV to the CZE capillary.
8. Switch the electrospray voltage back on. Monitor the CZE current (*see* **Note 8**) and start the normal electrophoresis when the stack limit current increased by 0.5 µA is reached by applying a voltage of 13 kV and a pressure of 10 mbar over the CZE capillary. Use the program made in **step 5**, for the CZE instrument to perform this switch automatically.

9. When starting the normal electrophoresis, immediately switch the Valco six-way switching valve to deliver the sheath-liquid from the first HPLC pump. Monitor the m/z value 59 on the mass spectrometers tuning page. After some time the signal at m/z 59 will reach its normal intensity (about 5 min after switching the valve, *see* **Note 13**). Start the data collection on the mass spectrometer.
10. Continue with **steps 17–19**, described in **Subheading 3.3.**

3.5. CZE-ESI-MS/MS

1. Once the m/z values of the products of interest (products resulting from DNA damage or DNA adducts) are determined using the technology described herein, the identity of these products can be determined using the collision induced dissociation (CID) possibilities of tandem mass spectrometers. In order to be able to inject sufficient amounts of sample on the CZE capillary to allow CZE-ESI-MS/MS analysis, the sample stacking technique will be used.
2. Perform the **steps 1–11** described in **Subheading 3.3.** and **step 3** described in **Subheading 3.4.**, if you have not yet done so.
3. Switch on the collision gas (Argon) in the collision gas cell and set the pressure to 3×10^{-3} mbar. Tune the mass spectrometer to attain maximum sensitivity and the wanted resolution on both quadrupole groups (*see* **Note 15**).
4. Position a vial containing the electrophoresis buffer: 0.1 *M* ammonium carbonate, pH 9.68, at the capillary inlet and rinse the CZE capillary with electrophoresis buffer by applying a pressure of 200 mbar (*see* **Note 9**).
5. Fill the CZE autosampler tray with the properly prepared samples (*see* **Subheading 3.1.**).
6. Prepare the data collection settings for the CZE-ESI-MS/MS analysis on the mass spectrometer. Set the first quadrupole group to select the ion (or ions) at the desired m/z value(s) and set the collision energy to an anticipated appropriate value for the selected compound. The collision energy will need optimization for each component (*see* **Note 16**).
7. Continue with **steps 4–9** described in **Subheading 3.4.**
8. Monitor extracted ion electropherograms of the m/z values of the mother ion and/or of expected daughter ions to follow the electrophoresis process.
9. When the analysis is completed, switch of the electrophoresis voltage and apply a pressure of 200 mbar for 5 min to rinse the CZE capillary with electrophoresis buffer. The system is now ready for the analysis of a new sample (repeat from **step 6**).
10. Find the product ion mass spectra in the CZE-ESI-MS/MS results by monitoring the extracted ion electropherograms of the m/z value of the mother ion and/or of expected daughter ions. Take the combined mass spectra of the peaks found in these extracted ion electropherograms. A typical product ion spectrum is shown in **Fig. 5**. Based on these results evaluate whether the collision energy was appropriate or not (*see* **Note 16**) and if necessary repeat the experiment with an adapted collision energy.
11. Based on the fragment m/z values present in the daughter ion spectrum of a selected compound try to identify the structure of the mother molecule.

4. Notes

1. When DNA is dissolved in water at the high concentrations as used here (10 mg/mL), it can result in a very viscous solution. It may take several hours before the DNA is totally dissolved. Take great care not to subject this solution to high-speed vortexing or violent shaking as this might induce shearing of the DNA strands. Pipet this solution slowly for the same reason.

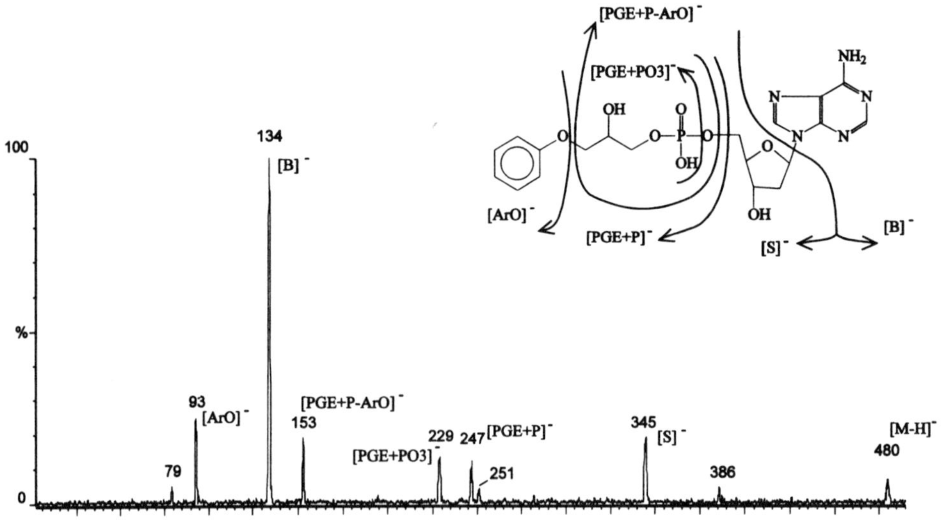

Fig. 5. Product ion spectrum of dAMP with a PGE alkylation on the 5'-phosphate moiety. Mother ion at m/z 480. The collision energy is set at 25 eV.

2. Normally, in DNA adduct research this ethanolic supernatant is thrown away, however we have shown that this supernatant can contain molecules shedding some light on the reactions that cause DNA strand breaks *(9)*. It is very important to evaporate the samples to dryness, because the presence of ethanol or methanol in the samples would disturb the purification on the Chromabond HR-P SPE columns.
3. It is of primary importance to remove the salts from the samples before CZE analysis. The presence of the vast amount of electrolytes (0.3 *M* sodium acetate) in the samples obtained after the precipitation of the DNA would otherwise preclude the use of the sample stacking procedure to concentrate the samples on the capillary.
4. The od of the capillary needs to be chosen in relation to the id of the inner sheath capillary of the ESI probe. The cut at both capillary ends should be straight, this is extremely important for the capillary end in the ESI probe. A nonclear cut can have a negative impact on the ESI process at the capillary tip. The cut can best be examined under a stereo microscope. The polyimide coating is removed by holding it in the flame of a cigarette lighter and cleaning of the burned coating with a cloth dampened with methanol.
5. In order to optimize/check the positioning of the capillaries at the ESI probe tip, the spray can be observed under a stereo microscope. For this purpose, a pressure of 10 mbar is applied to the CZE capillary in order to deliver electrophoresis buffer at the capillary end, a sheath flow of 10 µL/min, a nebulizing gas flow of 30 L/h, and an electrospray voltage of –4.5 kV is applied. The positioning of the capillaries may need slight adjustments in order to obtain a fine and stable electrospray mist. Great care needs to be taken to prevent the different capillaries to touch each other at the ESI tip, which has a detrimental effect on the electrospray.
6. Once the capillary is fully installed, it should be preconditioned before its first use. This procedure needs to be performed with the ESI probe **outside the source!** Rinse the capil-

lary with 0.1 *M* NaOH for 1 h applying a pressure of 1000 mbar, followed subsequently by a 15 min flush with HPLC water (1000 mbar) and then a rinse with the electrophoresis buffer (1000 mbar over 30 min).

7. This sheath liquid is optimal for the analysis of negatively charged species such as modified nucleotides. For the analysis of positively charged species, a sheath liquid of 50% (v/v) methanol in water is to be preferred. In order to deliver a flow rate of 7 μL/min a splitter (LC packings) may be needed. A lower sheath flow could increase the sensitivity though in our experience also led to an unstable electrospray, and subsequent MS signal.
8. In order to be able to monitor the electrophoresis process, it is important to be able to follow the current going through the CZE capillary. An indication that the electrophoresis process is disturbed is that the current is not stable at a certain level (using our conditions 180 μA) or drops to zero (e.g., bubbles in the capillary). In addition, the sample stacking procedure cannot be performed when the current cannot be monitored.
9. The CZE capillary is properly rinsed when there is no detectable signal at m/z 321 and 330. The pressure used to rinse the CZE capillary should never exceed 200 mbar, because a pressure higher than this value results in the production of electrolyte crystals in the ESI tip leading to a dramatic decrease in sensitivity.
10. When the electrospray voltage is not turned off during sample injection, the negative analytes (such as modified nucleotides) are prevented from entering the CZE capillary because of the electrostatic repulsion from the negative potential at the ESI tip and sensitivity drops dramatically. In order to perform normal injection (22 mbar for 0.2 min), the samples should be mixed with electrophoresis buffer and contain at least 25 m*M* buffer ions. The additional pressure of 10 mbar is applied in order to shorten the analysis times. This results in a slight decrease in resolution of the CZE separation. The analysis can be performed without the 10-mbar pressure to maintain the high resolution, however slow analysis times exceeding 1 h are to be expected. Data collection on the mass spectrometer can be started with a fixed delay (e.g., 10 min) in order to save disk space, since the modified nucleotides migrate past the detector only after the electroosmotic flow (EOF).
11. A dilution of dAMP/TMP with a concentration of 2.9×10^{-5} *M* can be injected (22 mbar, 0.2 min) and detected with a signal-to-noise ratio (S/N) of 5/1 on a conventional triple quadrupole instrument. A detection limit of 1.5×10^{-6} *M* (S/N 5/1) can be obtained on a Q-TOF instrument. It is a good idea to check the performance of the overall system by analyzing a series of dilutions of dAMP/TMP.
12. Performance of the "sample stacking procedure" demands that the electrophoresis buffer is present at the capillary end. This is necessary because a potential of –4.5 kV is applied over the CZE capillary during the "sample stacking" which causes an EOF in the direction of the injection side of the capillary, thus aspirating buffer into the capillary at the ESI probe end. If electrophoresis buffer is not present at the capillary end during sample stacking, the sample stacking process will fail to occur.
13. In order to shorten these times the flow rates of the HPLC pumps can be increased twofold during these steps.
14. Of course, the injection of a greater amount of sample increases the sensitivity of the method. In theory, the complete CZE capillary can be filled with sample solution for sample stacking concentration. In practice, we fill the CZE capillary up to 3/4 with sample (400 mbar, for 2 min.), and are able to obtain a detection limit for the dAMP/TMP solution of 1.45×10^{-7} *M* on the triple quadrupole (S/N 5/1), and a detection limit of 7.5×10^{-9} *M* (S/N 5/1) on the Q-TOF instrument. This is an enhancement in sensitivity with a factor 200X over normal injection.

15. In the MS/MS mode, the sensitivity increases with decreased mass resolution selected on the first quadrupole group. Using sample stacking (400 mbar, 2 min) on a triple quadrupole instrument, an acceptable daughter ion spectrum was obtained for TMP (collision energy 21 eV) at a concentration of 22×10^{-7} *M*. On the Q-TOF a detection limit of 1.2×10^{-7} *M* was obtained using the same conditions.
16. The optimal collision energies for the products resulting from the DNA damage varied between 15 and 35 eV. Optimization of the collision energy may require performing the experiment several times. It is best to start with moderate collision energy of 25 eV and vary the collision energy based upon the obtained results. When no mother ions are observed in the daughter ion spectrum, the collision energy should probably be lower, and when the daughter ion spectrum is dominated by mother ions, the collision energy should probably be increased.

References

1. Bartsch, H. (1996) DNA adducts in human carcinogenesis: etiological relevance and structure-activity relationship. *Mutat. Res.* **340,** 67–79.
2. van Zeeland, A. A. (1996) Molecular dosimetry of chemical mutagens. Relationship between DNA adduct formation and genetic changes analyzed at the molecular level. *Mutat. Res.* **353,** 123–150.
3. Farmer, P. B. (1994) Carcinogen adducts: Use in diagnosis and risk assesment. *Clin. Chem.* **40,** 1438–1443.
4. Chen, L., Devanesan, P. D., Higginbotham, S., Ariese, F., Jankowiak, R., Small, G. J., Rogan, E. G., and Cavalieri, E. L. (1996) Expanded analysis of benzo[a]pyrene-DNA adducts formed in vitro and in mouse skin: Their significance in tumor initiation. *Chem. Res. Toxicol.* **9,** 897–903.
5. Schmeiser, H. H., Bieler, C. A., Wiessler, M., van Ypersele de Strihou, C., and Cosyns, J.-P. (1996) Detection of DNA adducts formed by aristolochic acid in renal tissue from patients with Chinese herbs nephropathy. *Cancer Res.* **56,** 2025–2028.
6. Chiarelli, M. P. and Lay, J. O. (1992) Mass spectrometry for the analysis of carcinogen-DNA adducts. *Mass Spectrom. Rev.* **11,** 447–493.
7. Wolf, S. M. and Vouros, P. (1995) Incorporation of sample stacking techniques into the capillary electrophoresis CF-FAB mass spectrometric analysis of DNA adducts. *Anal. Chem.* **67,** 891–900.
8. Vanhoutte, K., Van Dongen, W., Hoes, I., Lemière, F., Esmans, E. L., Van Onckelen, H., Van den Eeckhout, E., van Soest, R. E. J., and Hudson A. J. (1997) Development of nanoscale liquid chromatography/electrospray mass spectrometry methodology for the detection and identification of DNA adducts. *Anal. Chem.* **69,** 3161–3168.
9. Deforce, D. L. D., Lemière, F., Esmans, E. L., De Leenheer, A., and Van den Eeckhout, E. G. (1998) Analysis of the DNA damage induced by phenyl glycidyl ether using capillary zone electrophoresis-electrospray mass spectrometry. *Anal. Biochem.* **258,** 331–338.
10. Deforce, D. L. D., Ryniers, F. P. K., Lemière, F., Esmans, E. L., and Van den Eeckhout, E. G. (1996) Analysis of DNA adducts in DNA hydrolysates by capillary zone electrophoresis and capillary zone electrophoresis-electrospray mass spectrometry. *Anal. Chem.* **68,** 3575–3584.
11. Deforce, D. L. D., Lemière, F., Hoes, I., Millecamps, R. E. M., Esmans, E. L., De Leenheer, A., and Van den Eeckhout, E. G. (1998) Analysis of the DNA adducts of phenyl glycidyl ether in a calf thymus DNA hydrolysate by capillary zone electrophoresis-electrospray mass spectrometry: evidence for phosphate alkylation. *Carcinogenesis* **19,** 1077–1086.

12. van der Vlis, E., Mazereeuw, M., Tjaden, U.R., Irth, H., and van der Greef, J. (1995) Combined liquid-liquid electroextraction-isotachophoresis for loadability enhancement in capillary zone electrophoresis-mass spectrometry. *J. Chromatogr. A* **712,** 227–234.
13. Dizdaroglu, M. (1994) Chemical determination of oxidative DNA damage by gas chromatography-mass spectrometry. *Method. Enzymol.* **234,** 3–16.
14. Lindemann, H. and Harbers, E. (1980) *In vitro* reaktion der drei alkylierenden pharmaka cyclophosphamid, ifosfamid und trifosfamid mit DNS und DNS bausteinen. *Drug Res.* **30,** 2075–2080.
15. Lindemann, H. (1984) Interaction of cyclophosphamide with DNA in isolated rat liver nuclei. *Anticancer Res.* **4,** 53–58.
16. Maccubbin, A. E., Caballes, L., Riordan, J. M., Huang, D. H., and Gurtoo, H. I. (1991) A cyclophosphamide/DNA phosphoester adduct formed *in vitro* and *in vivo*. *Cancer Res.* **51,** 886–892.
17. Singer, B., Sun, L., and Fraenkel-Conrat A. (1975) Effects of alkylation of phosphodiesters and of bases on infectivity and stability of tobacco mosaic virus RNA phosphate alkylated TMV RNA. *Proc. Natl. Acad. Sci. USA* **72,** 2232–2236.
18. Skibo, E. B. and Schulz, W. G. (1993) Pyrrolo[1,2-a]benzimidazole-based aziridinyl quinones. A new class of DNA cleaving agent exhibiting G and A base specificity. *J. Med. Chem.* **36,** 3050–3055.

32

Integration of Phosphodiesterase-Induced Degradation of Oligonucleotides with Capillary Polymer-Sieving Electrophoresis

Jan Saevels, Ann Van Schepdael, and Jos Hoogmartens

1. Introduction

Synthetic oligodeoxyribonucleotides (ODN) have been proposed as a class of potential therapeutic agents that can interact in a rational way with DNA or RNA, with the aim of inhibiting the expression of unwanted genetic information ***(1–3)***. One of the most critical questions in the evaluation of these molecules, is their stability toward enzymatic breakdown by [3' or 5']-exonucleases and endonucleases ***(4)***. In order to inhibit or at least limit the effect of these nucleases, chemically modified ODNs have been synthesized.

1.1. Stability Studies

In vitro stability studies are usually performed by incubating the oligonucleotide in serum, cell extracts or cell cultures, followed by extraction, desalting and off-line analysis ***(5–8)***. Here, sample cleanup is a laborious procedure, requiring a significant number of handling steps, such as protein digestion and phenol/chloroform extraction. In contrast, the use of the 3'-exonuclease Phosphodiesterase I can be justified as a model study, as 3'-exonuclease activity is the major cause of degradation of ODNs in serum ***(9)***. The use of a purified enzyme preparation abolishes the need for extensive sample preparation prior to analysis. In both cases, different manipulation steps are necessary, including one that can separate the starting material from the degradation products. Capillary gel electrophoresis (CGE) is such a high-resolution separation technique, proven to be suitable for the separation of oligonucleotides ranging from 10 to over 500 bases ***(10)***. These gels are usually chemically bound to the inner wall of the capillary in order to prevent electroosmotic flow (EOF) initiated disruption of their gel structure. Although CGE has proven to yield very high efficiencies in oligonucleotide analysis ***(11)***, it suffers from a major handicap, the crosslinked polyacrylamide sieving network in the capillaries is heat sensitive, and it becomes unstable over time.

From: *Methods in Molecular Biology, Vol. 162:*
Capillary Electrophoresis of Nucleic Acids, Vol. 1: Introduction to the Capillary Electrophoresis of Nucleic Acids
Edited by: K. R. Mitchelson and J. Cheng © Humana Press Inc., Totowa, NJ

These capillaries have not been reported to last for more than a few dozen runs due to this instability *(12)*, and possibly due to irreproducibility in the polymerization process and the fragile nature of the medium.

1.2. Capillary Polymer Sieving Electrophoresis

Recently, solutions of linear polymers have also received general acceptance for the analysis of DNA analogues, as well as fixed separation matrices. Replaceable entangled polymers such as linear polyacrylamide *(13)*, agarose *(14)*, poly(ethylene oxide) *(15)*, hydroxypropyl methyl cellulose *(16)*, and hydroxyethyl cellulose (HEC) *(17)* have been shown to be as effective as CGE for the analysis of numerous nucleic acids. The technique has been named capillary polymer sieving electrophoresis (CPSE). A dilute polymer solution can be reloaded into the capillary after each run, thereby extending column lifetime, as well as providing an essentially fresh medium for each analysis. Moreover, the quantitatively more interesting hydrodynamic injection mode of sample loading becomes feasible. This mode is impossible in CGE due to the extremely high viscosity of the gel. In the case of synthetic oligonucleotides, CPSE usually allows the separation of ODNs differing by only one nucleotide in length. This feature makes the CPSE technique particularly suitable for checking the purity (existence of failure sequences) of ODNs during chemical synthesis, as well as for undertaking studies on the stability of synthetic ODNs under various biological conditions.

1.3. Electrophoretically Mediated Microanalysis

CPSE can be applied for checking the quality of ODNs, as well as the resistance of some synthetic model ODNs to degradation. The enzymatic degradation is usually studied in an "off-line" analysis (enzymatic reaction physically detached from the product separation step), as well as in an "in-line" analysis format (enzymatic reaction and product separation steps are integrated). The continuous flow method that is used for these analyses is based on an electrophoretically mediated microanalysis (EMMA) methodology. During EMMA, the analyte and reagent(s) are introduced into the electrophoretic capillary as distinct zones *(18)*. These zones tend to merge together during the electrophoresis process on the condition that the zones have different electrophoretic mobilities. The enzymatic reaction proceeds during the mixing process in the presence, or absence of an electric field. Reaction products are then electrophoretically transported to the detector under the influence of an applied potential. Therefore, EMMA allows the reaction-based chemical or enzymatic analysis to be performed entirely "in-capillary" (*see* **Fig. 1**).

Since its introduction in 1992 *(19)*, EMMA took advantage of CE being able to perform each of the following tasks:

1. The transfer of different reagents by utilizing mobility differences.
2. The mixing of the reagents without disturbing the mixing zones.
3. Allowing sufficient time to carry out the reaction for quantitation of products.
4. The transport of identifiable and detectable species to the detection window.
5. Having sufficient detection sensitivity.

A great advantage in EMMA is that species are mixed by electrophoretic migration, rather than by convective flow. In this manner, mixing is achieved without either

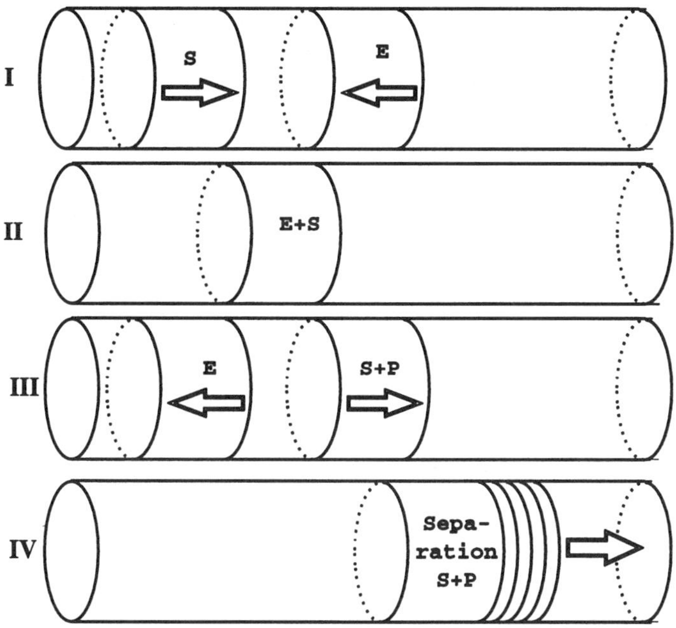

Fig. 1. Schematic representation of an EMMA of the interaction between an enzyme and an oppositely charged substrate. **I**) consecutive injections of the enzyme (E) and the substrate(S); **II**) the electric field drives the enzyme and the substrate in opposite directions: when the two zones meet, an enzymatic reaction takes place; **III**) the enzyme moves towards the capillary inlet, and the substrate and the reaction product (P) are driven in the opposite direction toward the detector; **IV**) the substrate is separated from the reaction products in the remainder of the capillary.

sample dilution or turbulence, which is important for dilute reagent solutions and for reproducibility between experiments, respectively. Potential applications of these microreactor techniques include, the analysis of samples for enzymatic activities, or for determining substrate concentrations, for evaluating molecules as substrates for enzyme-catalyzed reactions, for examining mixtures of enzymes for activity against a particular substrate, and for the study of enzyme-inhibitor interactions.

1.4. Overview

This chapter demonstrates how CPSE can be applied for the analysis of synthetic ODN. Antisense oligonucleotides are promising therapeutic agents for the treatment of viral infections and cancers. A critical question in the evaluation of these molecules is their stability toward nucleases. Numerous different chemical modifications of oligonucleotides have been proposed in order to limit breakdown by nucleases. Here, six homo-oligonucleotides of deoxyadenosine, both modified as well as unmodified are custom synthesized for subsequent analysis. A 4% solution of a low-viscosity grade HEC is used as the sieving agent for electrophoresis in a coated capillary, which allows for single-base resolution. Electropherograms reveal that the purity of the oligonucleotide ranges between 54.8% and 96.5%, suggesting that the reversed-phase liquid chromatography purification, used after synthesis to purify the synthetic ODNs from all their side products, is inadequate.

Phosphodiesterase I from snake venom is used to assess the susceptibility to 3'-exonuclease degradation of the ODNs, and these assays are constructed "off-line" as well as "in-line." "In-vial" degradation, followed by continuous injections into the capillary show nicely the degradation patterns and allow for the calculation of half-lives. "In-line" degradation and separation is accomplished using a unique EMMA concept, which is developed after opposite charges for enzyme and substrate have been determined. This novel approach is the first time that an enzymatic reaction has been integrated into a polymer-filled capillary, in which the polymer also acts as the sieving matrix for the reaction products. Results for the "in-capillary" degradation and separation are consistent with the "off-line" incubation and subsequent capillary separation, adding the extra advantage of reducing the number of sample manipulations steps, thus simplifying the procedure overall. This efficiency opens the possibility of screening the stabilities of large libraries of synthetic ODNs against degradation with purified nucleases using EMMA.

2. Materials

2.1. CE Instrumentation

1. A SpectraPHORESIS® 1000 CE system (Thermo Separation Products, Fremont, CA) is used for all experiments, operating under PC1000® software version 3.01 (Thermo Separation Products).
2. J&W DB-17 capillaries of 100 µm id coated internally with a 50/50 mixture of methyl- and phenylsilicone, are purchased from Alltech (Lokeren, Belgium). The capillaries are of 44.0-cm total length (*L*), and 36.0-cm effective length (*l*).
3. Before use, the capillaries are purged with methanol and water for no less than 5 min using the vacuum pump of the CE apparatus.
4. Oligonucleotide separation is monitored by "on-capillary" UV detection at 260 nm.
5. For sample injection, the vacuum system of the instrument applies a constant negative pressure of 0.75 psi.

2.2. Reagents

1. All reagents are of analytical grade.
2. TAPS (*N*-Tris(hydroxymethyl)methyl-3-aminopropanesulfonic acid) is from Sigma (St. Louis, MO).
3. Tris (Tris(hydroxymethyl)aminomethane) from Acros Organics (Geel, Belgium).
4. Urea is from ICN Biomedicals (Aurora, OH) and EDTA is from Carlo Erba (Milan, Italy).
5. The synthetic oligonucleotide mixture, $p(dA)_{12-18}$ is from Pharmacia Biotech (Roosendaal, The Netherlands) and is stored at –20°C as a 233 µg/mL solution in water.
6. HEC (Cellosize® EP09, technical grade) is a gift from Union Carbide (Antwerp, Belgium).
7. All buffer solutions are made with Milli-Q water (Millipore, Milford, MA) and are passed through a 0.2-µm filter, except for the polymer solution that is used without ultrafiltration.

Table 1
The Structures of the Oligonucleotides Under Study[a]

	Sequence (from 5' to 3')	5'	3'
dA_{21}	A-A-A-A-A-A-A-A-A-A-A-A-A-A-A-A-A-A-A-A-A	OH	OH
OLIGO 1	A-A-A-A-A-A-A-A-A-A-A-A-A-A-A-A-A-A-A-A-A	PO_4^-	OH
OLIGO 2	A-A-A-A-A-A-A-A-A-A-A-A-A-A-A-A-A-A-A-A-A	PO_4^-	NH_2
OLIGO 3	A-A-A-A-A-A-A-A-A-A-A-A-A-A-A-A-A-A-A-A-3dA	OH	H
OLIGO 4	A-A-A-A-A-A-A-A-A-A-A-A-A-A-A-A-A-A-A*s*A*s*A	OH	OH
OLIGO 5	A-A-A-A-A-A-A-A-A-A-A-A-A-A-A-A-A-A-A*m*A*m*A	OH	OH

[a]A stands for 2'-deoxyadenosine, 3dA is 3'-deoxyadenosine; - represents a phosphodiester linkage, *s* is a phosphorothioate bridge and *m* symbolizes a methylphosphonate group.

2.3. Compounds Under Study

1. Synthetic ODNs are custom made and are then purified using reversed-phase liquid chromatography (OliGold®, Eurogentec, Seraing, Belgium).
2. dA_{21}, an unmodified homo-oligomeric deoxynucleotide, is used as a model reference compound.
3. Five modified ODNs, all 21-mers of deoxyadenosine, are used. **Table 1** shows the structure of all oligonucleotides used in this work. Following the synthesis and lyophilization of the synthetic ODNs products, they are dissolved in 1X TE buffer (10 m*M* Tris-HCl, 1 m*M* EDTA, pH 7.5) and are stored at –20°C until use.

3. Methods

3.1. Preparation of the Polymer Solution

1. 25 mL of the polymer solution is prepared as follows: An appropriate amount of the low viscosity-grade HEC EP09 is added to 17 mL of water at 50°C.
2. After 10 min of continuous stirring, a clear solution is obtained, and 121.5 mg TAPS and 10.5 g urea are then added.
3. The mixture is stirred for another 10 min, then the pH is adjusted to 7.5 with solid Tris using a Consort P514 pH meter (Turnhout, Belgium). Finally, the viscous solution is made up to volume.

3.2. CPSE Method

1. Khan et al. ***(17)*** noted a relatively high concentration of a low viscosity-grade HEC may be used as a sieving agent in a CE system using a coated capillary. The preparation of the polymer solution for the separation of the oligonucleotides is described in **Subheading 3.1.**
2. Hydrodynamic injections (10 s at 0.75 psi) into the polymer-filled capillary are performed automatically, directly from the incubation vial at regular time intervals, with the first injection occurring immediately after vortex mixing.
3. Electrophoresis is run at –15 kV.
4. **Figure 2** shows excellent separation of the seven fragments of a standard oligonucleotide mixture is attained. In our hands, separation efficiencies ranging from 2×10^5 to 3×10^5 plates are achieved, with resolutions occurring between 1.8 and 2.1. The current generated is 22 μA at –15 kV and 25°C, which might seem high for a 20-m*M* TAPS buffer at pH 7.5. Indeed, inorganic impurities from the technical grade HEC may be up to 5%

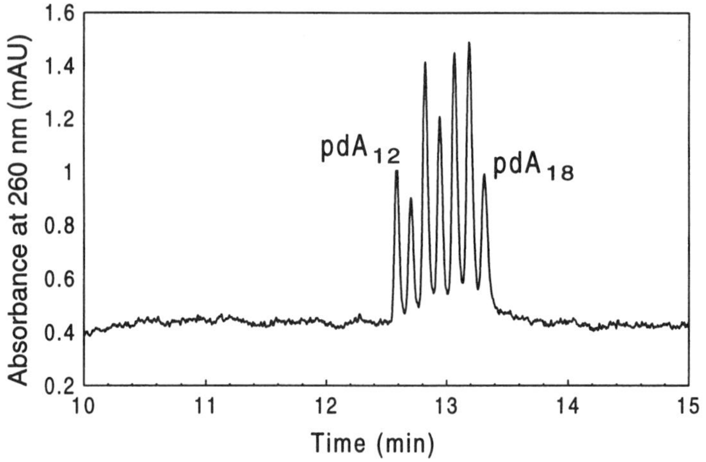

Fig. 2. Typical electropherogram of a 233 µg/mL $p(dA)_{12–18}$ mixture. Capillary: J&W DB-17 100 µm id, 44.0 cm *L*, 36.0 cm *l*; background electrolyte: 20 m*M* TAPS/Tris, pH 7.5, 4% HEC EP09 and 7 *M* urea; hydrodynamic sample injection for 5 s; run voltage is –15 kV (22 µA); temperature 25°C; UV detection at 260 nm.

sodium acetate *(20)*, which might contribute to an increase in the conductivity of the background electrolyte.

3.3. Purity Profiles (see Note 1)

1. All ODNs are subjected to a purity check using the CPSE method. The capillary is loaded with the HEC buffer for 5 min before each analysis, samples (109 µg/mL for dA_{21}, 125 µg/mL for OLIGO 1, and 100 µg/mL for OLIGOs 2 –5) are injected hydrodynamically and the electrophoresis is performed at reversed polarity.
2. CPSE of all six synthetic ODNs reveal a purity that strongly depends on the modification of the oligonucleotide *(21,22)*. The CPSE method has the ability to separate oligonucleotides from 12 to 24 bases in length with a single base resolution. As can be seen from **Fig. 3**, the synthesis failure lengths of individual preparations and of particular modification steps can be fairly significant.
3. The synthesized ODNs are in fact mixtures, consisting of the main component and of several related substances. Considering the absorption maxima between 255 and 265 nm and based on migration order, the compounds detected close to the main peak are identified as oligo (dA) nucleotides differing in length from the main compound, with the smaller molecules migrating faster through the pores of the polymer network. The relative content of the oligonucleotide fragments can be expressed as percentages (% m/m), because closely related homo-oligomeric ODNs all have similar extinction coefficient/molecular weight ratios.
4. The model compound dA_{21} and its 3'-deoxyadenosine counterpart OLIGO 3 are only 84.8 and 88.5% pure. The content of the main component is lowest for OLIGO 1 and 2 (70.1 and 54.8%, respectively), i.e., where a phosphate group has been incorporated in 5'-position. In the case of OLIGO 1, a substantial proportion of the oligonucleotide (20%) appears to be a nonphosphorylated molecule (peak at 14.8 min).

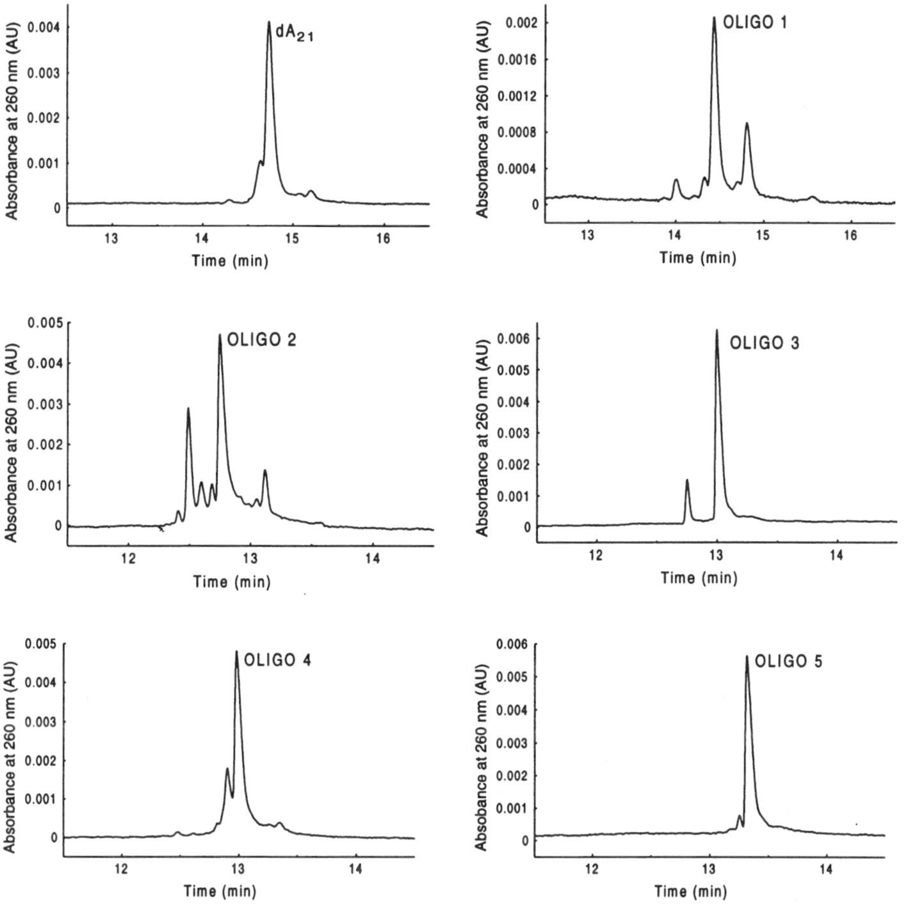

Fig. 3. Electropherograms of the ODNs. Sample injection is hydrodynamic (20 s for dA_{21}, 10 s for OLIGO 1 and 30 s for OLIGOs 2–5). The electrophoresis run is at –12 kV for dA_{21} and for OLIGO 1 (17 μA), while the run is at –15 kV for the other OLIGO's (22 μA); Electrophoretic conditions are the same as in **Fig. 2**.

5. An additional replacement step of the 3'-hydroxyl by a primary amine results in the detection of the highest number of related substances, and by far the lowest content of the desired compound. It is believed that these side products are impurities from ODN synthesis, because no further degradation is observed during storage as a lyophilized powder, or as a solution in water at –20°C.
6. Apparently, the reversed-phase liquid chromatography purification is inadequate to purify the synthesized ODN from all of its side products. Substitution of the last two phosphodiester bridges with phosphorothioate (OLIGO 4; 77.5%) or with methylphosphonate groups (OLIGO 5; 96.5%) gives a purity profile that is comparable to the reference compound dA_{21}. ODNs with n modified linkages such as OLIGO 4 and 5 are chiral at the phosphorous, and this nonstereospecific synthesis naturally leads to a mixture of 2^n dia-

stereo-isomers. Whether this CPSE method has the ability to separate stereoisomers is not known, because stereo-isomerically pure compounds are not available.

3.4. Procedure for the In-Vial Degradation

1. A stock solution of phosphodiesterase I (E. C. 3. 1. 4. 1) from *Crotalus adamanteus* venom (Pharmacia Biotech, Roosendaal, The Netherlands) is made at a concentration of 1 µg/µL in a 20-m*M* TAPS buffer, adjusted to pH 7.5 with Tris and containing 5 m*M* $MgCl_2$ (Merck, Darmstadt, Germany), identical to the incubation medium.
2. The procedure involves mixing of oligonucleotide and phosphodiesterase I in a 50-µL vial insert, to give final concentrations of 100 µg/mL for the substrate and 5 µg/mL for the enzyme.

3.5. In-Vial Degradation and Off-Line Analysis

1. For all synthesized ODNs, an off-line enzymatic degradation-separation-quantification was carried out (*see* **Subheading 3.4.**). Phosphodiesterase I catalyzes the hydrolysis of both DNA and RNA by processive exonucleolytic attack of the free 3'-hydroxyl terminus. Degradation of poly (dA) oligonucleotides by the snake venom exonuclease yields well-defined reaction products, namely only dA oligo's shorter than the starting product and the nucleotide dAMP.
2. The incubation of the model compound dA_{21} with phosphodiesterase I demonstrates that 3'-exonucleolytic breakdown results in a mixture of the nucleotide dAMP and deoxy-adenosine oligomers shorter than 21 bases in length. The polymer solution nicely separates the mononucleotide and the larger oligonucleotide fragments. The decrease in corrected peak area for the main peak, allows for the quantitative assessment of the degradation.
3. The half-life of dA_{21} under the stated conditions is 27.1 min as the mean of three degradation experiments, with an average standard deviation on the calculation of the slope of 3.5 min.
4. **Figure 4** shows half-lives of the five other ODNs, subjected to the same degradation process. Only OLIGO's 1 and 3 show substantial degradation within 12 h. For OLIGO 1, the mean half-life is 24.2 min with an average standard deviation of 2.4 min, which did not differ significantly (with a 98% confidence interval) from the half-life of the model compound. The only difference between both oligonucleotides is in a phosphate group in 5'-position. This phosphate does not affect the susceptibility to 3'-exonuclease degradation (*see* **Note 2**).
5. OLIGO 3 has 3-deoxyribose as the last sugar in the sequence, and is degraded somewhat slower than its 2-deoxyribose counterpart dA_{21}. The mean half-life for this ODN is 41.3 min with an average 2.9 min standard deviation on the calculation of the slope, showing some small effect of the 3'-OH on phosphodiesterase I to perform at optimal catalytic activity under these conditions.
6. These half-life times have only relative significance, because they have a strong dependence on the incubation conditions, such as the pH, the type of media, the presence of Mg^{2+} ions, the temperature, and the concentrations of reactants ***(11)***.
7. **Figure 5** depicts 4 stages in the degradation of OLIGO 3. A decrease of the main peak and the appearance of smaller fragments can be seen with increasing incubation time. 3'-Exonucleolytic breakdown of OLIGO 3 produces 2 mononucleotides, namely 3'-deoxyadenosine 5'-monophosphate (3'-dAMP) and dAMP, both of which migrate slower than the oligonucleotides, and one uncharged nucleoside. The slowest migrating mononucleotide is identified as 3'-dAMP. It is the first to appear and because complete

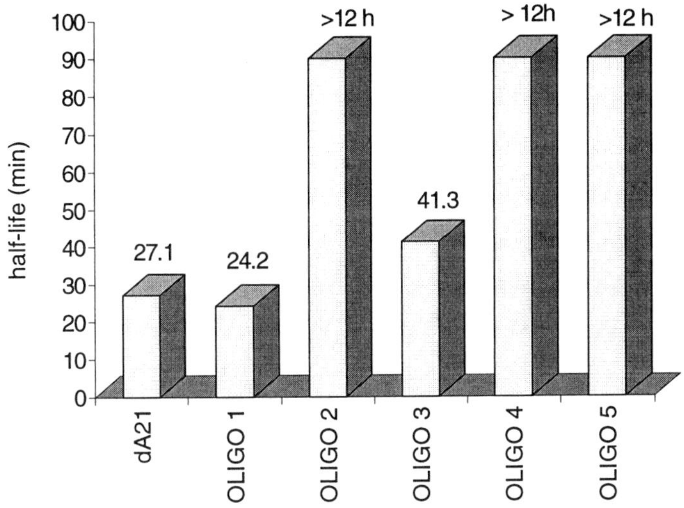

Fig. 4. Half-lives of the synthetic ODNs under the stated conditions.

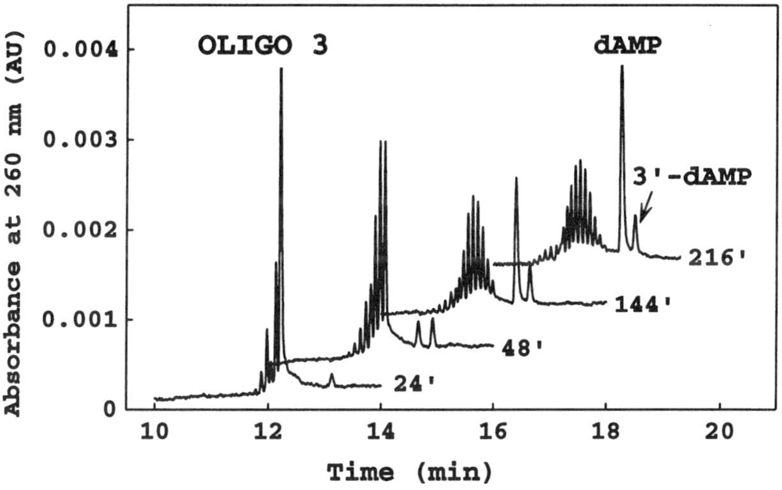

Fig. 5. Electropherogram runs taken at different stages of the "in-vial" degradation of OLIGO 3, which are indicated as the incubation time.

degradation yields a ratio of about 1–20 for this peak to the one migrating just before it, the theoretical composition is one 3'-dAMP for 19 dAMP nucleotides.

8. Contrary to expectations, dAMP migrates slower than the starting oligonucleotide. The charge to mass ratio for dAMP is equal to the ratio for all oligo (dA) nucleotides from 10 to 21 bases in length. Moreover, because of its smaller size, dAMP is expected to move faster than the larger oligo's through the pores of the matrix. When analyzing a mixture of

dA_{21} and dAMP in free solution electrophoresis, the same migration order as in HEC buffer is observed. The reason for the slow migration of the mononucleotide is not clear.

3.6. In-Line Degradation and Analysis

3.6.1. EMMA Procedure (see **Note 4**)

1. Although the injection, separation, and detection steps in the experiments described in **Subheading 3.5.** are already fully automated, the enzymatic degradation is not. The ideal system should be capable of performing each of the following tasks: mixing of the analyte and analytical reagents, reacting the analyte and reagents during a predefined time, quenching of the reaction, separation of all the reaction products formed and their individual detection and quantification.
2. The group of Regnier ***(18)***, have described how enzyme and substrate can meet inside a fused silica capillary and this has been named EMMA. Methods of transient engagement, such as EMMA, utilize differential electrophoretic mobility to merge distinct zones of substrate and enzyme under the influence of an electric field. The enzymatic reaction is then allowed to proceed for as long as the two zones overlap. After reaction, the remaining substrate is separated from the reaction product(s) by CE, thus allowing monitoring of both loss of substrate and appearance of product(s) simultaneously. The separation capability of the CE system can be of specific interest if more than one product results from the reaction.
3. The method described in this section is based on the transient engagement EMMA methodology and tries to adapt it to polymer filled capillaries where polymers actually act as sieving matrix for oligonucleotides. More precisely, it will be checked how the "plug-plug" reaction in transient engagement EMMA can be used to assess the resistance of synthetic ODNs to 3'-exonucleolytic breakdown by phosphodiesterase I, taking the off-line incubation and separation as a reference.
4. Initial electrophoresis experiments with the HEC buffer at pH 7.5 reveal a negative charge for oligonucleotides (*see* **Subheading 3.3.**). On the other hand, when phosphodiesterase I is injected under the same conditions, nothing could be detected between 200 and 300 nm. Using positive instead of negative polarity with the HEC buffer in a coated capillary did not produce any peak either. Two possible explanations are suggested: (a) the enzyme is too bulky to move through the pores of the polymer network, or (b) the background absorption of the HEC buffer in the lower UV region is too high to detect proteins. In order to deal with these problems, phosphodiesterase I is electrophoresed under CZE conditions using a nonviscous TAPS-Tris buffer in a coated capillary. The detector is set in its fast-scanning mode, and a protein peak is produced when electrophoresis is run at normal polarity, i.e., from positive to negative.
5. Negative charge for the substrate and positive for the enzyme prompted us to develop a unique EMMA concept. Up until now, EMMA methodologies have been devised for enzymes and substrates having different electrophoretic mobilities, but still migrating in the same direction. Here, oppositely charged enzyme and substrate call for another approach. Injection of enzyme prior to substrate will drive the oligonucleotide through the enzyme when working at negative polarity, whereby the enzyme moves toward the capillary inlet. In our setup, both enzyme and substrate are hydrodynamically injected as distinct zones from the cathodic side in a HEC-filled capillary. Both zones interpenetrate on condition that they have different electrophoretic mobility. More precisely, during electrophoresis at negative polarity, the oligonucleotide is driven from injection toward detection side, whereas phosphodiesterase moves in the opposite direction, leaving the

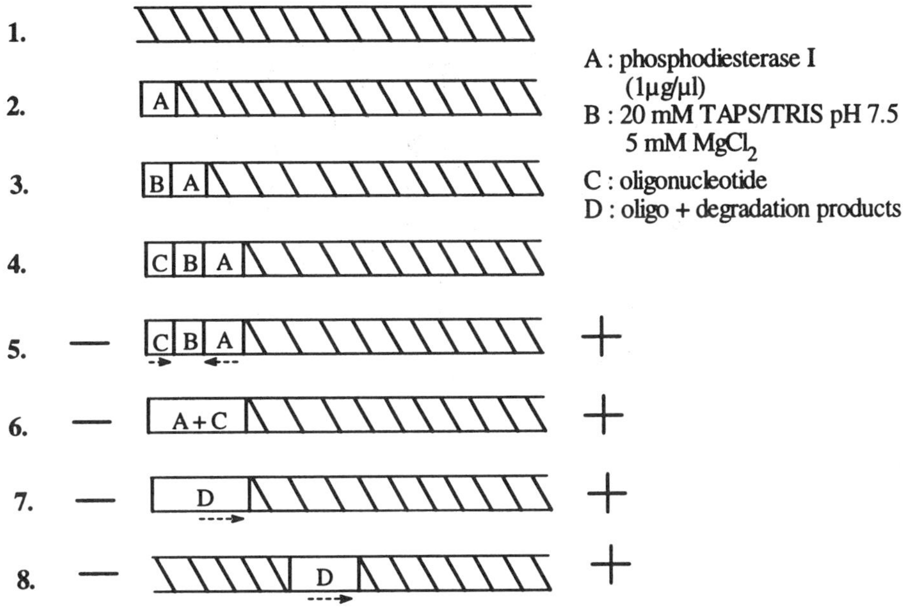

Fig. 6. Schematic representation of EMMA of oligonucleotides. *(1)* Capillary filled with sieving buffer. *(2–4)* Hydrodynamic injections of enzyme (A) for 5–60 s, buffer (B) for 10 s and then oligonucleotide (C) for 10 s. *(5)* Application of negative high voltage drives the oligonucleotide toward anode and the enzyme toward cathode. *(6)* An enzymatic reaction takes place. *(7)* The oligonucleotide has been degraded (D). *(8)* The separation of the oligonucleotide degradation fragments.

capillary inlet. This procedure makes the two zones meet. Thus, substrate and enzyme are brought into contact with each other, based on differences in their electrophoretic mobility. **Figure 6** shows the exact experimental conditions that are used.

6. An intermediate plug is injected between both reaction plugs, consisting of a nonviscous reaction buffer (20 m*M* TAPS, pH 7.5, 5 m*M* $MgCl_2$). The intermediate reaction buffer had a double function. First, it prevents the interpenetration of enzyme and substrate plugs due to diffusional band broadening prior to the application of an electric field. Second, it consists of the medium where enzyme can convert substrate in favorable conditions, namely a slightly alkaline buffer in the presence of Mg^{2+}. From **step 5** onward, a voltage gradient is applied in order to allow the enzymatic reaction to occur. Voltage rises in a linear way from 0 to –5 kV during the first 5 min of electrophoresis, followed by –12 kV or –15 kV during the remainder of the run. This voltage program assures that reaction is prolonged, such that a detectable amount of product can accumulate. After the enzymatic breakdown, the two plugs disengage, the fragments of different length of the degraded oligonucleotide are separated in the sieving buffer and swept to the detector. In between two runs, the HEC matrix in the capillary is replenished by a 5-min wash procedure with the same buffer.
7. Phosphodiesterase I degrades the oligonucleotide from the 3'-end for as long as the enzyme and substrate zones are in contact. After a voltage gradient for 5 min, no more

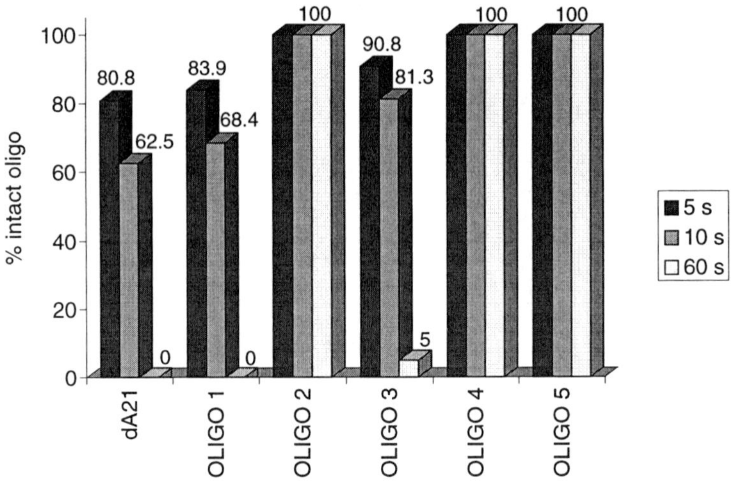

Fig. 7. Percentage of each ODN remaining intact after the injection of enzyme for time periods of 5, 10, and 60 s, respectively.

enzymatic reaction takes place, as experiments where the voltage program is prolonged reveal no additional degradation. Moreover, the relatively low voltage at the time enzyme and substrate meet (between 0 and –5 kV) is believed not to lead to any substantial Joule heating. The temperature inside the capillary, particularly during the first 5 min of electrophoresis, is taken to be equal to the thermostated capillary temperature, 25°C.

8. The extent of oligonucleotide degradation can be controlled by the length of injection time of the enzyme plug. If no voltage gradient is applied and enzyme is hydrodynamically injected for 5 s or less, no measurable degree of degradation takes place. A longer enzyme plug yields a longer reaction time when the voltage is applied, because of the longer time period the oligonucleotide needs to travel through the enzyme zone. Another hypothesis states that a longer injection time for the enzyme results in a more concentrated enzyme zone due to the stacking of the sample plug. In either case, the injection time for phosphodiesterase I seems proportional to the amount of degradation of the oligonucleotide.

3.6.2. Quantitative Assessment of Enzymatic Degradation

1. Enzymatic activity can be quantified by following the disappearance of the starting compound or by the appearance of degradation products, and more specifically by the appearance of the mononucleotide, dAMP. As with the "off-line" degradation, only dA_{21}, OLIGO 1 and OLIGO 3 can be degraded inside the capillary within a reasonable time period.
2. An enzyme injection time of 5 s yields some degradation of these compounds (19.2% for dA_{21}, 16.1% for OLIGO 1, and 9.2% for OLIGO 3 as a mean of 3 analyses).
3. However, a 10-s enzyme injection degrades 37.5% of the dA_{21}, 31.6% of OLIGO 1, and 18.7% of OLIGO 3 (*see* **Fig. 7**).
4. An enzyme concentration of 1 µg/µL is needed to obtain this fast rate of degradation, which is about 1000X higher than used by Bruin et al. ***(11)***. These authors incubated pdT_{20} with snake venom nuclease in a buffer of pH 9.0 at room temperature, and obtain a similar degradation pattern after 1–70-min incubation and a 20 min "off-line" CGE run.

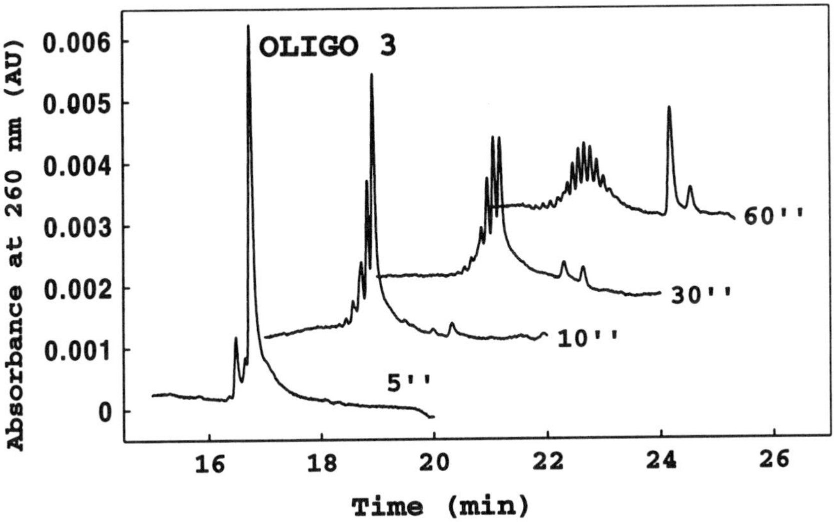

Fig. 8. Electropherogram runs of the "in-capillary" enzymatic degradation of OLIGO 3. The voltage gradient is applied after the injection of the enzyme plug (indicated time), the intermediate buffer plug (10 s), and the substrate plug (10 s).

5. In our experiments, enhancing the enzymatic activity by increasing to pH 8.9 of maximum activity *(23)* is not feasible because of instability of the capillary coating at this pH.
6. Raising the temperature to 37°C is another way to amplify the enzymatic activity *(24)* and although it was not tried out, this is believed to be compatible with the capillary electrophoretic system.
7. Electropherograms of the "in-capillary" degradation of OLIGO 3 are shown in **Fig. 8**. Comparing the results from the "off-line" and the "in-line" methods, OLIGO 3 appears to be about 1.6–1.9 times more stable than both dA_{21} and OLIGO 1, and no major deviation is seen in both methods.
8. For the other ODNs, substantial degradation is not seen with the integrated device, even with up to a 60-s injection of the enzyme solution, which is considered the maximum. This is in agreement with the findings from the "off-line" incubation, that OLIGO's 2, 4, and 5 are resistant against phosphodiesterase I-induced degradation.

4. Notes

1. The application of CPSE with a replaceable HEC matrix is a valuable tool in oligonucleotide analysis. Synthetic homo-oligomeric deoxynucleotides appear not to be unique molecules, but rather to be mixtures of the main compound and a number of related substances, in spite of a reversed-phase LC purification step after synthesis. The purity of the synthetic ODNs changes with nucleotide composition and with modification steps. Apparently, the incorporation of a phosphate group at the 5'-position results in more impure preparations.
2. Phosphodiesterase I-induced degradation is studied by incubation "'in-vial" on the one hand, and by "in-capillary" enzymatic reaction on the other hand. In both cases, CPSE

separates the uncut substrate from all the possible reaction products. The model compound dA_{21}, as well as the 5'-phosphorylated analogue are easily degraded. The modification of the sugar at the 3'-end by incorporation of a primary amine blocks 3'-exonucleolytic breakdown. In contrast, a 3'-deoxyadenosine instead of a 2'-deoxyadenosine only reduces the extent of degradation.

3. Replacement of two phosphodiester linkages with phosphorothioate or methylphosphonate also completely prohibits degradation. As with the ODNs containing mixtures of modified and natural linkages, exonucleases can sometimes skip over an isolated phosphonate or triester linkage to cleave the adjacent phosphodiester, but at a reduced rate *(25)*. In our experiments, two modified linkages are sufficient to block 3'-exonuclease activity.
4. The electrophoretically mediated enzymatic degradation of the homo-oligomeric ODNs can be achieved "in-capillary" and the results are consistent with findings of the "off-line" enzyme incubation and separation. This means that the classical assay sequence of: "sampling–mixing–incubating–quenching–sampling–separation–detection–quantification" can be fully automated, with all steps taking place in a single microliter reaction vessel in under 25 min. In fact, the total volume of the capillary is only 13.8 μL and injection volumes are in the nanoliter range. Moreover, in the capillary format, the EMMA procedure injects the two reactants into the same vessel and mixing of both takes place without any turbulence and is based exclusively on the differences in electrophoretic mobility. Integration of enzymatic reaction and separation in the same container reduces the number of sample manipulation steps. Possible errors and sources of contamination are eliminated and the procedure is simplified overall.

References

1. Stein, C. A. and Cheng, Y.-C. (1993) Antisense oligonucleotides as therapeutic agents—is the bullet really magical? *Science* **261,** 1004–1012.
2. Milligan, J. F., Matteucci, M. D., and Martin, J. C. (1993) Current concepts in antisense drug design. *J. Med. Chem.* **36,** 1923–1937.
3. De Mesmaeker, A., Häner, R., Martin, P., and Moser, H. E. (1995) Antisense oligonucleotides. *Acc. Chem. Res.* **28,** 366–374.
4. Stein, C. A., Subasinghe, C., Shinozuka, K., and Cohen, J. S. (1988) Physicochemical properties of phosphorothioate oligodeoxynucleotides. *Nucleic Acids Res.* **16,** 3209–3221.
5. Khan, K., Liekens, K., Van Aerschot, A., Van Schepdael, A., and Hoogmartens, J. (1997) Stability measurement of oligonucleotides in serum samples using capillary electrophoresis. *J. Chromatogr. B* **702,** 69–76.
6. Campbell, J.M., Bacon, T.A., and Wickstrom, E. (1990) Oligonucleoside phosphorothioate stability in subcellular extracts, culture media, sera and cerebrospinal fluid *J. Biochem. Biophys. Meth.* **20,** 259–267.
7. Eder, P. S., DeVine, R. J., Dagle, J. M., and Walder, J. A. (1991) Substrate specificity and kinetics of degradation of antisense oligonucleotides by a 3' exonuclease in plasma. *Antisense Res. Dev.* **1,** 141–151.
8. Akhtar, S., Kole, R., and Juliano, R. L. (1991) Stability of antisense DNA oligodeoxynucleotide analogs in cellular extracts and sera. *Life Sciences* **49,** 1793–1801.
9. Shaw, J.-P., Kent, K., Bird, J., Fishback, J., and Froehler, B. (1991) Modified deoxyoligonucleotides stable to exonuclease degradation in serum. *Nucleic Acids Res.* **19,** 747–750.
10. Guttman, A. and Ulfelder, K. J. (1998) Separation of DNA by capillary electrophoresis. *Adv. Chromatogr.* **38,** 301–340.

11. Bruin, G. J. M., Börnsen, K. O., Hüsken, D., Gassmann, E., Widmer, H. M., and Paulus, A. (1995) Stability measurements of antisense oligonucleotides by capillary gel electrophoresis. *J. Chromatogr. A* **709,** 181–195.
12. Shi, X., Hammond, R. W., and Morris, M. D. (1995) DNA conformational dynamics in polymer solutions above and below the entanglement limit. *Anal. Chem.* **67,** 1132–1138.
13. Heiger, D. N., Cohen, A. S., and Karger, B. L. (1990) Separation of DNA restriction fragments by high performance capillary electrophoresis with low and zero crosslinked polyacrylamide using continuous and pulsed electric fields. *J. Chromatogr.* **516,** 33–48.
14. Palm, A. and Hjertén, S. (1996) The resolution of DNA fragments in capillary electrophoresis in replaceable agarose gels. *J. Capillary Electrophor.* **3,** 173–179.
15. Fung, E. N. and Yeung, E. S. (1995) High speed DNA sequencing by using mixed poly(ethylene oxide) solutions in uncoated capillary columns. *Anal. Chem.* **67,** 1913–1919.
16. Cheng, J., Kasuga, T., Watson, N. D., and Mitchelson, K. R. (1995) Enhanced single-stranded DNA conformation polymorphism analysis by entangled solution capillary electrophoresis. *J. Capillary Electrophor.* **2,** 24–29.
17. Khan, K., Van Schepdael, A., and Hoogmartens, J. (1996) Capillary electrophoresis of oligonucleotides using replaceable sieving buffer with low viscosity-grade hydroxyethyl cellulose. *J. Chromatogr. A* **742,** 267–274.
18. Regnier, F. E., Patterson, D. H., and Harmon, B. J. (1995) Electrophoretically-mediated microanalysis (EMMA). *Trends Anal. Chem.* **14,** 177–181.
19. Bao, J. and Regnier, F. E. (1992) Ultramicro enzyme assays in a capillary electrophoretic system *J. Chromatogr.* **608,** 217–224.
20. Cellosize® HEC EP09 safety data sheet. (1997) Union Carbide, Antwerp, Belgium.
21. Saevels, J., Huygens, K., Van Schepdael, A., and Hoogmartens, J. (1997) In-line coupling of the enzymatic degradation of oligonucleotides with capillary polymer sieving electrophoresis. *Anal. Chem.* **69,** 3299–3303.
22. Saevels, J., Van Schepdael, A., and Hoogmartens, J. (1997) Phosphodiesterase susceptibility of modified oligonucleotides studied in an integrated capillary electrophoresis system. *J. Capillary Electrophor.* **4,** 167–172.
23. Björk, W. (1963) Purification of phosphodiesterase from *Bothrops atrox* venom, with special consideration of the elimination of monophosphatases. *J. Biol. Chem.* **238,** 2487–2490.
24. Richards, G. M., du Vair, G., and Laskowski, M. (1965) Comparison of the levels of phosphodiesterase, endonuclease, and monophosphatases in several snake venoms. *Biochemistry* **4,** 501–503.
25. Vandendriessche, F., Van Aerschot, A., Voortmans, M., Janssen, G., Busson, R., Van Overbeke, A., et al. (1993) Synthesis, enzymatic stability and base-pairing properties of oligothymidylates containing thymidine dimers with different N-substituted guanidine linkages. *J. Chem. Soc. Perkin. Trans.* **1,** 1567–1575.

33

Capillary Electrophoresis of Urinary Normal and Modified Nucleosides of Cancer Patients

Guowang Xu, Hartmut M. Liebich, Rainer Lehmann, and Sylvia Müller-Hagedorn

1. Introduction

RNA molecules contain a number of modified nucleosides, in addition to the normal ribonucleosides adenosine, guanosine, cytidine, and uridine. More than 90 different modified nucleosides with great structural diversity have been described in RNA *(1)*, most of them in tRNA, but some also in rRNA, mRNA and small nuclear RNA (snRNA). The modifications are formed posttranscriptionally within the polynucleotide molecule by numerous different modification enzymes, in particular by methyltransferases and ligases. Some of the modified nucleosides occur only in one species of RNA, whereas others occur in several species *(1,2)*. The function of the modified nucleosides of RNA is not quite clear. It has been suggested that tRNA modification plays a structural and regulatory role *(3)*.

During RNA turnover, free normal and modified nucleosides are created by the hydrolytic action of ribonucleases and phosphatases. Normal nucleosides can be either reutilized to form nucleotide triphosphates which are incorporated into nucleic acids, or further degraded to form uric acid and β-alanine *(4)*. Modified nucleosides in contrast, cannot be rephosphorylated and are generally not reincorporated during *de novo* RNA synthesis *(5,6)*. They circulate in the blood stream and, together with small amounts of normal nucleosides, they are excreted in the urine. Many of the modified nucleosides are not further metabolized and are virtually quantitatively excreted, e.g., 5,6-dihydrouridine (DhU), N^6-threonylcarbamoyladenosine (t^6A), pseudouridine (PseU), and N^1,N^2-dimethylguanosine (m^2_2G). Consequently, the levels of the modified nucleosides are a measure of RNA turnover in the organism *(2,7,8)*.

The whole-body tRNA turnover rates have been studied on the basis of various urinary modified nucleosides in animals *(2,9)* and in man *(2)*. In addition to tRNA turnover, the degradation of mRNA and rRNA have been investigated in man *(8,10)*. In general, RNA turnover rates in healthy individuals depend on several factors.

From: *Methods in Molecular Biology, Vol. 162:*
Capillary Electrophoresis of Nucleic Acids, Vol. 1: Introduction to the Capillary Electrophoresis of Nucleic Acids
Edited by: K. R. Mitchelson and J. Cheng © Humana Press Inc., Totowa, NJ

Whereas essentially no differences are found between female and male *(11–13)* nor between different ages during adulthood *(12)*, higher turnover rates are observed in children. The values decrease with increasing age of the children *(8,11)*. For some RNA metabolites, the excretion is about 6–10 times higher in infants than in adults and parallels the growth rate *(14)*. In preterm infants, the tRNA, mRNA, and rRNA turnover rates have been determined to exceed the rates in adults by factors of 3.1, 3.9, and 2.7, respectively *(8)*. The whole-body RNA turnover correlates quite well with the whole-body protein turnover *(15)*. Controversially, an effect of diet on the modified nucleosides and the RNA turnover has also been proposed *(11,16)*, as the metabolic state of protein appears to have a definite influence on the levels of modified nucleosides.

1.1. Rates of RNA Turnover

It has long been postulated that diseases may influence the rate of RNA turnover and thus be seen in the levels of excreted modified nucleosides. The most obvious alterations have been observed in malignant disease states. In their studies on tumor-bearing animals, Borek and coworkers *(9)* concluded that the elevated excretion of modified nucleosides results from increased tRNA turnover, rather than from cell death. The detailed molecular basis of the increased excretion is still unclear, but in extracts of tumor tissues aberrant tRNA methyltransferases *(17)* and enhanced tRNA methyltransferase activities *(18)* have been found, leading, in part, to abnormal tRNA in tumor tissue *(5)*. It has been suggested that the high turnover of tRNA is the result of the rapid degradation of aberrantly modified tRNA *(9,17)*.

Based on these biochemical findings, modified nucleosides have been frequently proposed, and partially evaluated, as tumor markers. A diagnostic benefit of the use of such markers has been suggested for a number of malignant diseases, including leukemia *(19–23)*, malignant lymphoma *(24,25)*, brain tumors *(21,26)*, nasopharyngeal cancer *(27)*, small cell lung cancer *(28–30)*, oesophagus cancer *(31)*, colorectal cancer *(32–35)*, mesothelioma *(36)*, hepatocellular carcinoma *(37)*, renal cell carcinoma *(38)*, and breast cancer *(39,40)*. Attempts have also been made to use modified nucleosides as diagnostic markers of disease development in patients with AIDS and individuals at risk of developing AIDS *(41,42)*.

1.2. Sample Materials and Analytical Methods

Several different types of sample material have been used for these analyses. Investigations on normal and modified nucleosides have been carried out in solid tumor tissues, e.g., in mucosa from stomach, colon and rectum obtained during surgery on patients with gastrointestinal and colorectal cancers *(34,35)*. Immunohistochemical analysis with monoclonal antibodies against pseudouridine and 1-methyladenosine have been performed with tissue sections from oesophageal cancer tissues *(31,43)*. In the majority of such studies urine is sampled, and in fewer cases serum is the sample. For the majority of the biochemical markers that occur in both blood serum and urine, measurement of serum levels is usually more reliable and of higher diagnostic value than measurement in urine, because errors may be involved during the collection and storage of the urine.

On the other hand, urinary excretion values supply a more integrated view of the metabolic situation, compared to serum values. Furthermore, because of the higher levels of modified nucleosides normally found in urine, better analytical precision can be expected. Most of the studies on modified nucleosides are based on the urinary excretions. It has been shown ***(32)*** that it is not necessary to work with urine from 24-h samples. Randomly collected urine samples can be used if the nucleoside levels are compared to creatinine levels, which is generally excreted in a quite constant manner within a period of 24 h. Additionally, sample collection is more convenient for the patients and the risk of sampling errors is reduced. Creatinine-normalized nucleoside levels in 24-h urines and randomly collected urines are found to be identical ***(32)***.

In most of the studies of the urine nucleosides, the analytes are isolated from the urine specimen by phenylboronate affinity gel chromatography and then separated by reversed-phase high-performance liquid chromatography (RP-HPLC) ***(13,44–47)***. The same procedure is also applied to serum nucleosides ***(12,23,30,37,48)***. However, because of their high protein content, serum samples are ultrafilter-purified prior to the nucleoside isolation on the phenylboronate columns. In addition to the RP-HPLC methods, radioimmunoassays ***(39)*** and enzyme immunoassays ***(16,22,40,43,49,50)*** have been developed and applied for the quantification of modified nucleosides in both urine and serum. Recently proton nuclear magnetic resonance (NMR) spectroscopy has been used to determine urinary modified nucleosides ***(10)***.

1.3. Capillary Electrophoresis Analysis

Capillary electrophoresis (CE) for the separation of standard mixtures of ribonucleosides has been introduced in 1987 ***(51)***. In the same year, another group reported the separation of deoxyribonucleosides ***(52)***. Because nucleosides are uncharged molecules, CE is applied in the mode of micellar electrokinetic capillary chromatography (MEKC). We have developed a CE method for the analysis of normal and modified nucleosides in urine, combining the isolation of the analytes by phenylboronate affinity gel chromatography, and the separation and quantification by MEKC ***(53–56)***.

2. Materials

2.1. Chemicals

1. Formic acid is purchased from Riedel-de Haen (Seelze, Germany).
2. Ammonium acetate, methanol, ammonia, sodium dihydrogenphosphate monohydrate ($NaH_2PO_4.H_2O$) are from Merck (Darmstadt, Germany).
3. Sodium tetraborate ($Na_2B_4O_7.10H_2O$) is from Sigma (Deisenhofen, Germany).
4. Sodium dodecyl sulfate (SDS) and Affi-gel 601 are obtained from Bio-Rad (Munich, Germany).
5. The nucleoside reference substances uridine (U), cytidine (C), adenosine (A), guanosine (G), inosine (I), xanthosine (X), pseudouridine (PseU), 5,6-dihydrouridine (DhU), 3-methyluridine (m3U), 5-methyluridine (m5U), N4-acetylcytidine (ac4C), N6-methyladenosine (m6A), 1-methylguanosine (m1G), 2-methylguanosine (m2G), 1-methylinosine (m1I), and the internal standard 3-deazauridine (3-DzU) are from Sigma.

2.2. Urine Samples

1. All urine samples are collected without preservative, immediately frozen after collection and stored at –20°C.

2. For the analysis of the nucleosides, the samples are thawed at room temperature.
3. Control urine samples (24-h urines) are collected from healthy individuals. In the examples shown here from 24 individuals (11 males and 13 females). Patient urine samples (24-h urines) are collected. In this example, from 25 patients with different kinds of cancer (16 males and 9 females, randomly collected urines) and 17 patients with breast cancer (randomly collected urines). The general cancer group was comprised of unselected patients in various stages of the disease and under different therapies. The breast cancer group included women one day before surgery.

3. Methods

3.1. Isolation of Nucleosides from Urine

1. The nucleosides are extracted from urine by phenylboronate affinity gel chromatography applying a modification of the method by Gehrke et al. *(45)* and Kuo et al. *(57)*.
2. The extraction is performed on Affi-gel 601 glass columns (140 × 14 mm) containing 500 mg of phenylboronate gel, which possesses a specific affinity for *cis*-diol structures.
3. After equilibration of the column with 35 mL of 0.25 *M* ammonium acetate, pH 8.5, 10 mL of centrifuged urine mixed with 0.3 mL of a 0.8 *M* aqueous internal standard (3-DzU) solution are loaded onto the column.
4. After loading the sample, the Affi-gel is washed with 20 mL of 0.25 *M* ammonium acetate and is then washed twice with 3 mL of 1:1 (v/v) methanol : water.
5. The nucleosides are eluted with 25 mL of 0.1 *M* formic acid in 1:1 (v/v) methanol:water.
6. The eluate is evaporated to dryness under vacuum at 40°C and is then redissolved in 1 mL of water to produce an extract, concentrated by a factor of 10 as compared to the original urine.
7. The same purification treatment is applied to the standard solutions that contain the 15 reference nucleosides and the internal standard.
8. Regeneration of the Affi-gel is carried out with 45 mL of 0.25 *M* ammonium acetate, pH 8.5. Under the described conditions the gel could be used for fifteen extractions.

3.2. Capillary Electrophoresis

1. All separations are performed on a Dionex (Idstein, Germany) CE system (CES I, with an automatic constant-volume sample injection system) in uncoated capillaries of 50 μm id × 50 cm, or 56.5 cm, with a distance of 65 mm between detection window and outlet of the column.
2. The separations are carried out in a buffer consisting of 300 m*M* SDS, 25 m*M* sodium tetraborate, 50 m*M* sodium dihydrogenphosphate buffer, adjusted to pH 6.7 with 5% hydrochloric acid. The buffer is filtered through a 0.45-μm membrane filter and is degassed in an ultrasonic bath for 10 min before use.
3. Other buffers used varied in the concentration of SDS: (a) 100 m*M* SDS, 25 m*M* sodium tetraborate, 50 m*M* sodium dihydrogenphosphate, pH 6.7, (b) 200 m*M* SDS, 25 m*M* sodium borate, 50 m*M* sodium dihydrogenphosphate, pH 6.7 (*see* **Note 1** and **Fig. 1**).
4. The samples are introduced by gravity injection at 100-mm head height for 45 s (*see* **Note 2**).
5. UV detection is performed at 260 nm and 210 nm in that sequence (*see* **Note 3**).
6. The applied voltage and current are 7.0 kV and 47–49 μA for the 500-mm capillary and 7.5 kV and 41 μA for the 565-mm capillary, respectively (*see* **Note 4**).
7. The voltage suitable for the separation is often determined on the basis of a "critical pair of substances," that is a pair of substances which is difficult to separate. Considering the

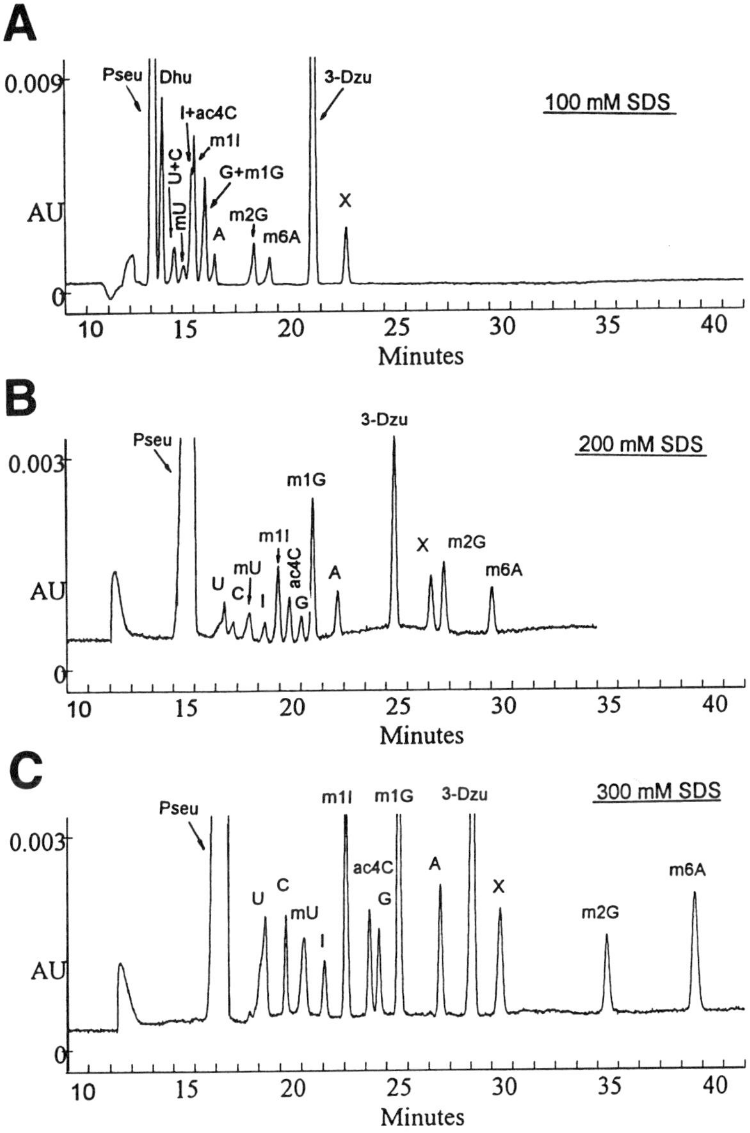

Fig. 1. Effect of SDS concentration on the separation of nucleosides in a standard solution. Capillary: 50 μm id X 50.0 cm (43.5 cm to the detection window); applied voltage: 7.0 kV; current: 47 μA; buffers and detection wavelengths: **(A)** 100 m*M* SDS, 25 m*M* sodium tetraborate, 50 m*M* sodium dihydrogenphosphate, pH 6.7, 210 nm; **(B)** 200 m*M* SDS, 25 m*M* sodium borate, 50 m*M* sodium dihydrogenphosphate, pH 6.7, 260 nm; **(C)** 300 m*M* SDS, 25 m*M* sodium borate, 50 m*M* sodium dihydrogenphosphate, pH 6.7, 260 nm; injection head height: 100 mm; load time: 45 s; peak identifications: PseU = pseudouridine, DhU = 5,6-dihydrouridine, U = uridine, C = cytidine, mU = 3-methyluridine+5-methyluridine, I = inosine, m1I = 1-methylinosine, ac4C = N4-acetylcytidine, G = guanosine, m1G = 1-methylguanosine, A = adenosine, 3-DzU = 3-deazauridine (internal standard), X = xanthosine, m2G = 2-methylguanosine, m6A = N6-methyladenosine.

pair ac4C/G, the optimal voltage is 7.0 kV when using the 50.0 cm column, and is 7.5 kV when using the 56.5-cm column. The currents for these electrophoretic separations are 47–49 μA and 41 μA, respectively.

8. Shortening the length of the capillary from 56.5 to 50.0 cm reduces the time taken for analysis by about 10%.
9. After completion of each electrophoresis run, the capillary is rinsed for 100 s with water, followed by 100 s with 0.1 *M* sodium hydroxide, 100 s with water, and finally 120 s with running buffer.

3.3. Identification and Quantification of Nucleosides

1. The peaks in the electropherograms are identified by comparing the migration times of the unknown components with those of the standard nucleosides eluted under the same conditions, and by spiking the urine sample with standard solution.
2. For the calibration procedure, three different volumes of an aqueous, standard stock solution containing the 15 nucleosides, to each of which was added 0.3 mL of the internal standard solution, are treated separately on the phenylboronate affinity gel.
3. Calibration curves are established at 260 nm and 210 nm for each of the 15 nucleosides with these three standard samples.
4. The nucleoside concentrations in the standard stock solution were 1280 μ*M* for PseU, 320 μ*M* for DhU, 64 μ*M* for m1G and m1I, 32 μ*M* for U, ac4C and m2G, 16 μ*M* for C, A, G, I, m3U and m6A, 8 μ*M* for m5U, and 3.2 μ*M* for X. The concentrations of the nucleosides in urine could be calculated based on the derived calibration curves.
5. The concentrations are then normalized by transforming the amounts of nucleoside into nmol/μmol of urinary creatinine. Urinary creatinine levels are determined by colorimetry using the conventional standard reaction between creatinine and picric acid ***(58)***.

3.4. Results

3.4.1. Urinary Nucleoside Excretion in Healthy Adults and in Cancer Patients

1. **Figure 2** shows the typical electropherograms of the nucleoside standard mixture.
2. **Figure 3** of the normal and modified nucleosides in urine of a healthy individual. The electropherograms were run under the described optimized analytical conditions.
3. **Table 1** shows the precision of the method when applied to the analysis of urine samples. Nine 10-mL aliquots are obtained from a randomly collected urine sample and extracted and analyzed under identical conditions. The coefficients of variation are determined to be between 5.6 and 16.8%, except for m6A with a coefficient of variation of 33%. The migration times have very high reproducibility. For the standard samples, the coefficients of variation are less than 1% for migration times, and less than 0.5% for migration times relative to the internal standard. The coefficients of variation of the peak areas as determined by analysis of nucleoside standard solutions are between 2–6%. The reproducibility of peak areas relative to the area of the internal standard is similar.
4. **Table 2** summarizes the reference values for the urinary excretion of normal and modified nucleosides in healthy individuals determined with the CE method. They are in good agreement with the reference values which are obtained with the RP-HPLC procedure ***(13)***.
5. Because the values found in this study and the values reported by other investigators ***(5,7,8,20,22,27,28,32,33)*** show very little variance, it is concluded that the excretion of nucleosides in urine is very constant in healthy, adult persons.

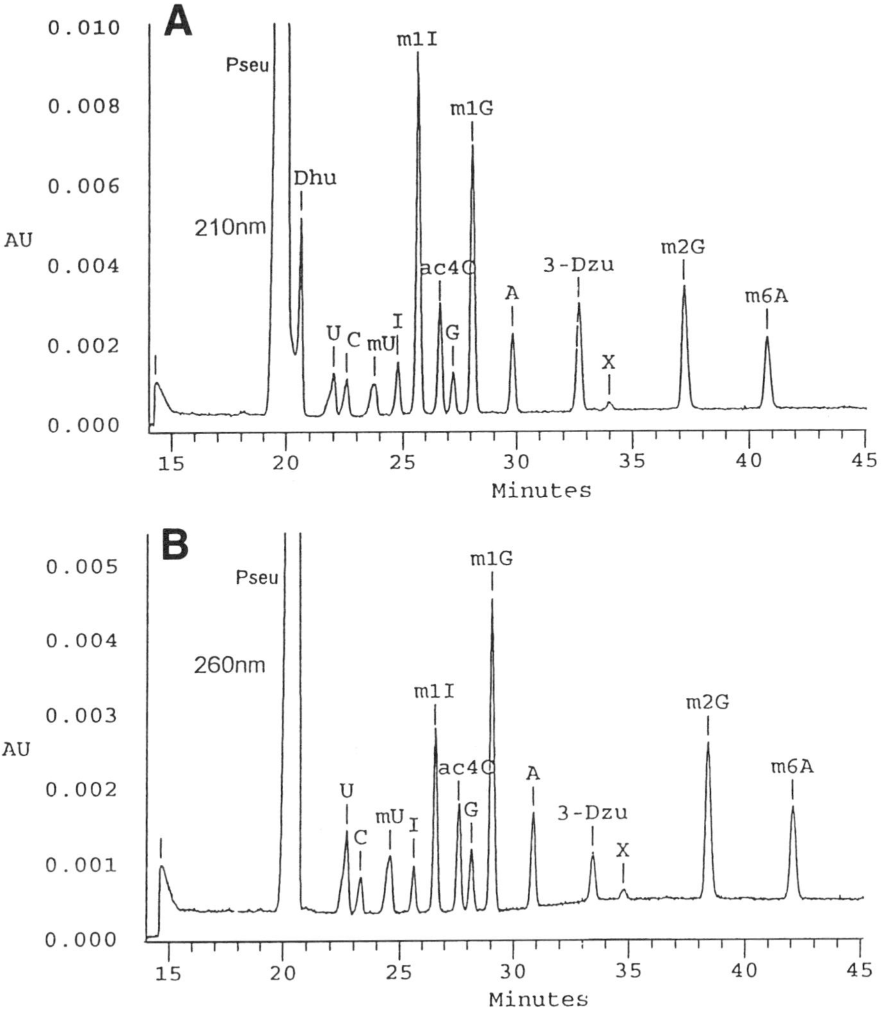

Fig. 2. Electropherograms obtained from an aqueous solution of standard nucleosides. Capillary: 50 μm id × 56.5 cm (50.0 cm to detection window); applied voltage: 7.5 kV; current: 41 μA; buffer: 300 m*M* SDS, 25 m*M* sodium borate, 50 m*M* sodium dihydrogenphosphate, pH 6.7; UV detection at: **(A)** 210 nm, **(B)** 260 nm; peak identifications are the same as described in **Fig. 1**.

6. It is notable that within the patients with different kinds of cancer, the levels of total nucleosides in urine are generally elevated, and further that the increase of modified nucleosides is more pronounced than that of normal nucleosides (**Fig. 4**). When a significant elevation of nucleosides is defined as a value higher than the average plus two standard deviations determined in normal urine, the concentrations of PseU, m1I, ac4C, m1G, and m2G in urine of patients with cancer are elevated significantly.

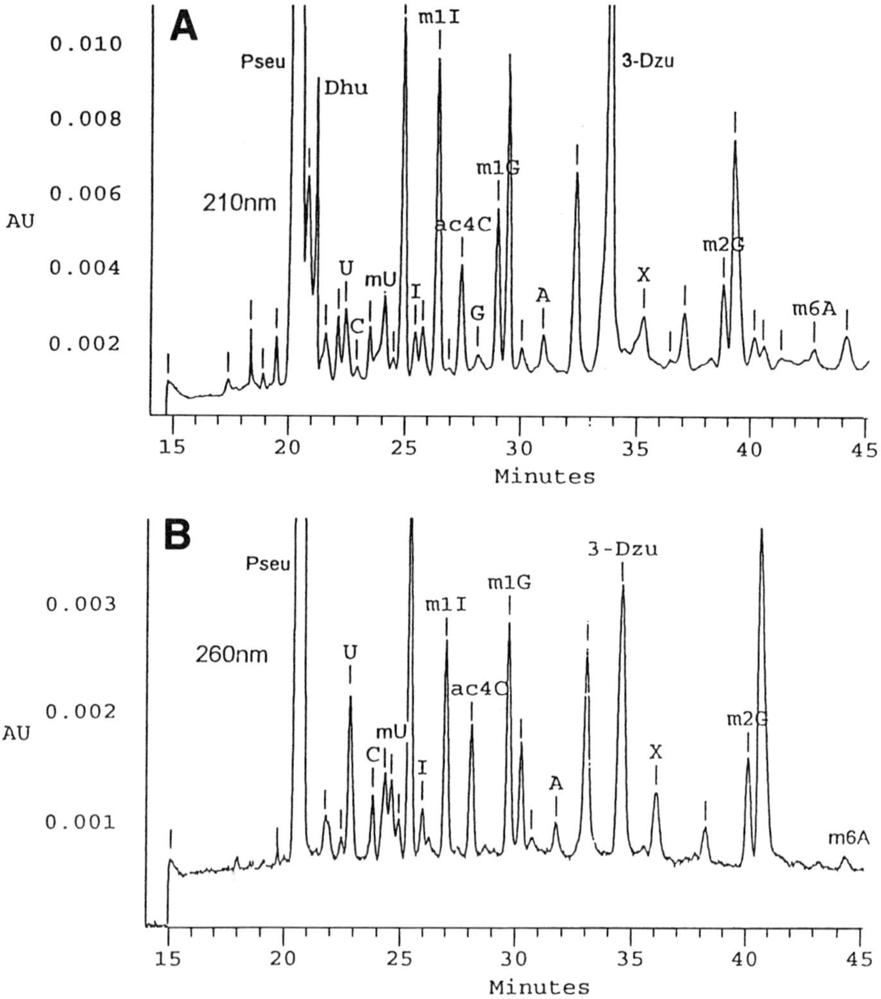

Fig. 3. Electropherograms obtained from urinary nucleosides in a healthy individual. **(A)** Detection at 210 nm; **(B)** detection at 260 nm. All other conditions are as described in **Fig. 2**.

7. These five modified nucleosides include PseU, which has been reported in the literature many times as being increased in urine of cancer patients, e.g., in patients with leukemia ***(19–22)***, malignant lymphoma ***(24,25)***, nasopharyngeal cancer ***(27)***, small cell lung cancer ***(28,29)***, oesophagus cancer ***(31)***, colorectal cancer ***(32,33)***, and breast cancer ***(40)***. In some of these reports, an increase of other modified nucleosides has also been described in addition to the elevation of PseU, including m1I ***(19,20,27,28,32,33,40)***, ac4C ***(32,40)***, m1G ***(19,20,27,28)***, and m2G ***(28)***.
8. Many studies on modified nucleosides in cancer patients demonstrate that in malignant diseases always several of the nucleosides are elevated, not just one nucleoside. In such a

Table 1
Precision of the Quantitative Determination of Normal and Modified Nucleosides in Urine by CE

Component	Concentration (nmol/μmol creatinine)	Standard deviation (nmol/μmol creatinine)	Coefficient of variation (%)
PseU	17.21	0.960	5.59
U	0.41	0.032	7.77
C	0.10	0.017	16.80
I	0.25	0.017	6.76
m1I	0.98	0.062	6.35
ac4C	0.58	0.040	7.01
G	0.05	0.006	11.81
m1G	0.64	0.036	5.61
A	0.38	0.022	5.80
X	0.25	0.038	14.96
m2G	0.26	0.020	7.66
m6A	0.04	0.013	33.10

Table 2
Mean Levels of Normal and Modified Nucleosides in Urine of Healthy Individuals ($n = 24$)

Component	Concentration (nmol/μmol creatinine)	Standard deviation (nmol/μmol creatinine)
PseU	25.32	10.320
DhU	4.25	1.105
U	0.47	0.190
C	0.07	0.095
mU	0.09	0.123
I	0.14	0.100
m1I	1.27	0.475
ac4C	0.60	0.384
G	0.01	0.021
m1G	0.82	0.298
A	0.18	0.172
X	0.45	0.260
m2G	0.39	0.197
m6A	0.01	0.023

multi-component alteration of the nucleoside levels, application of pattern recognition methods can reveal more information on differences between healthy individuals and cancer patients than the evaluation of single components (*see* **Note 5**).

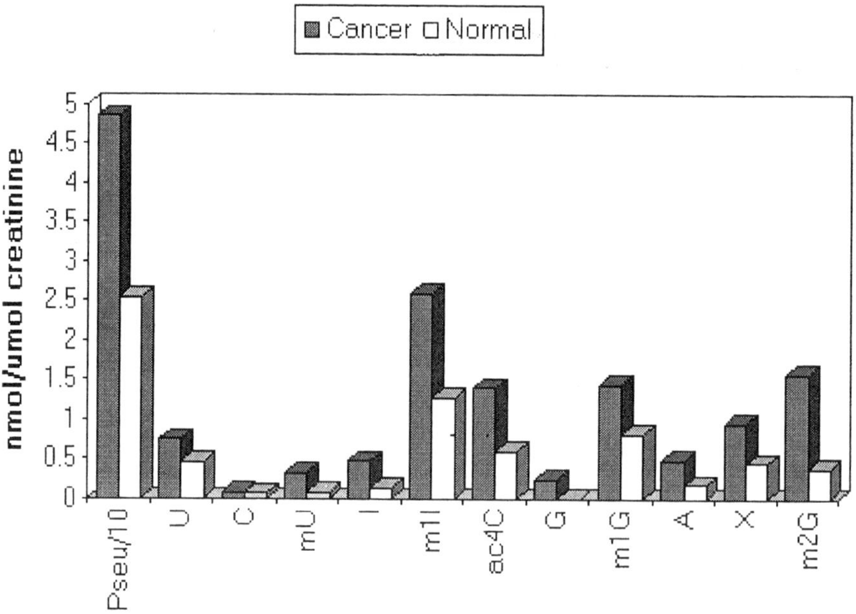

Fig. 4. Mean levels (nmol/μmol creatinine) of normal and modified nucleosides in urine of healthy individuals ($n = 24$) and patients with cancer ($n = 25$).

3.4.2. Analytical Characteristics

1. The calibration curves are characterized by a linear correlation between the nucleoside concentrations and the peak areas. The correlation coefficients are better than 0.99 for most of the nucleosides, when determined for standard nucleoside concentrations comparable to those of the urinary nucleosides. The correlation coefficients were 0.96 and 0.98 only for X and m6A, respectively.
2. The limits of detection, defined as the sample concentration that produces a peak with a height three times the level of the baseline noise, varied between 1.7 μ*M* and 9.2 μ*M* for the different nucleosides and were similar for detection at 260 nm and 210 nm. The limit of detection is higher by a factor of 10 only for PseU. These low micromolar detection limits are suitable for the analysis of the nucleosides in urine.

3.4.3. Comparison of the CE Method with the RP-HPLC Method

1. Most of the methods for analyses of nucleosides reported in the literature employ RP-HPLC for separation and quantification. The efficiencies of separation of the nucleosides by CE and RP-HPLC are comparable. The same holds true for the precision of the quantitative determination ***(13)***.
2. For certain nucleosides, differences in the separation quality are observed between the two methods. Ac4C and m2G are much better separated by CE than by RP-HPLC.
3. In contrast, 1-methyladenosine (m1A), that is proposedly one of the tumor markers among the modified nucleosides ***(22,40)***, which can be easily quantified by RP-HPLC could not be analyzed by CE under the analytical conditions applied in this study.

4. Because of the short UV-light pathlength available for on-line detection in CE, and the use of 10–15 nL of injection volume, the limits of detection are higher by one order of magnitude or more using the CE method, than the RP-HPLC method. However, for the analysis of nucleosides in urine the sensitivity of CE is sufficient.
5. Determination of nucleosides in serum would require the higher sensitivity of RP-HPLC be used, or necessitate an additional concentration step during the sample preparation steps when applying CE.
6. The CE method is a complementary technique to RP-HPLC and offers several advantages over RP-HPLC in the analysis of urinary nucleosides. CE uses uncoated capillary columns, which are cheaper and require less maintenance than RP-HPLC columns.
7. The CE capillaries usually have a longer lifetime than RP-HPLC columns. For the separations in this study the column, could be used more than 500 times.
8. Whereas RP-HPLC consumes considerable amounts of solvents as the mobile carrier, CE uses the electrical field for separation and does not consume solvents.
9. A further reduction in the costs of consumables results from a better degree of miniaturization of CE as compared to RP-HPLC. The CE method for the determination of normal and modified nucleosides is suitable for the analysis of large series of urine samples and is applicable in the clinical laboratory.

3.4.4. Relevance of Modified Nucleosides as Tumor Markers

1. Because of the increased levels of nucleosides in urine of cancer patients, the nucleosides have often been proposed as tumor markers. Even though a large number of tumor markers is available today in the clinical chemical laboratory, there is a continuing interest in additional markers with diagnostic relevance.
2. There are six areas of clinical questions and problems in which tumor markers may supply information: (a) therapy control, (b) prognosis, (c) differential diagnosis of tumor diseases, (d) screening for tumor diseases, (e) initial diagnosis of tumor diseases, and (f) tumor localization.
3. Presently, no ideal tumor markers exist which are of value for all six problems. Most of the tumor markers are informative and reliable in therapy control and prognosis, and only some are helpful in the other four areas of problems, e.g., prostate specific antigen (PSA) in tumor localization.
4. Even though detailed studies on the application of modified nucleosides as tumor markers have not yet been performed, many data suggest that the markers are useful for therapy control and as guides for prognosis (*see* **Note 6**).
5. Successful or unsuccessful therapy, remission or relapse are also reflected in the changes of the levels of the modified nucleosides ***(19,20,22,28,31)***, and highly elevated levels of the nucleosides are indicators for a bad prognosis ***(40)***.
6. It cannot be answered conclusively what additional relevance the modified nucleosides may have for specific diagnosis, because they are elevated in many kinds of cancer they can be considered as general markers for malignant diseases, and indicate the increased RNA turnover in these diseases. They have little tumor or organ specificity or cell autonomy, and therefore they certainly cannot be used for tumor localization (*see* **Note 6**).
7. Too little is known about the diagnostic sensitivity and specificity of the nucleosides. In the study on the women with breast cancer the modified nucleosides had a higher diagnostic sensitivity than carcinoembryonic antigen (CEA) and cancer antigen 15–3 (CA 15–3), which are conventionally used as tumor markers for breast cancer.
8. Using factor analysis, all 17 patients are classified correctly on the basis of the modified nucleosides, whereas only 3 patients had abnormal CEA and CA 15–3 levels in serum.

This is of particular interest, because at present no satisfactory tumor marker is available for breast cancer. Also cancer antigen 549 (CA 549) and mucin-like-cancer-associated antigen (MCA), which have been suggested as markers for breast cancer, have low diagnostic sensitivity. Not enough investigations have been carried out concerning the diagnostic specificity.

4. Notes

1. The composition of the buffer, e.g., components, ionic strength, pH value, amount of SDS, each influences the separation characteristics. At pH values near to neutral, the nucleoside molecules are uncharged and CE has to be performed in the MEKC mode. Micelles are formed by the addition of SDS to the buffer. The concentration of SDS in the buffer has a strong effect on the migration order, the resolution and the analysis time (**Fig. 1**). Pyrimidine nucleosides migrate in front of purine nucleosides, and normal nucleosides in front of their corresponding methylated analogs. Insufficient separation of nucleosides is obtained with 100 m*M* SDS. Although the nucleosides in the standard solution could be separated using a buffer with a concentration of 200 m*M* SDS, the optimal concentration for full separation of all components is 300 m*M* SDS. It is notable that m3U and m5U could not be separated at any of the SDS concentrations tested. The higher SDS concentration of 300 m*M* results in a longer analysis time, about 45 min for the 56.5-cm capillary and about 40 min for the 50.0-cm capillary.
2. The time applied in gravity injection defines the volume of sample that is introduced into the capillary. The increase in peak area is linear with injection time. However, when injection time is too long, the separation efficiency is reduced. At 100-mm head height, no obvious loss in resolution is observed using an injection time of 60 s, or less. A loading time of more than 100 s leads to a dramatic reduction in the peak resolution. A 45-s load time (sample volume 10–15 nL) was found to be optimal.
3. Nucleosides are usually monitored by UV detection at a wavelength between 254 nm and 280 nm. In this work, we chose 260 nm. However, it should be noted that DhU does not absorb to any significant degree at wavelengths above 240 nm and has its absorption maximum around 210 nm *(2)*. DhU should be included in the profile analysis of normal and modified nucleosides, because it is considered to be a good marker for the tRNA turnover *(2)*. Therefore, detection at both 260 nm and 210 nm are used, in sequence.
4. The quality of the electrophoretic separation and the time required for this separation depend on a number of analytical parameters, in particular on the voltage applied, the length of the column, the composition of the buffer, and the sample size introduced. Voltage control in CE offers an important parameter affecting analyte migration time and resolution efficiency. As voltage is increased, the migration velocity increases, and therefore the migration time decreases. Consequently, analysis time is shortened by increasing the voltage. Furthermore, the resolution increases as a function of the voltage up to a certain point under a given set of other analytical conditions. Above this point, the resolution efficiency may diminish.
5. The artificial neural network (ANN) method is applied for evaluation of the nucleoside levels in the group of patients with different kinds of cancer. Dividing the data from both healthy individuals and cancer patients in a training set and a predicting set, the ANN method results in a good classification of the healthy individuals, and the cancer patients of the predicting set, in two clusters. The recognition rate is 85% ***(56)***. The modified nucleoside levels are also elevated in the urine of the patients with breast cancer. In applying the factor analysis method for classification, a clear cut differentiation of the

breast cancer group and the healthy individuals in two clusters is obtained, without overlapping data *(59)*.

6. In order to use the modified nucleosides as tumor markers, it has to be clarified how their levels are affected by benign diseases occuring in the same organ, or in another adjacent organ, or by other hormonal or humoral influences. For instance, PseU which is increased in small-cell lung cancer, whereas the urinary excretion of PseU is at normal levels in several pulmonary infectious diseases *(29)*. On the other hand, modified nucleosides including PseU are elevated in patients who have undergone major surgery because of various nonmalignant diseases *(10)*. The elevation originates from an increased whole-body RNA turnover in conjunction with surgical stress and with continued catabolism. Because the whole-body RNA turnover correlates quite well with the protein turnover *(15)*, attention has to be paid to alterations in the levels of the modified nucleosides under various catabolic conditions other than that occurring during malignancies, e.g., during malnutrition, cardiac cachexia, endocrine abnormalities, alcoholism, infections, and stress. The CE method for the quantification of modified nucleosides described in this chapter is suitable for efficiently carrying out such wide-ranging studies.

Acknowledgments

This study has been supported in part by the Natural Science Foundation of China (No. 29775024) and the special foundation of the President of the Chinese Academy of Sciences. Dr. Guowang Xu gratefully acknowledges the fellowship from the Max-Planck-Institut of Germany. We are very grateful to Dr. Reichmann of the Radiologische Klinik Tuebingen and Dr. Zwirner of the Frauenklinik Tuebingen, Germany, for the donation of urine samples from cancer patients.

References

1. Limbach, P. A., Crain, P. F., and McClosky, J. A. (1994) Summary: the modified nucleosides of RNA. *Nucleic Acids Res.* **22,** 2183–2196.
2. Topp, H., Duden, R., and Schöch, G. (1993) 5,6-Dihydrouridine: a marker ribonucleoside for determining whole body degradation rates of transfer RNA in man and rats. *Clin. Chim. Acta* **218,** 73–82.
3. Björk, G. R., Ericson, J. U., Gustafsson, C. E. D., Hagervall, T. G., Jönsson, Y. H., and Wikström, P. M. (1987) Transfer RNA modification. *Ann. Rev. Biochem.* **56,** 263–287.
4. Simmonds, H. A., Fairbank, L. D., Morris, G. S., Webster, D. R., and Harley, E. H. (1988) Altered erythrocyte nucleotide patterns are characteristic of inherited disorders of purine or pyrimidine metabolism. *Clin. Chim. Acta* **171,** 197–210.
5. Borek, E., Sharma, O. K., and Waalkes, T. P. (1983) New applications of urinary nucleoside markers. *Recent Res. Cancer Res.* **84,** 301–316.
6. Borek, E. (1985) The morass of tumor markers. *Trends Biochem. Sci.* **10,** 182–184.
7. Nakano, K., Nakao, T., Schram, K. H., Hammargren, W. M., McClure, T. D., Katz, M., and Petersen, E. (1993) Urinary excretion of modified nucleosides as biological marker of RNA turnover in patients with cancer and AIDS. *Clin. Chim. Acta* **218,** 169–183.
8. Sander, G., Topp, H., Heller-Schöch, G., Wieland, J., and Schöch, G. (1986) Ribonucleic acid turnover in man: RNA catabolites in urine as measure for the metabolism of each of the three major species of RNA. *Clin. Sci.* **71,** 367–374.
9. Borek, E., Baliga, B. S., Gehrke, C. W., Kuo, C. W., Belman, S., Troll, W., and Waalkes, T. P. (1977) High turnover rate of transfer RNA in tumor tissue. *Cancer Res.* **37,** 3362–3366.

10. Marway, J. S., Anderson, G. J., Miell, J. P. Ross, R., Grimble, G. K., Bonner, A. B., Gibbons, W. A., Peters, T. J., and Preedy, V. R. (1996) Application of proton NMR spectroscopy to measurement of whole-body RNA degradation rates: effects of surgical stress in human patients. *Clin. Chim. Acta* **252,** 123–135.
11. Prankel, B. H., Clemens, P. C., and Burmester, J. G. (1995) Urinary excretion of nucleosides varies with age and protein metabolism. *Clin. Chim. Acta* **234,** 181–183.
12. Mitchell, E. P., Evans, L., Schultz, P., Madsen, R., and Yarbo J. W. (1992) Modified nucleosides in human serum. *J. Chromatogr. A* **581,** 31–40.
13. Liebich, H. M., Di Stefano, C., Wixforth, A., and Schmid H. R. (1997) Quantitation of urinary nucleosides by high-performance liquid chromatography. *J. Chromatogr. A* **763,** 193–197.
14. Schöch, G., Lorenz, H., Heller-Schöch, G., Baisch, H., and Clemens, P. (1981) Die Altersabhängigkeit der normalen und modifizierten Nucleobasen im Urin als Ausdruck der Wachstumsgeschwindigkeit. *Monatsschr. Kinderheilk.* **129,** 29–33.
15. Sander, G., Hülsemann, J., Topp, H., Heller-Schöch, G., and Schöch, G. (1986) Protein and RNA turnover in preterm infants and adults: a comparison based on urinary excretion of 3-methylhistidin and of modified one-way RNA catabolites. *Ann. Nutr. Metab.* **30,** 137–142.
16. Itoh, K., Aida, S., Ishiwata, S., Sasaki, S., Ishida, N., and Mizugaki, M. (1993) Urinary excretion patterns of modified nucleosides, pseudouridine and 1-methyladenosine, in healthy individuals. *Clin. Chim. Acta* **217,** 221–223.
17. Borek, E. and Kerr, S. J. (1972) Atypical transfer RNA's and their origin in neoplastic cells. *Adv. Cancer Res.* **15,** 163–190.
18. Borek, E. (1971) Introduction. *Cancer Res.* **31,** 596–597.
19. Heldman, D. A., Grever, M. R., and Trewyn, R. W. (1983) Differential excretion of modified nucleosides in adult acute leukemia. *Blood* **61,** 291–296.
20. Heldman, D. A., Grever, M. R., Speicher, C. E., and Trewyn, R. W. (1983) Urinary excretion of modified nucleosides in chronic myelogenous leukemia. *J. Lab. Clin. Med.* **101,** 783–792.
21. Graf, N., Bach, K., Frisch, B., Haas, H. J., and Sitzmann, F. C. (1989) Die klinische Bedeutung der Pseudouridinbestimmung im Urin von Kindern und Jugendlichen. *Klin. Pädiatr.* **201,** 154–162.
22. Itoh, K., Konno, T., Sasaki, T., Ishiwata, S., Ishida, N., and Misugaki, M. (1992) Relationship of urinary pseudouridine and 1-methyladenosine to activity of leukemia and lymphoma. *Clin. Chim. Acta* **206,** 181–189.
23. Pane, F., Savoia, M., Fortunato, G., Camera, A., Rotoli, B., Salvatore, F., and Sacchetti, L. (1993) Serum Pseudouridine in the diagnosis of acute leukaemias and as a novel prognostic indicator in acute lymphoblastic leukaemia. *Clin. Biochem.* **26,** 513–520.
24. Rasmuson, T. and Björk, G.R. (1995) Urinary excretion of pseudouridine and prognosis of patients with malignant lymphoma. *Acta Oncol.* **34,** 61–67.
25. Motyl, T., Traczyk, Z., Holska, W., Daniewska-Michalska, D., Ciesluk, S., Kukulska, W., Kaluzny, Z., and Podgurniak, M. (1993) Comparison of urinary neopterin and pseudouridine in patients with malignant proliferative diseases. *Eur. J. Clin. Chem. Clin. Biochem.* **31,** 205–209.
26. Marvel, C. C., Del Rowe, J., Bremer, E. G., and Moskal, J. R. (1994) Altered RNA turnover in carcinogenesis. The diagnostic potential of modified base excretion. *Mol. Chem. Neuropathol.* **21,** 353–368.
27. Trewyn, R. W., Glaser, R., Kelly, D. R., Jackson, D. G., Graham, W. P., and Speicher, C. E. (1982) Elevated nucleoside excretion by patients with nasopharyngeal carcinoma. *Cancer* **49,** 2513–2517.

28. Waalkes, T. P., Abeloff, M. D., Ettinger, D. S., Woo, K. B., Gehrke, C. W., Kuo, K., and Borek, E. (1982) Modified ribonucleosides as biological markers for patients with small cell carcinoma of the lung. *Eur. J. Cancer Clin. Oncol.* **18,** 1275–1283.
29. Tamura, S., Fujii, J., Nakano, T., Hada, T., and Higashino, K. (1986) Urinary pseudouridine as a tumor marker in patients with small cell lung cancer. *Clin. Chim. Acta* **154,** 125–132.
30. McEntire, J. E., Kuo, K. C., Smith, M. E., Stalling, D. L., Richens, J. W., Zumwalt, R. W., Gehrke, C. W., and Papermaster, B. W. (1989) Classification of lung cancer patients and controls by chromatography of modified nucleosides in serum. *Cancer Res.* **49,** 1057–1062.
31. Masuda, M., Nishihira, T., Itoh, K., Mizugaki, M., Ishida, N., and Mori, S. (1993) An immunohistochemical analysis for cancer of the esophagus using monoclonal antibodies specific for modified nucleosides. *Cancer* **72,** 3571–3578.
32. Gehrke, C. W., Kuo, K. C., Waalkes, T. P., and Borek, E. (1979) Patterns of urinary excretion of modified nucleosides. *Cancer Res.* **39,** 1150–1153.
33. Holstege, A., Pauw, M., Häring, R., Kirchner, R., Pausch, J., and Gerok, W. (1986) Die Wertigkeit einer erhöhten Urinausscheidung modifizierter Nucleoside als Tumormarker beim Kolonkarzinom. *Verh. dtsch. Ges. Inn. Med.* **92,** 114–120.
34. Nakano, K., Shindo, K., Yasaka, T., and Yamamoto, H. (1985) Reversed-phase liquid chromatographic investigation of nucleosides and bases in mucosa and modified nucleosides in urines from patients with gastrointestinal cancer. *J. Chromatogr.* **332,** 127–137.
35. Nakano, K., Shindo, K., Yasaka, T., and Yamamoto, H. (1985) Reversed-phase high- performance liquid chromatographic investigation of mucosal nucleosides and bases and urinary modified nucleosides of gastrointestinal cancer patients. *J. Chromatogr.* **343,** 21–33.
36. Fischbein, A., Sharma, O. K., Selikoff, I. J., and Borek, E. (1983) Urinary excretion of modified nucleosides in patients with malignant mesothelioma. *Cancer Res.* **43,** 2971–2974.
37. Amuro, Y., Nakaoka, H., Shimomura, S., Fujikura, M., Yamamoto, T., Tamura, S., Hada, T., and Higashino, K. (1988) Serum pseudouridine as a biochemical marker in patients with hepatocellular carcinoma. *Clin. Chim. Acta* **178,** 151–158.
38. Rasmuson, T., Björk, G. R., Hietala, S. O., Stenling, R., and Ljunberg, B. (1991). Excretion of pseudouridine as an independent prognostic factor in renal cell carcinoma. *Acta Oncol.* **30,** 11–15.
39. Vold, B. S., Kraus, L. E., Rimer, V. G., and Coombes, R. C. (1986) Use of a monoclonal antibody to detect elevated levels of a modified nucleoside, N-[9-(ß-D- ribofuranosyl) purin-6-yl-carbamoyl]-L-threonine, in the urine of breast cancer patients. *Cancer Res.* **46,** 3164–3167.
40. Sasco, A. J., Rey, F., Reynaud, C., Bobin, J.-Y., Clavel, M., and Niveleau, A. (1996) Breast cancer prognostic significance of some modified urinary nucleosides. *Cancer Lett.* **108,** 157–162.
41. Fischbein, A., Sharma, O.K., Valcuikas, J. A., Bekesi, J. G., Buschman, F., Apell, G., Kohn, M. A., Boesch, R. R., Teirstein, A., Selikoff, I. J., and Borek, E. (1985) Urinary excretion of modified nucleosides in patients with acquired immune deficiency syndrome and individuals at high risk of developing these diseases. *Cancer Detect. Prev.* **8,** 271–277.
42. Borek, E., Sharma, O. K., Buschman, F. L., Cohn D. L., Penley, K. A., Judson, F. N., Dobozin, B. S., Horsburgh, C. R., and Kirkpatrick, C. H. (1986) Altered excretion of modified nucleosides and ß-aminoisobutyric acid in subjects with acquired immunodeficiency syndrome or at risk for acquired immunodeficiency syndrome. *Cancer Res.* **46,** 2557–2561.
43. Itoh, K., Ishiwata, S., Ishida, N., and Mizugaki, M. (1992) Diagnostic use of anti-modified nucleoside monoclonal antibody. *Tohoku J. Exp. Med.* **168,** 329–331.

44. Uziel, M., Smith, L. H., and Taylor, S. A. (1976) Modified nucleosides in urine: selective removal and analysis. *Clin. Chem.* **22,** 1451–1455.
45. Gehrke, C. W., Kuo, K. C., Davis, G. E., Suits, R. D., Waalkes, T. P., and Borek, E. (1978) Quantitative high-performance liquid chromatography of nucleosides in biological materials. *J. Chromatogr.* **150,** 455–476.
46. Gehrke, C. W., Kuo, K. C., and Zumwalt, R. W. (1980) Chromatography of nucleosides. *J. Chromatogr.* **188,** 129–147.
47. Gehrke, C. W. and Kuo, K. C. (1989) Ribonucleoside analysis by reversed-phase high-performance liquid chromatography. *J. Chromatogr.* **471,** 3–36.
48. Pane, F., Oriani, G., Kuo, K. C. T., Gehrke, C. W., Salvatore, F., and Sacchetti, L. (1992) Reference intervals for eight modified nucleosides in serum in a healthy population from Italy and the United States. *Clin. Chem.* **38,** 671–677.
49. Reynaud, C., Bruno, C., Boullanger, P. Grange, J., Babesti, S., and Niveleau, A. (1991) Monitoring of urinary excretion of modified nucleosides in cancer patients using a set of six monoclonal antibodies. *Cancer Lett.* **61,** 255–262.
50. Ishiwata, S., Itoh, K., Yamaguchi, T., Ishida, N., and Mizugaki, M. (1995) Comparison of serum and urinary levels of modified nucleoside, 1-methyladenosine, in cancer patients using a monoclonal antibody-based inhibition ELISA. *Tohuku J. Exp. Med.* **176,** 61–68.
51. Cohen, A. S., Terabe, S., Smith, J. A., and Karger, B. L. (1987) High-performance capillary electrophoretic separation of bases, nucleosides, and oligonucleotides: retention manipulation via micellar solutions and metal additives. *Anal. Chem.* **59,** 1021–1027.
52. Row, K. H., Griest, W. H., and Maskarinec, M. P. (1987). Separation of modified nucleic acid constituents by micellar electrokinetic capillary chromatography. *J. Chromatogr.* **409,** 193–203.
53. Liebich, H. M., Xu, G., Di Stefano, C., Lehmann, R., Häring, H. U., Lu, P., and Zhang, Y. (1997) Analysis of normal and modified nucleosides in urine by capillary electrophoresis. *Chromatographia* **45,** 396–401.
54. Liebich, H. M., Lehmann, R., Di Stefano, C., Xu, G., and Voelter, W. (1997) Analysis of modified nucleosides in urine as potential tumour markers by HPLC and CE. *GIT Labor-Fachzeitschrift* **1997,** 92–93.
55. Liebich, H. M., Xu, G., Di Stefano, C., and Lehmann, R. (1998) Capillary electrophoresis of urinary normal and modified nucleosides of cancer patients. *J. Chromatogr. A* **793,** 341–347.
56. Zhao, R., Xu, G., Yue, B., Liebich, H. M., and Zhang, Y. (1998) Artificial neural network classification based on capillary electrophoresis of urinary nucleosides for the clinical diagnosis of tumors. *J. Chromatogr. A* **828,** 489–496.
57. Kuo, K. C., Phan, D. T., Williams, N., and Gehrke, C. W. (1990) Ribonucleosides in biological fluids by a high-resolution quantitative RPLC-UV method, in *Chromatography and Modification of Nucleosides* (Gehrke, C. W. and Kuo, K. C., eds.), Elsevier, Oxford, pp. C41–C113.
58. Bartels, H., Böhmer, M., and Heierli, C. (1972) Serum Kreatininbestimmung ohne Enteiweissen. *Clin. Chim. Acta* **37,** 193–197.
59. Xu, G., Schmid, H. R., Liebich, H. M., Zhang, Y., and Lu, P. (1999) Excretion pattern investigation of urinary normal and modified nucleosides of breast cancer patients by RP-HPLC and factor analysis method. *Manuscript in preparation.*

Index

C

D

Q

R

T

V

Z